A Study Book for the NEBOSH National Diploma
in occupational Health and Safety Practice

Hazardous agents in the workplace

Professional Membership (Grad IOSH)

Holders of the National Diploma may apply for Graduate membership (Grad IOSH) of the Institution of Occupational Safety and Health (IOSH), and on completion of a programme of Continual Professional Development (CPD) may apply for Chartered Safety and Health Practitioner status as a Chartered Member of IOSH (CMIOSH).
Chartered Member status reflects the competence demanded of professionals in health and safety management or leadership positions.

iosh

RMS Publishing
Suite 3, Victoria House
Lower High Street
Stourbridge
DY8 1TA

© RMS Publishing Limited.
First Published May 2005.
Second Edition August 2006.
Third Edition February 2008.
Third Edition November 2008 (reprint).
Third Edition April 2010 (reprint).
Fourth Edition July 2011.
Fourth Edition December 2012 (reprint).
Fifth Edition January 2014.
Fifth Edition November 2014 (reprint).
Fifth Edition April 2016 (reprint).
Sixth Edition August 2016.

Cover by Smudge Creative Design.
Printed and bound in Great Britain by CPI Antony Rowe.

ISBN-13: 978-1-906674-56-4

Contents

Preface

Publication users

This updated 6th Edition study book has been thoroughly updated in line with the current NEBOSH syllabus for the NEBOSH National Diploma 'Unit B Hazardous substances and agents' providing an excellent reference for those looking to undertake a career as a health and safety practitioner.

This study book Unit B provides a thorough grounding in all major aspects of managing hazardous substances and agents; topics covered include:

- Managing occupational health
- Identification, assessment and evaluation of hazardous substances
- Control of hazardous substances
- Monitoring and measuring of hazardous substances
- Biological agents
- Noise and vibration
- Radiation
- Mental ill-health and dealing with violence and aggression at work
- Musculoskeletal risks and controls
- Work environment risks and controls

It aims to prepare students for a career in health and safety by providing them with the ability to apply their knowledge and understanding of hazardous agent issues in the workplace.

The NEBOSH National Diploma is the qualification for aspiring health and safety professionals building directly upon the foundation of knowledge provided by the NEBOSH National General Certificate. It also provides a sound basis for progression to postgraduate study.

Syllabus

The study book has been structured to reflect the order and content of the NEBOSH National Diploma in Occupational Health and Safety syllabus 'Hazardous substances / agents'. In this way, the student studying for this qualification can be confident that this guide reflects the themes of the syllabus and forms an excellent study book for that purpose.

Each element of the study book has an element overview that sets out the learning outcomes, the contents and any connected sources of reference. In addition, the syllabus, and therefore this study book, is structured in a very useful way; focusing on hazards, their control and core management of health and safety principles which would be useful as reference for any health and safety practitioner.

Photographs and schematics

We have taken particular care to support the text with a significant number of photographs and schematics. The photographs have been selected to be illustrative of both good and bad working practices and should always be considered in context with supporting text. I am sure that students will find this a useful aid when trying to relate their background and experience to the broad based NEBOSH National Diploma syllabus. They will give an important insight into some of the technical areas of the syllabus for those who may not have a strong technical background.

Where diagrams/text extracts are known to be drawn from other publications, a clear source reference is shown and RMS wishes to emphasise that reproduction of such diagrams/text extracts within the Study Book is for educational purposes only and the original copyright has not been infringed.

Legal requirements

Legislation is referred to in context in the various elements that comprise the study book. This reflects the interest of the NEBOSH National Diploma syllabus and requirements to study new/amended legislation under the rule from NEBOSH that it has to have been in force for six months before it becomes examinable. In addition, the essential points of legislation relevant to this Unit of the Diploma syllabus are contained in the section of the study book under Relevant Statutory Provisions.

Decided cases have been included in Units B 'Hazardous substances / agents' and A 'Managing health and safety' and form part of the syllabus content; these cases are examinable.

All statistics shown throughout this publication are the latest available at time of going to press.

National Vocational Qualification

We are confident that those working to national vocational qualifications in occupational health and safety will find this Study Book a useful companion. For students working towards the S/NVQ Level 5 in Occupational Health and Safety Practice they will find a good correlation between the scope of the Study Book series for NEBOSH National Diploma and the domain knowledge needs at that level.

Higher Level Qualifications

The structure, level and content of this study book is appropriate for those involved in study of health and safety at university level. The NEBOSH National Diploma is recognised by IOSH as fulfilling the academic requirements for application for Graduate Membership (Grad IOSH) of the Institution of Occupational Safety and Health (IOSH – www.iosh.co.uk). This is the first step to becoming a Chartered Health and Safety Practitioner as a Chartered Member of IOSH (CMIOSH).Chartered Membership (CMIOSH) conferring the official title of Chartered Safety and Health Practitioner. It is also accepted by the International Institute of Risk and Safety Management (IIRSM) as meeting their requirements for full membership (MIIRSM).

Relationship to other RMS Study Books

Students with limited experience may find the foundation knowledge in the RMS Publishing Study Books for Certificate level particularly useful, in particular the study book for the NEBOSH National General Certificate in Occupational Health and Safety.

Production of the publication

Managing Editor: *Ian Coombes, Managing Director ACT, CMIOSH;* member of NEBOSH Council, member of NEBOSH Board of Trustees, member of NEBOSH Qualifications and Technical Council and former NEBOSH examiner. Former member of IOSH Professional Affairs Committee and chairman of the Initial Professional Development sub-committee. Member of the Safety Groups UK (SGUK) management committee.

Acknowledgements

RMS Publishing wishes to acknowledge the following contributors and thank them for their assistance in the preparation of the publication:

Geoff Littley, Principal Consultant ACT, CMIOSH, NEBOSH Diploma, CSPA; experienced health and safety advisor, including manufacturing, NHS Trusts, Local Authorities and transport industries. Lead tutor for NEBOSH Diploma and Certificate level courses, in particular for the NEBOSH National General Certificate and Construction Health and Safety qualifications. Provides training and mentor support for Principal Designers.

Barrie Newell, Director ACT, FCMI; Lead Auditor OHSAS 18001, former senior manager in the chemical industry with over 20 years' experience in the management of high risk facilities processing highly flammable and toxic chemicals, including HAZOP implementation. Implemented waste management systems including, waste reduction, recycling, reuse, incineration, including energy recovery and disposal to land fill.

Kevin Coley, CMIOSH; a NEBOSH examiner with many years' health and safety experience in the private and public sector. Kevin has been a production manager in a large foundry, so has a heavy engineering background as well as a senior safety manager for large government bodies.

Julie Skett, Design and Development Coordinator. Nick Attwood, Kris James, Alison Serdetschniy, Jack Aaron and Andy Taylor layout and formatting.

RMS Publishing also wishes to acknowledge the following contributors: Janice McTiernan.

Publications available from RMS:

Publication	Edition	ISBN
A Study Book for the NEBOSH National General Certificate in Occupational Health and Safety	Eighth	978-1-906674-44-1
A Study Book for the NEBOSH Certificate in Fire Safety and Risk Management	Fifth	978-1-906674-32-8
The Management of Construction Health and Safety Risk	Fourth	978-1-906674-37-3
The Management of Environmental Risks in the Workplace	Third	978-1-906674-24-3
The Management of Health and Well-being in the Workplace	First	978-1-906674-14-4
A Guide to International Oil and Gas Operational Safety	First	978-1-906674-19-9
A Guide to International Health and Safety at Work	Fifth	978-1-906674-47-2
Study Books for the NEBOSH National Diploma in Occupational Health and Safety:		
■ (Unit A) Managing health and safety	Sixth	978-1-906674-55-7
■ (Unit B) Hazardous substances/agents	Sixth	978-1-906674-56-4
■ (Unit C) Workplace and work equipment safety	Sixth	978-1-906674-57-1
Study Books for the NEBOSH International Diploma in Occupational Health and Safety:		
■ (Unit IA) Managing health and safety	Fourth	978-1-906674-52-6
■ (Unit IB) Hazardous substances/agents	Fourth	978-1-906674-53-3
■ (Unit IC) Workplace and work equipment safety	Fourth	978-1-906674-54-0
Controlling Skin Exposure (BOHS)	First	978-1-906674-00-7

Figure List (including tables and quotes)

Element B4

Element B5

Element B6

Element B7

Relevant statutory provisions

List of abbreviations

Legislation

BR	Building Regulations 2010
CAOR	Control of Artificial Optical Radiation at Work Regulations 2010
CAR	Control of Asbestos Regulations 2012
CDM	Construction (Design and Management) Regulations 2015
CLA	Criminal Law Act 1967, Section 3, (reasonable force)
CLAW	Control of Lead at Work Regulations 2002
CNWR	Control of Noise at Work Regulations 2005
COMAH	Control of Major Accident Hazards Regulations 2015
COSHH	Control of Substances Hazardous to Health Regulations 2002
CVWR	Control of Vibration at Work Regulations 2005
DDA	Disability Discrimination Act 1995
DPA	Data Protection Act 1998
DSE	Health and Safety (Display Screen Equipment) Regulations 1992
EA	Equality Act 2010
ERA	Employment Rights Act 1996
FAR	Health and Safety (First-Aid) Regulations 1981
HASAWA	Health and Safety at Work etc Act 1974
HSCER	Health and Safety (Consultation with Employees) Regulations 1996
IRMER	Ionising Radiation (Medical Exposure) Regulations 2000
IRR	Ionising Radiations Regulations 1999
LOLER	Lifting Operations and Lifting Equipment Regulations 1998
MAR	Health and Safety (Miscellaneous Amendments) Regulations 2002
MRRA	Health and Safety (Miscellaneous Repeals, Revocations and Amendments) Regulations 2013
MDA	Misuse of Drugs Act 1971
MHOR	Manual Handling Operations Regulations 1992
MHSWR	Management of Health and Safety at Work Regulations 1999
POA	Public Order Act 1986
PPER	Personal Protective Equipment at Work Regulations 1992
PSCPS	Prohibition of Smoking in Certain Premises (Scotland) Regulations 2006
PUWER	Provision and Use of Work Equipment Regulations 1998
REPPIR	Radiation (Emergency Preparedness and Public Information) Regulations 2001
REACH	REACH Enforcement Regulations 2008
RIDDOR	Reporting of Injuries, Diseases and Dangerous Occurrences Regulations 2013
RM(RT)	Radioactive Material (Road Transport) Regulations 2002
RSA	Radioactive Substances Act 1993
SD Act	Sex Discrimination Act 1974
SFPER	Smoke-free (Premises and Enforcement) Regulations 2006
SFPWR	Smoke-free Premises etc. (Wales) Regulations 2007
SRSC	Safety Representatives and Safety Committees Regulations 1977

| WHSWR | Workplace (Health, Safety and Welfare) Regulations 1992 |
| WTR | Working Time Regulations 1998 |

General

AC	Alternating Current
ACD	Allergic (sensitised) contact dermatitis
ACGIH	American Conference of Governmental Industrial Hygienists
ACOP	Approved Code of Practice
ACTS	Advisory Committee on Toxic Substances
ADI	Acceptable Daily Intake
AGIR	Advisory Group on Ionising Radiation
AGNIR	Advisory Group on Non-Ionising Radiation
AIB	Asbestos Insulating Board
AIDS	Acquired Immune Deficiency Syndrome
ALARP	As Low As is Reasonably Practicable
ANR	Active Noise Reduction
APF	Assigned Protection Factor
ART	Assessment of Repetitive Tasks
ARTP	Association for Respiratory Technology and Physiology
ASL	Approved Supply List
ATP	Adaptations to Technical Progress
BBV	Bloodborne Virus
BCS	British Crime Survey
BMGV	Biological Monitoring Guidance Values
BOHS	British Occupational Hygiene Society
BS	British Standards
BSC	British Safety Council
BSI	British Standards Institution
CA	Competent Authority
CAS	Co-operating with other member state
CEC	Co-operating with the European Commission
CET	Corrected Effective Temperature
CFM	Cubic Feet Per Minute
CGI	Convertible Gas Indicator
CGRO	The Compton Gamma Ray Observatory
CL	Control Limit
COHPA	Commercial Occupational Health Providers Association
CRE	Commission for Racial Equality
CSA	Chemical Safety Assessment
CSR	Chemical Safety Report
CTS	Carpal Tunnel Syndrome
dB	Decibel
DDT	Dichlorodiphenyltrichloroethane
DLBA	Direct Line Breathing Apparatus
DNA	Deoxyribonucleic Acid
DoE	Department of the Environment
DSD	Dangerous Substances Directive
DSE	Display Screen Equipment
EAV	Exposure Action Values
EC	European Community
ECD	Electron Capture
ECHA	European Chemicals Agency
ED	Effective Dose
EDS	Energy Dispersive X-Ray Spectrometry
EEA	European Economic Area
EFTA	European Free Trade Association
EH	Environmental Health
EHO	Environmental Health Office
EINECS	European Inventory of Existing Commercial Chemical Substances
ELV	Exposure Limit Values
EMAS	Employment Medical Advisory Service
EOC	Equal Opportunities Commission
ET	Effective Temperature
EU	European Union
FID	Flame Ionisation Detector
FOM	Faculty of Occupational Medicine
GHS	Globally Harmonised System
GIT	Gastro-intestinal Tract
GLC	Gas Liquid Chromatography
GM	Geiger Mueller
GP	General Practitioner
HAEO-1	The first High Energy Astrophysical Observatory

HAVS	Hand-arm Vibration Syndrome
HBV	Hepatitus B Virus
HEPA	High Efficiency Particulate Air
HGV	Health Guidance Value
HID	High-Intensity Discharge
HIV	Human Immunodeficiency Virus
HML	High, Medium and Low
HPA	Health Protection Agency
HPLC	High Performance Liquid Chromatography
HRA	Human Reliability Analysis
HRT	Hormone Replacement Therapy
HSC	Health and Safety Commission
HSE	Health and Safety Executive
HSG	Health and Safety Guidance
HSI	Heat Stress Index
IBS	Irritable bowel syndrome
ICD	Irritant contact dermatitis
ICNIRP	International Commission on Non-Ionising Radiation Protection
ICRP	International Commission on Radiological Protection
IOM	Institute of Occupational Medicine
IR	Infra Red
ISO	International Organisation for Standardization
LD	Lethal Dose
LEV	Local Exhaust Ventilation
LTEL	Long Term Exposure Limit
LVHV	Low volume high velocity
LX	Lux
MAC	Manual Handling Assessment Charts
MCE	Mixed Cellulose Ester
MDHS	Methods for Determinations of Hazardous Substances
MDI	Methylene Bisphenyl di-isocyanate
MELS	Maximum Exposure Limits
MF	Medium Frequency
MMMF	Man Made Mineral Fibres
MRSA	Methicillen-resistant Staphylococcus Aureus
MSD	Musculoskeletal disorders
NC	Noise Criteria
NDT	Non-Destructive Testing
NIHL	Noise induced Hearing Loss
NOAEL	No observed adverse effect level
NR	Noise Rating
NRPB	National Radiological Protection Board
NVQ	National Vocational Qualification
OECD	Organisation of Economic Co-operation and Development
OEL	Occupational Exposure Limit
OES	Occupational Exposure Standard
OHS	Occupational Health Service
OPCS	Office of Population Census and Surveys
OSHA	Occupational Safety and Health Administration
P4SR	Predicted 4-hour Sweat Rate
PCLM	Phase Contrast Light Microscopy
PID	Photo-ionisation Detector
PLM	Polarised Light Microscopy
PMT	Photo Multiplier Tubes
PMV	Predicted Mean Vote
PNS	Peripheral Nervous System
PPD	Percentage People Dissatisfied
PPE	Personal Protective Equipment
PTSD	Post Traumatic Stress Disorder
PVC	Polyvinyl Chloride
QEC	Quick Exposure Check
RF	Radio Frequency
RH	Relative Humidity
RL	Recommended Limits
RMS	Root Mean Square
ROSPA	Royal Society for Prevention of Accidents
RPA	Radiation Protected Advisors
RPE	Respiratory Protective Equipment
RRSAG	Radiation, Risk and Society Advisory Group
RSI	Repetitive Strain Injury
RULA	Rapid Upper Limb Assessment
RWL	Recommended Weight Limit

RXTE	Rossl X-Ray Timing Explorer
SAR	Specific Absorption Rate
SARS	Severe acute respiratory syndrome
SDS	Safety Data Sheets
SNR	Single Number Rating
SOM	Society of Occupational Medicine
SPHA	Special Health Authority
SSP	Statutory Sick Pay
SPL	Sound Pressure Level
STEL	Short Term Exposure Limit
SWORD	Surveillance of Work related and Occupational Respiratory Disease
TC	Thermal Conductivity
TDI	Toluene di-isocyanate
TLD	Thermoluminescent Dosimeters
TLV	Threshold Limit Value
TTS	Temporary Threshold Shift
TUC	Trades Union Congress
TWA	Time Weighted Average
UK	United Kingdom
ULD	Upper Limb Disorders
UV	Ultra Violet
VCM	Vinyl Chloride Monomer
VDU	Visual Display Unit
VHF	Very High Frequency
WATCH	Working on Action to Control Chemicals
WBGT	Wet Bulb Globe Temperature
WBV	Whole Body Vibration
WCI	Wind Chill Index
WEL	Workplace Exposure Limits
WHO	World Health Organisation
WRULD	Work Related Upper Limb Disorder
WRV	Work Related Violence
XRD	X-Ray Diffraction

Managing occupational health

Learning outcomes

On completion of this element, candidates should be able to demonstrate understanding of the content through the application of knowledge to familiar and unfamiliar situations. In particular, they should be able to:

B1.1   Outline the nature of occupational health.

B1.2   Outline the principles and benefits of the management of return to work including the role of outside support agencies.

B1.3   Outline the management of occupational health (including the practical and legal aspects).

Contents

Relevant statutory provisions

Equality Act (EA) 2010

Sources of reference

Reference information provided, in particular web links, was correct at time of publication, but may have changed.

A healthy return, Good practice guide to rehabilitating people at work, IOSH, Link to A healthy return, http://www.iosh.co.uk/~/media/Documents/Books%20and%20resources/Guidance%20and%20tools/A%20healthy%20return .pdf

Safe, Effective, Quality Occupational Health Service (SEQOHS), https://www.seqohs.org/

Workplace health: long-term sickness absence and incapacity to work, NICE guidelines (PH19), https://www.nice.org.uk/guidance/PH19

The above web links along with additional sources of reference, which are additional to the NEBOSH syllabus, are provided on the RMS Publishing website for ease of use - www.rmspublishing.co.uk.

B1.1 - Nature of occupational health

The relationship between occupational exposure to an agent and its associated diseases has been long established.

In the 19th century; for example:

- Felt hat makers showed symptoms of the effects on the central nervous system of exposure to the mercury salts used in the production process (hence the term 'as mad as a hatter').
- Coal miners have suffered respiratory disability due to the effects of coal dust on the lungs and painters have become unable to work in their chosen field due to occupational asthma. Despite this, occupational health medicine is a relatively junior science.
- An Austrian doctor attributed pulmonary troubles in one of his patients to the inhalation of asbestos dust.
- A report regarding the asbestos manufacturing process in England, where factories had been routinely inspected since 1833 to protect the health and safety of workers, cited 'widespread damage and injury of the lungs, due to the dusty surrounding of the asbestos mill'.

Thomas Legge was appointed as the first Medical Inspector of Factories in 1898, following his studies concerning working with lead. Lead poisoning was made a notifiable disease in 1899. Following his work as a Factory Inspector he published the results of his experiences in 1934 as 'Industrial Maladies'.

In 1906, the first documented death of an asbestos worker from pulmonary failure was recorded by Dr Montague Murray at London's Charring Cross Hospital. Two years later, in 1908, insurance companies in the U.S. and Canada began decreasing coverage and benefits, while increasing premiums, for workers employed in the asbestos industry.

Epidemiology is the study of categories of persons (populations) and the patterns of diseases from which they suffer so as to determine the events or circumstances causing these diseases. Thus the epidemiologist applies statistical techniques in the investigation of medical problems. One of the most famous examples is the Ghost Map (Johnson, 2008). The book describes how Dr Snow with the help from Rev Henry Whitehead undertakes an epidemiology study to discover the outbreak of cholera in Victorian London; he believes against popular opinion that the outbreak was waterborne.

Donald Hunter is regarded as the last great general physician and his book 'Diseases of Occupations' is still regarded as the current classic. Hunter, as with Ramazzini (a professor of medicine at the University of Modena in the 1700's), added another question to the list started by Hippocrates: *"Ask whether any similar disease has occurred in a fellow worker".* Hunter observed that: *"Many workers are intelligent, co-operative and good witnesses. Although some may be deaf, disconsolate, forgetful, obtuse, garrulous, monosyllabic, the worker is still the best witness to what happened."*

The meaning of health

The World Health Organisation (WHO) is a specialised agency of the United Nations, established in 1948, that is concerned with international public health. Their main areas of work include: health systems, promoting health, diseases, corporate services, preparedness, surveillance and response.

The WHO define health as a:

"State of complete physical, mental and social well-being and not merely the absence of disease or infirmity."

Figure B1-1: Definition of health. *Source: World Health Organisation.*

The meaning of occupational health

Occupational health is not simply about the effects of work on the health of workers, but also the effect of workers' health on work. Occupational health is a vital management issue, enabling employers to understand and comply with health and safety legislation and to ensure workplace risks are effectively managed.

The World Health Organisation (WHO) and the International Labour Organisation (ILO) issued this definition of occupational health jointly in 1995:

"Occupational health should aim at:

- *The promotion and maintenance of the highest degree of physical, mental and social well-being of workers in all occupations; the prevention amongst workers of departures from health caused by their working conditions.*
- *The protection of workers in their employment from risks resulting from factors adverse to health.*
- *The placing and maintenance of the worker in an occupational environment adapted to his physiological and psychological capabilities; and, to summarise, the adaptation of work to man and of each man to his job."*

Figure B1-2: Definition of occupational health. *Source: WHO and ILO.*

Note: This definition has been adopted by the International Commission on Occupational Health (ICOH) and features in the 2002 update of the International Code of Ethics for Occupational Health Professionals. The code is available at: http://www.icohweb.org.

The meaning of well-being

Although the concept of well-being is widely used, there is no commonly agreed definition of just what it is; the definition used by the Economic and Social Research Council (ESRC) from http://www.welldev.org.uk is:

> *"Well-being is a state of being with others, where human needs are met, where one can act meaningfully to pursue one's goals, and where one enjoys a satisfactory quality of life."*

Figure B1-3: Definition of well-being.   *Source: Economic and Social Research Council (ESRC).*

Key concepts relating to well-being are outlined by the ESRC as:

1) Well-being is complex and multifaceted. It is considered as a state and a process. It is a contested concept.
2) Well-being includes personal, interpersonal, and collective needs, which influence each other.
3) Well-being may take different forms, which may conflict across groups in society, requiring an overarching settlement. Well-being may also take different forms over the life-course of an individual.
4) Well-being is intimately intertwined with the physical, cultural and technological environment, and requires a global perspective.
5) Interventions to enhance well-being may take different forms. They should be conducted at individual, community, and societal levels, ideally in concert. Interventions need to recognise diversity and socio economic inequalities in society, and be concerned with the unintended as well as the intended consequences of action.

Categories of occupational health hazard

CHEMICAL

Chemical hazards dealt with in *Elements B1-B4* - examples include:

- *Acids and alkalis* - dermatitis.
- *Metals* - lead and mercury poisoning.
- *Non-metals* - arsenic and phosphorus poisoning.
- *Gases* - carbon monoxide poisoning, arsine poisoning.
- *Organic compounds* - occupational cancers, for example, bladder cancer.
- *Dust* - silicosis, coal worker's pneumoconiosis.
- *Fibres* - asbestosis.

BIOLOGICAL

Biological hazards dealt with in *Element B5* - examples include:

- *Animal-borne bacteria* - leptospirosis, E.coli.
- *Human-borne viruses* - viral hepatitis (hepatitis B).
- *Vegetable-borne fungi, moulds, yeasts* - aspergillosis (farmer's lung).
- *Environmental bacteria* - legionnaires' disease.

PHYSICAL

Physical hazards dealt with in *Elements B6, B7 and B10* - examples include:

- *Heat* - heat cataract, heat stroke.
- *Lighting* - miner's nystagmus.
- *Noise* - noise induced hearing loss (occupational deafness).
- *Vibration* - vibration induced white finger.
- *Radiation* - radiation sickness (at ionising wavelengths), burns, arc eye.
- *Pressure* - decompression sickness.
- *Environmental* - air, water pollution.

PSYCHO-SOCIAL

Psychosocial hazards dealt with in *Element B8* - examples include:

- Stress.
- Violence.
- Drugs.
- Alcohol.

ERGONOMIC

Ergonomic hazards dealt with in *Element B9* - examples include:

- *Manual handling* - musculoskeletal injuries (from poor handling techniques).
- *Job movements* - cramps (in relation to handwriting or computer data entry).
- *Friction and pressure* - bursitis, cellulitis, i.e. beat hand, traumatic inflammation of the tendons or associated tendon sheaths of the hand or forearm (tenosynovitis).

The prevalence of work-related sickness and ill-health

The Health and Safety Executive (HSE) reported the following data on ill-health for the period 2014/15:

ILL HEALTH

- Around 13,000 deaths each year from work-related lung disease and cancer are estimated to be attributed to past exposure, primarily to chemicals and dust at work
- In 2014/15 an estimated 2.0 million people were suffering from an illness (long standing as well as new cases) they believed was caused or made worse by their current or past work.
  - 1.2 million people who worked during the last year were suffering from an illness they believed was caused or made worse by their work.
  - A further 0.8 million former workers (who last worked over 12 months ago) were suffering from an illness which was caused or made worse by their past work.
- 0.5 million were new cases amongst those working in the last 12 months.
- Around 80 per cent of new work-related conditions were either musculoskeletal disorders or stress, depression or anxiety.
- 2,538 people died from mesothelioma in 2013 and thousands more from other work-related cancers and diseases such as COPD.
- There were 2,215 new cases of mesothelioma assessed for Industrial Injuries Disablement Benefit (IIDB) in 2014 compared with 2,145 in 2013.

WORKING DAYS LOST

- 27.3 million days were lost due to work-related ill-health or injury (15 days per case).
- 23.3 million days were lost due to work-related ill-health and 4.1 million due to workplace injury.

The Chartered Institute of Personnel and Development (CIPD) absence survey for 2015 reported that 8.3% of working time was lost due to work-related ill-health, equivalent to 6.9 working days per employee, resulting in a cost to the employer of £554 for each person annually.

Information that can assist in identifying health hazards can be sourced from internal data and external bodies.

INTERNAL DATA

Information may be derived from conducting formal health risk assessments for the range of hazards outlined in earlier text. The information gathered during accident/ill-health investigation, health surveillance and surveys of absence records can provide useful data on the types of hazards within a workplace.

Several cases of dermatitis can show a problem with a particular cleaning fluid, whereas one case, when analysed, may show the problem could have been caused by the person's own individual sensitivity or possibly pursuit of a leisure time hobby. Health professionals, such as an individual's GP or occupational physician, can provide information on health conditions related to workplace exposure.

EXTERNAL DATA

Health and Safety Executive (HSE)

A major source of information is the regulations produced in response to health hazards such as the Control of Noise at Work Regulations (CNWR) 2005 or Control of Asbestos Regulations (CAR) 2012. These regulations and their associated guidance aid the identification of conditions hazardous to health.

In addition, the HSE produce documents including the legal series, guidance notes, information sheets and leaflets, providing greater detail on certain topics. Specifically, the HSE guidance note EH40 provides information on chemical and biological hazards, including the lists of Workplace Exposure Limits (WEL) for use with the Control of Substances Hazardous to Health Regulations (COSHH) 2002 (as amended).

Other relevant bodies

Information can be sourced from agencies such as Public Health England, whose role is to provide an integrated approach to protecting UK public health through the provision of support and advice to the NHS, local authorities, emergency services and others, on infectious diseases and radiation protection.

Other bodies include:

- European Safety Agency.
- European Chemicals Agency (ECHA).
- Employment Medical Advisory Service (EMAS).
- Environment Agency.
- Fire Authority.
- Health and Safety Laboratory.
- World Health Organisation.
- Professional health organisations, for example, the Faculty of Occupational Medicine.
- Trade bodies.
- Insurance organisations.

The links between occupational health and general/public health

OCCUPATIONAL HEALTH AND GENERAL/PUBLIC HEALTH

Healthcare workers face a wide range of hazards at work; including needle-stick injuries, back injuries, latex allergy, violence, and stress. In these, and similar occupations, they are also exposed to general/public health hazards such as influenza, hepatitis and TB.

It is estimated that around 0.3% of the UK population are chronically infected with hepatitis B virus and about 0.4% similarly infected with hepatitis C virus. Chronic infection may lead to the development of chronic liver damage, cirrhosis or liver cancer.

Figure B1-4: Prevalence of hepatitis virus.   *Source: PHE, Corporate plan 2005/10.*

Although it is possible to reduce healthcare worker exposure to these hazards, healthcare workers are actually experiencing increasing numbers of occupational injuries and illnesses. Rates of occupational illness and injury to healthcare workers have risen significantly over the past decade.

Similarly, many other workers have experienced similar issues in their work with members of the public, such as:

- Police, fire and ambulance personnel.
- Care home workers.
- Social workers.
- Council workers engaged in: refuse collection; receipt of monies from tenants and rate payers.

It is expected, in the event of public health epidemics such as those arising from the flu virus, that many workers will be exposed to it through work-related contact with the public.

This makes it a significant occupational risk in this kind of work and other jobs that involve close contact with the public, for example, schools, nurseries and some retail activities.

Occupational exposure to such viruses could lead to mass worker illness and their absence from work. For critical public services this could be catastrophic and this has instigated government strategies for managing health in relation to work and arrangements to deal with epidemics.

New diseases can arise at any time as shown by the devastation of Human Immunodeficiency Virus (HIV) or more recently Severe Acute Respiratory Syndrome (SARS). Over the past two years the threat of an influenza pandemic has increased significantly and with the speed and extent of modern international travel the disease would spread much faster than in the past. The impact on the UK could be significant, and the Agency has plans in place to support the NHS in combating the threat.

Figure B1-5: PHE plans for epidemics.   *Source: PHE, Corporate plan 2005/10.*

ARRANGEMENTS TO DEAL WITH EPIDEMICS

Public Health England (PHE) is an independent UK organisation that was set up by the government in 2003 to protect the public from threats to their health from infectious diseases and environmental hazards.

It does this by providing advice and information to the general public, to health professionals such as doctors and nurses, and to national and local government.

The PHE identifies and responds to health hazards and emergencies caused by infectious disease, hazardous chemicals, poisons or radiation. It combines public health and scientific knowledge, research and emergency planning within one organisation.

This enables it to provide advice to the public on how to stay healthy and avoid health hazards, provides data and information to government to help inform its decision making, and advises people working in healthcare. It also makes sure the nation is ready for future threats to health that could happen naturally, accidentally or deliberately. The PHE also uses its research to develop new vaccines and treatments that directly help patients.

Governmental level arrangements to deal with epidemics have included establishing a system through various organisations, including the PHE, to monitor the occurrence of infectious diseases that could lead to an epidemic in the UK or pandemic, including monitoring overseas occurrences.

Arrangements also include:

- Provision of vaccines where they are currently available or development of them at an early point when an epidemic emerges.
- Early identification and clinical diagnosis of the presence of the disease in people in the UK.
- Provision of information and advice to public service organisations on the minimisation of the effects on workers and the maintenance of their provision of service.
- Requirement for public service organisations to establish contingency plans.

B1.2 - The principles and benefits of the management of return to work and vocational rehabilitation

The basic principles of the bio-psychosocial model and how it relates to the health of individuals

The bio-psychosocial model of health was developed in 1977 by George Engel, an American Psychiatrist, and refers to the concept that biological, psychological, and social factors coalesce to play a significant role in the way human beings function. The biological part of the model relates to the medical aspects of injury and illness, psychological, refers to the individual's personal beliefs and emotions. The social part of the model relates to the persons environment within which they live, including home and work. The health of individuals is best understood and managed in these terms.

> **Case Study - Contact centre operative**
>
> Jane is 35 years old and works in a contact centre as an operator. She does not enjoy her job and has been off sick with low back pain. She is still waiting for an appointment with a physiotherapist and is in constant discomfort. She has not been sleeping and has become very tired and emotional. Her partner has encouraged her to rest and not to take any exercise. Jane is convinced she has a serious problem and that she will never be well enough to return to work.

The above case study illustrates how the bio-psychosocial model can be used to explain the influence that individuals' belief and perceptions of illness can have upon the body and mind. Jane was genuinely concerned that she was suffering from a serious medical illness (biological). Her perception was that any activity would make her condition more serious and she lacks motivation (psychological). Jane does not enjoy her job as she finds it boring and does not have a good relationship with her manager (social).

Elements of the Equality Act 2010 that relate to well-being at work

The Equality Act 2010 legally protects people from discrimination in the workplace and in wider society. It replaced previous anti-discrimination laws with a single Act, making the law easier to understand and strengthening protection in some situations. It sets out the different ways in which it's unlawful to treat someone.

The Equality Act provisions set the basic framework of protection against direct and indirect discrimination, harassment and victimisation in services, public functions, work, education, associations and transport. The Act provides protection for people discriminated against because they are perceived to have, or are associated with someone who has, a protected characteristic.

The protected characteristics covered by the Equality Duty are:

- Age.
- Disability.
- Gender reassignment.
- Marriage and civil partnership (but only in respect of eliminating unlawful discrimination).
- Pregnancy and maternity.
- Race - this includes ethnic or national origins, colour or nationality.
- Religion or belief - this includes lack of belief.
- Sex.
- Sexual orientation.

The Equality Act 2010 defines disability as a physical or mental impairment that has a 'substantial' and 'long-term' negative effect on your ability to do normal daily activities. 'Substantial' is more than minor or trivial, for example, it takes much longer than it usually would to get dressed. 'Long-term' means 12 months or more, for example, a breathing condition that develops as a result of a lung infection.

An employer must make 'reasonable adjustments' to ensure that workplace requirements or practices do not disadvantage employees or potential employees with a disability. Reasonable adjustments should be made with the employee's involvement. They can often be simple and inexpensive. In law, adjustments have to be 'reasonable', and need not be excessive.

There are three elements to the approach taken to 'reasonable adjustments'. The first element involves changing the way things are done (equality law talks about where the disabled job worker is put at a substantial disadvantage by a provision, criterion or practice of their employer). The second element involves making changes to overcome barriers created by the physical features of a workplace. The third element involves providing extra equipment (which equality law calls an auxiliary aid) or getting someone to do something to assist you (which equality law calls an auxiliary service).

An example of a reasonable adjustment to a premises might be making a structural or other physical changes such as widening a doorway, providing a ramp or moving furniture for a wheelchair user; relocating light switches, door handles, or shelves for someone who has difficulty in reaching; or providing appropriate contrast in decor to help the safe mobility of a visually impaired person. A reasonable adjustment may be to allow a disabled person to work flexible hours to enable them to have additional breaks to overcome fatigue arising from their disability. A reasonable adjustment may also be to incorporate different working hours to avoid the need to travel in the rush hour if this is a problem related to impairment. A phased return to work with a gradual build-up of hours might also be appropriate in some circumstances.

When deciding whether an adjustment is reasonable an employer can consider:

- How effective the change will be in avoiding the disadvantage you would otherwise experience.
- Its practicality.
- The cost.
- Their organisation's resources and size.
- The availability of financial support.

The principles of fitness to work and fitness to work standards

Fitness to work standards are those standards set out by government, industry groups and organisations to establish the minimum health capabilities a worker would need to conduct a job. Standards are adopted or defined by organisations to assist them when selecting new employees, by using these standards during pre-employment health screening processes.

In the UK the Equality Act (EA) 2010 prevents the use of unreasonable health standards when selecting employees if this would unreasonably limit the ability of a disabled person from being selected. The fitness to work standards should be derived by careful analysis of the job, considering the hazards and demands of the job. The standards may relate to both the physical and psychological capabilities of the individual. Some jobs may have demands on physical dexterity, others may require high levels of mental reasoning.

The 'health risk assessment' used to derive the fitness to work standards needs to consider a number of factors:

- Nature of the work, and the work demands.
- How health of the worker might affect work, for instance, physical and psychological.
- Is there a recognised specific fitness standard.
- How the work might affect the health of the worker.
- The potential hazards.

With any fitness standard, there may be some room for 'professional clinical judgement' regarding new employees who, through ill-health or reduced fitness through ageing, may be able to do the job providing additional safe systems of work are put in place.

> **Case Study - Employee with poor eyesight and arthritis**
>
> A health risk assessment was used to develop control measures for an employee working for a local charity. She was based on the first floor and had lost sight in one eye, and had limited vision in the other. The health risk assessment concluded that she should be relocated on the ground floor - eliminating the risk of falling down stairs.

The role and benefits of 'pre-placement assessment'

The role of pre-employment (pre-placement) health assessment is to:

- Ensure someone is fit to do the job - assessed against specific fitness standards.
- Ensure that those with known health problems are not put at risk, or they put others at risk.
- Provide a reference profile (baseline) of health for the worker entering the workplace, for example, hearing, lung function, vision. This is important for the worker and organisation, especially if there has been previous exposure to noise or dust and the individual may already have impaired hearing or lung function. Both the worker and organisation need to know this, in case there are any future claims and to prevent further harm, for example, due to noise induced hearing loss or occupational asthma.
- Support an organisation's overall positive health strategy and provide an opportunity for early health education. It can also be used to inform workers of any specific health requirements and/or the positive health initiatives the organisation has in place that the workers can access.
- Provide an opportunity to review the current fitness standards by actively considering the standards in the context of assessing the health of real people. This enables fitness to work to be set that remain relevant and effective.

The benefit of pre-employment health assessment is that it enables an employer to:

- Judge a new worker's fitness to work against pre-defined standards.
- Not expose a new worker to hazards that may exacerbate a health condition.
- Make adjustments to work to accommodate a new worker's health condition.

A pre-employment health assessment should be undertaken once the individual has been offered the job and they have met all the other criteria for undertaking the work.

The role of 'Fit Note'

Research has shown that a person does not necessarily have to be 100% fit to perform satisfactorily in their work role. Being absent from work can be a factor in the condition getting worse, therefore, doctors will now review your condition and in considering your rehabilitation decide whether returning to work would promote your full recovery. The doctor will not automatically assess that you are not fit for work if you have a health condition. Instead, they will discuss with you how your health affects what you can do at work. The doctor will think about your fitness for work in general rather than just your current job. Research shows that work can be good for your physical and mental health, lowers the risk of experiencing financial difficulties, and improves your overall quality of life. The fit note is classed as advice from your doctor. Your employer can decide whether or not to accept it, and your doctor cannot get involved in any disputes between you and your employer.

Managing long-term and short-term frequent sickness absence/incapacity for work

Workplace health: long-term sickness absence and incapacity to work, was published in 2009 by the UK's National Institute for Health and Clinical Excellence (NICE) to produce public health guidance for primary care services and employers on the management of long-term sickness absence and incapacity for work. Other countries may have similar guidance.

The NICE guidance presents recommendations, based on evidence of effectiveness and cost effectiveness, for interventions that aim to:

- Prevent or reduce the number of employees moving from short-term to long-term sickness absence (including the prevention of recurring short-term sickness absence).
- Help employees on long-term sickness absence return to work.
- Reduce the number of employees who take long-term sickness absence on a recurring basis.
- Help people receiving incapacity benefit or similar benefits return to employment (paid and unpaid).

It is widely recognised that being employed can help improve a person's health and wellbeing and help reduce health inequalities. Conversely, unemployment is linked to higher levels of mortality and psychological morbidity.

> **Case Study - Health and safety practitioner with multiple sclerosis (MS)**
>
> After a diagnoses of MS, a health and safety practitioner chose to continue to work and study even though it had been suggested that since he had MS he should give up work. Whilst the MS affects his balance, he also occasionally drops things and suffers with chronic fatigue and pain. He continues to work as a consultant and trainer. He changed his car to an automatic, enabling him to continue to drive to clients' etc.

Long term medical conditions that may have an impact on work include:

- Arthritis.
- Cancer.
- Coronary heart disease.
- Mental ill-health - including anxiety, depression.
- Diabetes.
- Epilepsy.
- Hearing impairment.
- Motor neuron disease.
- Multiple sclerosis.
- Sight impairment.

Other conditions that may have an impact on work include:

- Pregnancy.
- New mothers.
- Young and older workers.

The effective management of absence is an important issue that requires a co-ordinated response, drawn from the occupational health and human resources disciplines. It is influenced by both employment law and health and safety law.

In the UK, the Advisory, Conciliation and Arbitration Service (ACAS), Health and Safety Executive (HSE) and Chartered Institute of Personnel and Development (CIPD) all have perspectives on how absence should be managed. There are many reasons why people take time off work. These can be categorised as:

- Short-term sickness absence.
- Long-term sickness absence.
- Unauthorised absence or persistent lateness.
- Authorised absences, for example, annual leave; maternity, paternity, adoption, or parental leave; time off for public or trade union duties, or to care for dependents; compassionate leave; educational leave.

Factors to be considered when managing long-term and short-term frequent sickness absence/incapacity for work are identifying the problem and developing a management strategy to address the concerns.

DEVELOPING AN ABSENCE MANAGEMENT STRATEGY

Developing an absence strategy involves developing an absence policy and establishing the 'building blocks' of absence practice:

- Clear absence procedures.
- Rigorous monitoring.
- Use of 'trigger points' for action, for example, the number or duration of absences.
- Defined roles and accountabilities.
- Effective management processes, for example, home visits.
- Options for incentives and sick pay.
- Preventative initiatives - flexible.
- Occupational health.

Meaning of vocational rehabilitation

The generally accepted definition for vocational rehabilitation is:

> *"A process to overcome the barriers a worker faces in returning to employment which result from their injury, illness or disability."*

Figure B1-6: Definition for vocational rehabilitation.   *Source: RMS.*

> *"A process that enables people with functional, psychological, developmental, cognitive and emotional impairments or health conditions to overcome barriers to accessing, maintaining or returning to employment or other useful occupation."*

Figure B1-7: Definition for vocational rehabilitation.   *Source: NHS Scotland, framework for rehabilitation.*

The definition encompasses the support a worker and employer needs to ensure the worker remains in or returns to work or can access work for the first time.

For workers suffering ill-health, it is an intervention to remove the causes of them leaving work or the barriers affecting their return to work, whilst providing therapeutic care. Vocational rehabilitation would be particularly relevant for workers with musculoskeletal or psychosocial illnesses.

The process of vocational rehabilitation is one of assessing the individual and work involved, setting goals, intervening to provide support, providing reasonable adjustment to the work and aids that make work easier to achieve, providing ongoing treatment, enabling the individual to self-manage the illness, then re-assessing to compare the situation after the provision of intervention.

At some point, the individual should exit this cycle and manage their own condition, but there will continue to be times when interventions may be required.

The Vocational Rehabilitation Association suggests that provision of vocational rehabilitation requires input from a number of disciplines, including medical professionals, health and safety professionals, human resources professionals, counselling professionals, management and the individual. The Vocational Rehabilitation Association has established professional standards for those working in the vocational rehabilitation field.

Effective vocational rehabilitation depends on communication and co-ordination between the worker concerned, the employer and healthcare professionals.

Benefits of vocational rehabilitation within the context of the employee and the employer

The detrimental impact that ill-health has on a worker is well documented and early intervention can be effective in avoiding long term absence from work. The workers will feel valued by the organisation if investment in them is focused on them returning to work.

There is strong evidence that provision of vocational rehabilitation provides a cost benefit for the organisation, in that it retains skilled workers in work, rather than having to replace them with someone who, whilst equally skilled, may take some time to achieve the level of performance required for the job.

The provision of vocational rehabilitation may also enable the retention of trusted and valued workers, which enables the organisation to benefit from their knowledge, experience and positive stability. If others can see how their employer works to support its employees in times of difficulty, they are likely to feel that their workplace is a positive environment in which to work. This can help with staff retention in general.

Knowing that it is possible to remain at work despite illness or injury, and that help is available to do so, may also make staff more likely to approach their employer, and others, for help before their condition worsens to the point at which work becomes impossible.

Overcoming any barriers to ensure that rehabilitation of the individual is effective

Overall, there is strong evidence that effective rehabilitation of the individual depends on two features:

- Accommodating workplaces.
- Healthcare that is work-focused.

One of the barriers to successful return to work can be the worker's negative perception about their illness, 'work will make their health worse'. In addition, they may have social barriers, for example, lack of child care, looking after elderly relatives and the job demands itself. Addressing these barriers is an important aspect of any rehabilitation programme, if it is to be successful. Organisations must support workplace interventions, and ensure they are tailored to support a safe and healthy return to work for the worker. It is the role of the occupational health function to monitor that this takes place and ensure that intervention to overcome barriers to work is effective.

The need to undertake or review risk assessments prior to return to work

It is important that the employee's first day back after extended absence is a positive experience for them. Returning to work after this time may be seen by the employee as a challenging experience and an informal visit before the return date may enable them to adjust, catch up with changes and re-orientate themselves to the workplace.

In addition to conducting a return to work interview, when the employee returns it is important that someone welcomes them back, eases their induction to the workplace and any adjustments that may have been made for their benefit. Fellow workers should be encouraged to make the return to work a positive, welcoming and encouraging experience. This may present challenges where it may be perceived that the employee returning to work is not likely to contribute fully to the workload, reduce bonuses and receive favourable treatment in the form of shorter hours. The negative feelings of the fellow workers can be reduced by explaining the return to work plan to them and encouraging a positive outlook.

The main findings of a CIPD absence survey conducted in 2012 included:

- The most commonly used approach to managing short-term absence was return to work interviews (65% of organisations), followed by trigger systems to review attendance (58%), and the provision of sickness absence information to line managers (20%) and the use of disciplinary procedures (27%).
- Return to work interviews were the most commonly used approach to managing long-term absence (85% of organisations), followed by risk assessments following return to work (75%), employee absence information for line managers (76%) and use of occupational health services (80%).

The involvement of occupational health professionals was rated the most effective approach for managing long-term absence by all the main employer sectors.

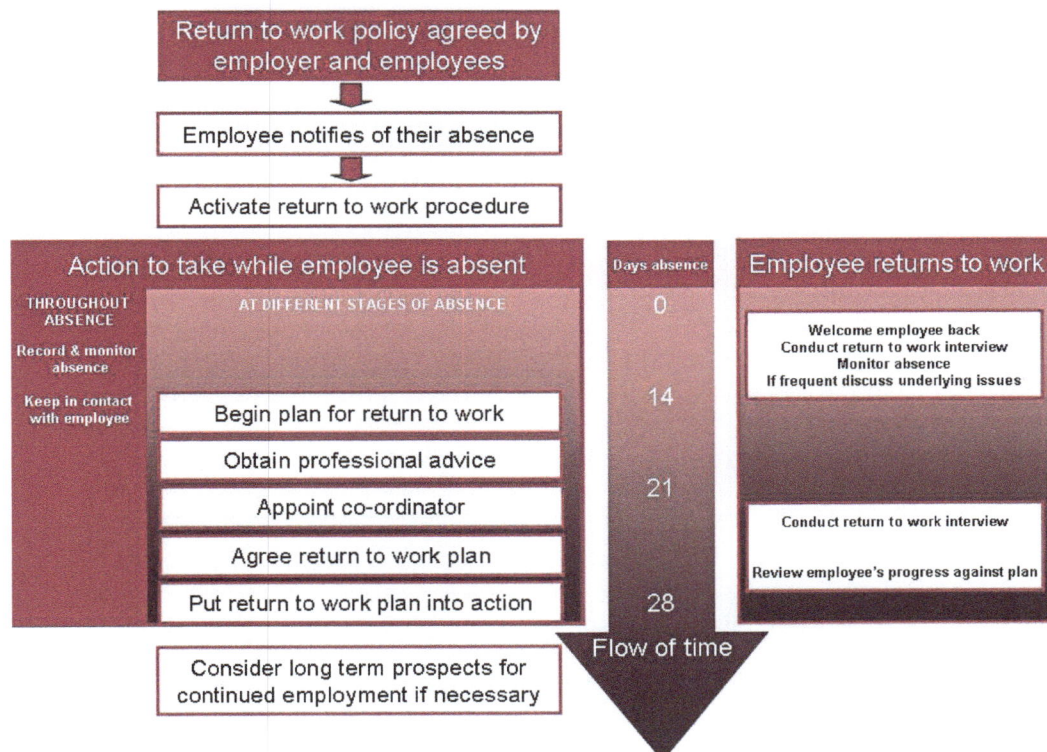

Figure B1-8: Managing sickness absence and return to work.   *Source: HSE, HSG249/RMS.*

Liaison with other disciplines in assessing and managing fitness for work

EXISTING HEALTH PROBLEMS

Many jobs do not require specific fitness standards and could be classed as low risk, for example, some office workers, and it may only be necessary to ask the employee to complete a basic health questionnaire that requires a declaration regarding their physical and mental health. The completion of this declaration may be organised by a Human Resources department and the results evaluated by an occupational nurse or doctor. For existing health problems that constitute a disability it may be necessary or helpful to consult outside organisations that are familiar with the particular health problem.

For more hazardous work, there may be a need to set minimum fitness standards to ensure that those with known health problems are not put at risk, or they do not put others at risk. This is important for the worker and organisation, especially if there has been previous exposure to noise or dust and the individual may already have impaired hearing or lung function. Both the worker and organisation need to know this in case there are any future claims (for example, for noise induced hearing loss or occupational asthma), and to prevent further harm.

Where existing health problems may affect a person's fitness for work, it may be necessary to refer the person to an occupational physician to determine the significance of the health problem and the degree to which it may affect their ability to work. The identification of an existing health problem may not prohibit a person from the intended work, but may mean that specific risk controls have to be established to minimise the effect. For example, where someone has partial deafness or is suffering from the early stages of HAVS, controls may have to be established to manage their exposure and prevent the condition worsening.

Some existing health problems may affect the ability of a person to do specific work, for example, a significant heart condition may prohibit a person driving or working at height. In order to assess the likely effects of a condition like this it may be necessary to obtain information from the person's general practitioner (GP) or a specialist consultant with detailed knowledge of the condition and how it may affect work.

With specific health conditions that may have arisen from work due to an overexposure to a harmful agent, such as lead or ionising radiation, it may be appropriate to consult the Employment Medical Advisory Service (EMAS), the specialist part of the HSE, who would have detailed understanding of the effects on the worker and their ability to continue in the work.

When employees are suffering from an illness that causes them to be absent from work it may be necessary to get independent specialist medical advice. Sometimes it is difficult to make sense of differing opinions from the employee, 'fit' notes, GP reports and other sources. It may be necessary to arrange for an independent consultant in occupational medicine to examine the employee, correspond with their doctor and produce a comprehensive medical report concerning their ability to work and any special arrangements that are necessary at work.

A medical report can provide the following information:

- The relationship between an underlying medical condition and work.
- An indication of the outlook for a return to work and the likelihood of further absence.
- The impact of the individual's condition on the health or safety of others.
- Any restrictions to the employee's duties or hours and the duration of the restrictions.
- The need for reasonable adjustments or resettlement into a different job.

FITNESS TO WORK STANDARDS

Fitness to work standards are those standards set out by government, industry groups and organisations to establish the minimum health capabilities a worker would need to conduct a job. Standards are set out by organisations to assist them when selecting new employees, by using these standards during pre-employment health screening processes. The fitness to work standards should be derived by careful analysis of the job, considering the hazards and demands of the job. The standards may relate to both the physical and psychological capabilities of the individual. Some jobs may have demands on physical dexterity; others may require high levels of mental reasoning.

The establishing of fitness to work standards should be as a result of a formal 'health risk assessment' and may involve a number of disciplines, for example, occupational health specialists, ergonomists, hygienists, engineers, health and safety practitioners, human resources, managers and worker representatives.

The 'health risk assessment' used to derive the fitness to work standards needs to consider a number of factors:

- Nature of the work, and the work demands, for example, physical and psychological.
- The potential hazards.
- How the work might affect the health of the worker.
- How health of the worker might affect work.
- Is there a recognised specific fitness standard?

With any fitness to work standard, there may be some room for 'professional clinical judgement' for employees who, through ill-health or reduced fitness through ageing, may be able to do the job providing additional safe systems of work are put in place. An example of this is where railway trackside worker, whose hearing has diminished with age, may be able to continue to work trackside if the team are aware of his limitations and the worker can be warned of danger by other means than sound, for example, lights, flags, buddy system. This type of decision should be made by an occupational health professional (doctor or nurse in conjunction with health and safety practitioners), who is aware of the work and hazards associated with it. This decision may be supplemented by actively monitoring the worker to ensure the measures in place provide safety and the decision was appropriate.

The following should be considered if there is not clear conformity with the fitness standards:

- Is a safe system of work required for this person to do the job?
- Are modifications to the workplace required?
- Is there alternative work?

DISCRIMINATION

The Equality Act (EA) 2010 prohibits, on the grounds of age, disability and sex, the discrimination of applicants or those in work by the use of unreasonable health standards. For example, it may be determined that it is essential that a train driver has a high standard of visual capability, which may be written into the fitness to work standard for the job, however to set high standards of fitness for someone carrying out a simple packing job may be considered unreasonable. In order to avoid discrimination of this type, particularly with regard to disability, it would be appropriate to liaise with other disciplines to ensure that assessments and arrangements to manage fitness for work are appropriate and effective. The EA 2010 further prohibits, on the grounds of discrimination, the consideration of the health of an applicant when determining who would be a suitable employee; it can only be considered as a confirmatory step after a job offer has been made. At this point, the employer is able to conduct appropriate assessment of the person's health to determine their suitability related to appropriate fitness to work standards and any adjustments to work that may be reasonably accommodated.

The employer may wish to liaise with specialist outside organisations that can provide advice and support on the assessment of disabled people, their suitability for particular work and any adjustments to work that may be needed to accommodate them. This might include specialist charities, such as MIND and RNIB, or government organisations, such as Access to Work.

Role of agencies that can support the employers and employees

PRIMARY CARE

The purpose of the primary care service framework is to:

- Equip commissioners, providers and practitioners with the necessary background knowledge.
- Service and implement details to safely deliver the service for people with long-term conditions.
- Improve patient's health and quality of life by providing patient-centred, systematic and on-going support.
- Reduce the reliance on secondary care services and increase the provision of care in a primary, community or home environment.

In January 2005, the Department of Health published 'Supporting people with long term conditions - an NHS and social care model', which described three levels of appropriate long-term care to help meet individual needs.

Level 3: Case management

Identifies the most vulnerable people, those with highly complex multiple long-term conditions, and uses a case management approach to anticipate, co-ordinate and join up health and social care.

Level 2: Disease-specific care management

Involves providing people who have a complex single need or multiple conditions with responsive specialist services, using multi-disciplinary teams and disease-specific protocols and pathways, such as the National Service Frameworks and Quality and Outcomes Framework.

Level 1: Self-care

Describes, the care and responsibility taken by the majority of individuals towards their own health and well-being and the support provided to them. It includes the actions people take for themselves, their children, and their families to stay fit and maintain good physical and mental health in order to ensure independence, self-worth and the ability to lead as near a normal life as is possible. It ensures people will have the necessary skills and education, information, tools and devices and support networks to manage their own health.

There are many support agencies such as the Shaw Trust and the Royal National Institute for the Blind that can provide targeted assistance. In addition, adaptations to the workplace may be paid for through the Government's 'Access to Work Scheme'.

FITNESS TO WORK SERVICE

From September 2015, all employers across the UK will be able to refer employees facing long-term sickness into the new support service, known as Fit for Work. Designed to help working people who face lengthy sickness absence return to work. Fit for Work provides the services of occupational health professionals to employed people if they have been, or are likely to be, off work for four weeks or more. The service is particularly helpful for small and medium-sized businesses (SMEs) employing a large proportion of the workforce but with no or limited occupational health support, and can often supplement the in house services of occupational health for larger employers. It is estimated that around 70% of employees do not have access to occupational health services.

ACCESS TO WORK

Access to Work is a Government initiative provided through Jobcentre Plus designed to keep in or get back to work those who may have a disability or health condition that prevent them being able to do parts of their job. There are Access to Work advisors who will provide advice and support to both employers and workers. Applicants may obtain payment towards:

- The equipment they need at work.
- Adapting premises to meet their needs.
- A support worker.
- The cost of getting to work if they cannot use public transport.
- A communicator at job interviews.

Access to Work may be available to those who are:

- In a paid job.
- Unemployed and about to start a job.
- Unemployed and about to start a work trial.
- Self-employed.

PATHWAYS TO WORK

The Pathways to Work programme helps people to get work if they are receiving allowances or benefits because of a condition or disability. The programme is tailored to the individual and may include:

- Work-focused interviews to identify needs, opportunities and support.
- Condition management programmes to help the person to better understand and manage their condition.
- Return to work credit to provide financial support.

WORK CHOICE

Work Choice helps people with disabilities whose needs cannot be met through work programmes like Access to Work or other workplace adjustments. This might be because they need more specialised support to find employment or keep a job once they have started work. It will also ensure that employers get the support they need to employ more disabled people.

THE SHAW TRUST

The Shaw Trust believes everyone should have the right to work. It is committed to supporting disabled and disadvantaged people into employment, and enabling them to live more independent lives. Every year the trust works with over 75,000 clients many of whom face barriers due to disability, ill-health or social circumstance. Thousands of employers and public sector organisations also benefit from the trust's range of services for business.

B1.3 - Managing occupational health

The role, function and benefits of occupational health services

The role of the occupational health service is to provide support to the management of an organisation in order that health at work is managed with a positive outcome.

The function of an occupational health service is to provide:

- Health promotion.
- Health assessment.
- Advice to management.
- Treatment services.
- Medical and health surveillance.

The benefits of occupational health services are that they will:

- Help protect the health of employees.
- Help employers make sure that they are complying with the legal requirements for a safe workplace.
- Enable the identification of health risks.
- Enable the establishment of workplace health standards.
- Enable detection of any adverse health effects at an early stage.
- Assist in the evaluation of control measures.
- Provide information useful in the detection of hazards and assessment of risks.
- Promote good health and well-being.
- Improve the level of general health of the workforce and reduce the level of ill-health absence.

Make-up and roles of a typical occupational health service and the importance of determining competency

THE OCCUPATIONAL HEALTH PHYSICIAN

They act in the role of consultant and advise organisations of likely health risks associated with their current practices. They make recommendations with respect to employee selection, i.e. to meet statutory requirements, for example, atopic screening - hay fever sufferers may not be ideal employees to work with known respiratory sensitisers such as flour or paints containing isocyanates.

The physician will be also carry out medicals on behalf of the employer, establish systems of employee examination and monitor the work of the Occupational Health Nurse.

THE OCCUPATIONAL HEALTH NURSE

The Occupational Health Nurse will be trained to identify symptoms resulting from exposure to specific workplace hazards and will carry out specific health monitoring, for example, those working with lead - blood tests; those working with compounds - nasal examination for 'chrome ulcer'.

OCCUPATIONAL HEALTH ADVISOR

The role of the Occupational Health Advisor (OHA) is to improve organisational effectiveness by providing a proactive Occupational Health Service. The OHA is required to develop the capability of the organisation to deal effectively with Occupational Health issues and legislative matters. This will require having expertise in supporting and enable people with a health condition or disability to work. The OHA will also work with management teams to increase productivity by lowering sickness absence. The OHA will also act as mentor for Occupational Health Technicians.

OCCUPATIONAL HEALTH TECHNICIAN

Occupational health technician is a relatively new discipline in occupational health. Qualified technicians may be trained to diploma level and have to pass theory and practical examinations in order to qualify. Many occupational health technicians receive in-house training, are able to carry out audiometry, spirometry, and venepuncture and have attended travel health, immunisation and hand arm vibration syndrome (HAVS) courses. Some have gained further qualifications in health related subjects. They are competent to carry out most practical procedures and some take on extra vocational training. They are not qualified to provide advice on sickness absence management policies, develop occupational health strategies and they must work under the supervision of a registered nurse or doctor.

THE OCCUPATIONAL HYGIENIST

The function of the occupational hygienist is to use science and engineering to identify risks to health in the workplace from physical, chemical and environmental hazards and assist with the establishment of controls to minimise these risks. For example, occupational hygienists take airborne samples and measurements of hazardous substances. They will identify when exposure is likely to be high and make suggestions in relation to control strategies to meet legal standards, in particular, regarding short-term exposure limits (STELs).

Typical services offered by an occupational health service

An occupational health service (OHS) may provide the following services:

HEALTH PROMOTION

Health promotions may reflect topical national issues related to lifestyle, for example, smoking, diet, exercise and weight, or work-related health issues that are specifically relevant to the organisation carrying out the promotion, for example, skin care promotion in a company where dermatitis is a problem.

HEALTH ASSESSMENT

Health assessment can identify fitness for work and certain health problems in workers before employment commences, which can help avoid placing workers in a work situation that could exacerbate a current health condition.

For example, an asthma sufferer should not be employed to work with or near respiratory sensitisers. Health assessment may also provide assessment of hearing loss in someone, who has previously worked in a noisy environment. This is important for health and safety reasons and to avoid any claims for compensation against the innocent company. It can also provide an assessment service that supports return to work and job related medical screening, such as is necessary for drivers and for pregnant workers.

ADVICE TO MANAGEMENT

Occupational health services can provide valid input to health risk assessments and development of policies, such as substance abuse and absence management. They can also provide specific advice regarding the management of health of individuals, which includes those returning to work from ill-health, disabled, older workers and those with specific medical conditions.

TREATMENT SERVICES

In smaller organisations, the occupational health service may not be employed full time and therefore may not be in a position to provide first aid, but the occupational health service may organise, co-ordinate and monitor first aid provision.

They may also provide an opportunity to review the first aid treatment through the provision of follow-up medical support. In larger organisations, the occupational health service may be more available and be involved with the direct provision of first aid. The occupational health service will be able to organise, and in some cases deliver, specific health treatments as part of an intervention process, for example, counselling, physiotherapy and rehabilitation services.

MEDICAL/HEALTH SURVEILLANCE

One of the main services provided by an occupational health service is the provision of medical and health surveillance. Medical and health surveillance helps to assure that workers continue to have the right health status for the work they are doing.

This will include periodic medical surveillance for such jobs as driving large goods vehicles (LGV) or fork lift truck, where risk of suffering a medical condition could cause immediate harm to the worker and others. Health surveillance will also assist in the identification of workers that are suffering a decline in their health, possibly due to workplace conditions.

The Management of Health and Safety at Work Regulations (MHSWR) 1999, regulation 6 deals with health surveillance and gives employers a duty to provide it where it is appropriate and links health surveillance in to the outcome of risk assessments.

Benefits of health needs assessment in relation to the planning of occupational health services

The benefits of conducting an occupational health needs assessment is that it will lead to the identification of health hazards relating to the work, the evaluation of current controls and identification of interventions to support health.

From this process it is possible to determine the level and types of occupational health services needed to meet the objectives identified. Meeting the objectives will require the development of an action plan and the assignment of individuals to enable the plan to be implemented.

If the organisation has a low level of health risk, such as an office environment, then the plan will be relatively simple. More complex organisations with complex risks to manage, such as within a chemical processing environment, will often choose to engage external assistance to complement their own occupational health department or contract the service to a third party.

If an occupational health service is to be provided it must be integrated into the organisation's operation. It should be clear how it will contribute to the success of the organisation and how this will be known.

Therefore the following should be addressed:

- Establish the purpose of the occupational health service.
- Identify the goals, objectives, which are based on the occupational health needs of the organisation.
- Determine how the performance of the occupational health service will be measured.
- Report the outcomes of the occupational health service to the top management.

This establishes direction for the occupational health service, defines the organisation better, and determines long term goals and what outcomes are to be achieved. It will set the context for development of the occupational health service, based on the needs of the organisation and its risks.

It should also be the basis of referral as OH is often regarded by staff as a punishment; this image needs to be changed and a more positive and enabling one created.

This requires a more enlightened vision of the role of OH by management, corporate identity and will make health and well-being more visible. The type of health services provided will depend on the resources available and the outcome of the occupational health needs assessment.

The importance of auditing against standards in occupational health provision

Safe, Effective, Quality Occupational Health Service (SEQOHS) is a set of standards and a voluntary accreditation scheme for occupational health services in the UK and beyond, launched in January 2010.

SEQOHS accreditation is the formal recognition that an occupational health service provider has demonstrated that it has the competence to deliver against the measures in the SEQOHS standards.

The scheme is managed by the Royal College of Physicians of London on behalf of the Faculty of Occupational Medicine (FOM). SEQOHS is available in Ireland and is endorsed by the Faculty of Occupational Medicine, Royal College of Physicians of Ireland.

An Occupational Health (OH) service must meet specific objectives; the SEQOHS has, with the FOM, set standards that will be audited:

A.	*Business probity*
A1	An OH service must conduct its business with integrity.
A2	An OH service must maintain financial propriety.
B.	*Information governance*
B1	An OH service must maintain adequate occupational health clinical records.
B2	An OH service must implement and comply with systems to protect confidentiality.
C.	*People*
C1	An OH service must ensure that its staff are competent to undertake the duties for which they have been employed.
C2	An OH service must ensure appropriate clinical governance.
D.	*Facilities and equipment*
D1	An OH service must conduct its business in facilities that are safe, accessible and appropriate for the services provided.
D2	An OH service must ensure that medical equipment is safe and appropriate for the services provided.
D3	An OH service must ensure that any medicines are handled appropriately.
E.	*Relationships with purchasers*
E1	An OH service must deal fairly and ethically with purchasers.
E2	An OH service must be customer-focused in its relationships with purchasers.
F.	*Relationships with workers*
F1	An OH service must ensure that workers are treated fairly and in line with professional standards.
F2	An OH service must respect and involve workers.

This page is intentionally blank

ELEMENT B1 - MANAGING OCCUPATIONAL HEALTH

This page is intentionally blank

Identification, assessment and evaluation of hazardous substances

Learning outcomes

On completion of this element, candidates should be able to demonstrate understanding of the content through the application of knowledge to familiar and unfamiliar situations. In particular, they should be able to:

B2.1 Explain the main routes of entry and the human body's defensive responses to hazardous substances.

B2.2 Explain the identification, classification and health effects of hazardous substances used in the workplace.

B2.3 Outline the factors to consider when undertaking assessment and evaluation of risks from hazardous substances.

B2.4 Outline the role of epidemiology and toxicology testing.

Content

Relevant statutory provisions

Control of Substances Hazardous to Health Regulations (COSHH) 2002 (and as amended 2004)

Control of Asbestos Regulations (CAR) 2012

Control of Lead at Work Regulations (CLAW) 2002 (and as amended 2004)

REACH Enforcement Regulations 2008 (as amended)

Control of Substances Hazardous to Health Regulations (Northern Ireland) 2003

Regulation (EC) No 1272/2008 on classification, labelling and packaging of substances and mixtures

Regulation (EC) No 1907/2006 concerning the Registration, Evaluation, Authorisation and Restriction of Chemicals (REACH), 18 December 2006 and subsequent amendments

Sources of reference

Reference information provided, in particular web links, was correct at time of publication, but may have changed.

COSHH Essentials www.coshh-essentials.org.uk

Asbestos Essentials, HSG120, ISBN 978-0-717662-63-0

Work with Materials Containing Asbestos. Control of Asbestos Regulations 2006, Approved Code of Practice and Guidance

Control of Lead at Work Regulations 2002. Approved code of practice and guidance, L132, ISBN 978-0-717625-65-9

Control of Substances Hazardous to Health, Approved Code of Practice and guidance, L5, HSE, ISBN: 978-0-717665-82-2, http://www.hse.gov.uk/pubns/priced/l5.pdf

COSHH Essentials, http://www.hse.gov.uk/coshh/essentials/

Globally Harmonized System of Classification and Labelling of Chemicals (GHS), http://www.unece.org/trans/danger/publi/ghs/ghs_rev06/06files_e.html#c38156

Managing skin exposure risks at work, HSG262, HSE Books, ISBN: 978-0-7176-6649-2, http://www.hse.gov.uk/pubns/priced/hsg262.pdf

The above web links along with additional sources of reference, which are additional to the NEBOSH syllabus, are provided on the RMS Publishing website for ease of use - www.rmspublishing.co.uk.

B2.1 - The routes of entry and the human body's defensive responses to hazardous substances

The structure and function of human anatomical systems and special sensory organs

THE HUMAN RESPIRATORY SYSTEM

Respiration is dependent on air being inhaled by one of two pathways; the nostrils (which pass through the nasopharynx) or through the mouth (the oral pharynx).

The air is warmed and moistened by vapour from the mucous membranes. It then passes through the pharynx (the throat) and enters the trachea though the glottis (which is closed by the epiglottis, a muscle flap, when swallowing). If the flap doesn't close properly, you can swallow 'the wrong way' into the lung - this is known as 'aspiration'.

The air proceeds down the trachea (windpipe) and divides through the right and left bronchi, which branch and re-branch (getting smaller and smaller) into the bronchioles. Each of the bronchioles terminates in a cluster of alveoli (not unlike a bunch of grapes in appearance), where the actual gas exchange takes place. Oxygen leaves the inhaled air and passes into the bloodstream. Carbon Dioxide leaves the bloodstream and is exhaled. The normal respiratory rate of an adult is 12-20 breaths/minute.

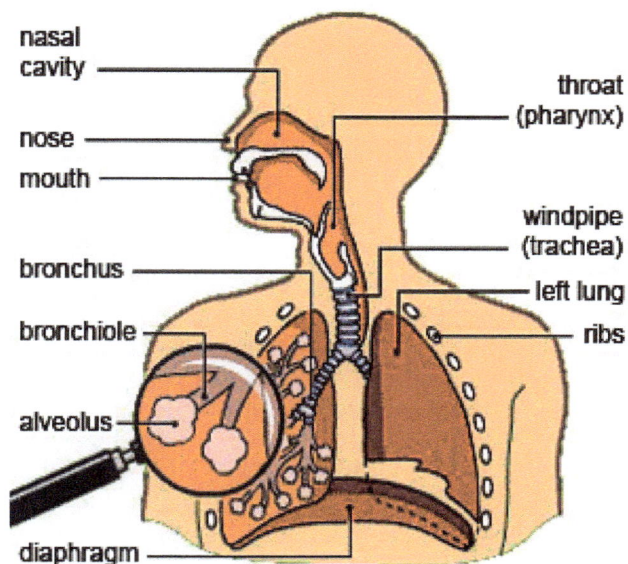

Figure B2-1: Respiratory system.   *Source: BBC.*

THE HUMAN DIGESTIVE SYSTEM

The human digestive system comprises the digestive tract, which is a series of hollow organs from the mouth to the anus. Organs that make up the digestive tract are: the mouth; oesophagus; stomach; small intestine; large intestine; rectum; and anus. Inside these hollow organs is a mucous membrane known as the mucosa. In the mouth, stomach, and small intestine, the mucosa contains tiny glands that produce juices to help digest food.

Digestion begins in the mouth, on chewing, when ptyalin (an enzyme in saliva) promotes the conversion of starches to sugars. Absorption is then completed in the small intestine.

The digestive tract also contains a layer of smooth muscle that helps break down food and moves it along the tract.

The liver and the pancreas produce digestive juices that reach the intestine through small tubes called ducts. The gall bladder stores the liver's digestive juices until they are needed in the intestine.

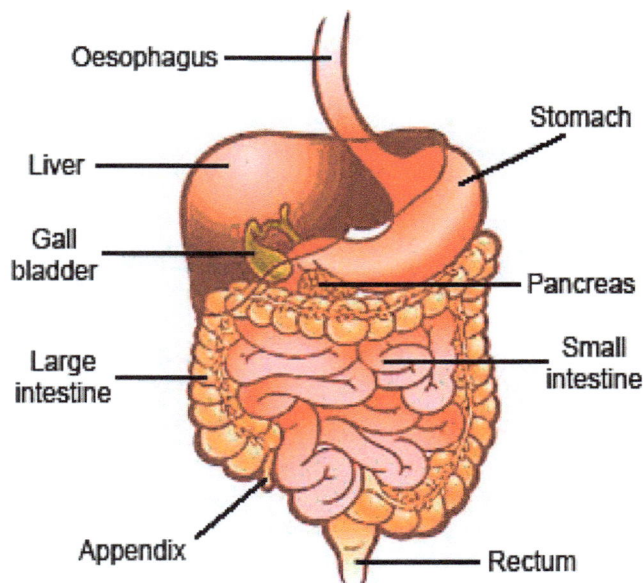

Figure B2-2: Digestive system.   *Source: STEM.*

THE HUMAN CIRCULATORY SYSTEM

The circulatory system supplies nutrients and oxygen to the body and removes any waste products, such as toxins and carbon dioxide. The circulatory system can be broken down into three main parts: the heart, the circulatory vessels and the blood.

The heart is a muscle that is divided by a septum into right and left sides acting effectively as two pumps to move blood throughout the body.

As the heart beats, the right side of the heart siphons in blood that has no oxygen and needs to be sent to the lungs for oxygenation. The left side of the heart pumps in the newly oxygenated blood from the lungs and sends it to the rest of the body.

The two main circulatory vessels are arteries and veins. There are also supporting vessels called capillaries.

Arteries are located throughout the body and carry oxygen-rich blood to all the muscles and tissues that need it.

Arteries have special walls that can respond to hormones and other chemicals that are released in the body.

This allows the walls to expand or contract, controlling the pressure and flow of the blood as it passes through the system. Branching off from these large arteries are smaller blood vessels called capillaries. Capillaries deliver oxygen-rich blood from the arteries to the tissues.

Once the oxygen and nutrients have been absorbed from the blood, by the tissues, it is returned to another major vessel called veins.

Veins return blood without oxygen back to the right side of the heart and onto the lungs.

Blood consists of two types of blood cells, red and white. Red blood cells contain haemoglobin, which absorbs oxygen and carries it to the tissues of the body. White cells are used to fight infection.

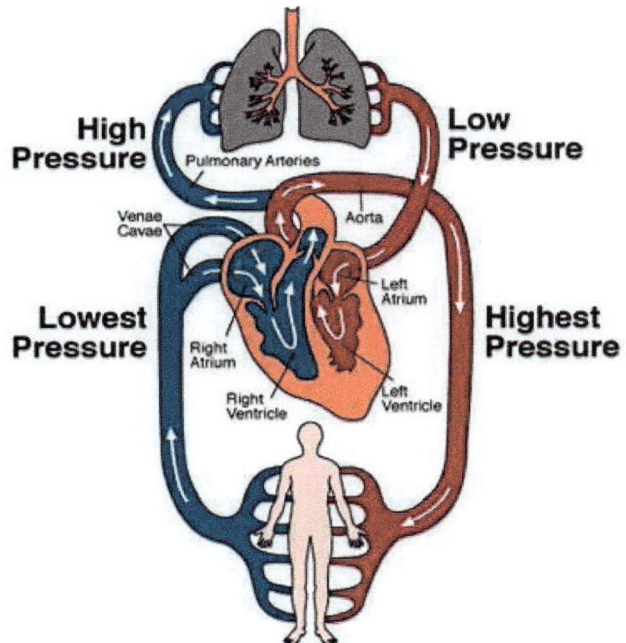

Figure B2-3: Digestive system. *Source: williamsclass.com.*

THE NERVOUS SYSTEM

The nervous system is divided into two main systems, the central nervous system (CNS) and the peripheral nervous system (PNS), **see figure ref B2-4.** The spinal cord and the brain make up the CNS. The function of the CNS is to get information from the body and send out instructions. The PNS is made up of all of the nerves. The brain sends and receives messages to the body through the spinal cord and nerves. It also controls the function of the body, and gives us awareness and personality.

The PNS is further divided into the Somatic and Automatic nervous systems. The Somatic nervous system relates to the voluntary control of body movements using skeletal muscles, such as using fingers to pick up an object. The Automatic nervous system controls automatic involuntary body functions, such as blinking.

The nerve system is made up of nerve cells known as neurons. Messages move from one neuron to another to keep the body functioning.

The axon of one neuron (the long extension shown in **figure ref B2-5**, which carries nerve impulses from the neuron) does not touch the dendrites of the next (the branches attached to the nucleus of the neuron, as shown in **figure ref B2-5**).

Nerve signals have to jump across a tiny gap. To get across the gap they have to change from electrical signals into chemical signals then back into electrical signals again.

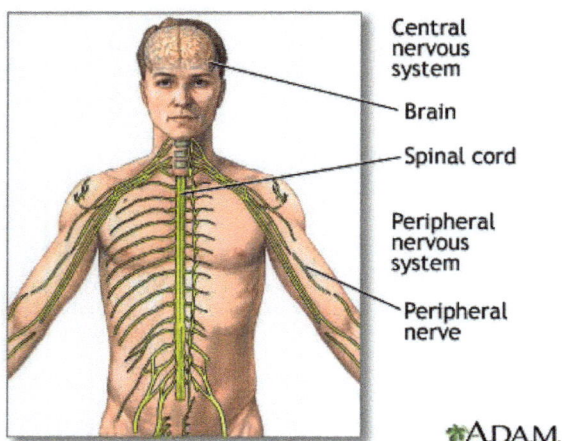

Figure B2-4: The nervous system.   *Source: A.D.A.M., Inc.*

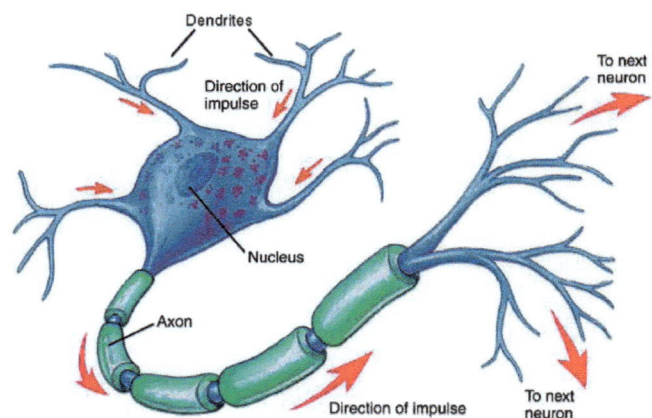

Figure B2-5: A neuron.   *Source: ibmmyositis.com.*

Neurons, unlike other body tissues, have a limited ability to repair themselves. Nerve cells cannot be repaired if damaged due to injury or disease.

The nervous system can act thousands of times faster than the endocrine system, which is a system of glands secreting hormones directly into the bloodstream.

Nerve impulses are often speedy and short lived, whereas the endocrine system is slower and longer lasting, for example, growth hormones.

In summary

```
                    ┌──────────────────────────┐
                    │     Nervous System       │
                    │       Structure          │
                    └──────────────────────────┘
                               │
              ┌────────────────┴────────────────┐
  ┌───────────────────────────┐    ┌───────────────────────────┐
  │  Central Nervous System   │    │ Peripheral Nervous System │
  │          (CNS)            │    │          (PNS)            │
  │    Brain and Spinal Cord  │    │     Nerves from CNS       │
  └───────────────────────────┘    └───────────────────────────┘
                                               │
                                 ┌─────────────┴─────────────┐
                    ┌───────────────────────────┐ ┌───────────────────────────┐
                    │         Somatic           │ │        Automatic          │
                    │ Touch, hearing, sight,    │ │ Automatically adjusted:    │
                    │ leg, finger arm muscles   │ │ pupils, sweat glands,      │
                    │ that move the body        │ │ saliva etc.                │
                    └───────────────────────────┘ └───────────────────────────┘
```

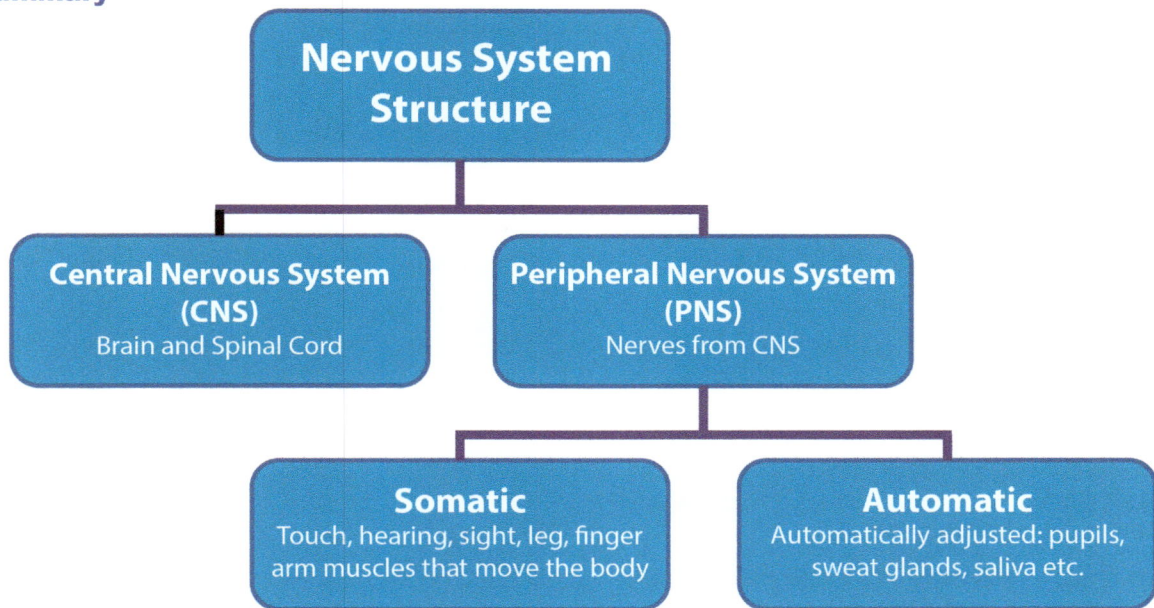

Figure B2-6: Nervous system structure.   *Source: RMS.*

Common problems of the nervous system include:

- Epilepsy - storms of abnormal electrical activity in the brain causing seizures.
- Meningitis - inflammation of the membrane covering the brain.
- Multiple sclerosis - the myelin sheaths protecting the electrical cables of the central nervous system are attacked.
- Parkinson's disease - death of neurones in a part of the brain called the midbrain. Symptoms include shaking and problems with movement.
- Sciatica - pressure on a nerve caused by a slipped disc in the spine or arthritis of the spine and, sometimes, other factors.
- Shingles - infection of sensory nerves caused by the varicella-zoster virus.
- Stroke - a lack of blood to part of the brain.

ENDOCRINE OR HORMONE SYSTEM

In addition to the nervous system, the body controls its functions by use of the endocrine system. The endocrine system consists of a series of glands that regulate hormone production, and their release into the blood stream to run the endocrine system.

The system utilises many glands to control such things as: the rate of growth in children (pituitary gland); the feelings of hunger; and our body temperature. The pancreas, ovaries (females) or testes (males), thyroid gland, parathyroid gland and the adrenal glands are some of the organs that run the endocrine system.

THE SKIN

The skin is the largest organ of the human body and has three main layers in its structure. It is soft, to allow movement, but still tough enough to resist breaking or tearing. It varies in texture and thickness from one part of the body to the next. For instance, the skin on the lips and eyelids is very thin and delicate, while skin on the soles of the feet is thicker and harder.

The skin provides:

- A waterproof covering for the entire body.
- The first line of defence against bacteria and other organisms (it is acidic, i.e. ph 6.5).
- A cooling system, via sweat and salt loss.
- A sense organ that gives us information about pain, pleasure, temperature and pressure.

The outer layer of the skin is called the epidermis. This protects the more delicate inner layers. The epidermis is made up of several layers of cells. The bottom layer is where new epidermal cells are made. As old, dead skin cells are lost from the surface by erosion, new ones replace them. The epidermis also contains melanin, the pigment that gives skin its colour and ability to darken to protect the skin from sunlight.

The next layer of the skin is the dermis (the true living skin), which is made up of elastic fibres (elastin) for suppleness and protein fibres (collagen) for strength. The dermis contains sweat glands, sebaceous glands, hair follicles, blood vessels and nerves.

The dermis is well supplied with blood vessels. In hot weather, or after exercise, these blood vessels expand to bring body heat to the skin surface.

Perspiration floods out of sweat glands and evaporates from the skin, taking the heat along with it. If the temperature is cold, the blood vessels in the dermis contract, which helps to reduce heat loss. Sebaceous glands in the dermis secrete sebum to lubricate the skin.

Both the dermis and epidermis have nerve endings. These transmit information on temperature, sensation (pleasure or pain) and pressure. The fingertips have more of these nerves than other parts of the body.

THE EYE

The eyeball is set into the orbit, an area of seven bones that make a pyramid shaped socket that points toward the back of the head. The eyeball is surrounded by a layer of fat to cushion it and allow smooth movement. The eyeball is encased in a white structure called the sclera; this is a tough material that protects the eyeball from damage.

The cornea is the clear part of the sclera that covers the iris, or the coloured part of the eye.

Light enters through the pupil, a hole in the middle of the eye. The iris is a series of muscles that allow it to adjust the size of the pupil's opening to regulate the amount of light that enters.

When there is not enough light, the iris adjusts the pupil's opening to receive the maximum amount.

When there is too much light, or the light is too bright, the iris contracts, or dilates the pupil's opening, so that it only takes in enough light to function.

The light passes through the pupil and enters the eye through the lens.

The purpose of the lens, which is a clear structure, is to focus the light toward the back of the eyeball. The lens is mobile; it is held in place by a network of fibres attached to the ciliary muscle.

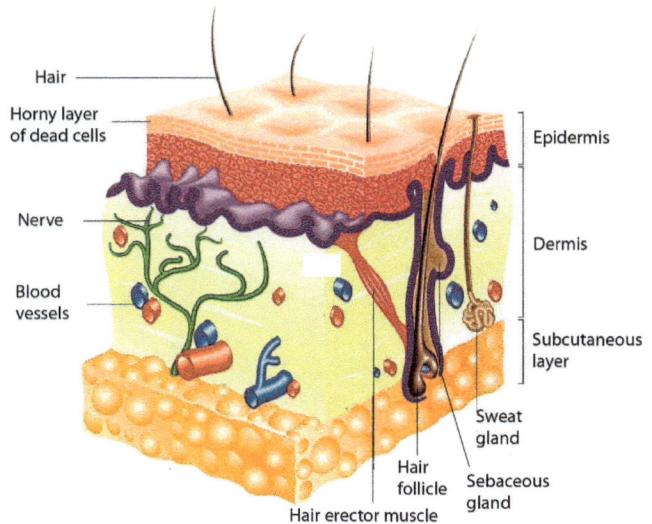

Figure B2-7: Skin layer.   *Source: SHP.*

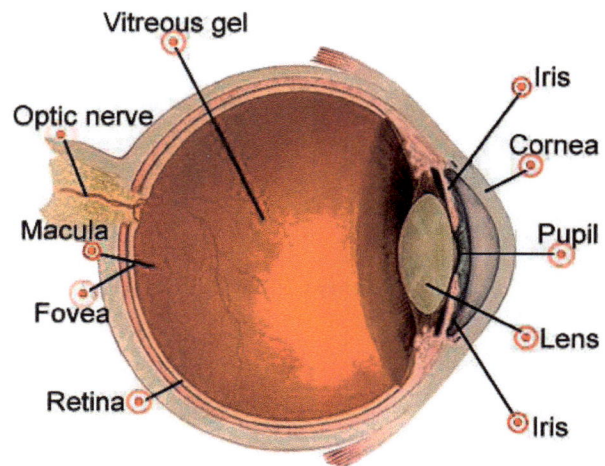

Figure B2-8: Eye diagram.   *Source: www.snezha.com.*

The ciliary muscle's function is to adjust the lens's thickness, depending on how it needs to be used. The ciliary muscle makes the lens thicker to view close items and much thinner to view distant objects.

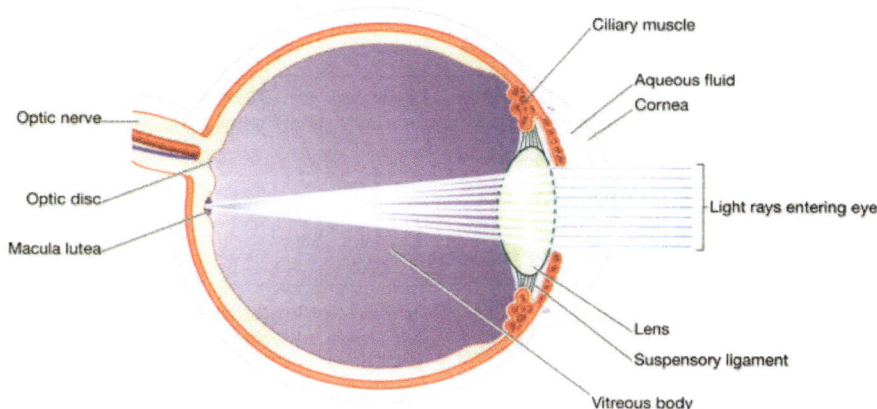

Figure B2-9: Section of the eye showing the focusing of light rays on the retina. *Source: www.selectspecs.com.*

After light passes through the lens, it passes through the vitreous humour, a gel-like substance made up of water and collagen. The vitreous humour gives the eye its shape and keeps the retina in place. The retina is a light-sensitive tissue consisting of rod and cone shaped cells that cover the back of the eye. Cones are responsible for detailed central vision and are located in the centre of the retina. Rods are located around the edges of the retina and are responsible for night and side vision but do not recognise colour. As light hits the optic disc is where the optic nerve joins the retina and is commonly known as the 'black spot', because no light receptor cells are located in this area.

The macula lutea, commonly called the 'yellow spot', is the area where the cone cells are most dense and sensitive. It is also the area where light is focussed by the lens, particularly when observing small close objects.

The conjunctiva is a thin delicate mucus membrane that covers the front of the eyeball and lines the inside of the eyelids. The eyeball is protected by the eyelid, which helps to prevent scratches, entry of dust and assists in the lubrication of the surface.

When the eyelid is closed, secretions from the tear glands are carried across the surface of the eyeball. There are several tear glands around the eyelids; each gland has a number of tear ducts. On blinking, the tears drain away via a small opening in the inner corner of the eyelid.

Main routes and methods of entry of hazardous substances into the human body

The main routes of entry into the body by substances (including toxic, corrosive and dermatitic substances, dusts and fibres) and agents are eyes, ears and skin. The nose and mouth are also routes of entry; they are covered in this section under inhalation and ingestion.

ENTRY THROUGH THE EYES

Some substances are water soluble, i.e. they dissolve in water, for example, ammonia gas. The mucous membrane (conjunctiva) of the eye will absorb the ammonia forming ammonium hydroxide, an alkali, which will irritate and eventually destroy the tissue.

Some substances will be absorbed by the mucous membrane, which allows the substance to pass into the eye and then gain a route into the body through the blood capillaries. Some viruses and bacteria can gain access this way, for example, the Brucella bacterium, which causes brucellosis; the leptospira bacterium, which causes leptospirosis (Weil's disease); and the Hepatitis B virus. The tear ducts that remove tears from the eye also provide a possible route of entry.

A possible method of entry is the use of a spray or aerosol. The eyes have many blood vessels that are close to the surface, if the spray is used at close range directly into the eye infiltration of the conjunctiva, cornea or sclera may occur. There may also be tearing of the cornea that allows the substance to enter the bloodstream.

ENTRY THROUGH THE EARS

Although the ear canal is coated with wax, which makes it a poor route of entry, it is possible for some substances, especially organic solvents, to pass through the skin of the canal or ear drum.

ENTRY THROUGH THE SKIN

Chemicals which pass through the skin are almost always in liquid form. Solid chemicals, gases and vapours do not generally pass through the skin without first being dissolved in moisture on the skin's surface. The skin is also waterproof, so water soluble chemicals must remain in contact with the skin for long time periods to gain entry.

The main methods of entry are:

- Inhalation (nose and mouth).
- Ingestion via the digestive tract (mouth).
- Skin pervasion (absorption).
- Injection.
- Aspiration.

This list represents the major routes of entry into the human body in their order of significance, inhalation being the most significant route, as more harmful substances enter the body through this route than any other.

INHALATION

The most significant industrial entry route is inhalation. It has been estimated that at least 90% of industrial poisons are absorbed through the lungs.

Harmful substances can directly attack the lung tissue, causing a local effect, or may pass through to the blood system, to be carried round the body and affect target organs such as the liver.

The effects of substances that enter the body through inhalation may be local or systemic.

Local effect

A local effect is where the hazardous substance has an effect where it first contacts the body. For example, silicosis, caused by inhalation of silica dust - where dust causes scarring of the lung, causing inelastic fibrous tissue to develop and reducing lung capacity

Systemic effect

A systemic effect is where the hazardous substance has an effect on a different site from where it first contacted the body. For example, anoxia, caused by the inhalation of carbon monoxide - the carbon dioxide replaces oxygen in the bloodstream, affecting the nervous system.

INGESTION

The ingestion route normally presents the least problem as it is unlikely that any significant quantity of harmful liquid or solid will be swallowed without deliberate intent. However, accidents/incidents will occur where small amounts of contaminant are transferred from the fingers to the mouth if eating, drinking or smoking is allowed in chemical areas or where a substance has been decanted into a container normally used for drinking.

The sense of taste will often be a defence if chemicals are taken in through this route, causing the person to spit it out. If the substance is taken in, vomiting and/or excretion may mean the substance does not cause a systemic problem, though a direct effect, for example, ingestion of an acid, may destroy cells in the mouth, oesophagus or stomach. Where hazardous substances are ingested, they may pass into the digestive system and be absorbed through the intestine to the blood system and may cause harm in another part of the body.

SKIN PERVASION (ABSORPTION THROUGH THE SKIN)

Substances can enter through the skin, via cuts or abrasions and through the conjunctiva of the eye; this is called absorption.

Solvents such as organic solvents, for example, toluene and trichloroethylene, can enter due to accidental exposure or if they are used for washing. The substance may have a local effect, such as de-fatting of the skin, resulting in inflammation and cracking of the horny layer, or pass through into the blood system, causing damage to the brain, bone marrow and liver.

INJECTION

Injection is a forceful breach of the skin, perhaps as a result of injury, which can carry harmful substances through the skin barrier. For example, handling broken glass that cuts the skin and transfers a biological or chemical agent.

On construction sites there are many items that present a hazard of penetration, such as nails in broken-up timber structures that might be trodden on and penetrate the foot, presenting a risk of infection from tetanus.

In addition, some land or buildings being worked on may have been used by intravenous drug users and their needles may present a risk of injection of a virus such as hepatitis. The forced injection of an agent into the body provides an easy route past the skin, which usually acts as the body's defence mechanism and protects people from the effects of many agents that do not have the ability to penetrate.

ASPIRATION

Pulmonary aspiration is the entry of material (such as secretions, food or drink, or stomach contents) from the oropharynx or gastrointestinal tract into the larynx (voice box) and lower respiratory tract (the portions of the respiratory system from the trachea (windpipe) to the lungs). A person may either inhale the material, or it may be delivered into the tracheobronchial tree during positive pressure ventilation.

When pulmonary aspiration occurs during eating and drinking, the aspirated material is often referred to as 'going down the wrong way', for example, vomit running back down the respiratory tract. This can also occur when substances are sucked directly into the lungs, which may occur during siphoning.

The concepts of target organs and target systems in relation to attack by substances

TARGET ORGANS AND TARGET SYSTEMS

A target organ or system is an organ or system within the human body on which a specified toxic substance exerts its effects.

Target organs include the lungs, liver, kidneys, brain, skin, bladder or eyes.

Target systems include the central nervous system, circulatory system, respiratory system, and reproductive system.

Examples of substances that have a systemic effect and their target organs/systems are:

- Alcohol - central nervous system, liver.
- Lead - bone marrow and brain.
- Mercury - central nervous system.
- Coal dust - lungs (pneumoconiosis).
- Mineral oils - skin.
- Benzidine - bladder.
- Leptospirosis - liver and kidneys.

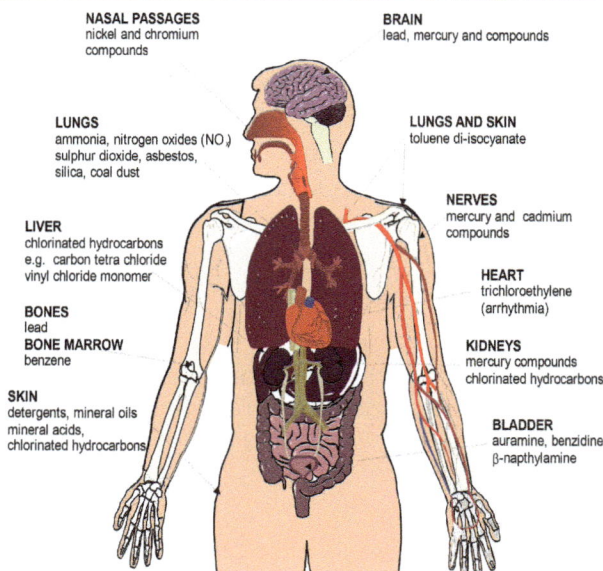

NASAL PASSAGES
nickel and chromium compounds

BRAIN
lead, mercury and compounds

LUNGS
ammonia, nitrogen oxides (NO$_x$) sulphur dioxide, asbestos, silica, coal dust

LUNGS AND SKIN
toluene di-isocyanate

LIVER
chlorinated hydrocarbons
e.g. carbon tetra chloride
vinyl chloride monomer

NERVES
mercury and cadmium compounds

HEART
trichloroethylene
(arrhythmia)

BONES
lead
BONE MARROW
benzene

KIDNEYS
mercury compounds
chlorinated hydrocarbons

SKIN
detergents, mineral oils
mineral acids,
chlorinated hydrocarbons

BLADDER
auramine, benzidine
β-napthylamine

Figure B2-10: Target organs/systems.   *Source: Mike Boyle.*

DIFFERENCE BETWEEN ACUTE AND CHRONIC HEALTH EFFECTS

The effect of a substance on the body depends not only on the substance, but also on the dose, and the susceptibility of the individual. No substance can be considered non-toxic; there are only differences in the degree of effect.

Acute effect

An acute effect is an immediate, or rapidly produced, adverse effect, following a single, or short-term, exposure to an offending agent, which is usually reversible (the obvious exception being death). Examples of acute effects are those from exposure to solvents, which affect the central nervous system, causing dizziness and lack of co-ordination; or carbon monoxide, which affects the level of oxygen in the blood, causing fainting.

Acute toxicity refers to those adverse effects occurring following oral or dermal administration of a single dose of a substance, or multiple doses given within 24 hours, or an inhalation exposure of 4 hours.

Chronic effect

A chronic effect is an adverse health effect produced as a result of prolonged or repeated exposure to an agent. The gradual or latent effect develops over time and is often irreversible. The effect may go unrecognised for a number of years. Examples of chronic effects are lead or mercury poisoning, cancer and asthma.

For example, lead:

- Health effects of *acute* lead toxicity include upset stomach (gastrointestinal problems), dullness, restlessness, irritability, poor attention span, headaches, kidney damage, hypertension and hallucinations.
- Health effects of *chronic* exposure to lead are blood disorder effects, such as anaemia, or neurological disturbances, including headache, irritability, lethargy, convulsions, muscle weakness, tremors and paralysis. Chronic lead exposure also causes cardiovascular and renal toxicity.

LOCAL AND SYSTEMIC EFFECTS

Local effects

Local effects are where a chemical causes harm at the point of first contact with the body. If a substance caused direct harm to the lungs when it was breathed in it would be said to have a local effect on the lungs. For example, if sulphur dioxide gas is inhaled, a dilute sulphurous acid forms on the lung lining, causing irritation and constriction of the lungs.

Systemic effects

Where a chemical enters via a route, such as the lungs, and has an effect elsewhere than its point of first contact, it is termed a systemic effect. For example, if carbon monoxide is inhaled it has no direct effect on the lungs; however, it acts as a systemic poison by replacing oxygen in the bloodstream thereby affecting cellular respiration.

It must be noted that many chemicals in use today can have both an acute and chronic effect. A simple everyday example is alcohol. The acute effect of drinking too much alcohol in a single evening may be vomiting and headache, whereas the chronic effect of drinking alcohol in smaller quantities, but over a prolonged period, is cirrhosis, a systemic effect with the liver as the target organ.

The body's defensive responses, in particular the respiratory system

The body's response against the invasion of substances likely to cause damage can be divided into:

- Respiratory (inhalation).
- Gastrointestinal (ingestion).
- Skin (absorption).
- Cellular mechanisms.

RESPIRATORY (INHALATION)

Nose

On inhalation, many substances and minor organisms are successfully trapped by nasal hairs, for example, larger wood and cement dust particles. Gases such as carbon monoxide are inhaled through the nose and easily defeat the defence mechanisms, the most superficial of which is smell. Carbon monoxide is highly poisonous to humans and has no smell. Vehicle exhausts are a common source of carbon monoxide in many countries.

Respiratory tract

The next line of defence against inhalation or substances harmful to health begins here, where a series of reflexes activate the coughing and sneezing mechanisms to forcibly expel the triggering substances. Silica and cement dust generated during construction activities may be trapped in the respiratory tract and ejected by coughing and sneezing.

Ciliary escalator

The passages of the respiratory system are also lined with mucus and well supplied with fine hair cells which sweep rhythmically towards the outside and pass along large particles.

The respiratory system narrows as it enters the lungs where the ciliary escalator assumes more and more importance as the effective defence. Smaller particles of agents, such as some silica particles, are dealt with at this stage. The smallest particles, such as organic solvent vapours and asbestos, reach the alveoli and are either deposited or exhaled.

Legionella bacteria that are inhaled can also defeat the ciliary escalator and reach the alveoli. Hardwood dusts (oak, mahogany, etc.), which are carcinogenic, if inhaled during cutting or machining operations, are usually trapped by the mucus in the escalator.

GASTROINTESTINAL (INGESTION)

Mouth

The mouth is used for the ingestion of substances in general. Saliva in the mouth provides a useful defence against hazardous substances that are not excessively acid or alkaline or present in large quantities.

Gastrointestinal tract

Acid in the stomach also provides a useful defence similar to saliva. Vomiting and diarrhoea are additional reflex mechanisms which act to remove substances or quantities that the body is not equipped to deal with.

SKIN (ABSORPTION)

Skin

The body's largest organ provides a useful barrier against the absorption of many foreign organisms and chemicals (but not against all of them). Its effect is, however, limited by its physical characteristics.

The outer part of the skin is covered in an oily layer and substances have to overcome this before they can damage the skin or enter the body. The outer part of the epidermis is made up of dead skin cells. These are readily sacrificed to substances without harm to the newer cells underneath. Repeated or prolonged exposure could defeat this.

When attacked by substances, the skin may blister in order to protect the layers beneath. Openings in the skin such as sweat pores, hair follicles and cuts can allow entry, and the skin itself may be permeable to some chemicals, for example, toluene.

Workers in the sewage industry or agriculture (animals) can come into contact with Leptospira bacteria, which can also pass through breaks in the skin. Other blood-borne viruses, such as Hepatitis, may defeat the skin by being injected into the bloodstream. Occupations at risk include those in health-care.

CELLULAR MECHANISMS

The cells of the body possess their own defence systems.

Scavenging action

A type of white blood cell called a macrophage attacks invading particles in order to destroy them and remove them from the body. This process is known as phagocytosis.

Secretion of defensive substances

Adrenaline is a hormone produced by the adrenal glands during high stress or exciting situations. This powerful hormone is part of the human body's acute stress response system, also called the 'fight or flight' response. It works by stimulating the heart rate, contracting blood vessels, and dilating air passages, all of which work to increase blood flow to the muscles and oxygen to the lungs. Histamine is a chemical that is released when the body is exposed to an allergen. Allergens may include airborne allergens (such as pollen and dust mites), certain foods (such as peanuts and shellfish) or insect venom. Histamine is released in an effort to protect the body from an allergen; however, sometimes an overload of histamine can result in life-threatening symptoms.

Prevention of excessive blood loss

Reduced circulation through blood clotting and coagulation prevents excessive bleeding and slows or prevents the entry of bacteria. Heparin is an anticoagulant (blood thinner) that prevents the formation of blood clots and is produced naturally in the lungs and liver.

Repair of damaged tissues

This is a necessary defence mechanism that includes removal of dead cells, increased availability of defender cells and replacement of tissue strength, for example, scar tissue caused by silica.

The lymphatic system

Acts as a form of 'drainage system' throughout the body for the removal of foreign substances. Lymphatic glands or nodes at specific points in the system act as selective filters, preventing infection from entering the blood system. In many cases, a localised inflammation occurs in the node at this time.

COMMON SIGNS AND SYMPTOMS OF ATTACK BY HAZARDOUS SUBSTANCES

Common signs of workplace harm to health are:

Lungs:	Wheezing, shortness of breath, coughing (possibly with presence of blood) and allergic asthma, with resultant, often permanent, reduction in lung functionality.
Skin:	Reddening of the skin and cracking blisters: common to dermatitis, often a chronic effect.
Nose and eyes:	Increased mucus, swelling and reddening, often acute effect on contact.
Digestive tract:	Nausea and vomiting, resulting from ingestion of a substance.

B2.2 - The identification, classification and health effects of hazardous substances used in the workplace

The influence of physical form and properties on routes of entry

The form taken by a hazardous substance is a contributory factor in its potential for harm. The form affects how easily a substance gains entry to the body, how it is absorbed into the body and how it reaches a susceptible site. Chemical agents take many forms, the most common being the primary forms or states - solids, liquids, gases, dusts, fibres, fumes, smoke, mists and vapours.

SOLIDS

Solids are materials that are solid at normal temperature and pressure. The atoms, molecules or ions that make up a solid may be arranged in an orderly repeating pattern or irregularly. Materials whose constituents are arranged in a regular pattern are known as crystals. Crystals that are large enough to see and handle are known as crystallites.

Other materials are called polycrystalline, which simply means they are composed of many crystallites of varying size and orientation. Almost all common metals, silicon and many ceramics are polycrystalline. Unlike a liquid, a solid object does not flow to take on the shape of its container, nor does it expand to fill the entire volume available to it like a gas. The risk from hazardous solids increases with reduction in particle size, particularly when it becomes a dust that can become airborne.

LIQUIDS

Liquids are substances that are liquid at normal temperature and pressure. Liquids have a definite volume but no fixed shape. Similar to a gas, a liquid is able to flow and take the shape of a container. Some liquids, such as water, resist compression, while others can be compressed.

Unlike a gas, a liquid does not disperse to fill every space of a container and maintains a fairly constant density. The density of a liquid is usually close to that of a solid and much higher than that of a gas.

> *"Liquid means a substance or mixture which:*
>
> *(i)   At 50ºC has a vapour pressure of not more than 300 kPa (3 bar).*
> *(ii) Is not completely gaseous at 20ºC and at a standard pressure of 101.3 kPa.*
> *(iii) Which has a melting point or initial melting point of 20ºC or less at a standard pressure of 101.3 kPa."*

Figure B2-11: Definition of liquid for purposes of classification and labelling.   *Source: Annex/CLP.*

GASES

Gases are formless fluids usually produced by chemical processes involving combustion or by the interaction of chemical substances. A gas will normally seek to fill the space completely into which it is liberated - for example, chlorine gas, carbon monoxide, methane.

> *"Gas means a substance which:*
>
> *(i) At 50ºC has a vapour pressure greater than 300 kPa (absolute).*
> *(ii) Is completely gaseous at 20ºC at a standard pressure of 101.3 kPa."*

Figure B2-12: Definition of gas for purposes of classification and labelling.   *Source: Annex/CLP.*

DUSTS

Dusts are solid airborne particles, often created by operations such as grinding, crushing, milling, sanding or demolition - for example, silica, cotton fibres, flour, cement. See later in this section for the distinction between inhalable and respirable dusts (and other particulates).

FIBRES

Dust may be created that is made up of tiny fibres, for example, mineral wool and asbestos.

FUMES

Fumes are solid particles formed by condensation from the gaseous state - for example, lead fume, welding fume.

SMOKE

Smoke is particles that result from incomplete combustion. They are a combination of gases and very small particles that can be in either solid or liquid state.

MISTS

Mists are finely dispersed liquid droplets suspended in air. Mists are mainly created by spraying, foaming, pickling and electro-plating - for example, mists from a water pressure washer, paint spray, pesticide, oil.

VAPOUR

Vapour is the gaseous form of a material normally encountered in a liquid or solid state at normal room temperature and pressure. Typical examples of vapours are those released from solvents - for example, trichloroethylene, which releases vapours when the container holding it is opened.

> "The terms 'dust', 'mist', and 'vapours' are defined as follows:
> **Dust:** Solid particles of a substance or mixture suspended in a gas (usually air).
> **Mist:** Liquid droplets of a substance or mixture suspended in a gas (usually air).
> **Vapour:** The gaseous form of a substance or mixture released from its liquid or solid state."

Figure B2-13: Definition of dust, mist and vapour. *Source: UN Globally Harmonised System of Classification and Labelling of Chemicals (GHS) part 3.*

Distinction between inhalable and respirable dust

The size of dust and other particles are usually described in microns (μm), a metric unit of measure where one micron is one-millionth of a meter. The eye can in general see objects > 40μm.

- **Inhalable dust** approximates to the fraction of airborne material which enters the nose and mouth during breathing; these are ≤ 100 μm diameter.
- **Thoracic dust** (course particles) will pass through the nose and throat, reaching the lungs; these are ≤ 10 μm diameter; they are referred to as PM10.
- **Repairable dust** (fine particles) will penetrate into the gas exchange region of the lungs. ≤ 2.5 μm diameter, or less; they are referred to as PM2.5.

Inhalable	up to 100 microns
Thoracic	up to 30 microns
Respirable	up to 12 microns

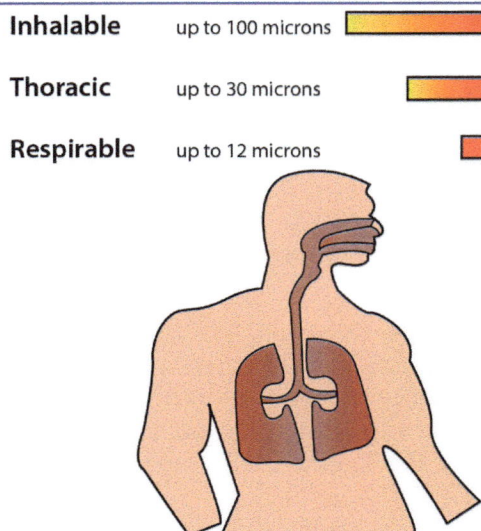

Figure B2-14: Entry through the nose and mouth. *Source: SKC.*

Purpose of classification and role of hazard and precautionary statements for hazardous substances

The Globally Harmonized System of Classification and Labelling of Chemicals (GHS) is the culmination of more than a decade of work. The purpose was to enable existing systems to be harmonised in order to develop a single, globally harmonised system to address classification of chemicals, labels, and safety data sheets (United Nations, 2011).

It uses three steps to outline the intrinsic hazardous properties of a chemical:

1) Identification of relevant data regarding the hazards of a substance or mixture.
2) Review of the data to determine the hazards associated with the substance or mixture.
3) A decision on whether the substance or mixture is classified as hazardous and the extent of the hazard when compared to an agreed hazard classification criteria.

A general principle adopted by GHS is that test data already generated for the classification of a chemical under the various existing systems should be accepted when classifying chemicals under GHS. This is to avoid unnecessary duplicative testing and unnecessary use of test animals.

Because GHS is not a treaty, and is not a legally binding international agreement, countries (or trading blocks) must create local or national legislation to implement the GHS. The European Union have implemented the GHS under European Regulation No 1272/2008 on Classification, Labelling and Packaging of Substances and Mixtures (CLP), which came into force on 20th January 2009 in all EU Member States.

Criteria for classifying chemicals have been established in the GHS and adopted in the CLP Regulations for the following health hazard classes:

- Acute toxicity.
- Skin corrosion/irritation.
- Serious eye damage/eye irritation.
- Respiratory or skin sensitisation.
- Germ cell mutagenicity.

- Carcinogenicity.
- Reproductive toxicity.
- Specific target organ toxicity - single exposure.
- Specific target organ toxicity - repeated exposure.
- Aspiration hazard.

A substance may be classified under more than one classification, although those with specific target organ toxicity only apply if other classifications do not. The various hazard classifications for substances harmful to health are defined in Annex 1 of GHS. The pictograms associated with the classifications are set out in tables in the annex.

The following section on hazard classifications and their associated pictograms are drawn from Annex 1 of GHS. The information is presented in a simplified format, for full details of the classifications the annex should be consulted. Text provided in brackets in a quotation is not part of the quotation from the GHS, but has been added so that it can be more easily understood.

Health hazard classes

ACUTE TOXICITY

'Acute toxicity refers to those adverse effects occurring following oral or dermal administration of a single dose of a substance, or multiple doses given within 24 hours, or an inhalation exposure of 4 hours.'

The classification is divided into 5 categories related to the level of toxicity of the chemical.

Acute oral toxicity - Annex 1 of GHS					
	Category 1	Category 2	Category 3	Category 4	Category 5
LD_{50}	≤ 5 mg/kg	>5 <50 mg/kg	>50 <300 mg/kg	>300 <2,000 mg/kg	>2,000 <5,000 mg/kg
Pictogram					No symbol
Signal word	Danger	Danger	Danger	Warning	Warning
Hazard statement	Fatal if swallowed	Fatal if swallowed	Toxic if swallowed	Harmful if swallowed	May be harmful if swallowed

The classification of categories and labelling for acute skin and inhalation toxicity are the same as oral exposure, except for slightly different hazard statements. An example of a substance with acute oral toxicity is arsenic, which is a systemic poison.

Another example is methanol, which is known to cause lethal intoxications in humans (mostly via ingestion) in relatively low doses and is a category 3 substance for acute oral toxicity.

SKIN CORROSION/IRRITATION

'Skin corrosion is the production of *irreversible* damage to the skin; namely, visible necrosis through epidermis and into dermis, following the application of a test substance for 4 hours. Corrosive reactions are typified by ulcers, bleeding, bloody scabs, and, by the end of observation at 14 days, by discolouration due to blanching of the skin, complete areas of alopecia, and scars.'

Skin corrosion/irritation - Annex 1 of GHS					
	Category 1A	Category 1B	Category 1C	Category 2	Category 3
Pictogram					No symbol
Signal word	Danger	Danger	Danger	Warning	Warning
Hazard statement	Causes severe skin burns and eye damage	Causes severe skin burns and eye damage	Causes severe skin burns and eye damage	Causes skin irritation	Causes mild skin irritation

The corrosion effect is divided into three categories related to duration of exposure necessary to create an effect.

'Skin irritation is the production of **reversible** damage to the skin following the application of a test substance for up to 4 hours.'

Sulphuric (battery) acid and sodium hydroxide (caustic soda) are examples of substances classified under skin corrosion.

SERIOUS EYE DAMAGE/EYE IRRITATION

'Serious eye damage is the production of tissue damage in the eye, or serious physical decay of vision, following application of a test substance to the anterior surface of the eye, which is **not fully reversible** within 21 days of application.'

'Eye irritation is the production of changes in the eye following the application of a test substance to the anterior surface of the eye, which are **fully reversible** within 21 days of application.' This generally relates to effects on the cornea, iris or conjunctiva.

Serious eye injury/eye irritation - Annex 1 of GHS			
	Category 1	Category 2A	Category 2B
Pictogram			No symbol
Signal word	Danger	Warning	Warning
Hazard statement	Causes severe eye damage	Causes severe eye irritation	Causes eye irritation

The difference between category 1 and 2 for eye injury is whether the harm to the eye is fully reversible within the observation period. A category 2B substance is one where it is considered to be mildly irritating to eyes and fully reversible within 7 days of observation.

RESPIRATORY OR SKIN SENSITISATION

'A respiratory sensitiser is a substance that will lead to hypersensitivity of the airways following inhalation of the substance. A skin sensitiser is a substance that will lead to an allergic response following skin contact.'

Sensitisation takes place in two stages; the first is the recognition stage, where the substance, following contact with the airways or skin, is recognised by the body as a pathogen (something that can cause harm to health). The second stage is the antibody generation and allergic response to further exposure to the substance.

Respiratory sensitisation - Annex 1 of GHS			
	Category 1	Category 1A	Category 1B
Pictogram			
Signal word	Danger	Danger	Danger
Hazard statement	May cause allergy or asthma symptoms or breathing difficulty if inhaled	May cause allergy or asthma symptoms or breathing difficulty if inhaled	May cause allergy or asthma symptoms or breathing difficulty if inhaled

The three categories of respiratory sensitisers relate to the type and level of evidence that identifies the substance as a sensitiser. Category 1 has been established by direct evidence that exposure will lead to specific hypersensitivity, for example, asthma, rhinitis/conjunctivitis and alveolitis, whereas Category 1B substances show a low to moderate frequency of occurrence of sensitisation. Flour dust and isocyanates are examples of respiratory sensitisers.

Skin sensitisation - Annex 1 of GHS			
	Category 1	Category 1A	Category 1B
Pictogram			
Signal word	Warning	Warning	Warning
Hazard statement	May cause an allergic skin reaction	May cause an allergic skin reaction	May cause an allergic skin reaction

As with respiratory sensitisers the three categories for skin sensitisers relate to the type and level of evidence that identifies the substance as a sensitiser. Category 1 substances are ones where there is strong documented evidence of causing allergic contact dermatitis. Nickel and epoxy resins are examples of skin sensitisers.

GERM CELL MUTAGENICITY

'This hazard class is primarily concerned with chemicals that may cause mutations in germ cells of humans that can be transmitted to the progeny (future generations, i.e. children). A mutation is defined as a permanent change in the amount of the genetic material in a cell.'

Germ cell mutagenicity - Annex 1 of GHS			
	Category 1A	Category 1B	Category 2
Pictogram			
Signal word	Danger	Danger	Warning
Hazard statement	May cause genetic defects	May cause genetic defects	Suspected of causing genetic defects

The different categories of mutagenicity reflect the degree of knowledge about the chemical and the indications that it may cause mutagenicity.

Category 1 substances are known or are presumed, because of related evidence, to cause an inherited mutation.

Whereas category 2 substances are ones where there is concern that they may induce heritable mutations. If a specific route of exposure is proven to be the only route causing harm, this route must be stated in the hazard statement.

CARCINOGENICITY

'The term carcinogen denotes a substance or mixture that induces cancer or increases its incidence. Substances and mixtures that have induced benign and malignant tumours in well performed experimental studies on animals are considered also to be presumed or suspected human carcinogens unless there is strong evidence that the mechanism of tumour formulation is not relevant to humans.'

Carcinogenicity - Annex 1 of GHS			
	Category 1A	Category 1B	Category 2
Pictogram			
Signal word	Danger	Danger	Warning
Hazard statement	May cause cancer	May cause cancer	Suspected of causing cancer

As with the different categories of mutagenicity, the categories for carcinogenicity reflect the degree of knowledge about the chemical and the indications that it may cause cancer.

Category 1 substances are known, or are presumed, because of related evidence, to induce cancer.

Whereas category 2 substances are ones where there is concern that they may induce cancer. If a specific route of exposure is proven to be the only route causing harm, this route must be stated in the hazard statement. Benzyl chloride is an example of a carcinogenic substance. Another example is benzene, which affects bone marrow, causing leukaemia.

REPRODUCTIVE TOXICITY

'Reproductive toxicity includes effects on sexual function and fertility in adult males and females, as well as developmental toxicity in the offspring. The genetically based inheritable effects in offspring come under the classification 'Germ cell mutagenicity'.'

The classification covers two main headings:

1) Adverse effects on sexual function and fertility - including alterations to the reproductive system, effects on the onset of puberty or the reproductive cycle, sexual behaviour, fertility and pregnancy outcomes.
2) Adverse effects on development of the offspring - including interference with the development of the foetus or child, before or after birth, resulting from exposure of either parent prior to conception or during development of the offspring.

Reproductive toxicity - Annex 1 of GHS				
	Category 1A	Category 1B	Category 2	Additional category on effects on or via lactation
Pictogram				No pictogram
Signal word	Danger	Danger	Warning	No signal word
Hazard statement	May damage fertility or the unborn child	May damage fertility or the unborn child	Suspected of damaging fertility or the unborn child	May cause harm to breast-fed children

The classification principally provides a warning for pregnant women, and men and women of reproductive capacity. Category 1 substances are known, or are presumed, because of related evidence, to be a human reproductive toxicant. Whereas category 2 substances are ones where there is concern that they may be a human reproductive toxicant. If a specific route of exposure is proven to be the only route causing harm, this route must be stated in the hazard statement.

The UN GHS advises that 'for many substances there is no information on their potential to cause adverse effects on the offspring via lactation. However, substances that are absorbed by women and have been shown to interfere with lactation, or which may be present (including metabolites) in breast milk in amounts sufficient to cause concern for the health of a breast-fed child, should be classified to indicate this property hazardous to breast fed babies.'

An example of a substance classified as reproductive toxic is 2-Ethoxyethanol, which has been implicated in impairing fertility. In addition, the effect of lead on the development of the brain of an unborn foetus has been long established.

SPECIFIC TARGET ORGAN TOXICITY - SINGLE EXPOSURE

This classification is for substances that produce specific, non-lethal, target organ toxicity, arising from a single exposure. The effects may be reversible or non-reversible, immediate or delayed, but are not covered in other classifications. Specific target organ toxicity can occur by any route and, therefore, includes oral, dermal and inhalation routes.

Specific target organ toxicity - single exposure - Annex 1 of GHS				
	Category 1	Category 2	Category 3	Category 3
Pictogram				
Signal word	Danger	Warning	Warning	Warning

Hazard statement	Causes damage to organs	May cause damage to organs	(Respiratory tract irritation) May cause respiratory irritation	(Narcotic effects) May cause drowsiness or dizziness

Category 1 substances are known to have produced significant toxicity in humans. Category 2 substances are presumed, because of related evidence, to produce significant toxicity in humans. Whereas category 3 substances cause transient target organ effects, that affect the respiratory tract or have a narcotic effect. Narcotic effects involve depression of the central nervous system, including drowsiness, loss of reflexes, lack of co-ordination, vertigo and reduced alertness. The symptoms may include severe headache, nausea, dizziness, sleepiness, irritability, fatigue, impaired memory function, perception, co-ordination and reaction time. Some solvents causing narcosis or central nervous system failure may take effect after a single exposure.

SPECIFIC TARGET ORGAN TOXICITY - REPEATED EXPOSURE

As with the similar classification for single exposure this classification is for substances that produce specific, non-lethal, target organ toxicity, but arising from repeated exposure. The effects may be reversible or non-reversible, immediate or delayed, but are not covered in other classifications. Specific target organ toxicity can occur by any route and, therefore, includes oral, dermal and inhalation routes.

Specific target organ toxicity - repeat exposure - Annex 1 of GHS		
	Category 1	Category 2
Pictogram		
Signal word	Danger	Warning
Hazard statement	Causes damage to organs through prolonged or repeated exposure	May cause damage to organs through prolonged or repeated exposure

Category 1 substances are known to have produced significant toxicity in humans. Category 2 substances are presumed, because of related evidence, to produce significant toxicity in humans. If a specific route of exposure is proven to be the only route causing harm, this route must be stated in the hazard statement. Similarly, if specific organs are known to be affected these organs must be stated.

ASPIRATION HAZARD

'Aspiration means the entry of a liquid or solid chemical directly through the oral or nasal cavity, or indirectly from vomiting, into the trachea and lower respiratory system'. Aspiration toxicity includes severe acute effects such as chemical pneumonia, varying degrees of pulmonary injury or death following aspiration.

Aspiration hazard - Annex 1 of GHS		
	Category 1	Category 2
Pictogram		
Signal word	Danger	Warning
Hazard statement	May be fatal if swallowed and enters airways	May be harmful if swallowed and enters airways

Category 1 substances are known or are regarded to cause aspiration toxicity in humans. Category 2 substances are those that cause concern of aspiration toxicity in humans.

Purpose of the European Regulation Registration, Evaluation, Authorisation and restriction of Chemicals (REACH)

REACH (Registration, Evaluation, Authorisation and Restriction of Chemicals) is the system for controlling chemicals in the EU. REACH is an EU Regulation that applies throughout the EU, including the UK. It has been adopted to improve the protection of human health and the environment from the risks that can be posed by chemicals.

In principle, REACH applies to all chemical substances; not only those used in industrial processes, but also used by the public, for example in cleaning products, paints as well as in articles such as clothes, furniture and electrical appliances. Therefore, the Regulation has an impact on most companies across the EU. It also promotes alternative methods for the hazard assessment of substances in order to reduce the number of tests on animals.

REACH provides a common EU system to gather hazard information, assess risks, classify, label, and restrict the marketing and use of individual chemicals and mixtures.

This is known as the REACH system:

R   egistration of basic information of substances to be submitted by companies, to a central database.

E   valuation of the registered information to determine hazards and risks.

A   uthorisation and restriction requirements imposed on the use of substances of very high concern.

CH   emicals.

REACH establishes procedures for collecting and assessing information on the properties and hazards of substances. Companies need to register their substances and to do this they need to work together with other companies who are registering the same substance. The European Chemicals Agency (the Agency; ECHA) receives and evaluates individual registrations for their compliance, and the EU Member States evaluate selected substances to clarify initial concerns for human health or for the environment. Authorities in member states and ECHA's scientific committees assess whether the risks of substances can be managed.

Authorities can ban hazardous substances if their risks are unmanageable. They can also decide to restrict a use or make it subject to a prior authorisation. If the risks cannot be managed, authorities can restrict the use of substances in different ways. In time, the most hazardous substances should be substituted with less dangerous ones.

REACH requires information, in the form of safety data sheets, to be provided to users by those that supply chemicals. 'Supply' means making a chemical available to another person. Manufacturers, importers, distributors, wholesalers and retailers are all examples of suppliers.

Aims, or purpose of REACH is:

- To provide a high level of protection of human health and the environment from the use of chemicals.
- To make the people who place chemicals on the market (manufacturers and importers responsible for understanding and managing the risks associated with their use).
- To allow the free movement of substances on the EU market.
- To enhance innovation in and the competitiveness of the EU chemicals industry.
- To promote the use of alternative methods for the assessment of the hazardous properties of substances, for example, quantitative structure-activity relationships (QSAR) and read across.

Hazardous substances

Hazardous substances are used and can be present in many workplaces, detailed in the following section are examples of hazardous substances, their routes of entry, target organs and likely acute/chronic health effect/s.

CARBON MONOXIDE

Carbon monoxide (CO) is an odourless, colourless gas that reduces the ability of haemoglobin to transport and deliver oxygen. Carbon monoxide is produced when organic material, such as coal, wood, paper, oil, gasoline, gas, explosives or any other carbonaceous material, is burned in a limited supply of air or oxygen. Naturally occurring sources produce 90% of atmospheric CO, and activity some 10%. Motor vehicles account for 55 to 60% of global man-made CO burden. The exhaust gas of gasoline-fuelled combustion engine (spark ignition) is a common source of ambient CO. The diesel engine (compression ignition) exhaust gas contains about 0.1% of CO when the engine is operating properly, but maladjusted, overloaded or badly maintained diesel engines may emit considerable amounts of CO.

Carbon monoxide competes with oxygen for the binding sites of the haemoglobin molecules. The affinity of human haemoglobin for CO is about 240 times that of its affinity for oxygen. The appearance of symptoms depends on the concentration of CO in the air, the exposure time, the degree of exertion and individual susceptibility. If the exposure is massive, loss of consciousness may take place almost instantaneously with few or no predictive signs and symptoms. Continued exposure will lead to death, separating acute conditions from chronic conditions is therefore, dependent on the concentration of the gas.

Exposure to concentrations of 10,000 to 40,000 ppm leads to death within a few minutes. Levels between 1,000 and 10,000 ppm cause acute symptoms of headaches, dizziness and nausea in 13 to15 minutes. Chronic symptoms of unconsciousness and death will follow if exposure continues for 10 to 45 minutes, the rapidity of onset depending on the concentrations. Below these levels the time before the onset of symptoms is longer: levels of 500 ppm cause headache after 20 minutes and levels of 200 ppm after about 50 minutes.

Carbon monoxide does not accumulate in the body. It is completely excreted after each exposure if sufficient time in fresh air is allowed. It is possible, however, that repeated mild or moderate poisonings which do not lead to unconsciousness would result in death of brain cells and ultimately lead to central nervous system damage with a multitude of possible symptoms, such as headache, dizziness, irritability, impairment of memory, personality changes and a state of weakness of the limbs.

ISOCYANATES

Isocyanates are a family of reactive low molecular weight chemicals and can be found in a range of construction industry products such as paints, coatings, foams, glues and flooring. They can also be found in the automobile industry in a range of products. They can be produced in different forms (solid crystals and liquids) dependent upon the type of isocyanate. The relative vapour density of most isocyanides is greater than 1 (air =1) so the vapours will pool at low levels. Methyl Isocyanate (MIC) will decompose on heating producing toxic gases (hydrogen cyanide, nitrogen oxides, carbon monoxide).

Isocyanates are powerful irritants to the mucous membrane, eyes, the gastrointestinal tract and the respiratory tract. Therefore routes of entry are likely to be inhalation and ingestion. Acute symptoms are likely to be a dry throat, sore eyes, coughing and shortness of breath. Continued exposure is likely to lead to sensitisation with shortness of breath leading to asthma which may be sufficiently severe as to cause death.

METAL WORKING FLUIDS

Metalworking fluids - sometimes referred to as suds, coolants, slurry or soap - are used during the machining of metals to provide lubrication and cooling, and to help carry away debris such as swarf and fine metal particles. They can also help to improve machining performance and prolong the life of the cutting tool, as well as provide corrosion protection for the surfaces of work pieces. Metalworking fluids are mostly applied by continuous jet, spray, or hand dispenser.

Routes of entry are therefore:

- Inhalation of the mist, aerosol or vapour generated during machining operations.
- Absorption through contact with the skin, particularly hands and forearms.
- Absorption by entering your body through cuts and abrasions or other broken skin.
- Ingestion when entering your body through the mouth if you eat or drink in work areas, or do not wash your hands before eating or smoking.

Neat oils in regular and prolonged contact with the skin can cause irritation of the hair roots. Also, fine microscopic metal particles, which are generated during machining, can damage the skin and make any existing irritation worse. Workers exposed to metalworking fluid, mist and vapour have an increased risk of developing acute health effect such as work-related asthma, bronchitis, irritation of the respiratory tract and breathing difficulties. It can also cause extrinsic allergic aveolitis, which can lead to chronic illnesses and cause increasingly severe breathing difficulties in recurrent episodes, following repeated exposure.

Acute conditions will also include irritation to the eyes, nose and throat. In the past, the use of unrefined mineral oils led to skin cancer affecting the exposed skin, often hands and forearms. Also, oil-soaked clothing and oily rags kept in overalls caused cancer of the scrotum. Today, the use of highly refined oils and the substitution of cancer-causing chemicals in metalworking fluids, as well as changes in work practices and improved personal hygiene, have reduced the risk of cancer.

USED ENGINE OIL

Motor oil picks up a variety of hazardous contaminants when used in engines and transmissions. These contaminants include lead, cadmium, chromium, arsenic, dioxins, benzene and polycyclic aromatics. If used motor oil and the contaminants it contains are disposed of inappropriately and released into the environment, they can harm humans, plants, animals, fish and shellfish.

Frequent and prolonged contact with used engine oil may cause dermatitis and other skin disorders, including skin cancer. Benzene is a natural component of both crude oil and petrol. It can be absorbed into your body through the skin or if you breathe it in and long-term exposure can lead to serious blood disorders such as anaemia and leukaemia (a form of cancer).

However, the risk of exposure to benzene from petrol is very small as its content in the EU is restricted to less than 1%. Sensible precautions when handling petrol (which is not only a fire and explosion risk but can defat the skin leading to a risk of dermatitis) should be more than adequate to protect against benzene. Used engine oils present a problem if they are absorbed by coming into contact with the eyes or skin, triggering an allergic skin reaction or eye irritation. Chronic affects from repeated contact include a higher risk of developing skin cancer.

SILICA

Silica is a natural substance found in most rocks, sand and clay and in products such as bricks and concrete. Silica is also used as filler in some plastics. In the workplace these materials create dust when they are cut, sanded down etc. The dust is too fine to see with normal lighting and some of this dust may be fine enough to reach deep inside the lung, this is known as respirable crystalline silica (RCS) and can cause harm to health. The primary route of entry for silica dust into the body is therefore through inhalation, and while silica does not have a 'target organ' particles lodging in the lungs will cause ill-health. Significant exposure to RCS can cause silicosis and lung cancer.

The acute effects of inhalation of silica dust can be asthma like symptoms and breathing difficulties leading to 'silicosis' where breathing is more difficult and increases the risk of lung infections. Silicosis usually follows exposure to RCS over many years, but extremely high exposures can lead rapidly to ill-health. Chronic effects include 'Chronic Obstructive Pulmonary Disease' (COPD). COPD is a group of lung diseases, including bronchitis and emphysema, resulting in severe breathlessness, prolonged coughing and chronic disability. It may be caused by breathing in any fine dusts, including RCS. It can be very disabling and is a leading cause of death. Cigarette smoking can make it worse. Heavy and prolonged exposure to RCS can cause lung cancer. When someone already has silicosis, there is an increased risk of lung cancer.

WOOD DUSTS

In addition to the tiny particles of wood produced during processing, wood dust can also contain bacteria, fungal and moss spores. The quantity and type of wood dust will depend on the wood being cut and the machine being used. The risk is from fine dust being inhaled, penetrating into the lungs where it will do the most damage.

Wood is classified into two general families, hardwood and softwood. The classification is botanical and depends on the fine structure of the cells in the wood species. It does not refer to the physical properties of the wood. Toxic activity is specific to a wood species, so knowing the exact species is important in establishing what the potential toxic effects may be.

Acute effects may be irritation to the skin from contact on the forearm, backs of the hands, the face (particularly eyelids), neck and scalp. Poor personal hygiene can also cause infection to the genitals. On average, they take 15 days to develop. Symptoms usually only persist as long as the affected skin site remains in contact with the source of irritation such as the wood dust or sap etc. Symptoms subside when contact with the irritant is removed. Sensitisation dermatitis is more challenging and is usually caused by skin exposure to fine wood dust of certain species. This is also referred to as allergic contact dermatitis and results in similar skin effects to those produced by skin irritants.

Respiratory and allied effects

Wood, especially inhalation of fine dust, can have many effects on the respiratory tract, including:

Nose:
- Rhinitis (runny nose).
- Violent sneezing.
- Blocked nose.
- Nose bleeds.
- Very rarely - nasal cancer (a recognised industrial disease associated with the inhalation of hardwood dusts).

Lungs:
- Asthma.
- Impairment of lung function.
- Rarely - extrinsic allergic alveolitis (a disease with 'flu-like' symptoms which can cause progressive lung damage), for example, when using western red cedar or iroko.

The International Agency for Research on Cancer (IARC) classifies wood dust as a carcinogen to humans (Group 1). This classification is based primarily on IARC's evaluation of increased risk in the occurrence of carcinomas of the nasal cavities and paranasal sinuses associated with exposure to wood dust.

Asthma is of particular concern as a chronic condition. Wood dusts can irritate the respiratory tract provoking asthma attacks in sufferers, although effective control of dust levels normally improves the problem. Wood dusts will also cause acute conditions to the eyes (conjunctivitis).

ASBESTOS

Asbestos is a term used to describe a group of naturally occurring fibrous minerals which are widely distributed throughout the world. The asbestos minerals fall into two groups - the serpentine group, which includes Chrysotile, and the amphiboles, which include Crocidolite, Tremolite, Amosite and Anthophyllite.

Chrysotile and the various amphibole asbestos minerals differ in crystalline structure, in chemical and surface characteristics, and in the physical characteristics of their fibres. When materials that contain asbestos are disturbed or damaged, fibres are released into the air. When these fibres are inhaled they can cause serious diseases. These often take a long time to develop, but once diagnosed, it is often too late to do anything.

Acute conditions from asbestos exposure will include irritation from contact with the skin (fibres may penetrate the skin leading to 'asbestos corns') or contact with the eyes (redness and sore). Asbestos can also cause gastrointestinal problems. Chronic conditions include; Asbestosis, asbestos-related pleural disease, malignant mesothelioma and lung cancer are specific diseases associated with exposure to asbestos dust. The fibrotic changes which characterise the pneumoconiosis, asbestosis, are the consequence of an inflammatory process set up by fibres retained in the lung. Asbestos target organs include the lungs, stomach and colon.

LEAD

Lead ores are found in many parts of the world. Metallic lead is used in the form of sheeting or pipes where pliability and resistance to corrosion are required, such as in chemical plants, it is also used for cable sheathing, as an ingredient in solder and is a valuable shielding material for ionizing radiations. It is used for metallising to provide protective coatings, in the manufacture of storage batteries and as a heat treatment bath in wire drawing.

Lead is present in a variety of alloys and its compounds are prepared and used in large quantities in many industries. About 40% of lead is used as a metal, 25% in alloys and 35% in chemical compounds.

The prime hazard of lead is its toxicity. Clinical lead poisoning has always been one of the most important occupational diseases. The main route of entry in industry is the respiratory tract. A certain amount may be absorbed in the air passages, but the main portion is taken up by the pulmonary bloodstream. The degree of absorption depends on the proportion of the dust accounted for by particles less than 5 microns in size and the exposed worker's respiratory volume. Increased workload therefore results in higher lead absorption.

Although the respiratory tract is the main route of entry, poor work hygiene, smoking during work (pollution of tobacco, polluted fingers while smoking) and poor personal hygiene may considerably increase total exposure mainly by the oral route. In the human body, inorganic lead is not metabolized but is directly absorbed, distributed and excreted. Once in the blood, lead is distributed primarily among three compartments: blood, soft tissue (kidney, bone marrow, liver, and brain), and mineralizing tissue (bones and teeth).

Whether lead enters the body through inhalation or ingestion, the biologic effects are the same; there is interference with normal cell function and with a number of physiological processes. The most sensitive target of lead poisoning is the central nervous system (CNS). Acute effects in the CNS are manifested by subtle behavioural changes, fatigue and impaired concentration.

Chronic exposure is likely to lead to recurrent (sometimes nearly continuous) episodes of complete disorientation with hallucinations, facial contortions and intense general somatic muscular activity with resistance to physical restraint. Such episodes may be converted abruptly into maniacal or violent convulsive seizures which may terminate in coma and death.

B2.3 - The assessment and evaluation of risk from hazardous substances

Information on substances or preparations/mixtures which have the potential to cause harm

PRODUCT LABELS

Hazardous chemicals have to be labelled so that workers are informed about their effects when they are exposed to them. The label should draw attention to the inherent hazards to those handling or using the chemical and provide information on precautions to prevent harm.

EUROPEAN REGULATION ON CLASSIFICATION, LABELLING AND PACKAGING OF CHEMICAL SUBSTANCES

The European Union (EU) Regulation on Classification, Labelling and Packaging of Chemical Substances and Mixtures (CLP) introduced throughout the EU a system for classifying and labelling chemicals based on the United Nations' Globally Harmonised System (GHS). The CLP regulation is concerned with the hazards of chemical substances and mixtures and how to inform others about them. Its purpose is to protect people and the environment from the effects of those chemicals by requiring suppliers to classify and provide information about the hazards of the chemicals and to package them safely.

It is the responsibility of manufacturers to establish what the hazards of substances and mixtures are before they are placed on the market, and to classify them in line with the identified hazards. When a substance or a mixture is hazardous, it has to be labelled in accordance with CLP so that workers are informed about its effects if it is used. Note that 'mixture' is the same as the term 'preparation', which has been used previously.

All substances available for use in the workplace should be labelled in accordance with CLP, including the use of the appropriate CLP hazard pictogram. Information on the use of symbols to indicate the harm related to various dangerous substance hazard classifications is provided earlier in this element.

The CLP regulation requires hazardous substances to carry a hazard label in a specified format, which is made up of specific symbols (known as 'pictograms'), warnings and precautions. These pictograms and the wording that supports them are defined in the CLP regulation, which requires chemical suppliers to use them where hazardous properties have been identified. Article 17 of the CLP Regulation requires that hazard labels include the following elements:

- Name, address and telephone number of the supplier(s).
- The nominal quantity of the substance or mixture in the package where this is being made available to the general public, unless this quantity is specified elsewhere on the package.

- Product identifiers.
- Hazard pictograms, where applicable.
- The relevant signal word, where applicable.
- Hazard statements, where applicable.
- Appropriate precautionary statements, where applicable.
- A section for supplemental information, where applicable.

A hazard **pictogram** is a graphical composition that includes a symbol plus other graphic elements, such as a border, background pattern or colour that is intended to convey specific information on the hazard concerned.

The pictogram forms an integral part of the label and gives an immediate idea of the types of hazards that the substance or mixture may cause.

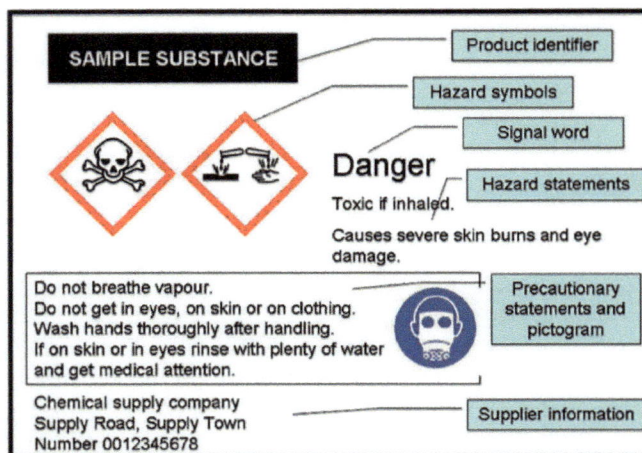

Figure B2-15: Sample GHS chemical labels. *Source: RMS.*

A **signal word** indicates the relative level of severity of hazardous substances to alert the potential reader of the significance of the hazard. More severe hazards are identified by the signal word 'danger', while less severe hazards are identified by the signal word 'warning'.

A **hazard statement** is a phrase assigned to a hazard class and category that describes the nature and severity of the hazard.

A **precautionary statement** is a phrase and/or pictogram that describe recommended measure(s) to prevent or minimise adverse effects resulting from exposure to a hazardous substance or mixture due to its use. A maximum of six precautionary statements is allowed under CLP requirements. Precautionary statements cover statements for prevention, response to problems or contamination, storage and disposal.

Hazard statements		*Precautionary statements*	
H300	Fatal if swallowed	P102	Keep out of reach of children
H320	Causes eye irritation	P271	Use only outdoors or in well ventilated area
H336	May cause drowsiness or dizziness	P331	Do not induce vomiting

Figure B2-16: Hazard and precautionary statements. *Source: EU, CLP Regulation, Annex III and IV.*

GUIDANCE DOCUMENTS

Guidance documents are used in many countries to set occupational exposure limit values for work activities. They are set by competent national authorities or other institutions acting nationally or internationally. Guidance documents set limits for concentrations of hazardous substances in workplace air.

Occupational exposure limit values for hazardous substances represent an important source of information for risk assessment and management of health risks. They provide valuable information for occupational health and safety activities involving hazardous substances and may carry legal status under national laws. As knowledge is gained about the health effects of substances, the occupational exposure limits are updated. Therefore, it is important to obtain current data on occupational exposure limits. There are a number of guidance documents that are updated regularly and provide useful sources of information, including the EU list of indicative occupational exposure limit values. There is also guidance in the form of the Health and Safety Executive (HSE) list of workplace exposure limits.

Occupational exposure limits related to a large number of chemicals are set out in the UK HSE Guidance Note EH40, which is prepared and published annually. The Guidance Note EH40 contains lists of occupational exposure limits, called workplace exposure limits (WELs) in the document, for use with the Control of Substances Hazardous to Health Regulations (COSHH) 2002. It also contains a description of the limit-setting process, technical definitions and explanatory notes.

In the EU, the list of indicative occupational exposure limit values (IOELVs), also commonly referred to as the List of Indicative Limit Values, contains human exposure limits to hazardous substances specified by the Council of the European Union, based on expert research and advice. They are not binding on member states but must be taken into consideration in setting national occupational exposure limits. The EU IOELVs are taken into account when the UK WELs are established. EH40 may contain WELs for substances that do not have an EU indicative occupational exposure limit value, because data may be established in the UK that enables a limit to be set before there is agreement in the EU.

SAFETY DATA SHEETS (SDS)

Under REACH, each EU manufacturer or importer of chemicals in quantities of one or more tonnes per year - around 30,000 substances - will have to register them with the European Chemicals Agency (ECHA), submitting information on properties, uses and safe ways of handling them. Health and safety information must be passed down the supply chain to 'downstream users', so that they know how to use the substances without creating risks for workers, consumers and the environment. REACH will restrict the use of the most dangerous substances, and for other substances of very high concern, use-specific authorisation will be required. This will only be granted to companies that can show that the risks are adequately controlled or if social and economic benefits outweigh the risks and where there are no suitable alternative substances or technologies. This is to encourage substitution for safer alternatives.

REACH requires that a Safety Data Sheet (SDS) be provided by the supplier for the following:

a) A substance or a mixture that is classified as hazardous under CLP.

b) A substance that is persistent, bio-accumulative and toxic (PBT), or very persistent and very bioaccumulative (vPvB) as defined in Annex XIII of REACH.

c) A substance that is included in the European Chemicals Agency's 'Candidate List' of substances of very high concern.

If a customer requests a SDS for a mixture that is not classified as dangerous under Directive 1999/45/EC, but contains either:

■ A substance posing human health or environmental hazards in an individual concentration of ≥ 1 % by weight for mixtures that are solid or liquids (i.e., non-gaseous mixtures) or ≥ 0.2 % by volume for gaseous mixtures.

■ A substance that is PBT, or vPvB in an individual concentration of ≥ 0.1 % by weight for mixtures that are solid or liquids (i.e., non-gaseous mixtures).

■ A substance on the 'Candidate List' of substances of very high concern, with an individual concentration of ≥ 0.1 % by weight for non-gaseous mixtures.

■ A substance for which there is Europe-wide workplace exposure limits.

■ A product listed as a 'special case' of CLP, for example, gas containers intended for propane, butane or liquefied petroleum gas.

A SDS does not need to be provided:

1) If the substances/mixtures are supplied in the UK and not classified as hazardous or considered PBT, vPvB or of equivalent concern (for example, endocrine disruptors).

2) For certain products intended for the final user, for example, medicinal products or cosmetics.

3) Suppliers of dangerous substances or mixtures to the general public if sufficient information is provided to enable users to take the necessary measures as regards safety and the protection of human health and the environment. However, a downstream user or distributor can request one is provided.

A SDS should be provided to the recipient free-of-charge, on paper or electronically, for example, by postal delivery, fax or email. The HSE states that 'a system that merely requires customers to obtain a SDS from a company's website or from a catalogue of SDS is not considered appropriate'. A SDS should be provided either before or at the time of first delivery of the substance or mixture. Where a customer re-orders substances or mixtures, then the supplier does not need to re-supply the SDS, unless the SDS contents have changed.

SDS Information

The safety data sheet shall be dated and shall contain the following headings and address the content under the headings:

1) Identification of the substance/mixture and the company/undertaking.

■ Product identifier, name of the substance.

■ Relevant identified uses of the substance or mixture and uses advised against, for example, "flame retardant".

■ Details of the supplier of the safety data sheet, name, address, telephone number and email address.

■ Emergency telephone number.

2) Hazards identification.

■ Classification of the substance or mixture, including the important hazards to man and the environment - hazard statements and risk phrases, referenced to section 16.

■ Label elements, hazard pictogram(s), signal word(s), hazard statement(s) and precautionary statement(s)

■ Other hazards, for example, dust explosion or freezing.

3) Composition/information on ingredients.

■ Sufficient information to allow the recipient to identify readily the associated risks.

4) First-aid measures.

■ Description of first aid measures for each relevant route of exposure.

- Advice on action of first aider, for example, whether immediate attention is required.
- Symptoms and effects including delayed effects.
- Indication of any immediate medical attention and special treatment is needed, for example, antidotes or medical monitoring.

5) Fire-fighting measures.

- Extinguishing media, suitable extinguishing media and extinguishing media that must not be used.
- Special hazards that may arise from combustion for example, gases, fumes etc.
- Advice for fire fighters, for example, special protective equipment for fire fighters.

6) Accidental release measures.

- Personal precautions such as removal of ignition sources, provision of ventilation, avoid eye/skin contact etc.
- Environmental precautions such as keep away from drains; need to alert neighbours etc.
- Methods for cleaning up for example, absorbent materials. Also, "Never use...."

7) Handling and storage.

- Precautions for safe handling, for example, measures to prevent aerosol, dust, fire and advice on occupational hygiene.
- Conditions for safe storage, including incompatibilities, for example, how to manage flammability hazards, how to control the effects of sunlight, ventilation requirements.
- Specific end use(s), detailed and operational.

8) Exposure controls/personal protection.

- Control parameters, exposure limit values.
- Exposure controls, for example, engineering controls taken in preference to personal protective equipment (PPE) and where PPE is required, type of equipment necessary for example, type of gloves, goggles, barrier cream etc.
- Environmental exposure controls.

9) Physical and chemical properties.

- Appearance, for example, solid, liquid, powder, etc.
- Odour (if perceptible).
- Boiling point, flash point, explosive properties, solubility etc.

10) Stability and reactivity.

- Reactivity.
- Chemical stability.
- Possibility of hazardous reactions.
- Conditions to avoid such as temperature, pressure, light, etc.
- Incompatible materials to avoid such as water, acids, alkalis, etc.
- Hazardous by-products given off on decomposition.

11) Toxicological information.

- Toxicological effects if the substance comes into contact with a person, for example, carcinogenic, mutagenic, toxic for reproduction etc.
- Likely routes of exposure.
- Symptoms related to physical, chemical and toxicological characteristics.
- Delayed, immediate and chronic effects from short and long-term exposure.
- Interactive effects.

12) Ecological information.

- Toxicity, effects, behaviour and environmental fate that can reasonably be foreseen.
- Persistence and degradability.
- Bioaccumulation.
- Mobility in soil.

13) Disposal considerations.

- Appropriate methods of disposal of the substance and any packaging for example, land-fill, incineration etc.

14) Transport information.

- Information on the transport classification for each of the UN Model Regulations: European Agreement concerning the International Carriage of Goods by Road (ADR) Regulations, International Carriage of Goods by Rail (RID) Regulations, International Carriage of Goods by Inland Waterways (ADN) Regulations; implemented by EU Directive 2008/68/EC.
- UN number.
- UN proper shipping name.
- Transport hazard class(es).
- Packing group.

- Environmental hazards.
- Special precautions in connection with transport or carriage.

15) Regulatory information.

- Safety, health and environmental regulations/legislation specific for the substance or mixture, relevant EU and UK regulatory provisions.
- Chemical safety assessment

16) Other information.

- Training advice, recommended uses and restrictions, sources of important data used to compile the data sheet.
- List of relevant R phrases, hazard statements, safety phrases and/or precautionary statements, written out in full if not provided earlier.
- Details of changes to previous safety data sheet.

Guidance on how to compile a SDS is detailed in Annex II of REACH (as amended). When REACH came into force it introduced some changes to the format of SDS. The main differences compared to the old (CHIP) style SDS are:

- An email contact address should be included in section 1, for competent person(s) to respond with appropriate advice.
- A SDS should be supplied in an official language of the Member State(s) where the substance or mixture is placed on the market (unless the relevant Competent Authority in the Member State(s) concerned has indicated otherwise).

CAS

CAS numbers identify the chemical, but not its concentration or specific mixture. For example, hydrogen chloride (HCL) has the CAS number 7647-01-0; this specific CAS number will appear on containers of anhydrous hydrogen chloride, a 20% solution of HCL in water, and a 2.0 molar solution of HCL in diethyl ether.

CAS Registry Number® is a Registered Trademark of the American Chemical Society all rights reserved. It is a good idea to file information on chemicals, not only by chemical name, but by CAS number. It is one way to avoid the problem of finding a substance name filed on a Safety Data Sheet (SDS) when the name of the substance might not be correctly catalogued or spelt. It is also a good idea to include a CAS number field if data is stored on a computer, especially if the electronic copy is used by a non-chemist to obtain data.

SAFETY ASSESSMENT/REPORTS

In order to register a substance under REACH it is necessary to conduct a chemical safety assessment (CSA). The purpose of the *Chemical Safety Assessment (CSA)* is to assess risks arising from the manufacture and/or use of a substance and to ensure that they are adequately controlled. A CSA has to be performed by registrants for substances manufactured and imported in quantities starting at 10 tonnes per year and by downstream users if their uses are not addressed by their supplier.

A CSA includes the following steps:

- Human health hazard assessment: determination of the classification and labelling of the substance, derivation of no effect levels (DNELs).
- Physicochemical hazard assessment: determination of the classification and labelling of the substance.
- Environmental hazard assessment: determination of the classification and labelling of the substance, derivation of predicted no effect concentrations (PNECs).
- Persistent, bioaccumulative and toxic (PBT) and very persistent and very bioaccumulative (vPvB) assessment (or substances of similar concern): comparison of the data on degradation, bioaccumulation and toxicity with the criteria available in Annex XIII of the REACH Regulation.

If the substance meets the criteria for classification as dangerous, or meets the PBT/vPvB criteria, the CSA shall also include:

- An exposure assessment for all identified and relevant uses of the substance and resulting life cycle steps, including the generation of exposure scenario(s).
- An exposure estimation of humans and the environmental compartments to the substance is performed from the conditions defined in the exposure scenarios.
- A risk characterisation, which is the final step in the chemical safety assessment. The risk characterisation identifies if the risks arising from manufacture/import and uses of a substance are adequately controlled. It consists of a comparison of the derived no effect levels (DNELs) and predicted no effect concentrations (PNECs) with calculated exposure concentrations respectively to human and the environment.

The CSA is a developmental process. Hazard assessment, exposure assessment and risk characterisation can be refined through the use of more precise information or improvement of measures taken until risks to human health or the environment are shown to be adequately controlled.

The output of the CSA will be exposure scenarios with operational conditions and risk management measures for adequate risk control. Exposure scenarios have to be documented in the *Chemical Safety Report (CSR)* and communicated to the downstream users as annexes to the Safety Data Sheets. The Chemical Safety

Report (CSR) is submitted to the European Chemical Agency (ECHA) as part of the registration dossier required by REACH must document the results of the CSA.

Factors to be considered in the assessment of risks to health from hazardous substances

'Hazardous substances' are substances that have the potential to harm human health. They may be solids, liquids or gases in a pure or mixed form. When used these substances often generate vapours, fumes, dusts and mists. A wide range of industrial, laboratory and agricultural chemicals are classified as hazardous and a systematic approach to safety is required. A systematic approach to safety requires an efficient flow of information from the suppliers to the users of chemicals on potential hazards and correct safety precautions.

The factors to be considered in the assessment of risks to health from hazardous substances include:

HAZARDOUS PROPERTIES

The hazardous properties of the substance, the most important hazards, including the most significant health, physical and environmental hazards, should be stated clearly and briefly, as an emergency overview on safety data sheets. The information should be compatible with that shown on the label of the substance. MSDS sheets include a section on 'toxicology', this section should give information on the effects on the body and on potential routes of entry into the body (the substance may become airborne as a dust or mist or fume and inhaled or remain as a liquid and be absorbed through the skin).

Substances that remain as solids (dusts) may also be ingested or come into contact with the eyes. The MSDS will also refer to acute effects, both immediate and delayed, and to chronic effects from both short and long-term exposure.

THE EFFECTS OF MIXTURES

The risk presented by the substance when mixed with other chemicals needs to be established. Some substances when mixed will have synergistic effects (2+2=5). In toxicology, synergism refers to the effect caused when exposure to two or more chemicals at the same time results in health effects that are greater than the sum of the effects of the individual chemicals.

As an example the risk of cancer from exposure to asbestos is significantly higher in people who smoke. Antagonism is the opposite of synergism. It is the situation where the combined effect of two or more compounds is less toxic than the individual effects (2+2= <4). Antagonistic effects are the basis of many antidotes for poisonings or for medical treatments. For example, ethyl alcohol (ethanol) can antagonize the toxic effects of methyl alcohol (methanol) by displacing it from the enzyme that oxidizes the methanol

QUANTITY AND CONCENTRATION

When evaluating the exposure to the substance a number of factors need to be taken into account such as consideration of the amount and concentration of the substance used. The higher the level of exposure to the substance, the more likely it is to be taken into the body, and the more likely that harm will arise.

Work Exposure Limits (WELs) are occupational exposure limits which are set in order to help protect the health of workers. WELs are concentrations of hazardous substances in the air, averaged over a specified period of time, referred to as a time-weighted average (TWA). The definition (given in The Control of Substances Hazardous to Health Regulations 2005) of a substance hazardous to health includes dust of any kind when present at a concentration in air equal to or greater than 10 mg.m-3 8-hour TWA of inhalable dust or 4 mg.m-3 8-hour TWA of respirable dust.

Legislation typically sets maximum long term exposure levels that cannot be exceeded, activities giving rise to short-term peak concentrations should receive particular attention when risk management is being considered due to the short term, high concentration, nature of the hazard. Where hazardous substances are stored in large quantities that could affect those in the vicinity of the plant there are requirements to reduce the possibility of accidental release and to mitigate the potential consequences.

OPERATING CONDITIONS

The risk to health is influenced by the nature of the task being assessed, including the specific methods being used, for example brush application or spray application of a substance. The nature of the task can influence other risk factors, such as:

- **Number of people exposed to the substance.** If a substance is sprayed instead of brush-applied, the small airborne particles of liquid have a higher chance of dispersing in the air and affecting a larger number of people, in addition to the person involved in the spraying operation.
- **Type of exposure.** The spraying operation provides small particles that may be more easily breathed in than when applied by brush and, as they are easily dispersed, may cause a wider range of parts of the body to come into contact with the substance particles. The operator of a spray application system may find they are enveloped in the spray, whereas someone applying the substance by brush may have exposure to their hands only.

- **Controls available.** The task the person is carrying out may influence the risk, because it may influence the controls available to limit exposure. For example, someone involved in a process using a substance may be protected by total enclosure of the operation; however, maintenance work may cause someone to have to enter the total enclosure area. This would mean that the high level of control provided is no longer available to the maintenance worker.
- **Morphology.** How a substance may change in a process. The substance may be relatively harmless until, during a task, it is mixed with another and then heated. What the substance morphs into may be extremely hazardous. For example, trichloroethylene in air at high temperatures degrades into phosgene. Trichloroethylene is a hazardous narcotic with a chronic effect on the heart rhythm; whereas phosgene, though hazardous, presents an acute effect on the pulmonary alveoli, causing chemical damage leading to asphyxiation.

THE RANGE OF USES

Hazardous substances are part of modern life, and we are likely to encounter them every day - from the chemicals used at work, to products in the home, such as paint and detergents and pesticides used in the garden. Workers may come into contact with chemicals as part of manufacturing, using, transporting or disposing of them.

Hazardous substances are used in many places and in many different ways, including in factories, shops, laboratories, offices, farms and in the home and garden. The substances used at work may include products purchased in your core business, or in maintaining equipment, or in general cleaning. The term hazardous substance could also include hazardous substances you create as part of your work processes, for example, dust created by cutting wood or stone.

NUMBER OF PEOPLE EXPOSED

One of the simplest factors influencing the level of risk is the number of people exposed to the hazardous substance. The higher the number of people exposed the higher the risk. Similarly, if the number of people exposed can be controlled and reduced this can greatly influence the risk.

TYPE AND DURATION OF EXPOSURE

It is important to consider the type of exposure to the substance. Substances that do not present a hazard through skin pervasion may not be a problem if small quantities are spilt on the hands, but the same substance may present a significant risk if volatile elements of it were breathed in while it was being applied. The duration of exposure is a significant factor influencing risk. Many substances have a cumulative effect and cause risk to health as exposure builds up over time. It is important to determine the level of exposure of specific groups of people affected as it may differ greatly due to the task being conducted.

Workers who directly use a substance may have a high level of exposure but benefit from a job rotation scheme that limits their duration of exposure. Whereas those who work in the area may have a lower level of exposure, but be exposed for the duration of their normal working day. Maintenance and cleaning staff may be exposed to the highest levels of exposure but only for a very short duration.

FREQUENCY OF EXPOSURE

A task involving a substance may be infrequent, because it is only conducted as part of a periodic cleaning/maintenance activity or the substance may be used daily as part of normal task operations. This, together with the type and duration of exposure can greatly influence risk. Exposure to some substances may be acceptable as they are infrequent, thus allowing the worker's body to recover from the exposure before additional exposures take place.

THE VARIETY AND NATURE OF TASKS AND THE METHODS USED

Particular activities, where exposure is likely to be unusually high, include those where there is an accidental release of the chemical, perhaps during maintenance work or failure of equipment.

Where substances are stored or handled, it is important to consider the risks arising from accidental release, due to spillage of the substances.

Other activities that present high exposure are maintenance and cleaning activities, which cause high levels of release of the substance, or cause the workers to do tasks where they are not protected by the usual controls in place, such as local exhaust ventilation (LEV). Those workers who service and clean local LEV, that has captured the hazardous substances in concentrated amounts, may be at risk of particularly high exposure to the substance. Tasks where workers are opening containers and decanting substances into process equipment may receive high exposure, for example emptying a bag of a substance into a process vessel.

THRESHOLDS OF EXPOSURE

The amount needed to cause harm may be a discernible threshold for some substances but others may have no safe level. Exposure to a substance does not automatically mean it is a high risk. It is important to take account of thresholds of exposure that are considered acceptable. Workplace exposure limits listed in EH40 will indicate what this may be.

THE CONSEQUENCES OF FAILURE OF EXISTING CONTROL MEASURES

The type of control measure used can greatly influence the level of risk, depending on its effectiveness. Whenever possible, priority to collective (shared by everyone in a group) controls should be given over individual controls, for example, effective LEV instead of reliance on the issue and use of respirators. The risk assessment should consider the actual effectiveness of existing control measures, by determining exposure levels when they are in use, where possible.

Otherwise, an estimate may be necessary, based on the manufacturer's data, for example when considering the protection provided by respiratory protection equipment. When determining exposure levels based on existing control measures, it is important to build in a margin, to take account of reduction of effectiveness over time.

THE RESULTS FROM HEALTH SURVEILLANCE AND EXPOSURE MONITORING

The results from relevant health surveillance and exposure monitoring will provide information on the effectiveness of controls. If workers are being exposed to higher than intended levels of the substance, this can provide an indication of actual risk to their health and indicate areas where controls are not effective. Health surveillance can provide information on actual harm to health and illustrate that there may be a higher level of risk than was expected; it can also help identify groups of people particularly at risk.

INDIVIDUAL SUSCEPTIBILITIES

The individual susceptibility of the people exposed may have an influence on the level of risk; this will include consideration of atopic people, women of child bearing capacity, age and sensitisation. **Atopic people** (typically those who have a genetic likelihood of sensitization) can be more susceptible to respiratory and skin irritants and include those people who naturally suffer from asthma, eczema and hay fever.

If a person is exposed to an allergen, or some other substance recognised as foreign by the body's immune system, this can lead to the process of **sensitisation.** Some substances can be more harmful to a specific gender.

Children have a less well developed blood-brain barrier and are, therefore, more susceptible to lead, as greater absorption across the barrier is likely. They are also more susceptible to harm from hazardous substances, because of their rapid growth rate and the ratio of their size to any exposure level.

LEVEL OF EXPOSURE TO THE SUBSTANCE

When evaluating the exposure to the substance, this will involve consideration of the amount and concentration of the substance used. The higher the level of exposure to the substance the more likely it is to be taken in to the body and the more likely that harm will arise.

The form of the substance will particularly influence how much can become airborne, the level of exposure and how much can be breathed in, for example, volatile liquids, dust from solids. It can also influence the likelihood of it coming into contact with the skin/eyes or being swallowed. The form the substance takes will often dictate its route of entry into the body and the measures needed to protect against exposure. A substance may have more than one state; thus a volatile solvent can be present in both liquid and vapour forms, leading to potential skin/eye contact and lung contact.

THE NATURE OF THE TASK

The risk to health is influenced by the nature of the task being assessed, including the specific methods being used, for example, brush application or spray application of a substance. The nature of the task can influence other risk factors, such as:

- **Number of people exposed** to the substance. If a substance is sprayed instead of brush-applied the small airborne particles of liquid have a higher chance of dispersing in the air and affecting a larger number of people in addition to the person involved in the spraying operation.
- **Type of exposure**. The spraying operation provides small particles that may be more easily breathed in than when applied by brush and, as they are easily dispersed, may cause a wider range of parts of the body to come into contact with the substance particles. The operator of a spay application system may find they are enveloped in the spray, whereas someone applying the substance by brush may have exposure to their hands only.
- **Controls available**. The task the person is doing may influence the risk because it may influence the controls available to limit exposure. For example, someone involved in a process using a substance may be protected by total enclosure of the operation; however, maintenance work may cause someone to have to enter the total enclosure area. This would mean that the high level of control provided is no longer available to the maintenance worker.
- **Morphology**. How a substance may change in a process. The substance may be relatively harmless until, during a task, it is mixed with another and then heated. What the substance morphs into may be extremely hazardous. For example, trichloroethylene in air at high temperatures degrades into phosgene. Trichloroethylene is a hazardous narcotic with a chronic effect on the heart rhythm; whereas phosgene though hazardous, presents an acute effect on the pulmonary alveoli causing chemical damage leading to asphyxiation.

PARTICULAR HIGH EXPOSURE ACTIVITIES

Particular activities where exposure is likely to be unusually high include those where there is an accidental release of the chemical, perhaps during maintenance work or failure of equipment. Where substances are stored or handled it is important to consider the risks arising from accidental release due to spillage of the substances.

Other activities that present high exposure are maintenance and cleaning activities that cause high levels of release of the substance or cause the workers to do tasks where they are not protected by the usual controls in place, such as local exhaust ventilation (LEV). Those workers that service and clean local exhaust ventilation equipment that has captured the hazardous substances in concentrated amounts may be at risk of particularly high exposure to the substance. Tasks where workers are opening containers and decanting substances into process equipment may receive high exposure, for example emptying a bag of a substance into a process vessel.

EFFECTIVENESS OF EXISTING CONTROL MEASURES

The type of control measure used can greatly influence the level of risk, depending on its effectiveness. Whenever possible, priority to collective (engineering) controls should be given over individual controls, for example, effective LEV instead of reliance on the issue and use of respirators. The risk assessment should consider the actual effectiveness of existing control measures by determining exposure levels when they are in use, where possible.

Otherwise, an estimate may be necessary, based on the manufacturer's data, for example, when considering the protection provided by respiratory protection equipment. When determining exposure levels based on existing control measures it is important to build in a margin to take account of reduction of effectiveness over time.

SURVEILLANCE AND MONITORING

The results from relevant health surveillance and exposure monitoring will provide information on the effectiveness of controls. If workers are being exposed to higher than intended levels of the substance this can provide an indication of actual risk to their health and indicate areas where controls are not effective. Health surveillance can provide information on actual harm to health and illustrate that there may be a higher level of risk than was expected; it can also help identify groups of people particularly at risk.

INDIVIDUAL SUSCEPTIBILITIES OF THE PEOPLE EXPOSED

The individual susceptibility of the people exposed may have an influence on the level of risk; this will include consideration of atopic people, women of child bearing capacity, age and sensitisation. *Atopic people* can be more susceptible to respiratory and skin irritants and include those people who naturally suffer from asthma, eczema and hay fever. If a person is exposed to an allergen or some other substance recognised as foreign by the body's immune system this can lead to the process of *sensitisation.*

Some substances can be more harmful to a specific sex. Mineral oil, for example, is associated with male cancer of the scrotum, while women are thought to be more susceptible to fat soluble toxins due to their greater fat to lean body ratio. Children have a less well developed blood-brain barrier and are therefore more susceptible to lead as greater absorption across the barrier is likely. They are also more susceptible to harm from hazardous substances because of their rapid growth rate and the ratio of their size to any exposure level.

Review of risk assessment

ASSESSING RISKS

In addition to the general risk assessment required under Regulation 3 of the Management of Health and Safety at Work Regulations (MHSWR) 1999 there are a number of other Regulations that require the assessment of risks to health from hazardous substances. These include:

- Control of Substances Hazardous to Health Regulations (COSHH) 2002 (as amended).
- Control of Lead at Work Regulations (CLAW) 2002.
- Control of Asbestos Regulations (CAR) 2012.

Although these Regulations require assessments to be made of substances hazardous to health, they do not specify the form the assessments should take. In the very simple and obvious cases, an assessment may not need recording as it can be readily explained. However, in most situations a record will be required in order to ensure compliance, accuracy and continuity.

THE RISK ASSESSMENT PROCESS

There are five major steps involved in making any assessment of hazardous substances:

1) Identify the hazards.
2) Decide who might be harmed and how.
3) Evaluate the risks and decide on precautions.
4) Record the findings and implement them.
5) Review and update as necessary.

B2.4 - The role of epidemiology and toxicological testing

Human epidemiological investigations

EPIDEMIOLOGICAL STUDIES

Occupational health and hygiene are concerned with recognition, measurement, evaluation and control. Epidemiology is the science which forms the basis of the four stages. At a basic level, epidemiology can be described as the explanation of the cause from observing the effects.

For example, if a number of workers on a particular process are suddenly affected by dermatitis, i.e. the effect, it is likely that the cause is of occupational origin, possibly a new raw material. However, the cause can often only be established in the light of current knowledge and understanding.

John Snow (1813-1858) flew in the face of conventional wisdom when he suggested that cholera was due to contaminated water rather than airborne gases (miasmas).

A semblance of the cause-effect relationship in occupational disease has been recognised since early times. *Hippocrates*, the father of medicine (born 460 BC), was probably among the first to recognise lead as a cause of colic; and *Pliny* (23-79 AD) describes mercurialism in writing of the diseases of slaves.

One of the greatest technical manuals ever written is De Re Metallica by *Agricola* (Georg Bauer), first published in Basle in 1556. He described, and illustrated in detail, the techniques and methods used by German mining engineers to win metallic ores and also covered aspects of occupational diseases (he was a medical doctor by profession). On miners in the Carpathian Mountains he comments:

1) *On ventilation:* "If a shaft is very deep and no tunnel reaches to it, or no drift from another shaft connects to it, or when a tunnel is of great length and no shaft reaches to it, then the air does not replenish itself. In such case it (the air) weighs heavily upon them (the miners), causing them to breathe with difficulty and extinguishing lamps. It is therefore necessary to install machines to enable the air to be renewed and for the miners to carry on their work."

2) *On accidents:* The Burgomaster "gives no-one permission to enter the mines after blasting until the poisonous vapours are cleared."

3) *On diseases of the lung:* "If the dust has corrosive qualities, it eats away the lungs and implants consumption in the body. In the mines of the Carpathian Mountains, women are found who have married seven husbands, all of whom this terrible consumption has carried off to a premature death."

To protect miners against the effects of the dust, Agricola advised purification of the air by ventilation and the use of loose veils over their faces.

Paracelsus (1493-1541) published the first monograph devoted entirely to the occupational diseases of mine and smelter workers *(Von der Bergsucht und Anderen Bergkrankheiten, 1567)*. Although he makes correct clinical observations, he then turns to alchemical theories to explain them: *"The lung sickness comes through the power of the stars, in that their peculiar characteristics are boiled out, which settle on the lungs in three different ways: in a mercurial manner like a sublimated smoke that coagulates, like a salt spirit, which passes from resolution to coagulation, and thirdly, like a sulphur, which is precipitated on the walls by roasting."*

If Hippocrates is the father of medicine then Bernardino Ramazzini (1633-1714) is the father of occupational medicine. His *De Morbis Artificum Diatriba* (1700) contains accounts of the occupational diseases suffered by miners of metals, healers by injunction (the rubbing in of oil or ointment for medicinal purposes), chemists, potters, tinsmiths, glass-workers, painters, sulphur-workers, blacksmiths, workers with gypsum and lime, apothecaries, cleaners of privies and cesspits, fullers, oil pressers, tanners, cheese-makers, tobacco-workers, corpse-carriers, midwives, wet-nurses, vintners and brewers, bakers and millers, starch-makers, sifters and measurers of grain, stone-cutters, launderers, workers handling flax, jute and silk, bath-men, salt-makers, workers who stand for long periods, sedentary workers, grooms, porters, athletes, runners, singers, preachers, farmers, fishermen, soldiers, learner men, priests and nuns, printers, scribes and notaries, confectioners, weavers, coppersmiths, carpenters, grinders of metals, brick-makers, well-diggers, sailors and rowers, hunters, and soap-makers; in short, most of the important occupations of the day.

As a result of his investigations he added an important question to the Hippocratic art, urging physicians to ask of their patients, 'what is your occupation?' He also 'urged physicians to leave their apothecaries shop, which is redolent with cinnamon and to visit the latrines where they may see the cause of ill-health'.

The results of an epidemiological study undertaken by Doll and Hill (1950) demonstrated the link between tobacco smoking and lung cancer. Their study began in 1947 where 20 London based hospitals were asked to record all patients admitted with carcinoma of the lung, stomach, colon or rectum. On receipt of the record a researcher visited the patient to conduct an interview with a series of pre-established questions. For each lung cancer patient interviewed, the interviewer also recorded the answers given by another patient who was not submitted suffering from lung cancer. This patient had to be the same sex as the target patient and within five years of the same age. Their study revealed a direct association between smoking and lung cancer.

DIFFERENT TYPES OF STUDY

Morbidity/mortality statistics

The relationship between occupational exposure to an agent and its associated diseases has been long established. In the 19th century, for example, felt hat makers showed symptoms of the effects on the central nervous system of the mercury salts used in the production process (hence the term 'as mad as a hatter'). In the 20th century coal miners became disabled by respiratory problems due to the effects of coal dust on the lungs and painters have become unable to work in their chosen field due to occupational asthma. This recognition and measurement of the relationship between an occupation and susceptibility to a disease comes from two main sources:

Morbidity data is the incidence of a disease in a community or population. This may be triggered through statistics published by government and health agencies, by questioning the occupational group involved or through observation of medical practitioners treating such groups. Once a disease is suspected both the group in question and a control group are studied to see if there is a statistically significant causal relationship.

Mortality data is extracted from death certificates using the International Classification of Diseases Injuries and Causes of Death (ICD) coding system. In this country, mortality statistics are compiled under the auspices of the Office of Population Census and Surveys (OPCS) from data extracted from death certificates. The antecedent cause of death and the occupation of the deceased are recorded. This has obvious drawbacks for people who changed jobs since contacting a particular disease or who are retired.

Once occupational exposure to an agent has been shown to present a hazard to health it must be evaluated and exposure eliminated or controlled. The criteria for doing so will depend on exposure levels and effects. This may result in banning the substance from use completely, for example, the Control of Substances Hazardous to Health Regulations (COSHH) 2002 (as amended) prohibits sand or other substance containing free silica from being used as an abrasive for blasting articles in any blasting apparatus due to the risk of silicosis.

Cross sectional surveys

The cross sectional survey is often used to assess the prevalence of acute or chronic health conditions in a population. The technique is used to measure exposure to risk and evidence of disease in groups, at the specific point in time of the survey. In a cross sectional survey a specific group is considered to see if a substance or activity is related to the health effect being investigated, for example, drinking alcohol and its relation to cirrhosis of the liver. If a significant number of alcohol drinkers have cirrhosis at the time of survey it could support the hypothesis that cirrhosis is caused by drinking alcohol.

It should be noted that it is not always possible to distinguish whether the exposure preceded or followed the disease, for example, a person may have developed cirrhosis prior to starting to drink alcohol. It is an easy study to conduct as it identifies current health conditions and the exposure to the substance or activity has already occurred. As such, this type of study is less expensive to conduct than a cohort study. The main disadvantage of the cross sectional survey is that it may not necessarily confirm causes of health conditions; the best it can usually do is to identify possible associations of substances or activities and suggest case-control or cohort studies as a follow-up.

Case control studies

This method may be used to investigate, for example, the frequency of arc welders in a 'case' group who have respiratory problems or lung disease and compare them against a 'control' group drawn from the general population. It is quicker and less expensive than a cohort study and is often used as the first step in an epidemiological investigation, to see if there may be an association between a suspected cause and a known effect. It is also a useful method for investigating a disease of low prevalence as it enables their study without having to study thousands of people. Unfortunately, case-control studies are generally less informative than cohort studies and spurious associations may occur. The epidemiologist needs to ensure that the case group and the control group have things in common, other than their work situation.

For example, if a case group of 50 arc welders exposed to welding fume is being studied with a control group of 50 office workers not exposed to welding fume, the results may show there is a strong association with exposure to welding fume and respiratory problems, but cannot prove causation. This could also be a spurious association if the 50 arc welders were all heavy smokers and the office workers were not. There are many variables that could affect the analysis. In addition, the results may not be considered valid if the number of subjects under study is not thought to be a large enough sample.

	50 welders	*50 office workers*
Respiratory problems	20	5
No respiratory problems	30	45

Figure B2-17: Example case group. *Source: RMS.*

Retrospective cohort studies

Retrospective cohort studies look back in time. In this situation the result (ill-health) is known, and the study tries to determine the cause. The 'cohort' is a number of individuals who have the health problem and who have been selected for study. The cohort may include individuals who have died as a result of the health problem being studied. Information is collected on the cohort's past lives and this is analysed to see what they had in common. If they all have, or had before they died, a respiratory problem, such as emphysema or lung cancer, the fact that they all spent some time working in the engineering industry would be of great interest.

This type of study is cheaper and quicker than a prospective cohort study. It allows the cohort to be selected purely on the basis of the ill-health condition of interest. Again, the number of individuals in the cohort will affect the validity of the results. Because it is a study of health conditions that have occurred, the epidemiologist does not have to wait for time to pass to gather the data necessary for the study. This is particularly useful when studying rare ill-health conditions, or chronic ill-health conditions that have a long latency period for their development. However, this type of study does have its limitations; there may be spurious associations made because the historic records analysed do not contain the full facts. Historic records may be very difficult to obtain, for example, a person's work record may not be available because the companies they worked for have gone out of business and no records were kept.

In addition, the person's ill-health may not have been caused by work but by their lifestyle, and reliable information on lifestyle is not always available. Individuals being interviewed as part of the study sometimes have difficulty with their recall of facts or they may tell the epidemiologist what they think they want to hear.

Prospective cohort studies

Prospective cohort studies look forward in time. They are considered the best method of obtaining epidemiological data, but they take a long time to conduct, a lifetime in some instances. The cohort, an identified group of people, is chosen and each individual is studied over time into the future. For example, the cohort may be a group of eight to ten-year-old children who use mobile telephones and are being studied for any health conditions which may be associated with mobile telephone use.

A further example may be a cohort made up of a group of workers using a chemical which has just come on to the market. The chemical will have gone through initial tests, and known information about the chemical would be made available in a health and safety data sheet, but the results of studying long term use of the chemical may change our knowledge of its effects. A prevalence of a particular health condition may be identified in the cohort group after years of use and as a result some of the data on the health and safety data sheet may have to be amended. These studies take a lot of resources, in particular time. The study may experience difficulties over time; the people in the cohort may leave the work activity or not be involved with the chemical anymore, because of career change or changing processes that cease its use in the workplace.

Furthermore, they may not want to be involved in the study any longer or some may die from totally unrelated causes, such as car accidents. The cohort group may become depleted in numbers to the point that the results of the study will not be valid. In some cases the outcomes of the research may not be what the sponsor of the research hopes for and the sponsor may choose to declare the results invalid. For example, a promoter of the use of mobile telephones might be carrying out research with the intent to prove that there are no ill-health effects associated with their use, but findings over time may suggest otherwise.

The largest cohort study in the UK is the Nurses' Health Study which started in 1976 and continues to gather data on the health of female nurses. This study has also produced landmark data on cardiovascular disease, diabetes and many other conditions. Most importantly, these studies have shown that diet, physical activity and other lifestyle factors can powerfully promote better health.

The role of toxicological testing

VERTEBRATE ANIMAL TESTING

A vertebrate animal is one that has a vertebral column, a stiff bone structure running through most of the length of the animal.

The use of animals for testing, to predict the effects of substances on humans, may be of limited value, unless the life form used has a physiology that is very close to that of humans (i.e. the primates). Rats and mice, for example, might be chosen because the effects over many generations may be observed in a relatively short period (life cycle 2 years, but reproductive capability within 9 to 12 weeks of birth). However, rats and mice do not retain urine within a bladder (as humans do) so would not be useful to determine the chronic effects of cancer of the bladder in relation to humans.

Over recent years public opinion has changed (historically, common testing procedures involved exposing rabbits to the effects of cigarette smoke exposure through inhalation, skin and eye contact with cosmetics) and vertebrate animal testing in many countries is subject to strict codes and licensing. Acute testing relates to tests lasting less than a month, tests lasting from one to three months are classed as sub-chronic and tests lasting greater than three months are classed as chronic tests.

There are several types of acute toxicity tests. The LD_{50} (lethal dose 50%) test is used to evaluate toxicity of a substance by determining the dose required to kill 50% of the test population. It is no longer an Organisation of

Economic Co-operation and Development (OECD) recommended method for acute toxicity tests and has mainly been replaced by the fixed dose procedure, which uses fewer animals and causes less suffering.

The Test Methods Regulation (EC440/2008) describes the fixed dose procedure as it applies to acute oral toxicity; this test method is equivalent to OECD TG 420 (2001). The fixed dose procedure is based on the observation of clear signs of toxicity at one of a series of fixed dose levels, instead of applying a dose that causes death.

Following UK and international *in vivo* (in living animals) validation studies, the procedure was adopted by the OECD as a testing method in 1992. The validation process demonstrated that the procedure is reproducible, uses fewer animals, causes less suffering than the traditional methods, and is able to rank substances in a similar manner to the other acute toxicity testing methods.

It is a principle of the method that, in the main, only moderately toxic doses are used and that administration of doses that are expected to be lethal should be avoided.

Also, doses that are known to cause marked pain and distress, due to corrosive or severely irritant actions, need not be administered. The process observes results of applied doses at a series of increasing fixed dose levels and, if it is clear from observations, that to apply the next dose would result in death, there is no need to apply the dose and cause unnecessary suffering.

In chronic toxicity testing the test substance is typically administered seven days per week, by an appropriate route, to several groups of experimental animals, one dose per group, for a major portion of their life span. During and after exposure to the test substance, the experimental animals are observed daily to detect signs of toxicity.

AMES TEST

The **Ames** test is based on the assumption that a mutagenic substance for bacteria will also be a carcinogen. Although this is strictly not always the case, the simplicity and cheapness of the Ames test makes it a valuable screening aid. A positive test indicates that the chemical might act as a carcinogen, as cancer is often linked to DNA damage.

The bacterium used carries a defective gene, making it unable to produce histidine. If the bacteria cannot make or obtain histidine they die. The mutagenic effect of the chemical the bacterium is exposed to causes many of the bacteria to regain the ability to grow, indicating they must have made their own histidine. The defective histidine gene has mutated back to its active version and, therefore, the chemical is a mutagen. If the bacteria die, then they have not made histidine and the chemical is not a mutagen.

QUALITATIVE/QUANTITATIVE STRUCTURE ACTIVITY RELATIONSHIP (QSAR)

Structure Activity Relationship (SAR) is a method designed to find relationships between the chemical structure and biological activity of compounds. The guiding principle of SAR is the assumption that the structure of a molecule (for example, its geometric, electronic properties etc.) contains the features responsible for its physical, chemical, and biological properties. SAR links the chemical structure to the chemical properties.

Quantitative SARs, referred to as (Q)SARs, use models of chemical structures to predict how the chemical will interact with the biochemistry of a human. (Q)SARs model mathematical relationships (often a statistical correlation), which relates a structural property to the presence or absence of another property, so they do not rely on information that is known on the substance.

Computer Assisted Evaluation of Industrial Chemical Substances According to Regulations (CAESAR) was an EU funded project that developed the existing (Q)SAR in silico models to specifically address the REACH legislation. In silico is an expression used to mean 'performed on computer or via computer simulation'. In silico research is thought to have the potential to speed the rate of discovery while reducing the need for expensive lab work and clinical trials. In silico models are not suitable for regulatory purposes, as they have not taken into account one or more factors essential for validation, quality assurance, or for a specific application of a given compound. The typical approach for in silico modelling is to develop a model and then propose its use. This is a generic approach, not addressing a specific application.

CAESAR models have been used to address bio concentration factors, skin sensitization, carcinogenicity, mutagenicity and developmental toxicity, which are all highly relevant for REACH.

'READ ACROSS' AND GROUPING

Grouping of substances and read-across is one of the most commonly used alternative approaches for filling data gaps in registrations submitted under REACH. This approach uses relevant information from analogous ('source') substances to predict the properties of 'target' substances. If the grouping and read-across approach is applied correctly, experimental testing can be reduced as there is no need to test every target substance.

For each standard information requirement that applies, registrants must indicate whether they are making an adaptation using read-across, and they must justify its use. *Further information with regards to submitting under REACH, can be found at http://echa.europa.eu/support/grouping-of-substances-and-read-across.*

The meaning of 'dose-response relationship', NOAEL, LD50, LC50

TOXIC AND VERY TOXIC

A toxic substance is one which if inhaled or ingested or it penetrates the skin, may involve serious acute or chronic health risks and even death, for example, arsenic, a systemic poison.

The toxicity of a substance is defined as its ability to cause death/serious health effects in a population. Thus toxicity is the property of being poisonous and is used to refer to the severity of adverse effects or illness produced by a toxin (for example, bacteria), a poison or a drug.

Highly toxic substances are classified as those which can cause death at relatively low concentrations. There are many measures of toxicity, for example, LD_{50} (lethal dose fifty) is the amount of a substance that will cause 50% of a standardised population of rats to die.

DOSE/RESPONSE RELATIONSHIPS

Toxins have very different effects on organisms. They include the minimum level at which an effect is detectable; the sensitivity of the organism to small increases in dose; and the level at which the harmful effect (most significantly, death) occurs. Such factors are indicated in the dose-response relationship, which is a key concept in toxicology:

Dose-response relationship

"All substances are poisons; there is none which is not a poison. The right dose differentiates a poison and a remedy."

Figure B2-18: Quote - toxicology. *Source: Paracelsus (1493-1541).*

Apart from the nature of the harm caused by a chemical, a key element in our ability to assess toxicity from animal studies or from human experience is the way the damage varies with the dose.

Definition of some terms

Dose	In some animal studies, this can be expressed as the amount administered (mg/kg bodyweight) via inhalation, oral or dermal routes with time. In humans, dose is the uptake of a chemical estimated from airborne measurements, biological monitoring or diet.
Dose-effect	Variation in the degree of effect with changing dose.
Dose-response	Proportion of the population that will demonstrate an effect or specified degree of effect, if graded, at a given dose. The dose-response relationship can be used to derive several values: LD_{50} - Lethal dose killing 50% of the test animals; provides indication or relative toxicity of a substance and possible target organ(s). LD_{90} - Lethal dose killing 90% of the test animals. LC_{90} - Lethal concentration killing 90% of the test animals.

NO-OBSERVED-ADVERSE-EFFECT LEVEL (NOAEL)

The NOAEL is the highest exposure level at which there is no statistically or biologically significant increases in the frequency or severity of adverse effect between the exposed population and its appropriate control; some effects may be produced at this level, but they are not considered adverse, nor precursors to other adverse effects. For some chemicals, for example, carcinogens, the dose at which no measurable effect occurs; used when setting workplace exposure limits, there may be no NOAEL.

The fixed dose procedure

The Fixed Dose Procedure (FDP), 1984, is a method to assess a substance's acute oral toxicity. In this procedure the test substance is given at one of the four fixed-dose levels (5, 50, 500, and 2000 mg/kg) to five male and five female rats. The objective is to identify a dose that produces clear signs of toxicity but no mortality.

Depending on the results of the first test, either no further testing is needed or a higher or lower dose is tested. In comparison to the older LD_{50} test developed in 1927, this procedure produces similar results while using fewer animals and causing less pain and suffering. As a result, in 1992 this test was proposed as an alternative to the LD_{50}. It is described in Annex V to the Dangerous Substances Directive and in the Organisation for Economic Co-operation and Development (OECD) guidelines for the Testing of Chemicals Number 420.

Lethal dose, lethal concentration

The toxicity of a substance is its potential to cause harm by reaction with body tissues. Measures of toxicity include lethal dose (LD_{50} and LD_{90}) and lethal concentration (LC_{50} and LC_{90}).

The LD_{50} is the single dose of a substance which, when administered to a batch of animals under test, kills 50% of them. It is measured in terms of mg of the substance per kg of body weight.

Typical degrees of LD_{50} toxicity are:

Extremely toxic	1 mg or less	*Slightly toxic*	>0.5-5g
Highly toxic	>1-50mg	*Practically non-toxic*	>5-15g
Moderately toxic	>50-500mg	*Relatively harmless*	15g or more

Figure B2-19: Degrees of LD_{50} toxicity.   *Source: RMS.*

In a similar way, the LC_{50} refers to an inhaled substance and is the concentration which kills 50% in a stated time. It should be appreciated that LD_{50} is not an exact value and in recent years there has been much discussion as to its usefulness and necessity in toxicology. The LD_{50} value may vary for the same compound between different groups of the same species of animal.

However, the value is of use in comparing how toxic a substance is in relation to other substances. The following table gives examples of LD_{50} values for a variety of chemical substances.

The dose-response curve for any substance will be typically S-shaped. The first point along the graph where a response above zero (or above the control response) is reached is usually referred to as a threshold-dose.

At first, the dose increase shows no ill effects. As soon as an effect is shown, i.e. death, the curve rises steeply.

The increase in the dose from the first deaths to 50% of the test population dying is very small.

The increase in the dose from 50% deaths to 90% deaths again is very small.

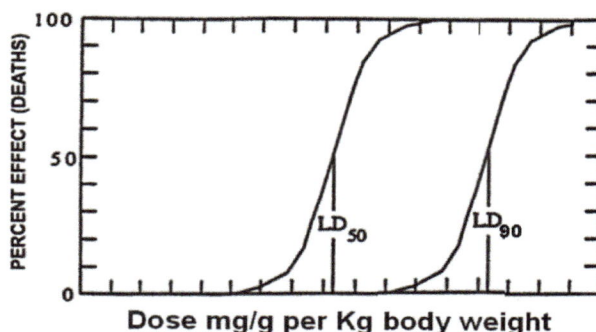

Figure B2-20: Dose/response curve.   *Source: RMS.*

Compound	LD_{50} (mg/kg)
Ethanol	10,000
DDT	100
Nicotine	1
Tetrodotoxin	0.1
Dioxin	0.001
Botulinus toxin	0.00001

Figure B2-21: Examples of LD_{50} values.   *Source: RMS.*

ED_{50} (effective dose for 50%) and TD_{50} (toxic dose for 50%) are related parameters, which indicate the dose at which a biological response is likely.

The effective dose is a pharmacological response, where an effect can be seen, and the toxic dose is the toxic effect which causes damage, but does not kill.

Comparison of ED_{50} with LD_{50} gives an indication of the margin of safety, between the dose which causes the first identifiable effect and that which is fatal.

Testing for carcinogenic potential is more complex, since there is no simple dose-response relationship. The toxicology of carcinogens is approached in a different way, but still involves exposing laboratory animals (usually rats and mice) to the chemical by oral, inhalation or skin contact techniques.

There are also short-term predictive tests available which are considered to simulate potential carcinogenicity in man. They are called short term, in contrast to the usual lifetime studies in rodents, which can take three to four years before a result is available.

Short term tests include those for:

■ Mutation (Ames test).
■ DNA damage.
■ Chromosomal damage.
■ Cell transformation.

Because of the extensive use of animals in carcinogenicity testing, there are at least 1,000 chemicals identified with this potential. However, not all chemicals which are carcinogenic in animals cause cancer in man, since there are wide differences in genetic constitution, metabolism and life span. It is essential, therefore, to interpret data on suspected carcinogens with great caution, when making judgements in an occupational setting.

Tests that can be carried out *'in vitro'* (in a test tube) have the advantage of being cheaper and quicker than *'in vivo'* (in living animals) and are more humane. A test developed by Bruce Ames for testing if a chemical was a mutagen is an *'in vitro'* test.

The Ames test is based on the assumption that a mutagenic substance for bacteria will also be a carcinogen. Although this is strictly not always the case, the simplicity and cheapness of the Ames test makes it a valuable screening aid. A positive test indicates that the chemical might act as a carcinogen, as cancer is often linked to DNA damage.

The bacterium used carries a defective gene, making it unable to produce histidine. If the bacteria cannot make or obtain histidine they die. The mutagenic effect of the chemical the bacterium is exposed to causes many of the bacteria to regain the ability to grow, indicating they must have made their own histidine. The defective histidine gene has mutated back to its active version and, therefore, the chemical is a mutagen. If the bacteria die, then they have not made histidine and the chemical is not a mutagen.

The control of hazardous substances

On completion of this element, candidates should be able to demonstrate understanding of the content through the application of knowledge to familiar and unfamiliar situations. In particular, they should be able to:

B3.1   Explain the principles of prevention and control of exposure to hazardous substances (including carcinogens and mutagens).

B3.2   Outline the specific requirements for working with asbestos.

B3.3   Explain the uses and limitations of dilution ventilation and the purpose and operation of local exhaust ventilation, including assessing and maintaining effectiveness.

B3.4   Explain the effectiveness of various types of personal protective equipment (PPE) and the factors to consider in selection of PPE.

Content

Relevant statutory provisions

Regulation (EC) No 1272/2008 on classification, labelling and packaging of substances and mixtures

Regulation (EU) 2016/425 on personal protective equipment at work

Control of Asbestos Regulations (CAR) 2012

Control of Asbestos Regulations (Northern Ireland) 2012

Control of Substances Hazardous to Health Regulations (COSHH) 2002 (and as amended)

Control of Substances Hazardous to Health Regulations (Northern Ireland) 2003

Inquiries into Fatal Accidents and Sudden Deaths etc. (Scotland) Act 2016

Personal Protective Equipment Regulations (PPER) 1992

Personal Protective Equipment at Work Regulations (Northern Ireland) 1993

Sources of reference

Reference information provided, in particular web links, was correct at time of publication, but may have changed.

Asbestos: The survey guide, HSG264, HSE Books, ISBN: 978-0-7176-6502-0, http://www.hse.gov.uk/pUbns/priced/hsg264.pdf

Controlling airborne contaminants at work; A guide to local exhaust ventilation (LEV), HSG258, HSE Books, ISBN: 978-0-7176-6415-3, http://www.hse.gov.uk/pubns/books/hsg258.htm

Control of Substances Hazardous to Health, Approved Code of Practice and guidance, L5, HSE Books, ISBN: 978-0-7176-6582-2, http://www.hse.gov.uk/pubns/books/l5.htm

COSHH Essentials, HSE, http://www.hse.gov.uk/coshh/essentials/

Managing asbestos in buildings: a brief guide, INDG223(rev5), HSE Books, http://www.hse.gov.uk/pubns/indg223.pdf

Managing and working with asbestos, Control of Asbestos Regulations 2012, Approved Code of Practice and guidance, L143, HSE Books, ISBN: 978-0-7176-6618-8, http://www.hse.gov.uk/pubns/books/l143.htm

Personal protective equipment at work, Guidance on Regulations, L25, HSE Books, ISBN: 978-0-7176-6597-6, http://www.hse.gov.uk/pubns/priced/l25.pdf

Personal protective equipment (PPE) at work; A brief guide, INDG174, HSE Books, http://www.hse.gov.uk/pubns/indg174.pdf

Respiratory protective equipment at work; A practical guide, HSG53, HSE Books, ISBN: 978-0-7176-6454-2, http://www.hse.gov.uk/pUbns/priced/hsg53.pdf

Working with substances hazardous to health, INDG136, HSE Books, http://www.hse.gov.uk/pubns/indg136.pdf

The above web links along with additional sources of reference, which are additional to the NEBOSH syllabus, are provided on the RMS Publishing website for ease of use - www.rmspublishing.co.uk.

B3.1 - The prevention and control of exposure to hazardous substances

Requirement for prevention and adequate control of exposure to hazardous substances (COSHH Regulation 7)

Hazardous substances or chemical agents are found in a wide range of workplaces and it is therefore important to determine whether hazardous substances are present. Risks to the health and safety of workers must be assessed and risk management measures established with the aim of reducing exposure to those risks.

Regulation 7 of the Control of Substances Hazardous to Health Regulations (COSHH) 2002 (as amended) sets out clear requirements for the control of substances hazardous to health, including carcinogens, mutagens and asthmagens.

> *"Every employer shall ensure that the exposure of his employees to substances hazardous to health is either prevented or, where this is not reasonably practicable, adequately controlled."*

Figure B3-1: Regulation 7(1).   *Source: The Control of Substances Hazardous to Health Regulations (COSHH) 2002 (amended).*

Regulation 7(7) specifies what may be considered adequate control of exposure in keeping with the requirement established in Regulation 7(1). This includes meeting the general principles of good practice, quite specific requirements for certain substances, such as asthmagens, and reducing exposure as low as is reasonably practicable.

"Exposure shall only be treated as adequate if:

a) *The principles of good practice for the control of exposure to substances hazardous to health set out in Schedule 2A are applied.*

b) *Any workplace exposure limit approved for that substance is not exceeded.*

c) *For a substance:*

(i) *Which carries the risk phrase R45, R46 or R49, or for a substance or process (to which the definition of a 'carcinogen relates') which is listed in Schedule 1.*

(ii) *Which carries the risk phrase R42 or R42/43, or which is listed in section C of HSE publication 'Asthmagen? Critical assessments of the evidence for agents implicated in occupational asthma' as updated from time to time or any other substance which the risk assessment has shown to be a potential cause of occupational asthma exposure is reduced to as low a level as is reasonably practicable."*

The amount of a substance that is needed to produce sensitivity and lead to asthma varies considerably between individuals. In addition only a minority of individuals at risk will actually develop asthma.

Once a person develops hypersensitivity as a result of exposure to a substance that causes asthma it is irreversible. However, people develop symptoms of asthma at much lower levels than those that will cause hypersensitivity and if people are removed from exposure to the substance as soon as they start to develop symptoms they are likely to make a complete recovery.

Principles of good practice (COSHH, Schedule 2A)

Regulation 7(2) of COSHH 2002 sets out a requirement that the employer comply with the requirement in Regulation 7 (1) by using the preferred method of prevention, i.e., substitution, thus replacing it with a substance or process that eliminates or reduces the risk.

Regulation 7(3) goes on to require the establishment of appropriate controls in circumstances where substitution is not reasonably practicable. These are set out in the following order of priority:

"(a) The design and use of appropriate work processes, systems and engineering controls and the provision and use of suitable work equipment and materials.

(b) The control of exposure at source, including adequate ventilation systems and appropriate organisational measures.

(c) Where adequate control of exposure cannot be achieved by other means, the provision of suitable personal protective equipment in addition to the measures required by sub-paragraphs (a) and (b)."

The eight principles of good practice for the control of exposure to substances hazardous to health are set out in Schedule 2A of COSHH 2002 (as amended in 2004), its ACOP and guidance. A summary of the main points is given here:

- Design and operate processes and activities to minimise emission, release and spread of substances hazardous to health.
- Take into account all relevant routes of exposure - inhalation, skin absorption and ingestion - when developing control measures.
- Control exposure by measures that are proportionate to the health risk.

- Choose the most effective and reliable control options which minimise the escape and spread of the substances hazardous to health.
- Where adequate control of exposure cannot be achieved by other means, provide, in combination with other control measures, suitable personal protective equipment.
- Check and review regularly all elements of control measures for their continuing effectiveness.
- Inform and train all employees on the hazards and risks from the substances with which they work and the use of control measures developed to minimise the risks.
- Ensure that the introduction of control measures does not increase the overall risk to health and safety.

Personal protective equipment (PPE) provided shall comply with the Personal Protective Equipment Regulations (PPER) 2002 (dealing with the supply of PPE).

Respiratory protection must be suitable and of a type or conforming to a standard approved by the Health and Safety Executive (HSE).

Control of hazardous substances with reference to the hierarchy quoted in 'Working with substances hazardous to health (INDG136)'

In the control of occupational health hazards many approaches are available, the use of which depends upon the severity and nature of the hazard. The main control strategies are outlined in the following section, in an order of priority.

ELIMINATION OF THE HARMFUL SUBSTANCE

This represents an extreme form of control and is appropriately used where high risk is present from such things as carcinogens, where it is usually achieved through the prohibition of use of these substances. Care must be taken to ensure that all stock is safely disposed of and that controls are in place to prevent their re-entry, even as a sample or for research.

Elimination of the harmful substance often means that a replacement substance will need to be used in its place; this may be a non-hazardous or less hazardous substance.

Substitution for a substance that does not have the hazard being eliminated may involve such things as the use of vegetable oils for metal machining processes to replace mineral oils (which may be carcinogenic), and glass fibre for asbestos (to eliminate the risk of asbestosis). Substitution may involve the use of a less toxic substance in place of a more highly toxic one, for example, toluene substituted for benzene.

If this level of control is not achievable then another must be selected.

CHANGE THE FORM OF THE SUBSTANCE

The form of the substance may be modified to reduce the risk of exposure to an effective level. For example, a paste or solution may be used to replace a dusty powder.

MODIFY THE PROCESS SO THAT IT EMITS LESS OF THE HAZARDOUS SUBSTANCE

A useful approach to control exposure to hazardous substances is to reduce the actual quantity of the substance which can become airborne; for example, prevention of large volumes of airborne vapours by the use of a brush to apply the substance rather than an aerosol can or spraying equipment. In the case of disposal of acids, the risk of acid burns or corrosion will be removed by neutralisation with a suitable alkali.

ENGINEERING CONTROLS

Isolation

This can involve the design and implementation of remote handling of hazardous substances, for example, the use of enclosure and automation for the bottling of bleach. This enables people to be isolated from the harmful substances.

Enclosure

This strategy is based on the containment of a hazardous offending substance to prevent its free movement in the working environment. It may take a number of forms, for example, glove boxes, pipelines, closed conveyors, laboratory fume cupboards.

Enclosure of process/plant, for example, the processing of harmful substances such as isocyanides used in the production of isocyanates.

Ventilation

See 'Element B3.3 - Ventilation' later in this Element.

MINIMISE THE NUMBERS OF WORKERS EXPOSED

Segregation is a method of controlling the risks from hazardous substances. It can take a number of forms:

By physical separation

This can be a relatively simple method such as where the minimum number of employees are working with biological or toxic substances and are distanced (segregated) from the general workforce. If the hazards cannot be enclosed close to their source it may be preferable that the workforce be segregated from the hazard by providing physical separation in the form of a refuge, for example, a control room of a chemical process.

A simple example would be, where herbicides are sprayed on the land others, including members of the public, are kept at a safe distance from the source.

By worker characteristics

The protection of young workers (aged 16 to18 years) in certain trades, for example, working with lead. The Control of Lead at Work Regulations (CLAW) 2002 excludes the employment of young persons in lead processes. There remains the possibility of gender linked vulnerability to certain hazardous substances, such as lead. It is a requirement of the CLAW 2002 that female workers of child bearing age are excluded from lead exposure through work. Segregation affords a high level of control in these circumstances.

By time

This is concerned with the restriction of certain hazardous operations to periods when the number of persons present is at its smallest, for instance at weekends. An example might be the spray painting of an underground railway station platform when the railway system is closed, from late evening until early morning.

PROVIDE PERSONAL PROTECTIVE EQUIPMENT

When adequate control of exposure cannot be achieved by other means, a combination of control measures and personal protective equipment (PPE) may be applied. Alternatively, PPE may be used on its own. It is important that other, more effective means of control be considered before resorting to the use of PPE.

Adequate control of carcinogens and mutagens

COSHH 2002, Regulation 7, states specific requirements for additional control measures for carcinogens and mutagens.

Regulation 7(5) states that where it is not reasonably practicable to prevent exposure to a carcinogen or mutagen:

"The employer must apply the following measures in addition to those required by Regulation 7(3):

a)   *Totally enclosing the process and handling systems, unless this is not reasonably practicable.*

b)   *The prohibition of eating, drinking and smoking in areas that may be contaminated by carcinogens.*

c)   *Cleaning floors, walls and other surfaces at regular intervals and whenever necessary.*

d)   *Designating those areas and installations which may be contaminated by carcinogens and using suitable and sufficient warning signs.*

e)   *Storing, handling and disposing of carcinogens safely, including using closed and clearly labelled containers."*

Ventilation may be by natural or mechanical means. Extract ventilation is an important control strategy in the prevention of occupational ill-health from exposure to dusts, fumes, mists, vapours and gases. It is referred to as such in the Control of Substances Hazardous to Health Regulations (COSHH) 2002 (as amended).

B3.2 - The specific requirements for working with asbestos

Identification of types of asbestos

Asbestos is a fibrous mineral extracted from the ground and mined throughout the world. 'Asbestos' is a general (generic) term for a group of fibrous minerals (known as silicates). There are three main types of asbestos:

1)   Chrysotile - 'white'.
2)   Amosite - 'brown'.
3)   Crocidolite - 'blue'.

The colours are of the mineral when mined. In the workplace the asbestos types *cannot* be identified just by their colour - laboratory analysis is required.

SERPENTINE FIBRE

Chrysotile - 'white' asbestos - typically found in cement products, floor tiles and gaskets. By far the most common type of asbestos.

AMPHIBOLES FIBRE

Amosite - 'brown' asbestos - typically found in wall panels, ceiling tiles and industrial (land and sea) boiler and pipe insulation.

Crocidolite - 'blue' asbestos - typically found in sprayed coatings (limpet), insulation industrial steam boiler systems and old textiles. Asbestos is made up of tiny sharp fibres that remain in the air easily. Fibres enter the body through the nose, mouth and skin. Fibres are easily inhaled and are difficult to breathe out because of their tiny sharp fibres. They may also be swallowed and affect the digestive system or may penetrate the skin.

Figure B3-2: Chrysotile 'white' asbestos. *Source: Forhealths.com.*

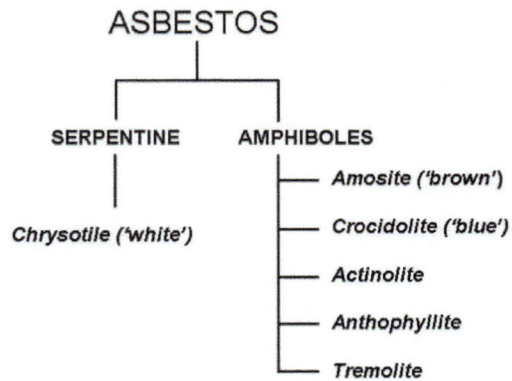

Figure B3-3: Two groups of asbestos types. *Source: RMS.*

Distinctions between licensed, notifiable non-licensed and non-licensed asbestos work

Under the Control of Asbestos Regulations (CAR) 2012, anyone carrying out work on asbestos insulation, asbestos coating or asbestos insulating board (AIB) needs a licence issued by the Health and Safety Executive (HSE) unless they meet one of the following exemptions set out in Regulation 3(2) of CAR 2012:

- If employee exposure is sporadic and low intensity, i.e. the concentration will not exceed 0.6 fibres per cm^3 measured over 10 minutes; and exposure will not exceed the 'control limit' (0.1 fibres per cm^3).
- The work involves short non-continuous maintenance activities, i.e., where one person works with the materials for no more than one hour in a seven-day period. The total time spent by all workers on the work should not exceed a total of two hours.
- The materials being removed have asbestos fibres firmly linked in a matrix, for example, asbestos cement.
- Encapsulation or sealing of asbestos-containing materials which are in good condition.
- Air monitoring and control, and the collection and analysis of samples to find out if a specific material contains asbestos.

Persons carrying out non-licensed asbestos work would include those whose work will knowingly disturb low risk asbestos containing materials, such as maintenance workers and their supervisors; and those who carry out asbestos sampling and analysis. All other work, for example, encapsulation or sealing asbestos-containing materials in poor condition, removal or demolition requires a licence.

Even if a licence is not required, the rest of the requirements of CAR 2012 to prevent exposure to asbestos must still be complied with.

Certain types of ancillary work will require a licence unless exempt under Regulation 3(2) of CAR 2012, this will include:

- Setting up and taking down enclosures for notifiable and licensed asbestos work.
- Putting up and taking down scaffolding to provide access for licensable work where it is foreseeable that the scaffolding activity is likely to disturb the asbestos.
- Maintaining air extraction equipment (which includes 'negative pressure' units).
- Work done within an asbestos enclosure, such as sealing an electric motor in polythene and installing ducting to the motor to provide cooling air from outside the enclosure.
- Cleaning the structure, plant and equipment inside the enclosure.

If the work is licensable, there are a number of duties:

- Notify the enforcing authority responsible for the site where the work is: HSE or the local authority, at least 28 days before work begins. A shorter time may be agreed by the enforcing authority, for example, in emergencies. The notification should be in writing and the particulars are specified in Schedule 1 of CAR 2012.
- Designate the work area *(see Regulation 18 for details).*
- Prepare specific asbestos emergency procedures.
- Pay for employees to undergo medical surveillance every two years.

CAR 2012 have introduced a new category of Notifiable Non-Licensed Work (NNLW). Whether a type of asbestos work is either licensable, NNLW or non-licensed work has to be determined in each case and will depend on the type of work being done, the type of material being worked on and its condition. The identification of the type of asbestos-containing material (ACM) to be worked on and an assessment of its condition are important parts of a risk assessment, which needs to be completed before work starts. It is the responsibility of the person in charge of the job to assess the ACM to be worked on and decide if the work is NNLW or non-licensed work. This will be a matter of judgement in each case, dependent on consideration of the above factors.

A decision flow chart is available from the HSE at, www.hse.gov.uk/asbestos/essentials/index.htm. If work is determined to be NNLW, the duties are:

- To notify the enforcing authority responsible for the site where the work is before work starts. (There is no minimum period).
- By 2015 all employees will have to undergo medical examinations which are repeated every three years.
- To have prepared procedures which can be put into effect should an accident, incident or emergency occur.
- To keep a register of all NNLW work for all employees.
- To record the significant findings of and comply with a risk assessment.
- To prevent or reduce exposure so far as is reasonably practicable and to take reasonable steps that all control measures are used.
- To ensure that adequate information, instruction and training is given to employees.

Typical locations where asbestos can be found

Asbestos has been widely used in building materials for a long time, though some countries have established programmes to phase out its use because of the risks to health.

As long as the asbestos-containing material (ACM) is in good condition, and is not being or going to be disturbed or damaged, there is negligible risk. But if it is disturbed or damaged, it can become a danger to health, because people may breathe in any asbestos fibres released into the air.

Workers who may be particularly at risk of being exposed to asbestos when carrying out building maintenance and repair jobs include:

- Construction and demolition contractors, roofers, electricians, painters.
- Decorators, joiners, plumbers, gas fitters, plasterers, shop fitters, heating and ventilation engineers, and surveyors.
- Anyone dealing with electronics, for example, phone and information technology (IT) engineers, and alarm installers.
- General maintenance engineers and others who work on the fabric of a building.

If asbestos is present that can be readily disturbed, is in poor condition and not managed properly, all people in the building could be put at risk.

Asbestos has been used in many parts of buildings, examples of uses and locations where asbestos can be found are shown in *figure ref B3-4*.

Asbestos products	What it was used for
Sprayed asbestos (limpet).	Fire protection in ducts and to structural steelwork, fire breaks in ceiling voids etc.
Lagging.	Thermal insulation of pipes and boilers.
Asbestos insulating boards (AIB).	Fire protection, thermal insulation, wall partitions, ducts, soffits, ceiling and wall panels.
Asbestos cement products, flat or corrugated sheets.	Roofing and wall cladding, gutters, rainwater pipes, water tanks.
Certain textured coatings.	Decorative plasters, paints.
Bitumen or vinyl materials.	Roofing felt, floor and ceiling tiles.
General uses.	Vehicle brake linings, woven fires, ropes used as high temperature gaskets for furnaces, jet engines, chemical pipelines. Electrical insulation for hotplate wiring, electrical fuse wire holders and in building insulation and sound absorption. Filters for cigarettes. Artificial (chrysotile) snow effects in Hollywood films made in the USA in the 1920's and 1930's.

Figure B3-4: Examples of uses and locations where asbestos can be found. *Source: RMS.*

Sprayed coatings, lagging and insulating board are more likely to contain blue or brown asbestos. In general, materials that contain a high percentage of asbestos are more easily damaged. Asbestos insulation and lagging can contain up to 85% asbestos and are most likely to give off fibres.

Work with AIB can result in equally high fibre release if power tools are used. Asbestos cement is of lower risk, since it contains only 10-15% asbestos.

The asbestos is tightly bound into the cement and the material will only give off fibres if it is badly damaged or broken, or is worked on, for example, if it is drilled, abraded or cut.

HIGH RISK MATERIALS

Figure B3-5: Asbestos pipe lagging.   *Source: HSE.*

Figure B3-6: Asbestos insulating board (AIB).   *Source: HSE.*

Figure B3-7: Textured decorative coating.   *Source: HSE.*

Figure B3-8: Window with AIB panel.   *Source: HSE.*

NORMALLY LOWER RISK MATERIALS

Figure B3-9: Asbestos cement roof sheeting.   *Source: HSE.*

Figure B3-10: Asbestos containing floor tiles.   *Source: HSE.*

The duty to manage asbestos

The duty to manage asbestos is covered by Regulation 4 of the Control of Asbestos Regulations (CAR) 2012. In many cases, the duty holder is the person or organisation that has clear responsibility for the maintenance or repair of non-domestic premises through an explicit agreement such as a tenancy agreement or contract.

Where there are domestic premises such as flats the duty holder will be responsible for common areas such as corridors or walkways.

The duty holder has the responsibility to:

- Carry out a survey to determine if there is asbestos in the premises (or assessing if ACMs are liable to be present and making a presumption that materials contain asbestos, unless there is strong evidence that they do not), its location and its condition.
- Identify and record the location and condition of any ACMs or presumed ACMs in the premises.
- Assess the risks from any material identified.
- Prepare a plan that sets out in detail how risks identified are to be managed.
- Establish a system for providing information on the location and condition of ACMs for anyone who is liable to work on or disturb it.

- Record the roles and responsibilities of those who manage asbestos in the organisation.
- Review and monitor the plan and the arrangements made to prevent risk.

There is also a requirement on anyone to cooperate as far as is necessary to allow the duty holder to comply with the requirements shown previously. If asbestos containing materials are in good condition they may be left in place and their condition monitored and managed.

ASBESTOS IDENTIFICATION

There are three main types of asbestos, which is a naturally occurring mineral. It can be amphibole asbestos which includes crocidolite (blue) and amosite (brown) asbestos, or serpentine asbestos which is chrysotile (white) asbestos. All three of these main types have been used in Great Britain at some time.

Types of survey

Management survey

This is the standard survey used to identify the presence and extent of any asbestos containing materials (ACMs) in the building which could be damaged or disturbed during normal occupancy, including foreseeable maintenance and installation, and to assess their condition.

A management survey might involve minor intrusive work and some disturbance which will vary between premises and depend on what is reasonably practicable for individual properties.

Management surveys can involve a combination of sampling to confirm asbestos is present or presuming asbestos to be present.

Refurbishment and demolition surveys

A refurbishment and demolition survey will be needed before any work of this type is carried out. This survey will be used to locate and describe all ACMs in the area where the work will take place or in the whole building if demolition is planned.

This type of survey is fully intrusive and involves destructive inspection to gain access to all areas, including those that may be difficult to reach.

A refurbishment and demolition survey may also be required in other circumstances, for example, when more intrusive maintenance and repair work will be carried out or for plant removal or dismantling.

Who can undertake them

An independent expert/specialist organisation must:

- Have adequate training and be experienced in survey work.
- Able to demonstrate independence, impartiality and integrity.
- Have an adequate quality management system.
- Carry out any asbestos survey work in accordance with recommended HSE guidance HSG264 'Asbestos: the survey guide'.

HSG264 states that organisations offering an asbestos survey service should be able to demonstrate their competence by holding United Kingdom Accreditation Service (UKAS) accreditation to ISO/IEC 17020 and individuals to ISO/IEC 17024.

Competence for assessment

The HSE document 'Managing and work with asbestos', the approved code of practice and guidance to the Control of Asbestos Regulations (CAR) 2012 clarifies competence requirements for those that carry out asbestos risk assessments.

The ACOP to Regulation 6 of CAR 2012, relating to assessment of work which exposes employees to asbestos, states:

> *"Whoever carries out the assessment should:*
> - *Have adequate knowledge, training and expertise in understanding the risks from asbestos and be able to make informed and appropriate decisions about the risks and precautions needed.*
> - *Know how the work activity may disturb asbestos.*
> - *Be familiar with and understand the requirements of the Regulations ACOP.*
> - *Have the ability and the authority to collate all the necessary and relevant information.*
> - *Be able to assess other non-asbestos risks on site.*
> - *Be able to estimate the expected level of exposure to decide whether or not the control limit is likely to be exceeded."*

Figure B3-11: Competence to carry out asbestos risk assessments.   *Source: HSE ACOP to CAR 2012.*

The use of specialist contractors for removal and disposal of asbestos

If ACMs need to be sealed, encapsulated or removed, then a licensed contractor needs to be employed if the materials are high risk (for example, pipe insulation and asbestos insulating panels). If the materials are lower risk (for example, asbestos cement sheets and roofing) then an unlicensed but competent contractor may carry out this work. A list of licensed contractors is available on the HSE asbestos licensing information community web page.

REQUIREMENTS FOR REMOVAL

Non-licensed work

Most asbestos work must be undertaken by a licensed contractor but any decision on whether particular work is licensable is based on the risk.

To be exempt from needing a licence the work must be:

- Sporadic and low intensity.
- Carried out in such a way that the exposure of workers to asbestos will not exceed the legal control limit of 0.1 asbestos fibres per cubic centimetre of air (0.1 f/cm^3).
- A short non-continuous maintenance task.
- A removal task, where the ACMs are in reasonable condition and are not being deliberately broken up.
- A task where the ACMs are in good condition and are being sealed or encapsulated.
- An air monitoring and control task to check fibre concentrations in the air, or it's the collection and analysis of asbestos samples to confirm the presence of asbestos in a material.

Notifiable non-licensed work

From April 2012, some non-licensed work, where the risk of fibre release is greater, is subject to three additional requirements:

1) Notification of work.
2) Medical examinations.
3) Record keeping.

This work is known as notifiable non-licensed work (NNLW). To decide if the work is NNLW, employers will need to consider the type of work that will be carried out, the type of material to be worked on and its condition:

- Decide what type of work will be carried out.
- Consider the asbestos type.
- Consider the material's condition.

Licensed work

Certain types of work with ACMs can only be carried out by those who have been issued with a licence by HSE. This is work which meets the definition of 'licensable work with asbestos' in Regulation 2(1).

That is work:

- Where worker exposure to asbestos is not sporadic and of low intensity.
- Where the risk assessment cannot clearly demonstrate that the control limit (0.1 f/cm^3 airborne fibres averaged over a four-hour period) will not be exceeded.
- On asbestos coating (surface coatings which contain asbestos for fire protection, heat insulation or sound insulation but not including textured decorative coatings).
- On asbestos insulation or AIB where the risk assessment demonstrates that the work is not sporadic and of low intensity, the control limit will be exceeded and it is not short duration work.

Short duration means the total time spent by all workers working with these materials does not exceed two hours in a seven-day period, including time spent setting up, cleaning and clearing up, and no one person works for more than one hour in a seven-day period.

If licensable work is to be carried out then the appropriate enforcing authority must be notified of details of the proposed work. Employers must carry out a risk assessment of the work. This assessment must be kept at the place where work is being carried out.

For licensable work, the plan of work *(see following section)* should be site specific and contain the following information:

- The scope of work as identified in the risk assessment.
- Details of hygiene facilities, transit route, vacuum cleaners, air monitoring, protective clothing, respiratory protection equipment (RPE), and communication between the inside and outside of the enclosure.
- Use of barriers and signs, location of enclosures and airlocks, location of skips, negative pressure units, cleaning and clearance certification, emergency procedures.

Notification and plan of work

If carrying out licensable work, it must be notified to the appropriate enforcing authority 14 days in writing before work commences (the authority may allow a shorter period in an emergency if there is a serious risk to health).

For any work involving asbestos, the employer must draw up a written plan of work. The plan of work must include the following information:

- Nature and duration of the work.
- Number of persons involved.
- Address and location of the work.
- Methods used to prevent or reduce exposure.
- Type of equipment used to protect those carrying out the work and those present or near the worksite.

Work must not take place unless a copy of the plan is readily available on site.

REQUIREMENTS FOR DISPOSAL

Licensed carrier

Health and Safety Executive (HSE) requirements on the licensed contractor has means that each company must have in place an up to date standard operating procedure manual, which should contain all of the details relating to the safe removal and disposal of asbestos.

This manual must also be sent to the local HSE inspector and constantly updated to reflect changes in guidance and legislation. Waste carriers licence involves a simple application and anyone looking to transport waste asbestos will need to be a Registered Waste Carrier.

Notification

Employers must notify the appropriate enforcing authority 14 days before carrying out any licensable work. This can be done on form FOD ASB5.

Licensed disposal site

Any facility looking to accept waste asbestos has to apply to the Environment Agency for a site licence. This will set out the range and volumes of wastes that can be accepted on the site, the site control and management systems, engineering and infrastructure, manning and qualification requirements and reporting and monitoring regimes.

Control limits for working with asbestos

ASBESTOS ESSENTIALS

'Asbestos Essentials', HSG210, is a task manual for building, maintenance and allied trades on **non-licensed** asbestos work. The manual is designed for all workers (employers, employees and self-employed) who may come into contact with asbestos; such as, electricians, builders, plumbers, carpenters and other trades.

The manual covers:

- Work with asbestos cement (AC) (non-licensed).
- Working with textured coatings (TC) containing asbestos (non-licensed).
- Strictly controlled minor work on Asbestos Insulating Board (AIB).
- Safe work with undamaged asbestos materials.
- Removal and replacement of other asbestos containing materials.
- Fly-tipped waste.
- Equipment and method sheets.

The manual headings are supplemented with 38-specific task sheets which cover most aspects of potentially hazardous work on, or near, asbestos materials including:

- Drilling, for example, holes in asbestos cement, textured coatings, asbestos insulation board.
- Removing, for example, asbestos cement or reinforced plastic product such as tanks, ducts and water cisterns.
- Repairing, for example, damaged asbestos cement.
- Painting, for example, undamaged asbestos insulating board.
- Enclosing, for example, undamaged asbestos materials to prevent impact damage.
- Cleaning, for example, debris from guttering on an asbestos roof.

The additional control measures relating to uncovering or damaging asbestos are covered by general advice given in stage-by-stage method sheets, such as the required training, vacuuming, damp wetting, decontamination and disposal.

THE CONTROL OF ASBESTOS REGULATIONS 2012

The **Control of Asbestos Regulations (CAR) 2012** place emphasis on assessment of exposure; exposure prevention, reduction and control; adequate information, instruction and training for employees; monitoring and health surveillance. The regulations also apply to incidental exposure. The section on prohibitions is now covered by REACH.

Identification and assessment

Before any work with asbestos is started the employer must ensure a thorough assessment of the likely exposure is carried out. Such an assessment must identify the type of asbestos involved in the work, or to which the employees are likely to be exposed. The assessment must also determine the nature and degree of any exposure and the steps required to prevent or reduce the exposure to the lowest level reasonably practicable. Assessments must be reviewed regularly and when there is reason to suspect that the original assessment is invalid or there is a significant change in the work to which the original assessment related. Assessments should be revised accordingly to take account of any such changes, etc.

Worker exposure must be below the airborne exposure limit (Control Limit). CAR 2012 has a single Control Limit for all types of asbestos of 0.1 fibres per cm^3. A 'Control Limit' (CL) is a maximum concentration of asbestos fibres in the air, averaged over any continuous 4 hour period, which must not be exceeded.

Short term exposures must also be strictly controlled. Worker exposure should not exceed 0.6 fibres per cm^3 of air averaged over any continuous 10 minute period; using respiratory protective equipment (RPE) if exposure cannot be reduced sufficiently using other means.

Plan of work

Employers must also prepare a suitable 'plan of work' before any work involving asbestos removal from buildings, structures, plant or installations (including ships) is undertaken. Such 'plans of work' must be retained for the duration of the work. The 'plan of work' should address the location, nature, expected duration and asbestos handling methods involved with the work, and the characteristics of the protection and decontamination equipment for the asbestos workers and the protection equipment for any others who may be affected by such work. The asbestos risk assessment and plan of work must be kept on site.

Information, instruction and training

Employees exposed to asbestos must be provided with adequate information, instruction and training to understand the risks associated with asbestos and the necessary precautions. Employees who carry out work in connection with the employer's duties under CAR 2012 must also be given adequate information, instruction and training to do their work effectively. Under Regulation 10, refresher training must also be provided.

Prevention or reduction of exposure

Wherever possible the employer must prevent exposure of asbestos to the employees. Where this is not reasonably practicable the employer must reduce the exposure to the lowest level reasonably practicable other than by using respiratory protective equipment and ensure the number of employees exposed at any one time is minimised.

Regulation 11 (2) requires:

"Where it is not reasonably practicable for the employer to prevent the exposure of his employees to asbestos in accordance with paragraph (1)(a)[of Regulation 11], the measures referred to in paragraph (1)(b)(i) [of Regulation 11] shall include, those in paragraph (2) [of Regulation 11], in order of priority:

a) *The design and use of appropriate work processes, systems and engineering controls and the provision and use of suitable work equipment and materials in order to avoid or minimise the release of asbestos.*

b) *The control of exposure at source, including adequate ventilation systems and appropriate organisational measures and the employer shall so far as is reasonably practicable provide the employees concerned with suitable respiratory protective equipment in addition to the measures required by sub-paragraphs (a) and (b)."*

In addition, to the provision of respiratory protective equipment (RPE), the employer must reduce the concentration of asbestos in the air inhaled by employees, taking account of the effect of RPE, to a level below the control limit and as low as is reasonably practicable.

Personal protective equipment (PPE) provided must comply with the Personal Protective Equipment Regulations (PPER) 2002 (dealing with the supply of PPE).

If the control limit is exceeded, Regulation 11(5)(b) requires the employer to:

"(i) Forthwith inform any employees concerned and their representatives and ensure that work does not continue in the affected area until adequate measures have been taken to reduce employees' exposure to asbestos to below the control limit.

(ii) As soon as is reasonably practicable identify the reasons for the control limit being exceeded and take the appropriate measures to prevent it being exceeded again.

(iii) Check the effectiveness of the measures taken pursuant to sub-paragraph (ii) by carrying out immediate air monitoring."

B3.3 - Ventilation

Uses and limitations of dilution ventilation for hazardous substances

This is often referred to as general ventilation and involves using natural air movement through doors and windows or using mechanical air movement to dilute any contaminated air, hence reducing or eliminating the airborne pollutants. Because dilution ventilation does not target any specific contamination source and relies on dispersal/dilution instead of specific contaminant removal, it is best suited to be used with nuisance contaminants that are themselves fairly mobile in air. Dilution ventilation systems will only deal with general contamination and will not prevent contaminants entering a person's breathing zone.

On its own dilution ventilation is not a reliable means of dealing with toxic or similarly hazardous contaminants, but supported by respiratory protection equipment it may be acceptable for some substances. Local exhaust ventilation is the preferred means of controlling exposure to toxic or similarly hazardous contaminants.

Dilution ventilation is a system designed to induce a general flow of clean air into a work area. This may be done by driving air into a work area, causing air flow around the work area, dilution of contaminants in the work area and then out of the work area through general leakage or through ventilation ducts to the open air.

A variation on this is where air may be forcibly removed from the work area, but not associated with a particular contaminant source, and air is allowed in through ventilation ducts to dilute the air in the work area. Sometimes a combination of these two approaches is used.

An example may be general air conditioning provided in an office environment. A particularly simple approach to providing dilution ventilation is to open a window and door and allow natural air flow to dilute the workplace air.

Dilution ventilation may only be used as the sole means of control in circumstances where there is:

- Non-toxic contaminant or vapour that is easily suspended in the workplace air (not dusts).
- Contaminant which is uniformly produced in small, known quantities.
- No defined point of release of the contaminant.
- No other practical means of reducing levels.

Dilution ventilation is often used to maintain a good working environment, enabling fresh, clean dilution air to be provided at a suitable temperature and humidity for comfort. This may be particularly useful for comfort in offices where a large number of people work closely together, such as a call centre, to reduce temperatures in workplaces where heat is generated by work processes, such as those involving molten metals, glass and ceramics or to reduce humidity in laundries and bakeries.

Roles and responsibilities in relation to local exhaust ventilation (LEV)

Local guidance and legislation may apply to the use of Local Exhaust Ventilation (LEV) systems. Key roles and responsibilities, in the UK for Local Exhaust Ventilation (LEV) systems, are outlined in 'Controlling airborne contaminants at work - a guide to local exhaust ventilation (LEV)':

- The employer is the 'system owner' and is the client for a new or redesigned LEV system. The employer should not apply LEV before considering other control options and using them where appropriate.
- The people who carry out routine checks of the LEV system are usually employees or supervisors, but may be service providers.
- LEV suppliers provide goods (an LEV system) and may then act as a service provider. Designers are responsible for interpreting the requirements of the employer and advising on an effective LEV system which is capable of delivering the required control.
- LEV installers work with commissioners to ensure the equipment supplied provides adequate control of the contaminant. The installer may be the design company, a service provider, or even the employer (if competent).
- Service providers offer services such as installation, commissioning, maintenance and thorough examination and tests.
- LEV commissioner's work with installers to ensure the equipment supplied provides adequate control of the contaminant.
- LEV maintenance and repair engineers are usually service providers, but sometimes an employee can carry out the work.
- LEV examiners' responsible for carrying out the thorough examination and test, are usually service providers but can be employees if competent.
- The requirement for competence for suppliers of goods and services means that the extent and depth of their knowledge and capability must be sufficient to assess and solve the problems they are likely to meet.

Typical components of an LEV and their function

Hazards that can be controlled by the effective use of LEV include: dust which could cause coughing, sneezing and various other respiratory diseases; chemicals which might cause sensitisation or other toxic effects; allergens which could aggravate asthmatic conditions; micro-organisms which can cause diseases; asphyxiants that can lead to breathing difficulties, unconsciousness and, ultimately, death; extreme heat which could result in heat exhaustion.

Various LEV systems are in use in the workplace, for example:

- Captor systems are used for welding, cutting and milling operations. *See figure ref B3-17,* which shows the fixed captor hood, flexible hose and rigid duct in use on a circular saw.
- Receptor systems, such as are used in fume cupboards and kilns.
- High velocity low volume flow systems, for example, as used on a grinding tool.

COMPONENTS OF A BASIC SYSTEM

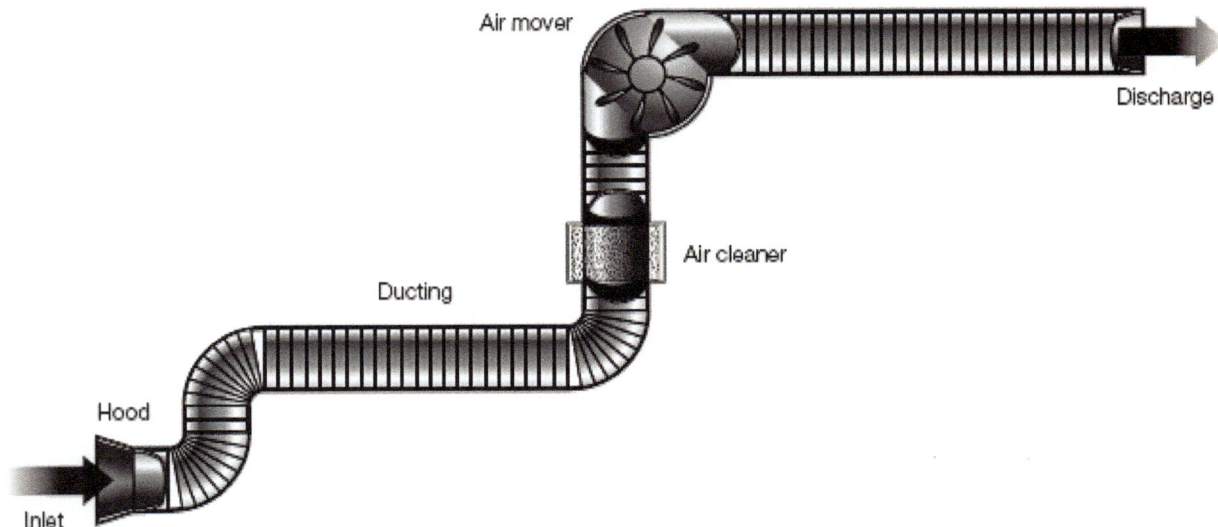

Figure B3-12: Components of a basic LEV system.   *Source: HSE INDG408.*

The objective of a LEV system is to remove contaminant at source, so that its range of contamination is minimised. Removal is usually achieved by mechanical air handling. A typical local exhaust ventilation system consists of the following major components:

- **Inlet** - this is the point at which the air is drawn to the LEV system from the atmosphere of a workplace, along with the airborne contaminant. It is located at or near the source of the contaminant and may be an opening in a hood, booth, canopy, cabinet or enclosure.
- **Hood(s)** - shaped to collect airborne contaminants and encourage them to flow through the inlet into the ducting. It may be part of a booth, canopy, cabinet or some other form of enclosure. **See later in this Element for more detailed explanation of design parameters**.
- **Ducts** - to carry the airborne contaminants away from the capture point. Ducting may contain bends, junctions, changes of section and dampers; it may be circular or rectangular in cross-section and be rigid or flexible. **Figure ref B3-15** shows the length of ducting with curves not corners and **figure ref B3-14** shows a self-contained unit which can be moved around the workplace.
- **Air cleaner** - to filter and clean the extracted air, for example, dust filter, wet scrubber, or solvent recovery device. Equipment designed as particulate dust and fume collectors and devices to remove mists, gases and vapours.
- **Air mover** - the equipment that moves the air through the system. Often this will be a fan; occasionally some other kind of air mover may be used, such as compressed air venturi. The air mover must be the right size and type to deliver sufficient 'suction' (negative pressure) to the hood. **Figure ref B3-18** shows the size of a fan and motor required for industrial scale LEV.
- **Discharge** - the safe release of cleaned, extracted air into the atmosphere. This may be back into the workroom via a diffuser, grille or open duct. Alternatively, this may be external to the workplace, via a stack.

Specific components and design features of the extraction system will depend upon the contaminant to be collected and parameters for its safe disposal.

Figure B3-13: Flexible hose and captor hood.   *Source: RMS.*

Figure B3-14: Portable self-contained unit.   *Source: RMS.*

Figure B3-15: Length of ducting with curves.   *Source: RMS.*

Figure B3-16: Shows a number of ports at the end of one duct - not all ports being used. *Source: RMS.*

Figure B3-17: Captor system on circular saw.   *Source: RMS.*

Figure B3-18: LEV fan and motor.   *Source: RMS.*

Source strength and capture zones

FACTORS IN LEV DESIGN THAT ENSURE EFFECTIVENESS

General points to be considered

The correct design of a LEV system is of vital importance to ensure minimum emission of contaminants into the workplace without the process being unduly affected. Successful control of contaminants for existing processes may well require modification of plant or layout to accommodate the design.

Consideration must be given to regulatory requirements and HSE guidance. When arranging installation of LEV it is vital that the pre and post ventilation contamination levels are specified and the required reduction should be part of the commissioning contract.

Successful control by ventilation may require a sound knowledge of processes, substances, plant, working methods and environmental factors.

It may require the application of more than one type of system:

Mini	On-tool extraction, typically portable, for example, fume extraction used by welders, air filtered and returned to workplace.
Midi	Local extraction, typically several inlets located close to the contaminant, air discharged to atmosphere.
Maxi	Comprehensive workplace extraction, typically involving a number of workrooms and operations with air removed from the workplace to a central point where it is cleaned and released to the external atmosphere.

Correct design of air inflow systems (midi and maxi systems) for make-up air may contribute greatly to the overall effectiveness of the LEV system. Process plant should be interlocked with the LEV system so that the system always operates when the process plant is active and the LEV system should give warning of defective operation.

Unauthorised modification, especially by operators should be prevented. LEV systems must not introduce other problems (for example, excess noise and explosion risk) and the design should not limit work rate. LEV systems should be designed to be reliable over a reasonable working life.

Their quality should be no lower than that of the process plant itself and a programme of thorough examination/planned preventive maintenance should be established.

Hood design

Hoods should be designed to confine or enclose the contaminant whenever possible and allow access for inspection and maintenance. Enclosing hoods establish a complete or partial enclosure around the contaminant preventing its escape into the workplace and allowing it to be drawn in to the LEV inlet.

Receptor hoods are used to direct contaminants that moves into the inlet due to characteristics of the process, for example upwards due to thermal air movement. Captor hoods are used to capture contaminant that would not naturally move into the inlet of the system, and are particularly important where the contaminant has high velocity and significant mass as the contaminant may quickly spread into the workplace. Captor hoods should, whenever possible, be positioned in line with normal contaminant travel. Examples of where captor hoods are used include rim/lip extraction, downdraught tables and LVHV (low volume high velocity) systems.

Wherever possible, flanges should be provided on hoods to eliminate the tendency of drawing air from ineffective zones (behind the hood) where no contaminant exists. Increasing hood effectiveness in this manner will usually result in a 25% reduction in the airflow required to achieve contaminant capture.

As a general rule, the width of the flange around a hood is equal to the hood diameter, or one side, but should not exceed 150mm. Flanges are especially important when enclosure of the process is impracticable, as the airflow pattern in front of the hood must be such that the selected capture velocity is maintained in the zone of contaminant generation, conveying it into the exhaust hood opening.

HOOD TYPE	DESCRIPTION
	Slot
	Flanged slot
	Plain opening
	Flanged opening
	Booth
	Canopy

Figure B3-19: Hood types.   Source: RMS.

Wherever possible, keep captor hoods close to the contaminant source (extract volume efficiency is proportional to distance squared).

Capture velocity of air

This is the velocity of air moving into the inlet, at any point in front of a captor hood, necessary to overcome opposing air currents and to capture the contaminated air by causing it to flow into the hood. If the actual velocity of air movement drops below the capture velocity for the contaminant in question the LEV system will not be effective. The value of the capture velocity necessary for effectiveness will depend upon:

■ The type of process involved.
■ The characteristics of the contaminant, including velocity and mass.
■ The mode of dispersion of the contaminant.
■ The distance from the captor to the contamination and the extent to which gravity may assist capture.

Transport velocity of air

This is the minimum air velocity to move the contaminant already in the air-stream the required distance through the system. If the transport velocity is not maintained, the particles will fall out of the air-stream. The velocity should be high enough to prevent settlement of the contaminant particles suspended in the air-stream, but low enough to minimise duct losses and consequently fan power consumption.

Air cleaners

The type of air cleaning device to be used within a LEV system depends upon the nature of the contaminant. Examples used to collect dusts and fume include:

■ Wet collectors.
■ Fabric filters.
■ Cyclone filters.
■ Electrostatic precipitators.

Mists, gases and vapours have a wide range of specialised devices and systems to remove them from the atmosphere, some examples of which include chemical absorption (scrubbers, for example, acidic exhausts may utilise alkaline scrubbers) and condensation. It is important that the design of the air cleaner does not unduly limit the air velocity and flow of the system. They must be regularly maintained to ensure the timely removal of the contaminant before it restricts air velocity and flow.

FACTORS THAT REDUCE A LEV SYSTEM'S EFFECTIVENESS

The efficiency of LEV systems can be affected by many factors including the following:

- Draughts in the workroom, influencing the ability for the system to collect the contaminant.
- Hood design, influencing the ability of the system to capture/receive the contaminant from a given effective field.
- Distance of the hood and inlet from the source. The ability of the hood and air movement to capture/receive the contaminant is reduced the further away it is from the source of the contaminant. High velocity low volume systems rely on the placement of the inlet very close to the source of the contaminant to increase its ability to capture it. *See figure ref B3-17* which shows how a captor hood can be repositioned to suit the work activity by the use of a flexible hose.
- Damaged ducting can lead to leaks, so that air is drawn into the system via unplanned inlets reducing the available suction and air movement around the contaminant making it less likely to capture the contaminant. Further down the system, damage could lead to contaminant being discharged back into the workroom.
- Leaving too many inlet ports open can lead to loss of suction and air movement at the inlet of any one of the many ports. This is particularly important in situations where the system has been designed to have a maximum number of available inlet ports open at a time. Where the design takes account of all the inlet ports being open this should not affect efficiency.
- Unauthorised alterations can lead to additional ducting and ports being added, thus reducing the overall effectiveness of the system.
- Process changes leading to overwhelming amounts of contamination. If the system has been designed to establish sufficient air movement to capture a given level of contaminant, changes to the process may render the performance of the system inadequate.
- Fan strength, air velocity achieved. The original air movement equipment may not have been designed to enable sufficient air velocity to deal with capture/reception of the contaminant or may be inadequate to transport the contaminant the distance it needs to be moved in order for it to be treated and returned to the atmosphere. This can lead to contaminant deposit in the ducting or insufficient suction to draw the air through the filter.
- Incorrect adjustment of fan. The cost of heating 'make up air' may encourage some employers to reduce extraction rates by adjusting the operating speed of the fan.
- Too many or too sharp bends in ducts can lead to the slowing down of air movement and deposition of contaminant on the bends.
- Blocked or defective filters will inhibit air flow through the system, reducing the amount of air for the fan to draw and to be drawn in at the inlet. This means less contaminant will be collected.

Thorough examinations of LEV

Regulation 7 of Control of Substances Hazardous to Health Regulations (COSHH) 2002 (as amended) requires that the exposure of employees to substances hazardous to health be either prevented or, where that is not reasonably practicable, be adequately controlled *(see also - HSE Guidance EH40)*. Schedule 3 of the regulation deals with the special provisions relating to biological agents. The regulation is supported by the COSHH Approved Code of Practice (paragraph 33) which lists ways in which control can be achieved and makes specific mention of enclosure, partial enclosure with LEV, LEV and sufficient ventilation.

MAINTENANCE, EXAMINATION AND TESTING OF LEV

Regulation 9 of COSHH 2002 requires that any control measure taken to comply with Regulation 7 must be maintained in an efficient state, in efficient working order, good repair and in clean condition. LEV systems should be examined and tested at least once every 14 months (more frequently for those processes listed in schedule 4 of COSHH 2002). Regulation 9 also specifies that records shall be kept of the results of the tests including details of any repairs carried out as a result of the examinations and tests. These records shall be kept for at least five years. There is a duty (Regulation 8) on the employee to use the LEV provided and report any defects observed. Both COSHH 2002 and Management of Health and Safety at Work Regulations (MHSWR) 1999 require that those who carry out duties under these regulations should be competent to do so.

OTHER REGULATIONS

The MHSWR 1999 require an employer to make appropriate arrangements for the effective planning, organisation, control, monitoring and review of the preventative and protective measures, which includes LEV systems. The Control of Asbestos Regulations (CAR) 2012 and the Control of Lead at Work Regulations (CLAW) 2002 also impose specific requirements for the provision of, and maintenance, examination and testing of LEV. The Workplace (Health, Safety and Welfare) Regulations (WHSWR) 1992 also require the maintenance of general ventilation systems. *See also - 'Relevant statutory provisions' section.*

GENERAL POINTS

All control measures require monitoring and maintaining to ensure their effectiveness and efficiency.

There are a number of reasons why this approach should be adopted:

- Statutory obligations.
- To comply with WELs.
- Provision of information to employees.
- Indicate the need for health surveillance.
- For insurance purposes.
- To develop in-house exposure standards.

Inspection/thorough examination can be divided into two types:

1) *Initial commissioning inspection/thorough examination* - to ensure design criteria has been achieved (before use).

2) *Regular monitoring inspection/thorough examination* - to ensure design parameters are still being met (in use).

Inspection/thorough examination of performance may be conducted using qualitative means, including:

- Visual inspection of the system.
- Smoke tests with the system running, with observation of smoke leakage, eddying and smoke proximity to the breathing zone of the worker.
- High intensity lamp observation for dust or mist contaminants.
- Observation of the contamination sources and worker activities.

Inspection/thorough examination of performance may also be conducted quantitatively, including:

- Measuring the air flow rates at various points, including hood faces, hood ducts and main ducts.
- Measuring static pressure in various parts of the system including hood ducting and the pressure drop across filters and fans.
- The fan speed, motor speed and power consumption.

VISUAL INSPECTION

A competent person with appropriate knowledge and experience should periodically inspect ventilation used as an engineering control for any visual irregularities or defects that may be present.

The frequency of the inspection of the equipment will depend on its criticality in providing health and safety and on the conditions of use that could cause its degradation.

The equipment should be checked for damage, leaks or corrosion on any part or component and also all auxiliary supplies, such as electrical cabling and switching. It should be noted that the majority of components or sections of such equipment are often hidden from normal viewing; some contaminants present may also be invisible to the naked eye. Significant dust deposits within the entrances to the duct may also suggest extraction is not adequate at the points inspected.

THE EQUIPMENT REQUIRED FOR TESTING

Smoke tubes

Smoke tubes are manufactured to produce a small amount of non-toxic white smoke as a single cloud. Some produce an acidic mist derived from smoking sulphuric acid located in the tubes. When the top of the tube is opened, a small rubber ball (or similar) is used to pump air through the tube. This creates visible white smoke, which is carried on any existing air flow. For larger scale tests a battery operated fog generator may be used.

These usually use an alcohol mixture that is heated in the generator. When the vapour contacts ambient air it condenses into a white fog. The smoke or fog patterns and direction of flow are observed to assess air flow movements associated with ventilation systems and leakage tests.

Figure B3-20: Smoke tube used to test a ventilation system.
Source: Draeger.

Figure B3-21: Fog generator used to test air movement.
Source: Draeger.

Dust lamp

By shining a powerful beam of light those dust particles within the respirable range, less than 10 micrometers, become visible to the naked eye. The phenomenon is often termed the 'Tyndall effect', after the British scientist (John Tyndall) who first investigated it.

Although this does not indicate the exact quantity of dust present it does give a qualitative assessment *(see figure refs B3-24 and B3-25)*. The lamp can be used near the capture source to indicate the effectiveness of LEV equipment at capturing the dust and to check for leaks along its path of travel.

Commonly used types of lamp are either mains or battery powered. The mains operated type is fitted with a parabolic reflector unit, which uses fixed focusing to enable a parallel beam of light. They are fitted with reflectors which have safety glass screens attached.

There is a range of bulbs available up to 2,000 watts. Battery operated versions are often easier to use and will provide a 12-volt output for approximately 60 minutes. They have output powers of between 100 to 250 watts.

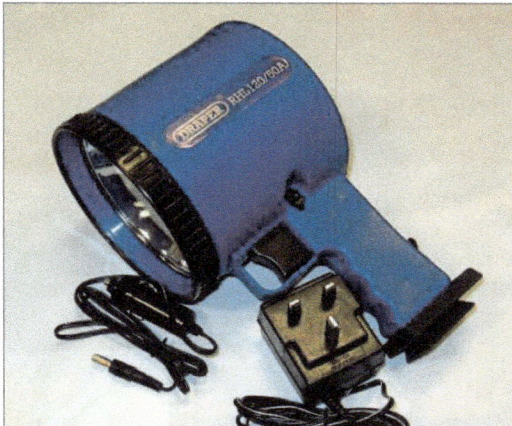

Figure B3-22: High powered dust lamp.   *Source: HSE.*

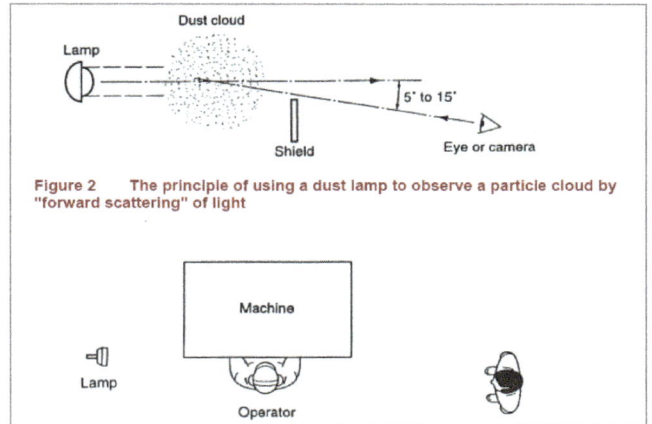

Figure B3-23: Use of dust lamp.   *Source: HSE.*

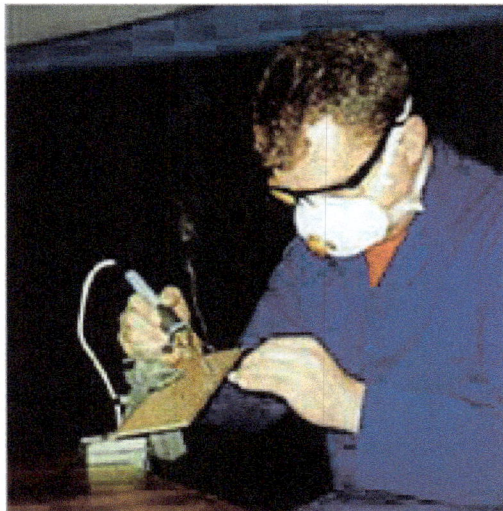

Figure B3-24: Soldering without Tyndall illumination.
Source: HSE, HSG258.

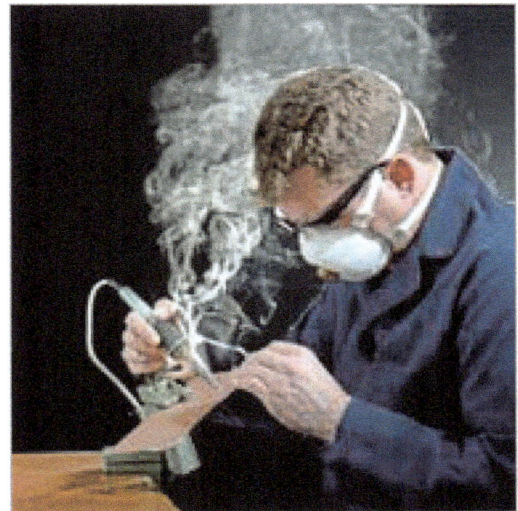

Figure B3-25: Soldering with Tyndall illumination.
Source: HSE, HSG258.

INSTRUMENTS FOR MEASURING AIR VELOCITY

There are three main instruments used for measurement of air velocity in ventilation systems:

1) Hot wire anemometer. *(See figure ref B3-26)*.
2) Swinging vane anemometer. *(See figure ref B3-27)*.
3) Rotating vane anemometer. *(See figure ref B3-28)*.

Hot wire anemometer

The hot-wire anemometer is the most well-known thermal anemometer, and measures a fluid (in this case air) velocity by noting the heat reduction in the wire, caused by convection as the fluid passes over the wire.

The core of the anemometer is an exposed hot wire (usually made of platinum or tungsten) either heated up by a constant current or maintained at a constant temperature. In either case, the heat lost to fluid convection is a function of the fluid velocity.

By measuring the change in wire temperature under constant current or the current required to maintain a constant wire temperature, the heat lost can be obtained.

The heat lost can then be converted into a fluid velocity in accordance with convective theory.

Due to the tiny size of the wire, it can be fragile and is best suited for clean fluid flows of the type experienced when conducting measurements under test conditions rather than actual operation.

Swinging vane anemometer

This consists of a box containing a hinged vane. By mounting the vane on low friction bearings as air passes through the box it deflects the vane and a measurement can be achieved by reproducing the movement on a pointer which moves across a scale.

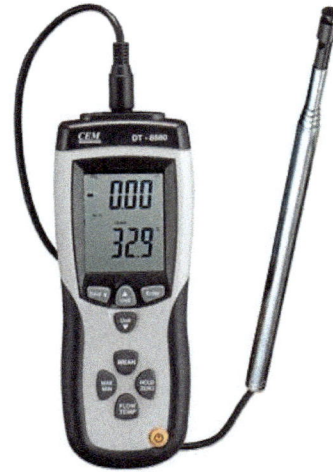

Figure B3-26: Hot wire anemometer. *Source: ATP Instrumentation Ltd.*

Rotating vane anemometer

This instrument has a series of light vanes mounted on radial arms, which rotate on a common spindle. As the spindle rotates it moves a set of counters on the instrument via a set of gears.

This variety is not a direct reading instrument as the counter only records the linear air movement in a given period of time. However some of the more modern electronic versions can give continuous direct readings.

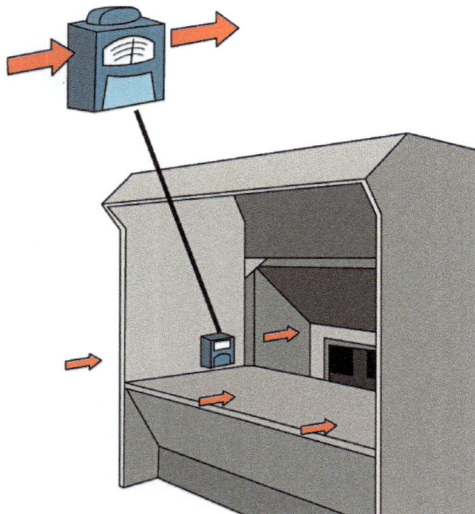

Figure B3-27: Swinging vane anemometer. *Source: HSE.*

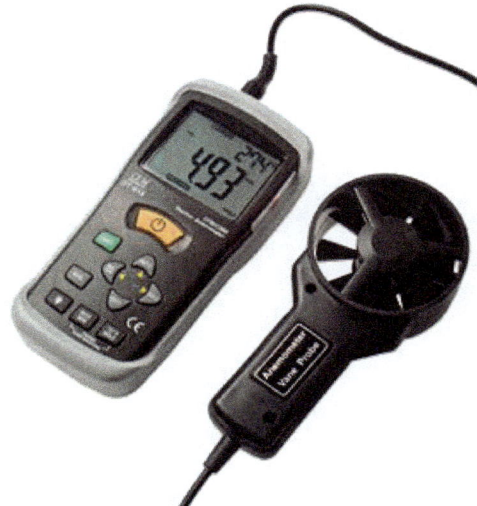

Figure B3-28: Rotating vane anemometer. *Source: HC Slingsby Plc.*

CAPTURE VELOCITY

The *capture velocity* is the air speed necessary to overcome opposing air currents and draw a contaminant into a hood or intake. Acceptable capture velocity depends on the mass of the particulate being captured, the prevailing air currents outside the hood, and thermal properties of the contaminant and the velocity of the particulate or gas relative to the hood flow.

There should be a standard procedure for determining the effectiveness of the LEV system in capturing a contaminant, i.e. the *capture efficiency*. Such procedures should be used as part of the initial or periodic LEV system effectiveness test.

Measurements of the capture velocity are taken at numerous points in the vicinity of the intake, in the zone of capture. The *'zone of capture'* is the area around the intake in which contaminants may be captured and is visualised by releasing smoke from a smoke tube around the intake to the LEV.

Before taking readings, it is important to observe and record the conditions in the workplace that may affect the result, for example, windows open/shut; status of the room heating, ventilation and air-conditioning systems; traffic/movement of people/equipment around the system; permanent/temporary storage of equipment around the system.

■ The velocity-measuring meter's probe should be placed at various locations around the duct inlet and the reading recorded at each point tested in a systematic way, noting air velocity readings at each point of reference in relation to distance and position from the centre of the zone of capture. The reading should be compared with the design parameter requirements and any deviance should be noted and investigated.

FACE VELOCITIES

During the measurement of face velocities all other equipment within the vicinity should be operated normally. The velocity should then be measured at a number of different collection points. By sub-dividing the face area an accurate measurement drawn on a grid can be taken. A value should be recorded as the mean face velocity. If at any point the reading is significantly different from the design mean face velocity value further investigation should be made.

TRANSPORT VELOCITIES

Transport velocity is the speed at which air is transported through the ducting of a LEV system. Transport velocity is measured at a number of points in the system by taking measurements inside the ducting with the system operating.

It is important to measure transport velocity with all the inlets that are normally in use open and working at the time of measurement. Measurements may be taken in the branch of the LEV system under evaluation and in the main branch(es) that take the contaminated air to the air cleaner.

STATIC PRESSURES

Air inside a container presses outward, creating pressure on the walls of the container. This is called the *static pressure*. This pressure exists even if the container has not been 'pumped up'. The pressure in this case is simply the atmospheric pressure.

The static pressure of the LEV system at the point being measured is obtained by using a manometer to measure the absolute pressure inside the system and comparing this with atmospheric pressure to determine the difference - the static pressure.

The static pressure in a LEV system is a useful indicator of how much negative pressure the fans are creating and how effective they will be at drawing fresh air in through the inlets. An overly high static pressure can also indicate that there are not enough inlets in the LEV system.

Measurements are taken at various points along the ducts, including behind the fan and near the outlet. The absolute pressure before the fan (on the suction side) is lower than atmospheric pressure so the static pressure is negative.

On the exhaust side of the fan the absolute pressure is higher than atmospheric pressure so the static pressure is positive.

Static pressure is measured using a *pitot tube;* this instrument consists of two concentric tubes; the inner tube has a front opening, which is pointed into the flow. The outer tube has annular openings at right angles to the airflow.

Pitot tubes are simple to use, inexpensive and robust. The tube does not require calibration and is suitable for measurements above $3ms^{-1}$. (For example, wood dust produced from circular saw cutting requires velocities of $>20ms^{-1}$ for removal).

By connecting a pressure gauge, manometer, to the Pitot tube the static pressure of the air stream can be measured.

Measurements at lower air speeds can be made if a more accurate digital micromanometer is used. For accurate measurements, pitot static tubes should only be used to measure velocities in ducts where the flow direction is well defined.

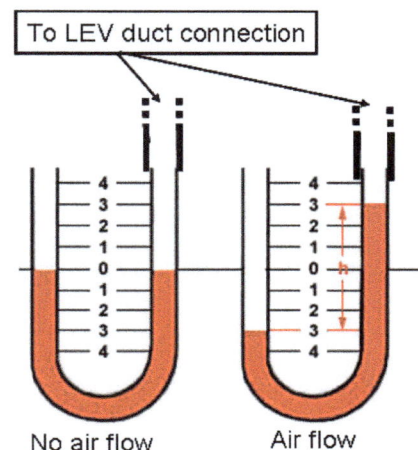

Figure B3-29: Manometer measuring static pressure. *Source HSE.*

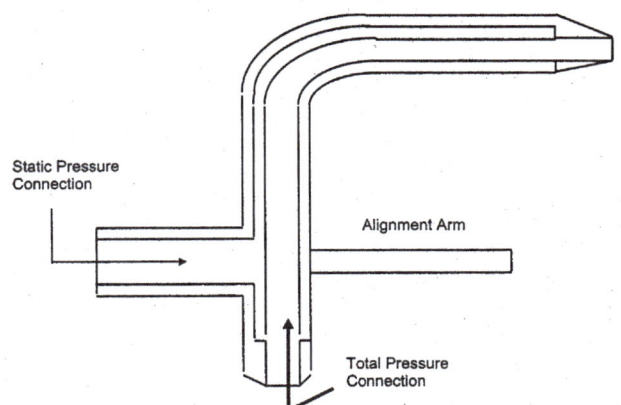

Figure B3-30: Pitot tube.   *Source RMS.*

EMISSIONS TO ATMOSPHERE AFTER INSTALLING LEV SYSTEMS

The design will need to take into consideration any exhaust materials from the LEV system after filtration or cleaning. Discharges should not be positioned in the area of any air intake systems, such as with air conditioning, other openings such as windows and doors or onto neighbouring properties or public spaces.

Permits will be required if it is intended to release contaminated air into the atmosphere and the environment agency should be informed of the proposed installation to ensure permissions to discharge will be granted, before the design is implemented.

THE PURPOSE OF AIR CLEANING DEVICES

The purpose of air cleaning devices in LEV systems is to remove contaminants from the air-stream to enable it to be returned to the workplace or exhausted into external atmosphere without risk to humans or the environment. The type of air cleaner must be matched to the nature of the contaminant. They include devices for collecting particulates, such as fabric/cyclone filters, electrostatic precipitators, chemical absorption scrubbers and specialist condensation units.

Many employers buy local exhaust ventilation (LEV) to protect workers' health, but it may not suitable for the purpose. This may be because it's the wrong type or because it's not properly installed, used or maintained, INDG408 provides advice on sourcing the correct LEV and using and maintaining it properly.

FREQUENCY OF TESTING

Engineering control measures should be thoroughly examined and tested at suitable or specified intervals, to ensure that they are continuing to perform as originally intended. The intervals and content of the thorough examination should be in accordance with current legislation, or criteria specified in national or international standards, approved or recognised by the competent authority, taking into account the extent of the risk in the event of failure of the control measure. The results of each thorough examination and test should be compared with the assessment of risks and control measures. Any defects disclosed, as a result of the examination or test, should be remedied as soon as possible, or within such time as the examiner directs. A suitable record of each thorough examination should be kept.

REPORT ON LEV TESTING

The examination and test for LEV should typically provide the following information:

- Name and address of the employer responsible for the plant.
- Identification and location of the LEV plant, and the process and hazardous chemicals concerned.
- Date of last thorough examination and test.
- Conditions at time of test: normal production or special conditions (for example, maximum use).

Information about the LEV plant which shows:

- Its intended operating performance for controlling the hazardous chemicals.
- Whether the plant still achieves the same performance.
- If not, the repairs required to achieve that performance.
- Methods used to make assessments in respect of the above (for example, visual, pressure measurements, air flow measurements, dust lamp, air sampling and filter integrity tests).
- Date of examination and test.
- Name, designation and employer of the person carrying out the examination and test.
- Signature or authentication of the person carrying out the examination and test.
- Maintenance of engineering control measures; details of repairs to be carried out - to be completed by the employer responsible for the LEV plant.
- The effectiveness of repairs carried out should be determined by a retest.

B3.4 - Personal protective equipment

Personal protective equipment (PPE) is equipment that will protect the user against health, safety risks at work. It can include items such as safety helmets, hard hats, gloves, eye protection, high-visibility clothing, safety footwear and safety harnesses.

The requirements of the Personal Protective Equipment at Work Regulations (PPE) 1992

The Personal Protective Equipment Regulations (PPE) 1992 does not apply where the following five sets of regulations require the provision and use of PPE against these hazards:

- The Control of Lead at Work Regulations (CLAW) 2002.
- The Ionising Radiations Regulations (IRR) 1999.
- The Control of Asbestos Regulations (CAR) 2012.
- The Control of Substances Hazardous to Health Regulations (COSHH) 2002 (as amended).
- The Control of Noise at Work Regulations (CNWR) 2005.

The above regulations each use an hierarchical approach to dealing with hazards, i.e. avoid or eliminate; substitute or reduce; use engineering controls, for example, COSHH 2002 when dealing with volatile materials or dust LEV; the use of PPE in addition to engineering controls may also be appropriate. PPE should only be used when it is not reasonably practicable to use other engineering controls.

Types of PPE for use with hazardous substances

Whenever possible employees should be protected by collective measures such as LEV, and PPE should be regarded as the last choice. Often PPE may be cumbersome, uncomfortable and impair both mobility and dexterity. PPE will only give protection to the user, and not the rest of the workforce.

Whatever control measures are used, they must be adequately maintained. Ventilation equipment must be tested regularly by a competent person. PPE must be maintained in an efficient state and replaced when worn out, time expired or damaged.

It is an employer's duty to provide PPE when necessary and it is the employees' duty to wear, or use PPE provided when required. PPE must be used when engineering controls are not adequate to reduce risk to an acceptable level.

In general terms this will be when the provision and use are:

■ A condition of a work permit.
■ A requirement of a risk assessment.
■ Where mandatory signs are displayed indicating the need to wear PPE, for example, hearing protection.

In addition, the PPE 1992 require that people, who are expected to work in the open air in low temperature or cold stores, should be provided with adequate and suitable protective clothing. PPE must be provided free of charge to employees, and the employer must provide training in use, care, limitations and maintenance.

It is important to consider the following factors when choosing personal protective equipment:

■ Type of protection required.
■ Level of protection required.
■ Compatibility.
■ Individual issues, for example, beards and spectacles.

■ Wearer acceptability/comfort.
■ Fit.
■ Information, instruction and training requirements.

SUPERVISION/MANAGEMENT CONTROLS

The PPE 1992 state the following PPE should:

■ Be appropriate for the risks involved and the conditions where it is to be used.
■ Consider the ergonomics of the user including comfort and acceptability, and take account of any health conditions they may have, for example, provision of breathing apparatus, may not be suitable for use by an asthmatic person.
■ Be capable of fitting the user correctly, after any adjustments have been made.
■ Not increase the overall risks to the user, for example, those with beards or spectacles may not be able to obtain a protective fit.

In addition:

■ If the presence of additional hazards make it necessary to wear more than one piece of PPE, the equipment used must be compatible and not give rise to additional risk.
■ The user must be given adequate information and training in the use and limitations of the PPE provided; the safe removal if it becomes contaminated and how to report any faults or defects.
■ The method for cleaning and sterilisation after use should be specified and understood by the user.
■ Employees must be provided with adequate PPE storage facilities.

There are several areas of special importance in respect of the provision, wearing and care of personal protective equipment. They include protection for ears, eyes, skin, whole body and lungs.

This Element is concerned with PPE specifically with regard to:

■ Respiratory protection.
■ Eye protection.

■ Skin protection.

Respiratory protective equipment (RPE)

TYPES OF RESPIRATORY PROTECTIVE EQUIPMENT

RPE is available in a wide range of varieties. They can be split into six basic types and two categories.

The two categories of respiratory protection are:

1) Respirators - made up of 3 different types.
2) Breathing apparatus (BA) - made up of 3 different types.

Category 1 - respirators

Particle filtering face pieces

These are commonly known as disposable facemasks **(see figure ref B3-33 'A')**.

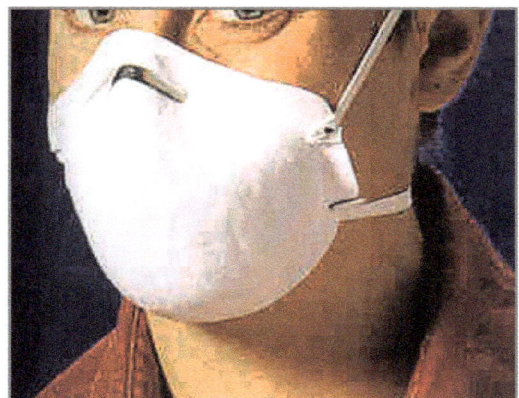
Figure B3-31: Paper filter respirator. *Source: Haxton Safety.*

They have a relatively low filter breathing resistance but are easy to use and should be replaced at least at the end of every working period if not more frequently.

Mask and filter respirators

This type of respirator comprises a face piece connected to a suitable filter cartridge or canister for protection against particulates or chemicals.

The face piece can just cover the mouth and nose (half mask/orinasal) *(see figure ref B3-33 'B')* or whole face (full-face mask) *(see figure ref B3-33 'C' and 'D')*. This type is not to be used in atmospheres containing less than 18% oxygen.

It can take a great effort to breathe through the filters and they should be changed on a frequent basis. For these reasons the users require a high level of training and discipline.

Power assisted respirator

This is a similar type of respirator to the mask and filter variety.

Figure B3-32: Breathing apparatus. *Source: Haxton Safety.*

The main difference is that it is fitted with a battery-powered fan or blower, which draws the contaminated air through the filter and supplies clean air to the user *(see figure ref B3-33 'E' and 'F')*. The fresh air can be ducted to the user via a visor or hood, or alternatively through to a full-face mask for a higher level of protection.

Typical examples of respirators:

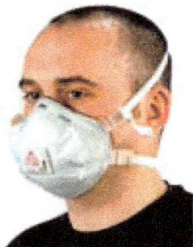

Type 'A'

Type 'B'

Type 'C'

Type 'D'

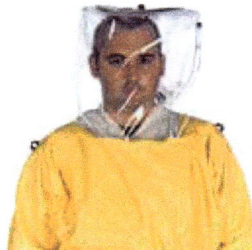

Type 'E'

Type 'F'

Figure B3-33: Category 1 respirators.

Source: HSG53.

Category 2 - breathing apparatus

Compressed airline apparatus

Commonly known as 'Direct Line Breathing Apparatus' (DLBA). This system uses an airline connecting the facemask to a supply of breathable air generated by a compressor. The type of compressor used should ideally be designed for this purpose and be of a non-oil type.

The face mask/headpiece is supplied with air at pressure, which creates a positive pressure in the mask thus preventing the contaminants entering the breathing zone of the user.

Self-contained open circuit air (BA)

The main difference between this variety and the compressed airline apparatus is that the supplied air is stored at 300 bar in cylinders mounted on the users back or on a portable trolley.

Self-contained closed circuit air (BA)

This type uses a supply of air either in compressed, liquid or chemically bonded form usually held in a container within the apparatus.

As it is a closed circuit the air is expelled and chemically cleaned or 'scrubbed' and mixed with a supply of clean air and thus can remain in use for longer periods of time.

Typical examples of breathing apparatus:

Figure B3-34: Category 2 - typical example of breathing apparatus.   *Source: HSG53.*

BA systems are not for general use. They are expensive to purchase and require specialised training in fitting use and limitations and regular health checks for users.

THEIR APPLICATIONS AND LIMITATIONS

Respirators

Respirators filter the air breathed but do not provide additional oxygen. There are a number of types of respirator that provide a variety of degrees of protection from dealing with nuisance dusts to high efficiency respirators for solvents or asbestos. Some respirators may be nominated as providing non-specific protection from contaminants whereas others will be designed to protect from a very specific contaminant such as solvent vapours.

There are five main types of respirators:

1) Filtering face piece.
2) Half mask respirator.
3) Full face respirator.
4) Powered air purifying respirator.
5) Powered visor respirator.

Advantages

Some of the advantages of using respirators include:

- Unrestricted movement.
- Often lightweight and comfortable.
- Can be worn for long periods.

Limitations

Purify the air by drawing it through a filter to remove contaminants. Therefore, it can only be used when there is sufficient oxygen in the atmosphere.

Other limitations include:

- Requires careful selection by a competent person (must fit the face of the user - face piece fit).
- Requires regular maintenance.
- Requires understanding by the user when a cartridge is at the end of its useful life.
- Requires separate storage facilities from personal clothing.
- Can give an enclosed/claustrophobic feeling.
- Relies on user for correct fit/use etc.
- May be incompatible with other forms of PPE.
- Performance can be affected by beards, long hair and spectacles.
- Interferes with other senses, for example, sense of smell.

Breathing apparatus

Breathing apparatus provides a separate source of supply of air (including oxygen) to that which surrounds the person. Because of the self-contained nature of breathing apparatus it may be used to provide a high degree of protection from a variety of toxic contaminants and may be used in situations where the actual contaminant is not known or there is more than one contaminant.

There are three types of breathing apparatus:

1) Fresh air hose apparatus - clean air from uncontaminated source.
2) Compressed air line apparatus from compressed air source.
3) Self-contained breathing apparatus - from cylinder.

Advantages

- Supplies clean air from an uncontaminated source; therefore, can be worn in oxygen deficient atmospheres.
- Has high assigned protection factor (APF). Therefore may be used in an atmosphere with high levels of toxic substance.
- Can be worn for long periods if connected to a permanent supply of air.

Limitations

- Can be heavy and cumbersome which restricts movement.
- Requires careful selection by competent person.
- Requires special training.
- Requires arrangements to monitor/supervise user and for emergencies.
- Requires regular maintenance.
- Requires correct storage facilities.
- Can give a 'closed in'/claustrophobic feeling.
- Relies on user for correct fit/use etc.
- May be incompatible with other forms of PPE.
- Performance can be affected by, for example, facial or long hair.
- Interferes with other senses, for example, sense of smell.

SELECTION AND SUITABILITY

There are a number of issues to consider in the selection of respiratory protective equipment (RPE) not least the advantages and limitations shown above. One of the important factors is to ensure that the equipment will provide the level of protection required.

This is indicated by the assigned protection factor given to the equipment by the manufacturers - the higher the factor the more protection provided. With a little knowledge it is possible to work out what assigned protection factor (APF) is needed using the following formula.

APF   =   Concentration of contaminant in the workplace

Concentration of contaminant in the face-piece

It is important to understand that this factor is only an indication of what the equipment will provide. Actual protection may be different due to fit and the task being conducted.

Use

Every employee must use any personal protective equipment provided in accordance with the training and instructions that they have received.

THE SIGNIFICANCE OF ASSIGNED PROTECTION FACTORS

The PPE 1992, Regulation 4, states that PPE is to be used only after all other possible control measures have been considered or have been put in place.

To protect against dusts, vapours, particulates and dusts from inhalation, the type of PPE that is used is **respiratory protective equipment (RPE)**.

Items of RPE have an **assigned protection factor (APF)**, decided on after research and testing by the manufacturer. Ten volunteers wear the RPE in laboratory conditions. The amount of contaminant in the atmosphere is measured and then, after a set period of time, the amount of contaminant in the face-piece is measured. The APF is the amount of contaminant in the atmosphere divided by the amount in the face piece. The APF will be a number such as 10 or 40, for example.

In practice, RPE does not fit the face as well in the workplace as it does in laboratory conditions because the worker may be bending and twisting during their normal work. A worker's face may be sweaty or greasy and in the case of a male may have facial stubble as the day progresses. For these reasons, a higher APF would be needed than the result of a mathematical formula would indicate.

Example:

- Amount of airborne contaminant measured in the workplace = 20ppm.
- Workplace exposure limit value (from EH40) = 2ppp.
- Minimum theoretical APF required would be: 20/2 = 10.
- A higher value of APF would be recommended to allow for real conditions.

The HSE use a scheme to identify the capabilities of various forms of RPE and provide advice on its selection and use.

Advice sheet	APF	Type of RPE
R1	4	■ Filtering half-mask, EN 149. ■ Filtering half-mask with valve, EN 405. ■ Filtering half-mask without inhalation valves EN 1827. ■ Half mask EN 140 and filter. ■ Full face mask EN136 and filter. ■ Any of the above devices incorporating a low efficiency P1 particulate filter. ■ Caution: these are not suitable for use in confined spaces.
R2	10	■ Filtering half-mask, EN 149. ■ Filtering half-mask with valve, EN 405. ■ Filtering half-mask without inhalation valves EN 1827. ■ Half mask EN 140 and filter. ■ Full face mask EN136 and filter. ■ Any of the above devices incorporating a medium efficiency P2 particulate filter, gas filter, or gas and P3 filter. ■ Powered hood model TH1 EN 146/EN 12941. ■ Power-assisted mask model TM1 EN 147/EN 12942. ■ Caution: these are not suitable for use in confined spaces.
R3	20	■ Filtering half-mask, EN 149. ■ Filtering half-mask without inhalation valves EN 1827. ■ Half mask EN 140 and filter. ■ Any of the above devices incorporating a high efficiency P3 particulate filter. ■ Full face mask EN136 and gas filter. ■ Powered hood model TH2 EN 146/EN 12941. ■ Power-assisted mask model TM2 EN 147/EN 12942. ■ Caution: these are not suitable for use in confined spaces.
R4	40	■ Full face mask EN 136 and P3 filter. ■ Powered hood, helmet or blouse model TH3 EN 146/EN 12941. ■ Power-assisted full face mask model TM3 EN 147/EN 12942. Caution: these are not suitable for use in confined spaces.
R5	40	■ Fresh air hose BA EN 138/269. ■ Compressed airline BA hood/helmet/visor LDH3 EN 1835. ■ Constant flow compressed airline BA hood EN 270/271 or mask EN 139 or 12419. ■ Constant flow compressed airline BA full face mask EN 139. ■ Constant flow compressed airline BA LDM3 mask EN 12419.
R6	2000	■ Positive demand compressed airline BA with full face mask EN 139. ■ Positive demand full face mask self-contained BA (SCBA) EN 137.

Figure B3-35: COSHH Essentials: RPE guidance. *Source: HSE.*

HSE RPE SELECTOR TOOL (REF HSG53) TO AID SELECTION

The HSE have produced a practical guide, HSG53 which provides essential guidance for the correct selection and use of respiratory protective equipment (RPE) for use in the work place.

The guidance is designed to assist employers to comply within the law.

The guide explains the use of the HSE's selector tool for RPE. The RPE selector tool uses a step-by-step generic approach and has been specifically designed to help small and medium-sized enterprises.

Following this approach is not compulsory, and alternative approaches may be used to comply with the law.

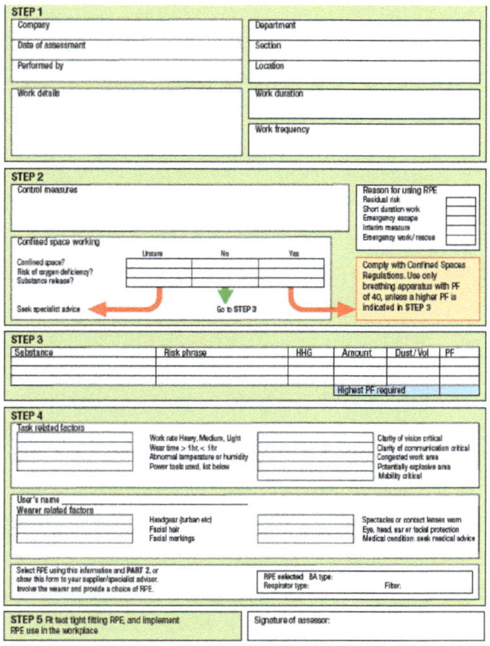
STEP 1
Company
Department
Date of assessment
Section
Performed by
Location
Work details
Work duration
Work frequency
STEP 2
Control measures
Reason for using RPE
Residual risk
Short duration work
Emergency escape
Interim measure
Emergency work/rescue
Confined space working
Confined space?
Risk of oxygen deficiency?
Substance release?
Unsure
No
Yes
Seek specialist advice
Go to STEP 3
Comply with Confined Spaces Regulations. Use only breathing apparatus with PF of 40, unless a higher PF is indicated in STEP 3
STEP 3
Substance
Risk phrase
HHG
Amount
Dust/Vol
PF
Highest PF required
STEP 4
Task related factors
Work rate Heavy, Medium, Light
Wear time > 1hr < 1hr
Abnormal temperature or humidity
Power tools used, list below
Clarity of vision critical
Clarity of communication critical
Congested work area
Potentially explosive area
Mobility critical
User's name
Wearer related factors
Headgear (turban etc)
Facial hair
Facial markings
Spectacles or contact lenses worn
Eye, head, ear or facial protection
Medical condition: seek medical advice
Select RPE using this information and PART 2, or show this form to your supplier/specialist adviser. Involve the wearer and provide a choice of RPE.
RPE selected BA type:
Respirator type:
Filter:
STEP 5 Fit test tight fitting RPE, and implement RPE use in the workplace
Signature of assessor:

Figure B3-36: HSE RPE selector tool. *Source: HSG 53.*

FACE FIT TESTING

To ensure that the selected RPE has the potential to provide adequate protection for individual wearers, the ACoPs supporting COSHH, CAR and CLAW stipulate that tight-fitting RPE must be fit tested as part of the selection process.

Ill-fitting face pieces can create inward leakages of airborne contaminants.

A tight-fitting face piece is a full face mask, a half mask, or a filtering face piece (commonly referred to as a disposable mask). The performance of the face piece, irrespective of whether they are used in negative pressure respirators, power assisted respirators or compressed air supplied breathing apparatus, relies heavily on the quality of fit of the face piece to the wearer's face.

An inadequate fit will significantly reduce the protection provided to the wearer. The presence of facial hair in the region of the face seal will significantly reduce the protection provided.

RPE fit testing should be conducted by a competent person. Competence can be demonstrated through achieving accreditation under the 'Fit2Fit RPE Fit Test Providers Accreditation Scheme'. This Scheme has been developed by the British Safety Industry Federation (BSiF) together with industry stakeholders and is supported by the HSE. The scheme is not compulsory and employers are free to take other action to comply with the law.

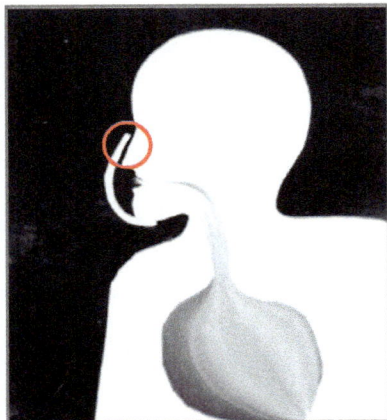

Figure B3-37: If the respirator does not fit tightly. *Source: BSiF.*

Figure B3-38: Contaminated air will enter the mask. *Source: BSiF.*

Prior to using or issuing RPE the following questions must be asked:

- Is it possible to control the hazard at source?
- Is the RPE designed to protect against the particular hazard?
- Will the RPE reduce the exposure to an acceptable level?
- Will it protect in practice?
- Have face fit tests been carried out and recorded?
- What training is required?
- Have provisions been made for cleaning, storage and maintenance?
- Does it increase exposure to any other hazards?
- Is it fully compatible with any other PPE issued?
- Are adequate rescue and resuscitation facilities established for the user?

Skin and eye protection

SKIN PROTECTION

Dermatitis is a major cause of absenteeism, accounting for over half of all working days lost through industrial sickness. It is an inflammatory skin condition (not infectious or contagious) caused by certain irritants contained in many industrial materials, or an allergy caused by dermatitic sensitisers. When other control measures have been used, but there is still a risk of skin exposure, skin protection will have to be used. The most likely exposure is to the hands and there are many types of protective gloves available. The type chosen will depend on the chemical the operator is exposed to and how much dexterity and feeling is necessary to carry out the task.

Latex gloves were once widely used as they allow dexterity and feeling, but latex rubber is both a dermatitic and a respiratory sensitiser. They have been replaced with nitrile, a man-made equivalent.

There are other hand and arm protective gloves and they should be chosen according to the chemical being handled. If the wrong ones are chosen, the chemical may pass directly through the glove and be held against the skin. The hands may also sweat, allowing the chemical to be more easily absorbed by the skin or to be held against the skin causing irritation or corrosion. Some gloves will break down very quickly if wrongly used with acidic or alkaline substances. The protective gloves manufacturer will advise on the correct type as will the catalogues they produce.

Protective gloves often have a powder inside for ease of pulling on and taking off, but unfortunately some users may have an allergic reaction to the powder, depending on the type of powder used.

The hands and arms can be protected with gauntlets, which reach high up the arm. Occasionally, whole body protection may be necessary where the substance can be absorbed through intact skin. These suits protect all the skin, but can be restrictive and hot.

In additions to personal protective equipment, barrier substances and skin cleansers could be used when contact is unavoidable. Barrier creams protect against either water soluble or solvent soluble irritants. To be effective the correct type must be used and it must cover the whole hand surface. After 2-3 hours it may need to be re-applied and re-application will be necessary after washing. These creams are not effective against skin sensitisers, for example, epoxy resin dermatitis. Also important, is the provision of adequate skin cleansers so as to not leave glues or resins to harden on the skin, and the use of after work creams may be necessary to replace some of the natural oil the skin has lost.

EYE PROTECTION

When selecting suitable eye protection, some of the factors to be considered are:

- Type and nature of hazard (impact, chemical, ultra violet (UV) light, etc.).
- Type/standard/quality of protection.
- Comfort and user acceptability issues.
- Compatibility.
- Maintenance requirements.
- Training requirements.
- Cost.

Figure B3-39: Eye and ear protection.   *Source: Speedy Hire plc.*

Figure B3-40: Arc welding visor - UV reactive.   *Source RMS.*

Types	Advantages	Limitations
Spectacles	Lightweight, easy to wear. Can incorporate prescription lenses. Do not 'mist up'.	Do not give all round protection. Relies on the wearer for use.
Goggles	Give all round protection. Can be worn over prescription lenses. Capable of high impact protection. Can protect against other hazards, for example, dust, molten metal.	Tendency to 'mist up'. Uncomfortable when worn for long periods. Can affect peripheral vision.
Face shields (visors)	Gives full face protection against splashes. Can incorporate a fan which creates air movement for comfort and protection against low level contaminants. Can be integrated into other PPE, for example, head protection.	Require care in use, otherwise can become dirty and scratched. Can affect peripheral vision. Unless the visor is provided with extra sealing gusset around the visor, substances may go underneath the visor to the face.

Figure B3-41: Advantages and limitations of eye protection.   *Source: RMS.*

There are several types of eye and face protector and it is important to select the correct type to give the required protection. Some of the hazards for which protection is required are:

- **Impact of solids.** (Grade 1 face shield or goggles that are manufactured to the highest impact standard. Normal safety glasses are made to Grade 2 Impact Standard).
- **Ingress of liquid, dust or gas.** (Goggles or respirator required).
- **Splashes of molten metal.** (Face shield).
- **Exposure to glare/ultra-violet radiation.** (Face shield/lens filter).
- **Lasers.** (Special eye protection equipment required). The normal eye protection is no protection against some types of laser beams.

In all cases, Codes of Practice, Standards or Operational procedures will determine the type of protection required.

SPECIFICATION AND STANDARDS

The Personal Protective Equipment Regulations (PPE) 1992 applies unless there are more specific regulations that apply. Even where they do, Regulation 5 requiring all PPE to be compatible always applies. Regulations which are more specific to the use of PPE include:

- ■ *Control of Lead at Work Regulations (CLAW) 2002* which aim to protect people at work exposed to lead by controlling that exposure. PPE may be used to protect all routes of entry, i.e. protection from inhalation, ingestion and absorption through the skin.
- ■ *Control of Asbestos Regulations (CAR) 2012.*
- ■ *Control of Substances Hazardous to Health Regulations (COSHH) 2002.*

The above regulations are covered in more detail - *see relevant statutory provisions section.*

Storage and maintenance of PPE

STORAGE

Secure and clean storage, separately housed from personal clothing, should be provided for RPE issued to users. RPE, along with other PPE, should not be worn in areas set aside for meal breaks; local temporary, clean storage facilities should be provided as required adjacent to canteens, and welfare facilities to prevent contamination of 'clean areas'.

MAINTENANCE

Where respiratory protective equipment (other than disposable respiratory protective equipment) is provided the employer must ensure that a thorough examination, and where appropriate testing, of that equipment is carried out at suitable intervals. All maintenance work should be carried out by properly trained people, using spare parts supplied by the manufacturer of the RPE.

Thorough maintenance, examination and tests should be carried out at least once a month. If the RPE is used only occasionally, an examination and test should be made before use and in any event the interval should not exceed three months.

Emergency escape-type RPE should be examined and tested in accordance with the manufacturer's instructions.

The need for training in the correct use of PPE

Regulation 9 of the Personal Protective Equipment Regulations (PPE) 1992 requires the employer to ensure that adequate training and instruction, including when and where to use PPE is provided.

- ■ Make sure anyone using PPE is aware of why it is needed, when to use, repair or replace it, how to report it if there is a fault and its limitations.
- ■ Train and instruct people how to use PPE properly and make sure they are doing this. Include managers and supervisors in the training, they may not need to use the equipment personally, but they do need to ensure their staff are using it correctly.
- ■ It is important that users wear PPE all the time they are exposed to the risk. Never allow exemptions for those jobs which take 'just a few minutes'.
- ■ Check regularly that PPE is being used and investigate incidents where it is not. Safety signs can be useful reminders to wear PPE, make sure that staff understand these signs, what they mean and where they can get equipment, for example, for visitors or contractors.

In particular, the information and training must be sufficient and suitable for workers to know:

- ■ How to use the PPE, and especially RPE. For example, the correct method of removing and refitting gloves and masks and determining how long protective gloves should be worn before any liquid contamination is liable to permeate them.
- ■ The cleaning, storage and disposal procedures they should follow, why they are required and when they are to be carried out, for example, cleaning contaminated PPE with water or a vacuum fitted with a high-efficiency particulate arrester (HEPA) filter, and not with an airline, or the risks of using contaminated PPE.
- ■ Training should include elements of theory as well as practice. Training in the use and application of control measures and PPE should take account of recommendations and instructions supplied by the manufacturer.

The monitoring and measuring of hazardous substances

Learning outcomes

On completion of this element, candidates should be able to demonstrate understanding of the content through the application of knowledge to familiar and unfamiliar situations. In particular, they should be able to:

B4.1 Explain how workplace exposure limits are used in the workplace.

B4.2 Outline the methods for sampling of airborne contaminants.

B4.3 Outline the principles of biological monitoring.

Content

Relevant statutory provisions

Control of Asbestos Regulations (CAR) 2012

Control of Lead at Work Regulations (CLAW) 2002 (and as amended 2004)

Control of Substances Hazardous to Health Regulations (COSHH) 2002 (and as amended 2004)

Ionising Radiations Regulations (IRR) 1999

Control of Asbestos Regulations (Northern Ireland) 2012

Control of Lead at Work Regulations (Northern Ireland) 2003

Control of Substances Hazardous to Health Regulations (Northern Ireland) 2003

Ionising Radiations Regulations (Northern Ireland) 2000

Sources of reference

Reference information provided, in particular web links, was correct at time of publication, but may have changed.

Control of lead at work, Approved Code of Practice and guidance, L132, HSE Books, ISBN: 978-0-7176-2565-6, http://www.hse.gov.uk/pUbns/priced/l132.pdf

Control of Substances Hazardous to, Health, Approved Code of Practice and guidance, L5, HSE Books, ISBN: 978-0-7176-6582-2, http://www.hse.gov.uk/pubns/books/l5.htm

COSHH Essentials, HSE, http://www.hse.gov.uk/coshh/essentials/

EH40/2005 Workplace exposure limits (as amended), HSE Books, ISBN: 978-0-7176-6446-7, http://www.hse.gov.uk/pUbns/priced/eh40.pdf

HSE's Health Surveillance Cycle, http://www.hse.gov.uk/health-surveillance/

Methods for the Determination of Hazardous Substances (MDHS) guidance, http://www.hse.gov.uk/pubns/mdhs/

Monitoring strategies for toxic substances, HSG173, HSE Books, ISBN: 978-0-7176-6188-6, http://www.hse.gov.uk/pubns/books/hsg173.htm

The above web links along with additional sources of reference, which are additional to the NEBOSH syllabus, are provided on the RMS Publishing website for ease of use - www.rmspublishing.co.uk.

B4.1 - Workplace exposure limits (WELs)

Concept of WELs

UK STANDARDS

The Health and Safety Executive (HSE) Guidance Note EH40, containing data on occupational exposure limits, has been published since the late 1960s, first by Her Majesty's Factory Inspectorate and later their successors, the HSE. At first these limits were entirely based on data supplied by the American Conference of Governmental Industrial Hygienists (ACGIH) under a copyright agreement. The US Threshold Limit Values (TLVs) were published as guidance only. The UK started to evolve their own standards in respect of substances such as cotton dust and asbestos which differed from those set by the ACGIH. The differing needs of industry, the trades unions and government made the need for UK specific occupational hygiene standards apparent.

The first UK standard was a Control Limit (CL) for asbestos, which was followed by other substances, including lead. Less hazardous substances were assigned HSE Recommended Limits (RLs). The difficulty of assigning a 'safe limit' for substances like genotoxic carcinogens led to the setting up of a framework of Occupational Exposure Limits (OEL) which encompassed both Maximum Exposure Limits (MELs) and Occupational Exposure Standards (OESs). The framework has now been replaced by one type of occupational exposure limit, called Workplace Exposure Limits (WELs). These are listed in EH40 and the document is regularly reviewed and often amended each year. The HSE now take into account standards set at European level and other hygiene standards when setting national occupational exposure limits.

EUROPEAN STANDARDS

The European Union has established lists of Indicative Occupational Exposure Limit Values (IOELV) through the introduction of commission directives, a recent one being Council Directive 2009/161/EU which was the third list produced in keeping with Council Directive 98/24/EC on the protection of the health and safety of workers from the risks related to chemical agents at work.

IOELVs are health-based, non-binding values, derived from the most recent scientific data available and taking into account the availability of measurement techniques. They set threshold levels of exposure below which, in general, no detrimental effects are expected for any given substance after short-term or daily exposure over a working life time. They constitute European objectives to assist employers in determining and assessing risks, in accordance with Article 4 of Directive 98/24/EC.

For any chemical agent for which an IOELV is established at community level, Member States are required to establish a national occupational exposure limit value taking into account the Community limit value, but may determine its nature in accordance with national legislation and practice.

OTHER HYGIENE STANDARDS

As with the UK, most standards do not claim to be absolute safety standards. The ACGIH say of their TLVs that:

"Threshold limit values refer to airborne concentrations of substances and represent conditions under which it is believed that nearly all workers may be repeatedly exposed day after day without adverse effect. Because of wide variation in individual susceptibility however, a small percentage of workers may experience discomfort from some substances at concentrations at or below the threshold limit: a smaller percentage may be affected more seriously by aggravation of a pre-existing condition or by development of an occupational disease."

Figure B4-1: Quote - ACGIH hygiene standards. *Source: ACGIH.*

A US body, the Occupational Safety and Health Administration (OSHA), say of their standard that:

"No employee will suffer material impairment to health or functional capacity even if such employee has regular exposure."

Figure B4-2: Quote - OSHS hygiene standards. *Source: OSHA.*

This compares with the old Soviet Union which stated that 'no detectable changes of any kind in a test organisation should occur' in respect of exposure.

Figure ref B4-3 below compares the USSR's approach with that of the US:

Contrasts in approaches to standard-setting in the USA and the former USSR (Calabrese, 1978)

USA	Former USSR
Minor physiological adaptive changes are permitted.	Maximum allowable concentration will not permit the development of any disease or deviation from normal.
Economic and technological feasibility are important considerations in the development of standards.	Standards should be based entirely on health and not on technological and economic feasibility.

USA	Former USSR
Values are time-weighted averages.	Concentrations are maximum values.
Research emphasis is on pathology.	Research emphasis is on nervous system testing.
Except for carcinogens, goals of near zero exposure are not widely adopted.	The goal is a level of exposure which does not strain the adaptive and compensatory mechanisms of the body.

Figure B4-3: Contrasts of standard-setting in the USA and former USSR.   Source: RMS.

Clearly, the inference from the process shown above is that these standards must be intended for different purposes. The USSR standards were more stringent and were those which should be **aimed** at. ACGIH-type standards are pragmatic standards, **achievable now** by a reasonable employer and are potentially enforceable.

The real danger, with both the WELs in EH40 and with ACGIH Standards, is that users believe that they are set entirely on medical and scientific knowledge related to health and therefore afford **total** protection.

Meaning of WELs

EH40 contains the lists of Workplace Exposure Limits (WEL) for use with the Control of Substances Hazardous to Health Regulations (COSHH) 2002 (as amended).

*"**Workplace exposure limit** for a substance hazardous to health means the exposure limit approved by the Health and Safety Commission for that substance in relation to the specified reference period when calculated by a method approved by the HSE, as contained in HSE publication **EH/40 Workplace Exposure Limits/Year** as updated from time to time."*

Figure B4-4: Definition of WEL.
Source: The Control of Substances Hazardous to Health (Amendment) Regulations 2004 (which came into force on 17/01/05 and 06/04/05).

How WEL's are established ACTS and WATCH

Workplace Exposure Limits are set on the recommendation of the Advisory Committee on Toxic Substances (ACTS) following assessment, by the Working Group on Action to Control Chemicals (WATCH), of the toxicological, epidemiological and other data. The committees have to consider at what concentration the limit should be set. Each substance is first reviewed by WATCH, which considers what value should be recommended to ACTS. WATCH comes to a decision based on a scientific judgement of the available information on health effects and ACTS makes recommendations to the HSE. The decision is also influenced by whether the European Union has established an IOELV for the substance in question; if they have this must be taken into account.

WELs is derived by using the following criteria:

1) The value would be set at a level at which no adverse effects on human health would be expected to occur, based on known or predicted effects.
2) If this value is not identifiable with reasonable confidence then the value would be based at a level corresponding to what is considered to be good control, taking into account the likely severity of health hazards and the cost and effectiveness of controls.
3) The WEL should not be set at a level where there is evidence of adverse effects on human health.

The status and use of EH40

EH40 contains the list of workplace exposure limits (WEL) for use with the Control of Substances Hazardous to Health Regulations 2002 (as amended).

The **list is legally binding,** as it reproduces the WEL values approved by the Health and Safety Executive and COSHH requires the employer to work to the WEL.

COSHH states that exposure to hazardous substances should be prevented where it is reasonably practicable. Where this cannot be done by, for example, changing the process, substituting it for something safer or enclosing the process, exposure should be reduced by other methods.

COSHH Regulation (7) sets out the following specific requirement:

"Without prejudice to the generality of paragraph (1), where there is exposure to a substance hazardous to health, control of that exposure shall only be treated as being adequate if:

(a) The principles of good practice for the control of exposure to substances hazardous to health set out in Schedule 2A are applied.

(b) Any workplace exposure limit approved for the substance is not exceeded."

EH40 also contains biological monitoring guidance values (BMG).

Specific issues relating to the setting and applying of exposure limits are also documented alongside matters such as calculation methods, exposure monitoring and mixed exposures.

EH40 is primarily concerned with airborne contaminants, which includes vapours, fumes and dusts. Certain substances listed have a skin annotation, **'sk'**, to denote that they can be absorbed through the skin. This warns that additional controls, in addition to respiratory protection, may be required.

Some substances are not listed in EH40 because their use is prohibited, and some substances are listed, but certain uses are prohibited. COSHH Regulation 4(1) states that those substances described in Column 1 of Schedule 2 are prohibited to the extent set out in the corresponding entry in Column 2 of that Schedule. An extract from Schedule 2 is shown below:

	Description of substance	*Purpose for which the substance is prohibited*
1)	2-naphthylamine; benzidine. 4-aminodiphenyl. 4-nitrodiphenyl; their salts and any substance containing any of those compounds, in a total concentration equal to or greater than 0.1 per cent by mass.	Manufacture and use for all purposes including any manufacturing process in which a substance described in Column 1 of this item is formed.
2)	Sand or other substance containing free silica.	Use as an abrasive for blasting articles in any blasting apparatus.
3)	A substance - Containing compounds of silicon calculated as silica to the extent of more than 3 per cent by weight of dry material, other than natural sand, zirconium silicate (zircon), calcined china clay, calcined aluminous fireclay, sillimanite, calcined or fused alumina, olivine. Composed of or containing dust or other matter deposited from a fettling or blasting process. Use as a parting material in connection with the making of metal castings.	Use as a parting material in connection with the making of metal castings.
4)	Carbon disulphide.	Use in the cold-cure process of vulcanising in the proofing of cloth with rubber.

Figure B4-5: Extract from Schedule 2. *Source: Control of Substances Hazardous to Health Regulations (COSHH) 2002.*

The significance of short-term and long-term exposure limits and time-weighted average values

WELs are occupational exposure limits set under COSHH to protect the health of persons in the workplace. They are concentrations of airborne substances averaged over a period of time known as a Time Weighted Average (TWA). The two periods that are used are 8-hours and 15-minutes. The 8-hour TWA is known as a LTEL (long-term exposure limit), used to help protect against chronic ill-health effects. The 15-minute TWA is known as STELs (short-term exposure limits) and are to protect against acute ill-health effects such as eye irritation, which may happen within minutes or even seconds of exposure.

Where there are different exposures to a substance throughout the day, the 8 hour TWA of the inhaled substance can be expressed mathematically as follows:

$$\frac{C_1T_1 + C_2 T_2 + C_3T_3 + \dots C_nT_n}{8}$$

Where C_1 is the occupational exposure and T_1 is the associated exposure time in hours in any 24-hour period. When working out the 8 hour TWA, there may be no exposure for certain time periods such as the lunch break, other breaks and time spent on another job. This will reduce the TWA.

B4.2 - Strategies, methods and equipment for the sampling and measurement of airborne contaminants

The atmosphere has a relatively fixed composition, 78.09% nitrogen, 20.95% oxygen, 0.93% argon, 0.03% carbon dioxide, and insignificant amounts of neon, helium and krypton and traces of hydrogen, xenon, oxides of nitrogen and ozone which may be mixed with up to 5% water vapour. Any of these gases in a greater proportion or any other substance present in the atmosphere is regarded as a contaminant.

Contaminants are often classified according to their physical state at normal temperature and pressure, for example.

- Gas.
- Liquid.
- Solid.

Though this classification has subdivisions that describe the form of the contamination more precisely, for example dust, fibre, fume and vapour.

The role of the occupational hygienist

Occupational hygienists work at the interface of people and their workplaces. They use science and engineering to prevent ill-health caused by the work environment - specialising in the assessment and control of risks to health from workplace exposure to hazards.

Hygienists help employers and employees to understand these risks and to minimise or eliminate them. Occupational hygienists can come from many backgrounds - chemists, engineers, biologists, physicists, doctors, nurses and others who have chosen to apply their skills to improving working practices and conditions.

With good occupational hygiene science and practice, some occupational health risks have been eliminated, others brought under control. So, it is possible, today, to be a healthy miner. Also, the ill-health effects of working with or near to asbestos, and how to avoid them, are now understood.

Occupational hygiene has enabled the risk of silicosis to be eliminated in pottery workers who used to die from this lung disease. These are some of the major achievements of occupational hygiene and its scientists and practitioners.

The role of the occupational hygienist is to identify measure, assess and control atmospheric contaminants that can harm the health of workers or others. This will involve consideration of the introduction of new substances or their creation by process changes. It will include the measurement of actual levels of contaminant or prediction of likely levels based on scientific data.

They will develop contaminant sampling strategies for identification and monitoring purposes. Occupational hygienists will be involved with the development of suitable control strategies and the confirmation of their effectiveness. In some cases the occupational hygienist may be involved in the establishment of in-house standards for exposure to substances.

COMPETENCE OF HYGIENISTS

There are no specific legal requirements relating to the competence of occupational hygienists but the Management of Health and Safety at Work Regulations (MHSWR) 1999 (as amended), Regulation 7(5) sets out a general requirement for those providing assistance to the employer in complying with health and safety:

> *"A person shall be regarded as competent for the purposes of paragraphs (1) and (8) where he has sufficient training and experience or knowledge and other qualities to enable him properly to assist in undertaking the measures referred to in paragraph (1)."*

Figure B4-6: Regulation 7(5) of MHSWR 1999.   Source: Management of Health and Safety at Work Regulations (MHSWR) 1999.

When looking to use the services of an occupational hygienist, to ensure competency, it may be appropriate to contact the only professional society representing qualified occupational hygienists in the UK, the British Occupational Hygiene Society (BOHS), with the largest number of members in Europe.

The BOHS has a high level award, the Diploma of Professional Competence in Occupational Hygiene and other awards that help establish the competence of those conducting specific aspects of occupational hygienist work, for example, measurement of hazardous substances and specific ones related to asbestos.

Interpreting a hygienist's report

It is important to ensure the *strategy, approach and methods* used by the hygienists are suitable for the contaminant to be measured, for example, the suitability of absorbant used in vapour sample tubes. If the contaminant is to be collected by absorbtion and then flushed out by a solvent for analysis, will all the absorbed material be released from the chosen absorbant? This is essential to ensure that the results are *valid, reliable* and not understated. The hygienist should state in the report what methods were used and for what purpose. Typically they should refer to the relevant HSE 'Methods for Determining Hazardous Substances' (MDHS) guidance.

All measurement should be *representative* of the actual risk of exposure to the worker. Personal measurements should be carried out where there is an occupational exposure limit. In a similar way, results of inspections of local exhaust ventilation equipment should be stated, together with the method used, frequency and the reasons why they were used. It is important that the report consider measurements taken *relative to any workplace exposure standards* and express the significance of this in a manner that is understandable to the reader.

Monitoring strategy (ref HSG173)

Before any monitoring is carried out it is important to know the typical size of the particles to be studied. *Figure ref B4-7* illustrates the different sizes of particles from gas molecules up to very large grit particles.

Small particles of less than 10 microns (μm) have a slow falling velocity, and therefore any dust that is generated can remain in the atmosphere with little or no air movement. The particulate size of interest in occupational hygiene sampling is the total inhalable dust and respirable dust.

Measurement of these fractions is important to determine the amount of the size of dust that exerts harmful effect on the three areas of the respiratory tract: the naso-pharynx, the trachea and bronchial tree and the pulmonary region. COSHH 2002 defines a 'substantial concentration of dust of any kind' as:

- 10 mg/m³, as a time weighted average over an 8 hour period, of total inhalable dust.
- 4 mg/m³, as a time weighted average over an 8 hour period, of respirable dust.

Note that this applies only to dusts not assigned an occupational exposure limit or classified as very toxic, toxic, harmful, corrosive or irritant which are already defined as hazardous substances. There are specific limits for both asbestos and lead.

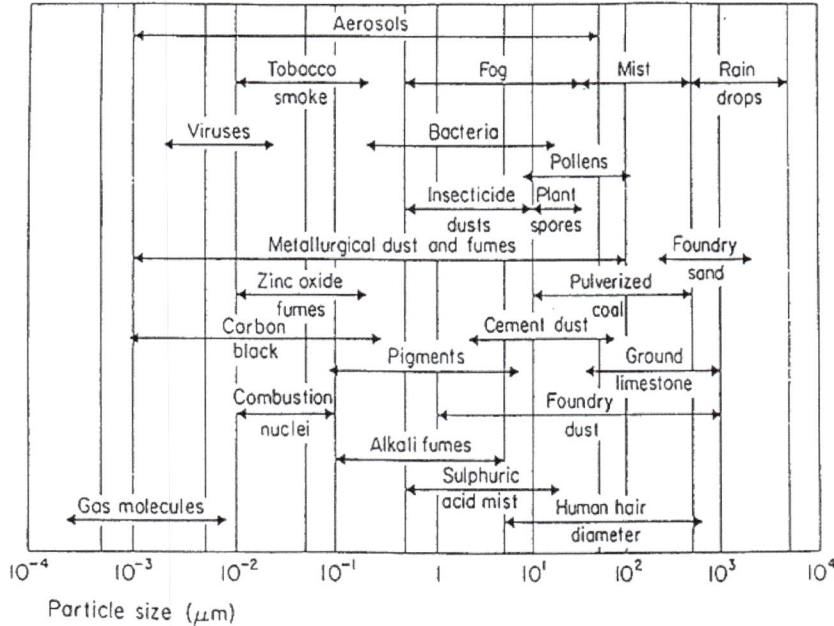

Figure B4-7: Particle size diagram.   Source: Ambiguous.

Guidance Note HSG173 'Monitoring strategies for toxic substances' advises on the methodology to be used when investigating the nature, extent and control of exposure to substances hazardous to health. It is possible to determine the scope of exposure without taking any measurements (i.e. deciding that exposure is under adequate control).

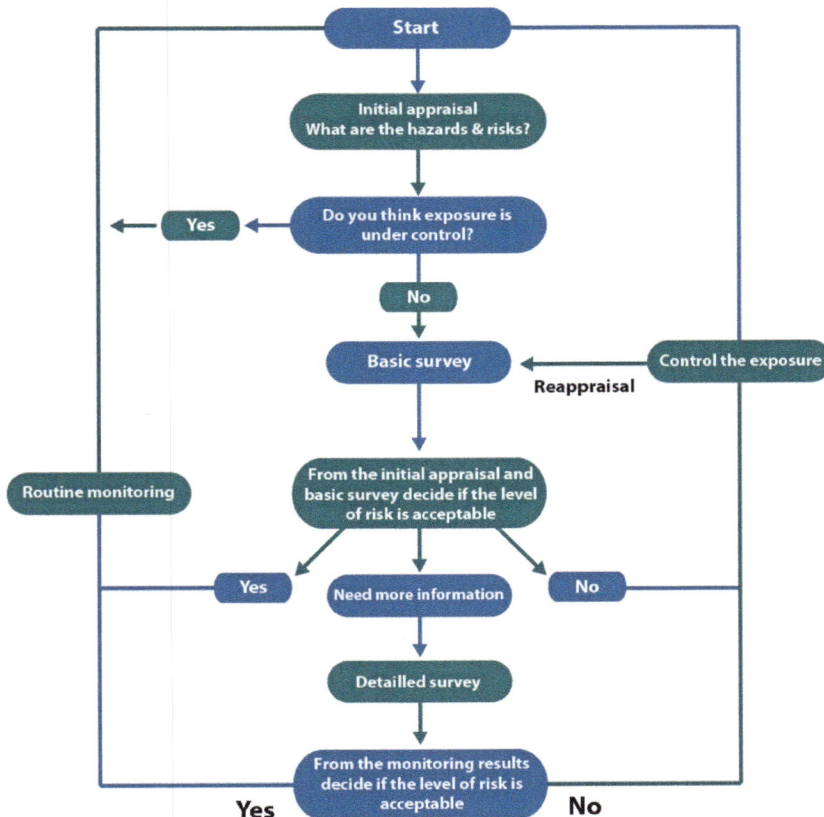

Figure B4-8: Monitoring strategy.   Source: HSG173/RMS.

Having identified the need to take measurements then a three stage strategy is advocated:

1) Initial appraisal.
2) Basic survey.
3) Detailed survey.

Depending on the results, these may be followed by:

- Re-appraisal.
- Routine monitoring.

INITIAL APPRAISAL

The aim of the initial appraisal is to establish the need for and extent of monitoring required. The initial appraisal will provide information about the type of the hazards and the potential risks involved. It will also identify the need to obtain further information that will influence exposure monitoring.

During the initial appraisal information is gathered on the substance(s) that workers are exposed to and its hazards, physical properties and airborne forms. In order to appraise the significance of this, information will also be required on:

- Controls already in place to limit exposure.
- Duration and pattern of exposure.
- Individuals and groups of workers most likely to be exposed.
- Number, type and location of the sources.
- Work conducted where exposures are likely to occur.
- Working practices that influence exposure.
- Respiratory protective equipment and/or other personal protective.
- Workplace exposure limits (WELs) set out in EH40 that relate to the substance, limits from other bodies or in-house standards for the substance involved.

This information can be obtained from a number of sources, including labels, safety data sheets, HSE and other technical publications.

Simple methods can then be used to determine the extent of the risk. Qualitative tests can aid this process; for example:

- Smell can also be used as an indicator of contamination (this is, however, an unreliable method).
- Smoke tubes to highlight air movement under the influence of draughts or general and local exhaust ventilation systems.
- Use of a dust lamp (for example, a Tyndall beam) which allows very fine airborne particles, which are invisible under normal light, to be seen.

Based on the information established during the initial appraisal it may be decided that the level of exposure relating to inhalation is acceptable. Therefore it may not be necessary to carry out exposure monitoring if this is the case.

It should be borne in mind that levels of exposure to hazardous substances can change and the need for exposure monitoring should be reviewed as often as necessary.

BASIC SURVEY

A basic survey should be carried out if the initial appraisal suggests that:

- Risk remains uncertain.
- A new process is introduced.
- A new WEL or in-house standard has been set.
- Significant changes have taken place since the last assessment.
- Unusual work activities that may influence exposure are planned.

The basic survey provides an estimation of personal exposure and an indication of the effectiveness of controls.

Before monitoring, the employees who are likely to be at significant risk of exposure should be identified, together with the conditions which give rise to them. Semi-quantitative methods, such as stain indicator tubes that give a rough numerical estimate of exposure, can be used to estimate personal exposure. These semi-quantitative methods can be comparatively inexpensive and easy to use.

Some instruments that give a specific direct reading of the level of exposure to a substance may be used, where these are available for the substance in question. Alternatively, more complex methods may have to be used which require specialist knowledge. HSG 173 suggests that these may include:

- Computer exposure modelling.
- Organic vapour analysers such as photo-ionisation detectors, portable gas chromatographs and infra-red analysers.
- Validated laboratory based sampling and analytical techniques.

Simple measuring instruments, such as air velocity meters, can be used to assess local exhaust ventilation systems to ensure that they are performing in accordance with the design specification. Also, the qualitative methods used in the initial appraisal can also be used again.

Based on data gathered during the basic survey and information gathered during the initial appraisal, it may be concluded that the control of inhalation exposure is acceptable.

However, the basic survey may identify limitations in the control strategies being used. If the conclusion is not certain, there are two alternatives:

1) Take direct action to control exposure.
2) Carry out a detailed survey and take remedial action as necessary.

DETAILED SURVEY

A detailed survey may be required when:

- The parameters of exposure would not be accurately determined by the basic survey.
- There is unexplained difference in measured exposure of employees doing similar tasks.
- Exposure to carcinogenic substances (risk phrase R45) or respiratory sensitisers (risk phrase R42) is involved.
- The initial appraisal and basic survey identify that:
    - The time-weighted personal exposure is close to the WEL.
    - The costs of improvements need further justification.
    - The work involves complex processes, such as major maintenance or plant decommissioning.

A detailed survey is likely to involve techniques used in the initial appraisal and basic survey, together with more detailed monitoring.

A detailed survey will require a greater depth of assessment, for example, consideration of:

- Control measures in use and their suitability.
- Health surveillance results.
- Information, instruction and training provided.
- Maintenance procedures.
- Personal protective equipment provided.
- Previous monitoring results.
- Work practices.

A detailed survey usually requires the person(s) conducting it to have more specialist knowledge. For certain substances a detailed survey may need to include the use of biological monitoring.

REAPPRAISAL

Once remedial action identified by the survey(s) has been carried out the situation should be re-appraised. If the risks to health are judged to be high, additional exposure monitoring may be required.

For example, when:

- Carcinogens are used.
- Patterns of exposure are very variable.

ROUTINE MONITORING

When the risk to employees has been adequately controlled, a routine monitoring programme may need to be set up in order to ensure that the control measures stay effective.

Factors that determine the necessity and frequency

Routine monitoring is **mandatory** under COSHH 2002 (as amended) Regulation 10 for the processes listed at column 1 of Schedule 5 of COSHH 2002 (as amended) at least at a frequency as that specified in column 2, for example, the continuous monitoring for vinyl chloride monomer and monitoring every 14 days for sprays given off from vessels during electroplating processes involving hexavalent chromium.

Routine monitoring can be time-consuming and expensive and it may be more cost-effective to invest in better control measures that reduce the need for monitoring and improve the health risks in the workplace.

Monitoring need not be complex as there are some simple and inexpensive instruments available that can provide information on the effectiveness of controls, for example:

- Smoke tubes.
- Dust lamps.
- Pressure sensing devices fitted to ventilation systems.

Routine monitoring results should be compared with those obtained from the earlier surveys and previous monitoring conducted. This allows progress to be monitored. However, in order to make comparison valid care should be taken to consider:

- The similarity of the processes and tasks monitored compared with at the time of previous monitoring.
- Where monitoring was carried out on this and previous occasions.
- The method of collecting and analysing the samples in each case.

The frequency of routine surveys will vary depending on the level of risk related to the substance, the degree of control of the risk in place and any changes affecting the way work is done. The nearer the measured exposure is to the WEL the more often monitoring will be required. A scheme for determining the frequency of routine monitoring is given in 'BS EN 689 Workplace Atmospheres: Guidance for the Assessment of Exposure by Inhalation to Chemical Agents for Comparison with Limit Values and Measurement Strategy'.

DIFFERENCE BETWEEN STATIC AND PERSONAL MONITORING

Instruments that are used for long-term sampling are broadly of two types - personal samplers and static sampling systems:

- **Personal samplers** - devices attached to the person and may take a number of forms, for example, gas monitoring badges, filtration devices and impingers (limit devices).
- **Static sampling systems** - sampling systems stationed in the work area. They sample continuously over a period of time, for example, a working shift of 8 hours, or over longer periods if necessary. They operate using mains or battery-operated pumps.

IMPORTANCE OF USING STANDARD METHODS (MDHS SERIES) FOR MONITORING

Methods for the determination of hazardous substances (MDHS) guidance on analysis

The Methods for Determining Hazardous Substances (MDHS) are a specialised guidance series of HSE publications that advise on general methods of monitoring and methods for monitoring exposure to various substances. The series does change as techniques are amended; the HSE website informs of the status of all documents in the MHDS series, this should be consulted to ensure documents are current.

Examples include:

MDHS No.	Category	Method
6/3	Lead and organic compounds of lead	Laboratory method using flame or electro thermal atomic absorption spectrometry.
14/3	General	General methods for sampling and gravimetric analysis of respirable and inhalable dust.
54	General	Protocol for assessing the performance of a pumped sampler for gases and vapours.
56/2	Hydrogen cyanide in air	Laboratory method using an ion-selective electrode.
59	Man-made mineral fibre	Airborne number concentration by phase contrast light microscopy.
70	General	General methods for sampling airborne gases and vapours.
94	Pesticides in and/or on surfaces	Methods for the sampling and analysis of pesticides in air/or on surfaces using pumped filters/sorbent tubes and gas chromatography.
96	Volatile organic compounds in air	Laboratory method using pumped solid sorbent tubes, solvent desorption and gas chromatography.
101	Crystaline silica	In respirable airborne dusts.

Figure B4-9: MDHS Examples.   *Source: RMS.*

Sampling techniques

Sampling is the continuous or intermittent sampling of air in the working environment with a view to detecting the presence of contaminants and can be undertaken on a long-term or short-term basis.

SHORT-TERM SAMPLING TECHNIQUES

Generally known as 'spot', 'snap' or 'grab' sampling - taking an immediate sample of air from the workplace and passing it through a particular chemical agent which responds to the chemical being monitored or using a direct reading instrument to determine the concentration of the substance in air.

LONG-TERM SAMPLING TECHNIQUES

Instruments that are used for long-term sampling are broadly of two types - personal samplers and static sampling systems.

PRELIMINARY CONSIDERATIONS FOR SAMPLING

The following must be considered when deciding an appropriate sampling strategy and methods:

Type of air contamination - this involves a review of the materials, processes and operating procedures being used within a work activity. Safety data sheets are also of use. The initial appraisal should determine these things and establish what needs to be measured.

People affected - this will depend on the size and diversity of the group affected. The individuals selected to assist with the provision of personal sampling exposure must be representative, but selecting those with the highest exposure is a reasonable starting point. If the group is large, then random sampling may have to be employed. The group being considered should be made aware of the reason for the sampling.

Frequency of measurement - this will depend upon the hazard, whether it is acute or chronic, exposure limits and results of prior measurements.

Technique for measurement - the particular type is based on the hazard presented and purpose of the sample being taken; an approach is outlined in the following table:

Measurements required to determine	Suitable types of measurement
Chronic hazard	Continuous personal dose measurement
	Continuous measurements of average background levels
	Spot readings of containment levels at selected positions and times
Acute hazard	Continuous personal monitoring with rapid response
	Continuous background monitoring with rapid response
	Spot readings of background contaminant levels at selected positions and times
Environmental control status	Continuous background monitoring
	Spot readings of background contaminant levels at selected positions and times
Whether area is safe to enter	Direct reading instruments

Figure B4-10: Sampling strategies.   *Source: RMS.*

Direct reading instruments

Direct reading instruments were developed as early warning devices for use in industrial settings, where a leak or an accident could release a high concentration of a known chemical into the ambient atmosphere.

Today, some direct reading instruments can detect contaminants in concentrations down to one part contaminant per million parts of air (ppm), although quantitative data are difficult to obtain when multiple contaminants are present.

Some direct reading instruments are:

- *Combustible gas indicator (CGI)* - used to monitor combustible gases and vapours.
- *Flame ionisation detector (FID)* - used to monitor many organic gases and vapours.
- *Gamma radiation survey instrument* - monitors gamma radiation.
- *Portable infrared (IR) spectro-photometer* - used to monitor many gases and vapours.
- *Ultraviolet (UV) photo-ionisation detector (PID)* - monitors many organic and some inorganic gases and vapours.
- *Oxygen meter* - monitors oxygen.

ADVANTAGES

Direct reading instruments provide information at the time of sampling, enabling rapid decision making. Such instruments may be used to rapidly detect flammable or explosive atmospheres, oxygen deficiency, certain gases and vapours, and ionising radiation.

The information provided by these instruments can be used to:

- Institute appropriate protective measures (for example, PPE, evacuation).
- Determine the most appropriate equipment for further monitoring.
- Develop optimum sampling and analytical procedures/protocols.

DISADVANTAGES

- Usually detect and/or measure only specific classes of chemicals.
- Generally not designed to measure and/or detect airborne contaminants below 1 ppm.
- Many of the instruments that have been designed to detect one particular substance also detect others (interference) and consequently may give false readings.
- Direct reading instruments must be operated (and their data interpreted) by qualified individuals who are thoroughly familiar with the particular device's operating principles and limitations and who have obtained the device's latest operating instructions and calibration curves.

STAIN TUBE (COLOURMETRIC) DETECTORS

A colour change detector, also called 'colourmetric' detector, is chemical detection equipment that incorporates an inert supporting material which is made to contain chemical reagents(s) that react in the presence of a target gas, liquid or vapour, resulting in a specific colour change.

These detectors may be either 'paper' or glass tubes. The sampling method may be either passive (diffusion/time) or via a pump (volumetric). Intensity of colour change or length of stain (for glass tube) is directly related to concentration of the substance.

See figure ref B4-11 detailing the accuro® Pump which draws a calibrated 100 ml sample of air through the Draeger-Tube® with each stroke.

Limitations

A limitation of this equipment is that the measured concentration of the same substance may vary among different manufacturers' tubes. Another limitation is that similar chemicals can interfere with the reaction, causing false readings.

Also, whilst minimal operator training and expertise is required to actually use this piece of equipment, the greatest sources of error are the user judging the stain's end-point and the tube's limited accuracy. The instrument may also be affected by humidity.

Figure B4-11: The accuro® Pump. *Source: Draeger.*

For further information, see 'Sampling equipment for vapours', earlier in this Element, which includes details on use of stain tube (colourmetric) detectors.

General methods for sampling and gravimetric analysis of dust

GENERAL APPROACH

There are four critical factors that may influence the health impact of dusts - each of these four factors are interrelated and influence sampling and analysis methods.

1) The nature of the dust in question.
2) Particle size.
3) Duration of exposure time.
4) The airborne concentration of the dust in the breathing zone of the exposed person.

Nature of the dust

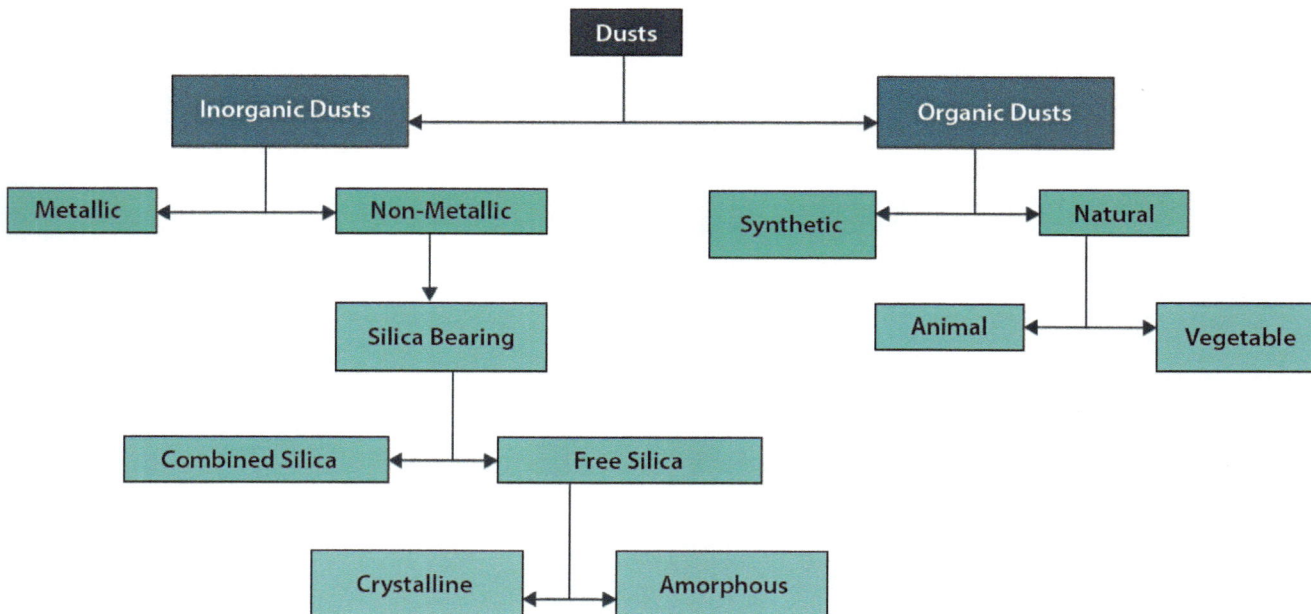

Figure B4-12: Classification for sampling and evaluating respirable dusts.

Source: www.safetyline, Hazardous Substances Management, Sampling of Airborne Contaminants (this figure is based on a similar diagram, which was adapted from a US Bureau of Mines Circular 8503, Feb. 1971 [Alpaugh, 1988, p.124]).

Particle size

This is critical in determining where particulates will settle in the lung. Larger particles of dust will settle in the bronchi and the bronchioles and will not tend to penetrate the smaller airways found in the alveolar region.

These are termed *inspirable* particles. Those smaller particles that can penetrate to the gas exchange region of the lungs, the alveolus, are termed *respirable* particles.

The term *inhalable* dust describes dust that is hazardous when deposited anywhere in the respiratory tree, including the nose and mouth. When carrying out sampling and gravimetric analysis of dust it is important to measure in such a way that the amount of inhalable and respirable dust is determined.

Inhalable dust approximates to the fraction of airborne material that enters the nose and mouth during breathing, and is therefore available for deposition in the respiratory tract. Respirable dust approximates to the fraction of airborne material that penetrates to the gas exchange region of the lung.

Figure B4-13: Inhalable and respirable dusts.   *Source: HSE, MDHS 14/3.*

Duration of exposure

This may be ***acute*** (i.e. minutes, hours, but usually no longer than a day or two at the most) or ***chronic***, in that the duration of exposure may be measured in months or years, over a full working lifetime. Some airborne dust may exert a toxic effect after a single acute exposure, for example, beryllium.

Other dusts may exert a toxic effect following a longer period of exposure; maybe several days to weeks, for example, lead. Such exposures could be termed ***sub-chronic***. Chronic lung conditions, for example, mesothelioma, may follow prolonged exposure to crocidolite (blue asbestos).

Airborne concentration of the dust in the breathing zone of the exposed person

The worker's breathing zone is described by a hemisphere of 300 mm radius extending in front of their face and measured from the mid-point of an imaginary line joining the ears. The sampling method should determine the inhalable and respirable airborne concentration of the dust. Various methods of sampling may be used.

SAMPLING DUST PARTICULATES

The principle of sampling and gravimetric analysis is expressed in MDHS 14/3:

A measured volume of air is drawn through a collection substrate such as a filter mounted in a sampler, and the mass of dust collected is determined by weighing the substrate before and after sampling.

Figure B4-14: Principle of sampling and gravimetric analysis.   *Source: HSE, MDHS 14/3.*

Filtration sampling

Typically, a filtration sampling system consists of:

Figure B4-15: Typical sampling train.   *Source: RMS.*

Filters

The filter diameter, type and pore size will vary depending on the dust being sampled and will be specified in the sampling method. An air sampling filter may contain millions of pores, ranging from 0.5 to 30 microns diameter.

Sampling heads

A sampling head is a device that acts as a filter holder and may be designed to overcome problems of sampler orientation or wind speed.

The selection of the sampling head is determined by the nature of the airborne contaminant and the particle size range. ***See later in this Element for more information.***

Tubing

The length and diameter of the tubing should be appropriate to ensure smooth flow of the sampled air.

Pump

The pump must be capable of maintaining smooth flow at the specified rate throughout the sampling period. It is best to use pumps that automatically maintain a constant flow rate to collect samples, since it is difficult to observe and adjust pumps while wearing gloves, respirators and other PPE.

For personal sampling, the pump must be light and portable, and a belt may be required if the pump is too large to fit in the worker's pocket. Pumps should be protected with disposable coverings, to make decontamination easier.

Personal and background sampling

Filtration sampling may be personal, where the equipment is attached to the worker, or background, where the equipment is located in a fixed point.

Personal sampling

The selective monitoring of high-risk workers, i.e., those who are closest to the source of contaminant generation, is highly recommended.

This approach is based on the rationale that the probability of significant exposure varies with distance from the source.

If workers closest to the source are significantly exposed, then other workers away from the source will not be significantly exposed and probably do not need to be monitored.

When carrying out personal sampling, the sampler is attached to the wearer within his or her breathing zone, and the pump is usually connected to it by a length of flexible tubing and worn on a belt, harness, or in a pocket. The breathing zone is the space around the worker's face from where the breath is taken, and is generally accepted to extend no more than 30 cm from the mouth.

Personal sampling instruments are normally mounted therefore on the upper chest, close to the collar-bone. As personal monitoring samples should be collected within the breathing zone, if workers are wearing respiratory protection equipment (RPE) this means outside the face-piece.

These samples represent the actual inhalation exposure for workers who are not wearing RPE and the potential exposure of workers who are wearing respirators. Personal monitoring may require the use of a variety of sampling media.

Figure B4-16: Personal sampler. *Source: SKC Inc.*

Background sampling

Fixed point sampling may be used to determine background levels of dust in the workplace. However, the HSE caution in MDHS 14/3 states that:

> *"It is not appropriate to compare fixed point (background) samples with the exposure limit, because the distribution of dust in the workplace is not uniform. In addition to this discrepancy, because of aerodynamic effects, background samplers will not exhibit the same characteristics as when mounted on the body, and will usually underestimate the inhalable dust concentration."*

Figure B4-17: HSE caution - background sampling. *Source: MDHS 14/3.*

Personal samplers are sometimes used for background sampling, and in this situation they should be positioned at approximately head height, away from obstructions, fresh air inlets or strong winds.

General equipment and methodology for personal sampling of solid particulates

SAMPLING HEADS

Protected head samplers

There are three main types of protected head samplers used to determine *inhalable* dust concentrations. These are the multi-orifice (7-Hole) sampler, the Institute of Medicine (IOM) sampler and the conical sampler. In particular, the IOM inhalable dust sampler is designed to overcome problems of sampler orientation or wind speed and has been determined to give the best performance in workplace conditions and the HSE in MDHS 14/3 state that its use is the preferred method. The protected head is used in preference to the traditional *open faced sampler,* which has a tendency to over-sample in environments where particulates of large sizes are present.

Figure B4-18: Multi-orifice (7-hole) sampler. *Source: SKC Inc.*

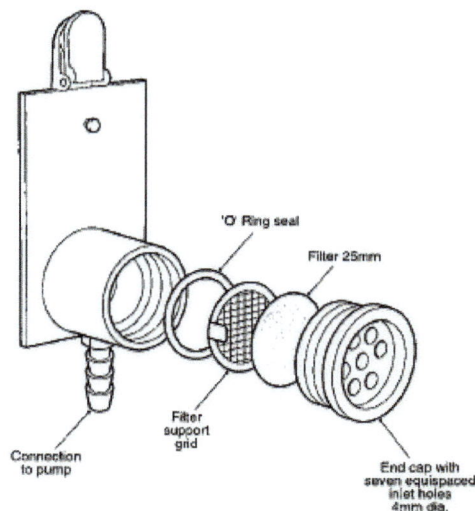

Figure B4-19: Multi-orifice (7-hole) sampler. *Source: MDHS 14/3, HSE.*

The IOM personal inhalable sampler is a conductive plastic sampling head that houses a reusable 25-mm filter cassette with specified filter for the collection of inhalable airborne particles. The sampling head is attached to a personal sampling pump operating at 2 l/min and clipped near a worker's breathing zone, near the nose and mouth, to closely simulate the manner in which airborne workplace particles are inhaled.

Because both the cassette and the filter are pre-and post-weighed as a single unit, all particles collected (even larger ones) are included in the analysis. The cassette can be cleaned, reloaded with a new filter, and reused.

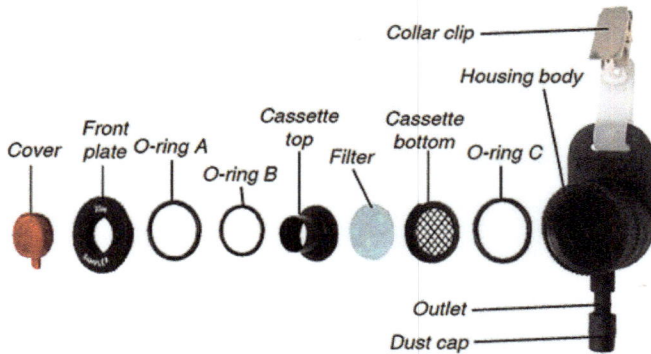

Figure B4-20: IOM inhalable sampler.   *Source: MDHS 14/3, HSE.*

Figure B4-21: IOM inhalable dust sampler. *Source: SKC Inc.*

Cyclone head samplers

A *cyclone head* is used for *respirable* dust sampling. Cyclones are designed to simulate the collection characteristics of the nose and mouth, its name is derived from the rotation of air that takes place within the sampler chamber.

The cyclone functions on the same principle as a centrifuge: the rapid circulation of air separates particles according to their equivalent aerodynamic diameter.

The respirable particles are small enough to remain airborne in the cyclone and pass out of the chamber collecting on the filter, while larger and heavier particles fall into the grit pot. Depending on the type of cyclone used, a two-or three-piece cassette may be needed.

Figure B4-22: Conical inhalable sampler.   *Source: MDHS 14/3, HSE.*

Recommended sampling rates: of 1.7 l/min or 2.2 l/min (depending on the type of cyclone used), for a maximum of 400 litres.

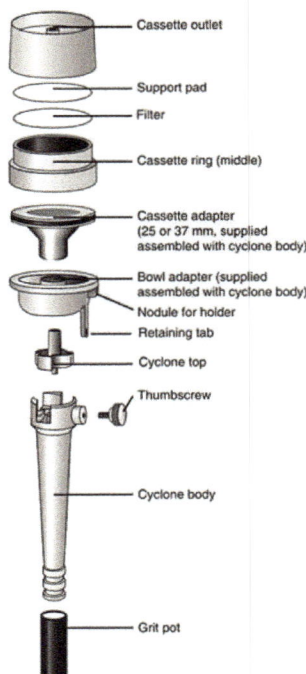

Figure B4-23: Cyclone respirable sampler. *Source: IOM.*

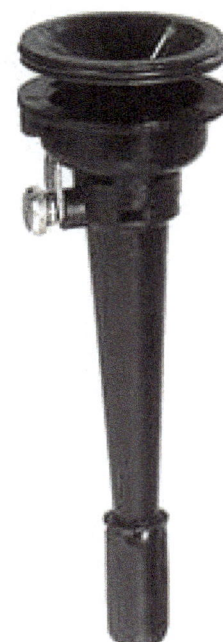

Figure B4-24: Cyclone sampling head. *Source: IOM.*

Respirable dust and sampling efficiency curves

Respirable dust refers to particles that settle deep within the lungs that are not ejected by exhaling, coughing or expulsion by mucus. Since these particles are not collected with 100% efficiency by the lungs, respirable dust is defined in terms of a sampling efficiency curve. These curves are sometimes referred to in terms of the 50% sampling efficiency or cut-point at a certain flow rate.

The term 50% cut-point is used to describe the performances of *cyclones* and other particle-size selective heads or devices. The 50% cut-point is the size of the dust that the device collects with 50% efficiency. Particles smaller than the 50% cut-point will be collected with efficiency greater than 50%.

Particles larger than the 50% cut-point will be collected with efficiency less than 50%. To reach world-wide consensus on the definition of respirable dust in the workplace, a compromise curve was developed with a 50% cut-point of 4 microns.

Cowl head sampler

This type of sampler head is used to sample fibres. This sampler head is an open-faced filter holder fitted with an electrically conducting cylindrical cowl extending between 33 mm and 44 mm in front of the filter, and exposing a circular area of filter at least 20 mm in diameter.

This type of head is intended to protect the filter, whilst still permitting a uniform deposit. In correct use, the cowl will point downwards. A suitable design is shown in *figure ref B4-25*. The head is used in position shown, with the opening downwards.

The HSE state in MDHS59 'Man-Made Mineral Fibre Sampling' that as this type of sampler head is used to collect fibres, the membrane filter must be of mixed esters of cellulose, of pore size 0.8 to 1.2µm, and 25 mm in diameter, with a printed grid.

The HSE advise that the pump give a smooth flow, be capable of having its flow set to within 5% and of maintaining the chosen flow rate through the membrane filter to within ±10% during the period of sampling.

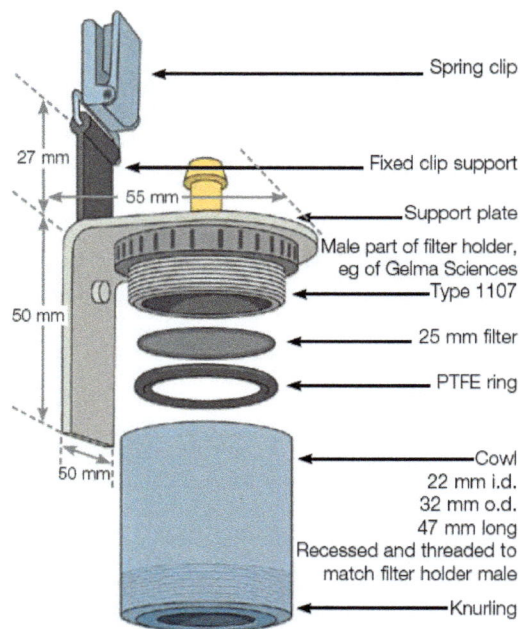

Figure B4-25: Cowl sampling head. *Source: HSE, HSG248.*

SAMPLING PUMPS

A wide range of personal sampling pumps is now commercially available, most of which are small and light enough to be comfortably worn on the belt or carried in a pocket.

Various pumping techniques are used, the most common are:

- Piston pumps.
- Rotary vane pumps.
- Diaphragm pumps.

Typical designs include a flexible cylinder which is alternately compressed rapidly and then allowed to expand slowly over a few seconds to draw atmosphere into it.

Piston and compressed chamber pumps

The volume sampled is measured by counting the number of strokes during the sampling period. The volume sampled at each stroke is determined by the geometry of the pump and this technique is a very reliable method of measuring sampled volume. Compressed chamber pumps cannot be used for dust sampling.

Rotary vane pumps

The volume sampled is then obtained by multiplying the flow rate by the time. It is necessary to measure the rate of flow through the pump, and this is often achieved with an integral flow meter.

Variations in flow rate can occur due to malfunctions of the pump during the sampling time, although it is general practice to measure the flow rates at the beginning and end of the sampling period and assume that any variations between them have been linear during the sampling period.

Diaphragm pumps

The volume sampled is measured by counting the number of movements of the diaphragm during the sampling period. The volume sampled at each stroke is determined by the geometry of the diaphragm.

Some pumps are double acting, drawing air in on the backward movement of the diaphragm as well as the outward movement. This enables smoother operation of the pump and more consistent air flow.

Many airborne substances may cause potentially flammable atmospheres. It is for this reason that a sampling pump chosen should be of a suitable design to not be a source of electrical ignition.

The battery capacity of the pump should be sufficient to cover the sampling time chosen and, in general, pumps should be capable of operating continuously for at least eight hours. Poor battery choice or quality may result in marked drops in sampling rates during the sampling period. The sampling pump is the most expensive item in the sampling system and is probably the biggest source of error in hygiene monitoring.

It is essential that it is serviced regularly. This servicing should include cleaning of the pump and relevant parts of the flow system and careful checks on battery capacity under sampling conditions.

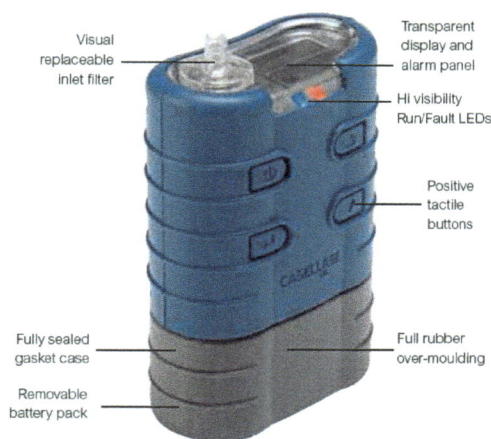

Figure B4-26: Double acting diaphragm pump. *Source: Casella.*

With most pumps used for sampling gases the flow rate can be varied over a range of pre-set values, generally between 10 ml/min and 200 ml/min. The techniques used to vary flow rates include by-pass valves on rotary vane pumps, variable motor speeds on piston pumps and a range of standard orifices on compressed chamber pumps.

The flow rate for personal sampling is normally 2 litres per minute. When high volume samples are taken, flow rates up to 100 litres per minute are used.

Sampling pumps used for dust sampling should have as a minimum the following features:

■ Automatic flow control, which keeps the volumetric flow rate constant to within ±0.1 litre/min in the case of changing back pressure.
■ Malfunction indicator (which indicates that air flow had been reduced or interrupted during sampling) or an automatic cut-out (which stops if the air flow is reduced or interrupted).
■ Adjustment of flow rate, with restriction on the ability to adjust requiring a tool (for example, a screw driver) or knowledge (for example, passworded software).

For cyclone samplers, pulsation damped flow is particularly important and an external pulsation damper must be used if the pump does not contain an integral damper. In addition, the HSE recommends in MDHS 14/3 compliance with the requirements of BS EN1232 'Workplace atmospheres. Pumps for personal sampling of chemical agents. Requirements and test methods'.

FILTERS

The use of filters is the most common form of measurement for dust concentration in the atmosphere. The process involves a known volume of air being drawn through a special filter or membrane by a pump. There is a wide variety of filters available in various sizes ranging from 13 mm (personal sampling) to 50 mm diameter (background/static monitoring).

Types

The main filter types are:

Media	Advantages	Disadvantages
Paper (cellulose fibre)	Inexpensive and easy to handle.	Absorbs water vapour.
Glass fibre	Inexpensive, high particle retention, low water vapour absorption.	Sheds fibres easily, highly variable manufacturing tolerances.
Silver	Low chemical interference (particularly x-ray diffraction) and low water vapour absorption.	Expensive.
Membrane (plastic)	Good optical properties under microscope and precise pore sized.	Poor adhesion, high static, pores plug easily due to surface collection.
PVC	Low chemical interference, low water vapour absorption and precise pore size (particularly silica IR).	Poor adhesion, high static, pores plug easily due to surface collection.

Figure B4-27: Filters. *Source: RMS.*

Due to the variety of methods for analysing filters available, the appropriate filter must be chosen for the particular analysis method that will be used.

For sampling asbestos fibres, membrane filters must be of mixed esters of cellulose or cellulose nitrate, with a pore size of 0.8 to 1.2 microns, optically clear grade, and should be 25 mm in diameter with a printed grid.

Pre-weighed and match weight filters

There are two main formats of filter: pre-weighed and matched weight.

Pre-weighed filter analysis starts when the filters are equilibrated (consistent temperature and humidity) and weighed, or 'tared'. They are then placed in cassettes, sealed with colour-coded bands, and labelled with a unique identification number. After sampling, they are again equilibrated and post sampling weight is determined. The difference between the initial weight and the final weight is the number reported (along with the air volume sampled and the resulting mg/m^3 calculation). Pre-weighed filters more than six months old should not be used. Pre-weighed filters should always be returned to the same laboratory that prepared them, which will prevent slight variations in analytical balances which will affect the results. Blank filters in their cassettes should always be submitted when using pre-weighed filters. Blank filters are considered field blanks and help confirm the integrity of your samples during transport and storage.

Matched weight filters do not require the initial weighing, or 'tare', step. A matched weight filter cassette contains two filters of equal weight, one on top of the other. During sampling, dust is collected on only the top filter; the bottom filter acts as a blank (equivalent to a pre-weighed). Once back in the laboratory, both filters are weighed and the difference between them is the number you see on your report.

Weight in grams must be accurate to 5 decimal places, as air humidity can affect the weight of some filters; preconditioning may be required.

METHOD OF USE

When samples are collected for comparison with control limits, it is essential that the methods should achieve a high degree of reproducibility and specificity. The methods referred to in this Element are expected to meet these requirements.

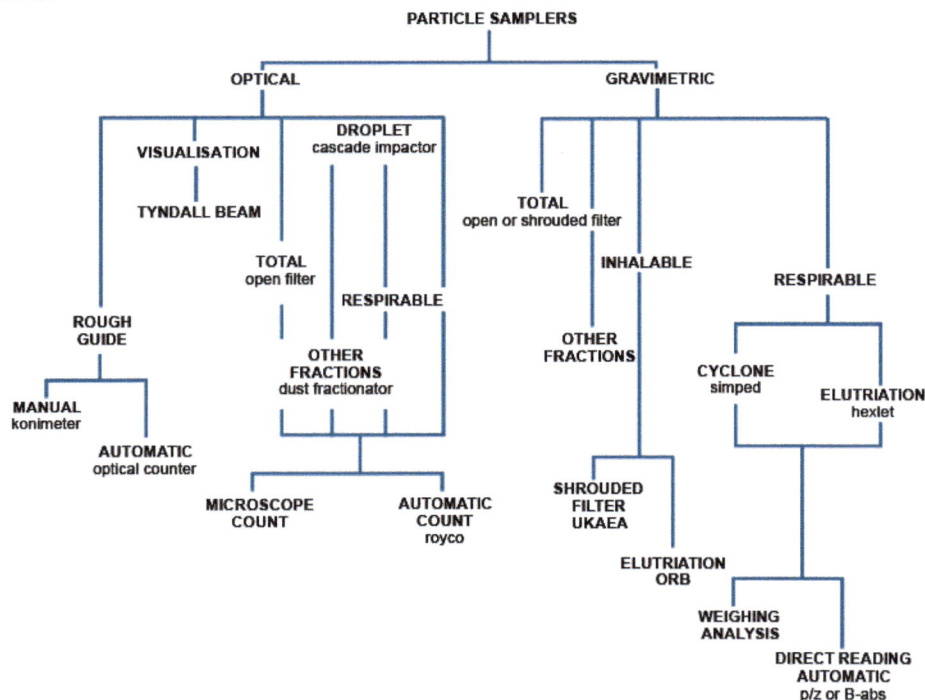

Figure B4-28: Classification of the methods for the sampling of particulates.   *Source: RMS.*

The method of sampling should be selected based on what is to be sampled and what is to be measured. The main methods are designed to focus on the fact that there are two basic categories of dust that are sampled and measured:

1) Inhalable - under 100 microns in diameter.
2) Respirable - under 10 microns in diameter.

Both types of dust exist in the air we breathe; the only difference between them is the diameter of the particle. The collection method varies, depending upon the type of dust to be evaluated. It will be necessary to use one of the types of protected head samplers, for example, an IOM sampler, when sampling for inhalable dust; this collects dust up to 100 microns. When sampling for respirable dust, 10 microns or less, a cyclone will need to be used, which excludes dust particles larger than the respirable size. The cowel head sampler should be considered when sampling for fibres.

The method of conducting sampling based on MDHS 14/3 is summarised as:

1) Decide on the type of sampler head needed.
2) Select a clean and dry sampler head.
3) Prepare the sampler head with the appropriate filter in a dust free area, label and seal.

4) Attach sampler heads to the pumps and set the pump flow rate, using a calibrated flow meter.
5) Prepare and retain enough blank loaded samplers to represent the batch, one in ten and a minimum of three.
6) Treat the blank loaded samplers as the others, but do not draw air through.
7) Select and attach the sampler equipment to the worker, ensure head is in the breathing zone.
8) Remove protective covering from sample head and record the time and volumetric flow rate.
9) Check periodically, if necessary adjust flow rate and look for fault conditions.
10) At the end of sample period record the flow rate and sample time.
11) Place protective covering over sample head and remove carefully from worker.
12) Ensure cyclone sample heads are kept upright.
13) Remove the filter and prepare for transportation and analysis.

MEASUREMENT PRINCIPLES - DUSTS AND FIBRES

Dusts

Gravimetric analysis

Gravimetric analysis enables the weight of the dust collected to be determined. This involves weighing the filter before and after sampling in order to determine the net weight gained. This will require the use of weighing equipment that has been validated as accurate at intervals by the use of a calibrated standard weight and independently verified as accurate periodically.

For improved accuracy the filters should be conditioned before weighing to allow moisture in the filter to reach equilibrium with that of the weighing environment. This involves them being placed in individual, labelled, clean tins (or other suitable containers) and left with the lids slightly ajar in the weighing environment overnight. Any electrostatic charge should be eliminated by passing the filter over a static eliminator before weighing.

When the net weight has been determined the dust concentration can be calculated by comparing it with the amount of air flow that created the weighed dust sample. The volume of air is calculated by multiplying the mean volumetric flow rate of the pump in cubic metres per minute by the sampling time, in minutes. It is useful to note that if the flow rate is given in litres per minute, flow rate litre/min = 1,000 x flow rate in m^3/min, so the flow rate in litres/min is divided by 1,000 to obtain the flow rate in m^3/min. To obtain the dust concentration in mg/m^3 the net weight gain in mg is divided by the volume of air sampled in m^3.

Physical

Samples that have been collected on a filter may be analysed using a variety of physical techniques:

- X-Ray Fluorescence Spectroscopy.
- X-Ray Diffraction.
- Infra-Red Techniques.
- Atomic Absorption.
- High Performance Liquid Chromatography (HPLC).

Silica exposures can be evaluated by infrared (IR) or X-ray diffraction (XRD) techniques.

Chemical analysis

Metals analysis of dust samples may be important. Ideally, samples for metals analysis should be collected on preweighed MCE filters. These filters are easily digested in acid, whereas often the PVC filters fail to digest completely. The residues left after chemical digestion can be further examined by a variety of chemical means, for example, gravimetric i.e. direct weighing.

Fibres

Microscopy for fibres

If a membrane type filter made of cellulose acetate is used, it can be made transparent by adding a clearing fluid, allowing the dust to be examined under a microscope and the particles or fibres may be counted if necessary.

This process is particularly important if fibrous matter, such as asbestos, is present.

Microscopes can range in power from a simple single lens instrument (i.e. a magnifying glass) through to high-powered scanning electron microscopes and transmission electron microscopes capable of up to approximately five million times magnification.

An optical or light microscope can magnify up to about 1,500 times and is used extensively for analytical work.

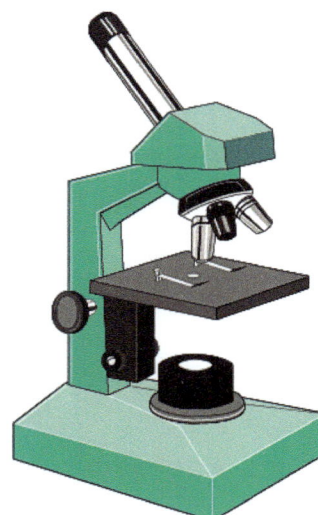

Figure B4-29: Microscope. *Source: RMS/Corel Draw.*

The compound light microscope consists of:

- An eyepiece or viewing lens.
- A secondary lens in the body tube.
- An objective lens.
- An optical condenser that concentrates light (from a mirror or more usually a built-in illuminator).
- A stage which holds the specimen in place.

Phase contrast and interference types of microscope are types of light microscopes that have modified illumination and optical systems, which make it possible for unstained transparent specimens to be seen clearly.

Phase contrast light microscopy (PCLM) is a technique for revealing the structural features of microscope transparent objects with varying, but invisible differences in thickness, which result from varying differences in the phase of transmitted light. These phase differences are converted to visible intensity differences when part of the transmitted light has its optical path changed by about ¼ of a wavelength. In addition to fibre analysis, they are very useful for the examination of living tissues and cells.

Asbestos fibre concentration is estimated by use of PCLM using the method described in Appendix 1 of HSG248 'Asbestos: The analysts' guide for sampling, analysis and clearance procedures'. Accurate asbestos measurement is difficult to achieve because of the difficulty of fibre counting. Laboratories carrying air sampling and analysis must be accredited to ISO/IEC 17025 and those conducting visual inspection procedures need to conform to ISO 17020. Phase contrast microscopy is used to enhance the contrast between the fibre on the filter and the background. Although the type of fibre cannot be positively distinguished, such as between asbestos and some other fibres such as gypsum and machine-made mineral fibres (MMMF), an experienced analyst can comment on whether the fibres are visually consistent with asbestos. The microscope used is fitted with a Walton-Beckett graticule with a network of fine lines that can be viewed through the eyepiece. A sample is taken and mounted on a slide. The operator examines 100 fields or 200 fibres whichever occurs first.

A respirable fibre is considered to be:

- ≥ 5 µm long.
- ≤ 3 µm in diameter.
- A length to diameter ratio of $\geq 3:1$.

Samples can also be examined using polarised light techniques in order to determine whether the fibres are crystalline, glass or mineral. Here two polarising filters can be used to distinguish visually identical crystals by highlighting their individual colour and tone patterns (known as their birefringence). A rotating polarising filter is positioned above the sample and one fixed below. Observing the optical properties of the sample can provide a unique identification for thousands of compounds. The sample's unique refractive indices, birefringence, and dispersion can all be quantitatively determined. A specific technique for fibres is described in Appendix 2 of HSG248 'Asbestos: The analysts' guide for sampling, analysis and clearance procedures'. Asbestos can also be positively identified using electron microscopy and energy dispersive x-ray analysis. The aim with these techniques is to eliminate non-asbestos fibres from the count.

Test results given should include:

- The type(s) of asbestos identified.
- The sample number.
- The sample site.
- The sampling duration (in minutes).
- The flow rate (in litres per minute).
- The fibres counted.
- The number of graticule areas counted.
- The calculated result, expressed as fibres per millilitre of air (f/ml).
- The name of the analyst and date of analysis.

General equipment and methodology for personal sampling of vapours

PASSIVE AND ACTIVE DEVICES

Passive personal samplers

These are devices that collect a sample without the use of a pump. Gas is absorbed or adsorbed by the collecting medium at a rate of diffusion across a well-defined diffusion path. They rely upon a concentration gradient across a static layer of air, which is proportional to the concentration in the atmosphere outside the sampler.

A passive sampler is attached close to the worker's breathing zone and the total exposure time is noted. Analysis of the sample gives the total concentration and from this the Time Weighted Average (TWA) can be calculated. Some passive samplers, like gas badges, are generally fitted to the lapel and change colour to indicate contamination.

The 2 main types of passive sampler are:

Badge		Activated charcoal is useful for collecting common solvents. Sampling rate 50ml/min.
Tube		Sampling rate of 1ml/min.

Figure B4-30: Types of passive samplers. *Source: RMS.*

Passive samplers tend to be light, compact and easy to use. For this reason they are very well received by users amongst the workforce.

Active personal samplers

Active personal samplers are designed to be worn by the worker whilst carrying out their normal work routine. The sample device consists of a pump, fitted to a belt positioned at the waist.

The pump is connected by a flexible tube to the collector which is attached near to the collar of the worker, as close to the respiratory zone as possible.

The collector may contain a *liquid or solid sorbent* to collect vapours or gases. Analysis is carried out after completion of the work period.

ACTIVATED CHARCOAL TUBES AND PUMPS

Tubes

In these devices a powdered sample of solid charcoal with particle sizes in the range 0.1 - 1 mm is contained in a short glass or plastic tube.

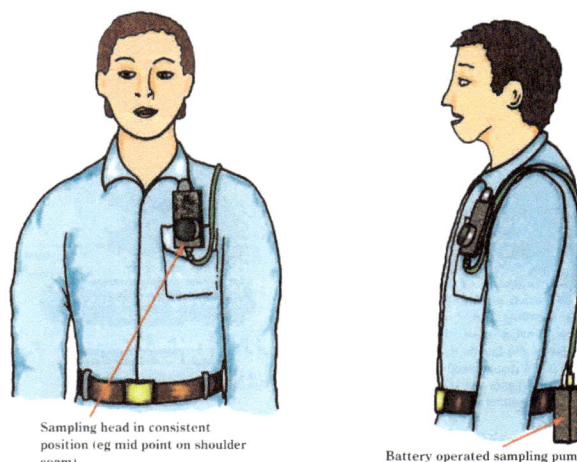

Sampling head in consistent position (eg mid point on shoulder seam)

Battery operated sampling pump

Figure B4-31: Example of an 'active personal sampler'.
Source: RoSPA OS&H.

Charcoal is a suitable adsorbent for a very wide range of organic vapours and a short length of charcoal will effectively remove vapours from an air stream. Often two plugs of adsorbent material are used in the tube so that the presence of vapour on the second downstream plug indicates breakthrough and that the sample will be giving a low value. Solid adsorbents other than charcoal can be used in such devices, for example, the porous polymers tenax and poropak Q.

These alternative adsorbents are important for vapours that are so strongly adsorbed on charcoal that they cannot subsequently be easily desorbed for analysis.

After sampling, the tubes are sealed with plastic caps before the tube is sent for analysis. Obviously, many variations in geometry and packing are possible. Larger amounts of adsorbent will enable larger concentrations of vapours to be assessed over longer time periods. Similarly, polar materials, such as silica gel, can be used in special applications for polar gases.

The simplicity of the solid adsorbent sampling tube means that two or more of them can be used simultaneously with a simple splitting device so that the pump is pulling the same flow rate through two tubes mounted in a suitable adapter, enabling duplicate samples to be obtained.

In general, adsorbent tubes have more than sufficient capacity to sample organic vapours at their workplace exposure limit continuously for eight hours. Problems can sometimes occur where high humidity (> 90% RH) is present, when the adsorbent capacity can be reduced. Such limitations are indicated in the literature for the substance and by the manufacturer of the absorbent tubes.

The sampling pump must operate consistently within a defined range to ensure accurate sampling is achieved. For gas and vapour determination, a knowledge of the volume of atmosphere pulled through the sampler in a given time is essential, but variations in flow rate within that time are not important.

In fact, very large variations in flow rate are admissible, provided the time over which the variations take place is small compared with the time over which variations in gas and vapour concentration take place.

An accuracy of ± 5% in volume sampled is sufficient for most purposes.

It is important that the sampling technique selected for a particular application meets the following criteria:

- It should have a known collection efficiency for the contaminant, which should be at least 75% with a reproducibility of ±5%.
- It must provide sufficient material for analysis. If the contaminant is at its control limit, the amount collected, over the sampling period, should be at least five times the detection limit of the analytical technique used.
- Collection of the sample should be compatible with the analytical technique and must ensure the minimum amount of degradation prior to analysis.
- Sampling should involve the minimum of on-site preparation.
- The sampling equipment and medium should not present a hazard to the wearer or the environment.

The sampling procedure selected should be as simple and straightforward as possible. It is advisable to have a written protocol which includes procedures for numbering and mounting the sampling head, testing for leaks, and handling and packing of the used sampling heads. Records of time of sampling, sampling period, flow rate, place and other relevant information must be kept.

There are several different manufacturers of detector tubes, and it is most important that the literature provided with the pumps and tubes is strictly followed.

Types of tube construction:

1) Commonest is the simple stain tube (colour metric) detector, but it may contain filter layers, drying layers, or oxidation layers.
2) Double tube or tube containing separate ampoules, avoids incompatibility or reaction during storage.
3) Comparison tube.
4) Narrow tube to achieve better resolution at low concentrations.

The above illustrates the main types of tubes. There are more variations and therefore it can only be re-emphasised that the manufacturer's operating instructions must be read and fully understood before tubes are used.

Pumps

There are three types of hand operated pumps:

1) Hand operated bellows pump.
2) Hand operated piston pump.
3) Hand operated ball pump.

In addition to the hand operated pumps, a variety of battery operated pumps are available - *see earlier in this element - 'Sampling equipment for solid particulates'*.

Never mix pumps and tubes of different manufacturers.

How to use tubes

- Choose tube to measure substance of interest and expected range.
- Check tubes are in date.
- Check leak tightness of pump.
- Read instructions to ensure there are no limitations due to temperature, pressure, humidity or interfering substances.
- Break off tips of tube, prepare tube if necessary and insert correctly into pump. Arrows normally indicate the direction of air flow.
- Draw the requisite number of strokes.
- Immediately evaluate stain tube, unless manufacturers' operating instructions say otherwise. If there is any doubt when reading tube, always accept the higher figure.
- Remove tube and discard according to instructions. Many of the tubes are hazardous once opened (COSHH procedures will need to be considered as will disposal arrangements).
- Purge pump to remove any corrosive contaminants from inside the pump.

In terms of accuracy, manufacturers claim a relative standard deviation of approximately 20% or less, i.e. 1 part per million (ppm) error in 5 ppm.

Long-term tubes

Detector tubes have been developed that indicate contaminant mass over extended periods of time. The tubes are designed for a flow rate of 10 - 20 ml/minute, much lower than that of most short-term tubes, which sample 100 ml in about 10 seconds. The total sample is also greater, up to 10 litres in 8 hours.

MEASUREMENT PRINCIPLES - CHEMICAL AND PHYSICAL ANALYSIS

In order to carry out chemical or physical analysis of the contaminant it may be necessary to remove it from the material that absorbed it. The simplest way of removing adsorbed vapour is to wash it off with a suitable solvent. For efficient removal, the partition of the vapour between the solid absorbent and washing solvent must lie heavily on the solvent side - otherwise excessive amounts of solvent will need to be used with undesirable dilution of the sample for analysis.

Carbon disulphide is widely used, although care has to be exercised as this solvent is highly toxic and flammable. A convenient alternative for desorption is to heat the solid in a gas stream, so that the vapour is desorbed into the gas. The stream can then be analysed or the vapour can be re-trapped and flash heated and passed into a suitable analytical device. Solvent desorption dilutes the original sample and some losses can occur during 'wetting', but the resulting solution can be repeatedly analysed. No dilution occurs with thermal desorption, but the analyst must get the conditions right first time.

Chemical analysis

Certain procedures may have to be performed on the adsorbed contaminant during analysis. For instance, a specific reaction may be carried out by adding reagents that produce a compound, which can be estimated visually or using other analytical methods, including:

- Spectroscopy - the quantative measurement is known as spectrophotometry.
- Gas Liquid Chromatography.
- High Performance Liquid Chromatography.
- Thin Layer Chromatography.

Spectrophotometry

There are a number of instruments available to determine concentrations of contaminants.

Infrared spectrometry

Infrared spectrometry is where a beam of infrared light is passed through the sample. Examination of the transmitted light reveals how much energy was absorbed at each wavelength of the infrared spectrum. Analysis of these absorption characteristics reveals details about the molecular structure of the sample. Infra-red analysis is limited to covalent bonded substances such as organic chemicals, for example, methanol.

Ultraviolet spectrometry

Ultraviolet spectrometry measures the intensity of light passing through a sample (I), and compares it to the intensity of light before it passes through the sample (Io). The ratio I / Io is called the transmittance and is usually expressed as a percentage (%T). Samples for UV/Vis spectrometry are usually used to measure transition metals in solution, although the absorbance of gases and some translucent solids can also be measured.

Atomic absorption spectrometry

Atoms absorb energy at the same frequency as they emit it. If a substance lies in a radiation path its atoms will absorb certain of the energy quanta. This will produce dark lines or absorption spectra. This can then be analysed by a spectrometer. A sample in solution is atomised in a flame and the absorption of radiation by these atoms is measured. The degree of absorption is proportional to the quantity of the element within the solution. By also examining the absorption of known quantities of the substance (standard solutions) we can determine the actual concentration of the substance in the sample.

One common tool used in laboratory analysis is energy dispersive x-ray spectrometry (EDS). The analytical sample is placed in an electron microscope to which a (EDS) capability has been added. In the microscope, the bombardment by the energetic electron beam induces the emission of x-rays at energies that are characteristic of the elements present in the sample. The resultant EDS spectrum is analysed. The peaks indicate the presence of the associated element. The height of the peaks is related to the concentration of the element, so that quantitative elemental analysis is possible.

Chromatography

Chromatography *(writing in colour)* was first discovered in 1906 when a Russian Botanist (Mikhail Semenovitch Tswett) found that he could separate plant pigments, which were chemically similar, by washing them down a column of powdered limestone with a solvent. Each moved at a different rate because each differed in strength of adhesion to the powder. This resulted in the pigments separating into a series of bands, each of a different colour. With continued washing with the solvent (for example, petroleum) the separated substances trickled out separately at the base of the column - one after the other. Chromatography was later developed as a method of separating, often complex, mixtures.

Gas liquid chromatography (GLC)

Gas liquid chromatography is a technique whereby solid and liquid samples are vaporised before introduction on to a packed column. The various components are separated out in a gas stream on the packed column where they are detected. This detection is normally carried out by flame ionisation (FID), thermal conductivity (TC), photo ionisation (PID), or electron capture (ECD). There are many thousands of compounds that can be detected by gas liquid chromatography. The very sensitive detectors used allow the analysis of sub-microgram amounts of materials.

X-ray diffraction

A sample is exposed to an intense X-ray source. As the X-rays pass through a solid crystal they are diffracted according to the distance between the layers of the atoms. The diffraction pattern is then photographed and the spot densities and positions analysed. The spectrum produced will be characteristic of a particular compound.

Calculation of 8 hour equivalent TWA exposures from gathered data

DUSTS

The filter is weighed prior to use and again after the end of the sampling period. The difference between the two weights represents the weight of dust collected.

This, divided by the total volume of air that passed through the filter during the sampling period, gives the average concentration of dust over the sampling period:

$$\frac{\text{Weight gain in filter mg}}{\text{Total volume of air sampled m}^3} = \text{concentration in mg/m}^3$$

Thus, if air was drawn through the filter for 8 hours at a rate of 2 litres per minute and the weight increase of the filter is 0.9 mg then:

Total volume of air sampled   =   2 x 60 x 8 = 960 litres = 0.96 m^3

Airborne concentration   = $\frac{0.9 \text{ mg}}{0.96 \text{ m}^3}$ = 0.937 mg/ m^3

FIBRES

The fibres are counted using a graticule to divide up the area of the filter. The number of fibres found is compared with the area of the graticules analysed. Not the entire filter is analysed for the number of fibres; a representative sample is taken.

The airborne concentration of fibres is given by the following formula:

- $C = 1{,}000\, N D^2 / n\, d^2\, V$ fibres per millilitre (f/ml).

Where:

- N   is the number of fibres counted.
- D (mm)   is the diameter of the exposed filter area.
- N   is the number of graticule areas examined.
- d (μm)   is the diameter of each graticule.
- V (litres)   is the volume of air sampled.

VAPOURS

The amount of vapour desorbed from the sampler can be expressed as a mass concentration or a volume concentration.

The mass concentration is expressed by the formula:

$$\text{Concentration in air (mg/ m}^3) = \frac{1{,}000\,(m - m_{blank})}{DE \times U \times t}$$

Where:

- m   =   weight (μg) of organic vapour on sample tube.
- M$_{blank}$ =   weight (μg) of organic vapour on blank tube.
- DE   =   desorption efficiency, as read from the DE curve, taking m as the weight recovered.
- U   =   uptake rate (cm^3 min^{-1}).
- t   =   exposure time (min).

The volume concentration is similarly expressed by the formula:

$$\text{Concentration in air (ppm)} = \frac{1{,}000\,(m - m_{blank})}{DE \times U^v \times t}$$

Where:

- m   =   weight (μg) of organic vapour on sample tube.
- m$_{blank}$ =   weight (μg) of organic vapour on blank tube.
- DE   =   desorption efficiency, as read from the DE curve, taking m as the weight recovered.
- U^v =   uptake rate (ng ppm^{-1} min^{-1}).
- t   =   exposure time (min).

Calculation of 8-hour time weighted average

The occupational exposures in any 24-hour period are treated as equivalent to a single uniform exposure for 8 hours, the 8-hour time-weighted average (TWA) exposure.

The eight-hour time weighted average concentration can be calculated using the following formula:

8 hour TWA = $\dfrac{C_1T_1 + C_2T_2 + \ldots C_nT_n+}{8}$

Where C_1 is the occupational exposure and T_1 is the associated exposure time in hours in any 24-hour period.

COMPARISON WITH LTEL

The calculated exposure limit should be compared with any published or in-house long-term exposure limit (LTEL) in order to determine its significance and if any further action is needed. If the calculated exposure is over the LTEL then the reasons for this must be investigated and measures put in to prevent further exposures above the LTEL. Those affected may require medical surveillance and will need to be informed of their exposure.

COSHH 2002 Regulation 12 (2) (d) requires that:

> *"The results of any monitoring of exposure in accordance with Regulation 10 and, in particular, in the case of a substance hazardous to health for which a workplace exposure limit has been approved, the employee or his representatives shall be informed forthwith, if the results of such monitoring show that the workplace exposure limit has been exceeded."*

Figure B4-32: Regulation 12 (2) (d) of COSHH 2002. Source: Control of Substances Hazardous to Health Regulations (COSHH) 2002.

Where the calculated exposure is near to the LTEL it is recommended that similar action be taken to the above. If the calculated exposure is below 80% of the LTEL then further reasonably practicable action should be taken and the work continue to be monitored frequently.

Qualitative dust monitoring (Tyndall beam)

THE TYNDALL EFFECT

The Tyndall beam is a strong beam of light scattered by colloidal particles. This is known as the Tyndall effect (after its discoverer, the 19th-century British physicist John Tyndall), and is a special instance of diffraction.

Colloid - a combination of two or more substances so that very small particles of each are suspended throughout the others. The particles in a colloid are larger than a molecule, but small enough to remain in suspension permanently and be homogeneous. The particles in a colloid are large enough to act as tiny mirrors and reflect or scatter light. If a beam of light is passed through a colloid, the light beam will be scattered. The Tyndall effect can be visualised in air when strong sunlight comes in through a window and the dust and other particles can be seen floating in the air; however, the air molecules are too tiny to be seen.

The Tyndall effect occurs when the dimensions of the particles that are causing the scattering are larger than the wavelength of the radiation that is scattered. Tyndall noticed the effect, but did not know how to explain it.

The explanation was given by Gustav Mie in 1908. So the scattering of light by particles larger than the wavelength of the light is called 'Mie scattering'.

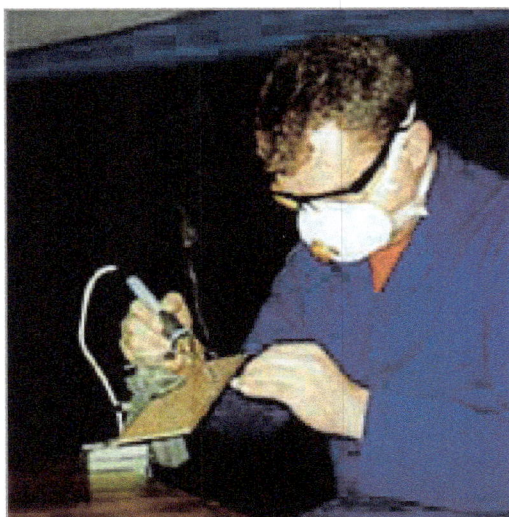

Figure B4-33: Soldering without Tyndall illumination.
Source: HSE, HSG258.

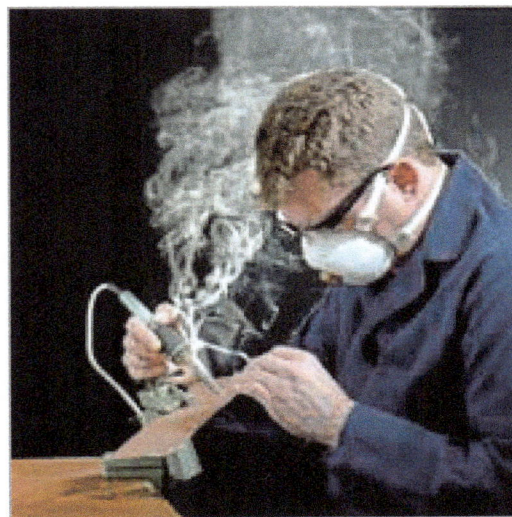

Figure B4-34: Soldering with Tyndall illumination.
Source: HSE, HSG258.

USE AND LIMITATIONS

Very fine dust may not be visible to the naked eye so the Tyndall beam can be used in the workplace to provide a qualitative measurement to show where the dust is, where it is escaping from and where it is at its worst. For example, dust escaping from the ducting in a local exhaust ventilation system used for removing fine hardwood dust from the process will show up in the beam.

The Tyndall beam can be used as a first step to assess a perceived problem with dust by determining where the problem is and how much of a problem the dust is. Showing up the dust with the beam can give an indication of how much dust there is, but other methods must be used to give an exact concentration in the atmosphere.

It is a visual qualitative measurement and though it provides immediate information it does not measure the size of the particles, so methods of dust collection and laboratory analysis will be necessary to show if the dust is inhalable and what part of it is respirable.

The beam can be used by anyone with a small amount of training and after the initial purchase does not require extensive maintenance. They are also fairly sturdy.

B4.3 - Biological monitoring

Distinction between general health assessment and health surveillance

Health assessment is conducted in order that a person may be considered for being put to work in a given job or activity, for example, fitness for work health assessments offered to night workers under the Working Time Regulations (WTR) 1998.

This will also include pre-employment, return to employment and change of status situations like pregnancy and wider issues such as those related to health promotions, sickness absence, drug or alcohol misuse and the general health of critical workers.

Health surveillance is the systematic, regular provision of appropriate programmes to detect early signs of work related ill-health among workers exposed to certain health risk. This may relate to a specific duty to provide health surveillance or to an organisation's own practices to provide effective control of risks.

It is important that the employee and employer know the difference between health assessment and health surveillance.

Elements of the HSE health surveillance cycle

The diagram in *figure ref IB4-35* below provides an overview of the health surveillance cycle. The employer has a central role in every aspect with involvement from employees to ensure effective implementation.

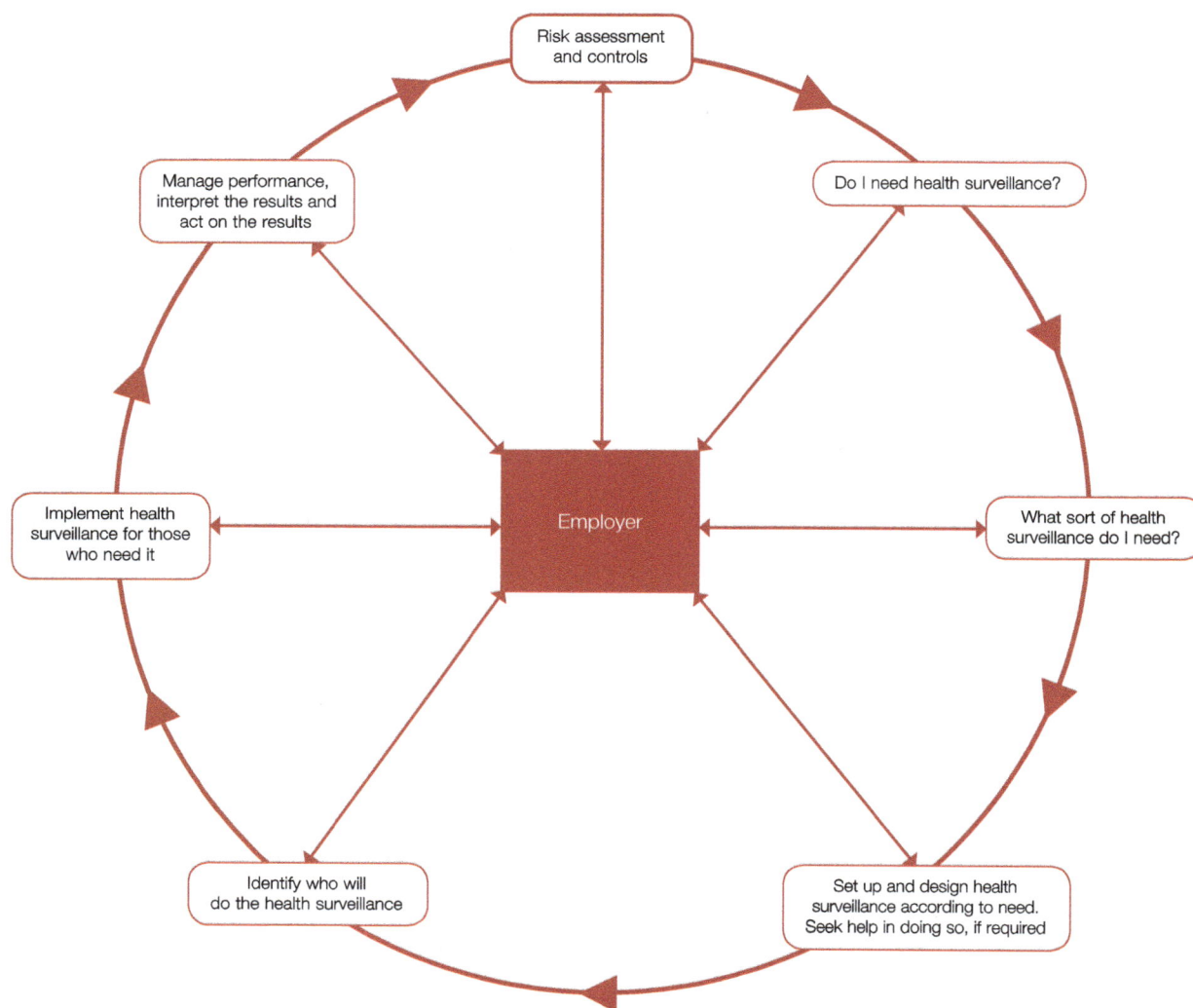

Figure B4-35: Health surveillance cycle. *Source: HSE, http://www.hse.gov.uk/health-surveillance/assets/documents/health-surveillance-cycle.pdf.*

Legal requirements for medical surveillance

GENERAL DUTIES RELATED TO MEDICAL SURVEILLANCE

Employees to whom the Control of Substances Hazardous to Health Regulations (COSHH) 2002 (as amended), Control of Lead at Work Regulations (CLAW) 2002, Control of Asbestos Regulations (CAR) 2012 and Ionising Radiation Regulations (IRR) 1999, apply must present themselves for medical examination and provide any information concerning their health as the doctor may require, at the cost of the employer and during their working hours. The regulations require that the medical surveillance can only be conducted if it is of low risk to the employee and records must be kept for 40 years.

CONTROL OF SUBSTANCES HAZARDOUS TO HEALTH REGULATIONS

Regulation 11, of the Control of Substances Hazardous to Health Regulations (COSHH) 2002 (as amended), places a duty on employers to carry out medical surveillance, for the protection of employees' health where they are liable to be exposed to a substance hazardous to health. Surveillance is appropriate where the employee is exposed to one of the substances specified in Column 1 of Schedule 6 to the regulations and is engaged in a process specified in Column 2 of that Schedule, *see figure ref B4-36*. Where there is a reasonable likelihood that:

- An identifiable disease or adverse health effect may be related to the exposure.
- There is a reasonable likelihood that the disease or effect may occur under the particular conditions of his work.
- There are valid techniques for detecting indications of the disease or effect.
- The technique of investigation is of low risk to the employee.

Column 1 - Substances for which health surveillance is appropriate		Column 2 - Processes
Vinyl Chloride Monomer (VCM).		In manufacturing, production, reclamation, storage, discharge, transport, use or polymerization.
Nitro or amino derivatives of phenol and of benzene or its homologues.		In the manufacture of nitro or amino derivatives of phenol and of benzene or its homologues and the making of explosives with the use of any of these substances.
1-Napthylamine and its salts. Orthotolidine and its salts.	Dianisidine and its salts. Dichlorbenzidene and its salts.	In manufacture, formation or use of these substances.
Auramine.	Magenta.	In manufacture.
Carbon Disulphide. Disulpher Dichloride. Benzene, including benzol.	Carbon Tetrachloride. Trichloroethylene.	Process in which these substances are used, or given off as a vapour, in the manufacture of indiarubber or of articles or goods made wholly or partially of indiarubber.
Pitch.		In manufacture of blocks of fuel consisting of coal, coal dust, coke or slurry with pitch as a binding substance.

Figure B4-36: Substances requiring medical surveillance.   *Source: HSE, COSHH ACOP.*

Medical examinations related to substances listed in Schedule 6 must be carried out at intervals of not greater than 12 months, or a shorter interval specified by a doctor. Where a doctor specifies in a health record following, medical surveillance, work restrictions to minimise exposure of the person to the substances, the employer must follow these restrictions.

CONTROL OF LEAD AT WORK REGULATIONS

Regulation 10, of the Control of Lead at Work Regulations (CLAW) 2002 places a duty on employers to ensure that employees who may be exposed to lead are subject to suitable medical surveillance by a medical inspector of the HSE's Employment Medical Advisory Service (EMAS) or a registered medical practitioner (an appointed doctor) where:

"1) a) *The exposure of the employee to lead is, or is liable to be, significant.*

   b) *The blood-lead concentration or urinary lead concentration of the employee is measured and equals or exceeds the levels detailed in paragraph (2).*

   c) *A relevant doctor certifies that the employee should be under such medical surveillance.*

2)   *The levels referred to in paragraph (1)(b) are:*

   a)   *A blood-lead concentration of:*

      i)   *In respect of a woman of reproductive capacity, 20 µg/dl.*

      ii)   *In respect of any other employee, 35 µg/dl.*

   b)   *A urinary lead concentration of:*

      i)   *In respect of a woman of reproductive capacity, 20 µg Pb/g creatinine.*

      ii)   *In respect of any other employee, 40 µg Pb/g creatinine."*

Where reasonably practicable the medical surveillance must be conducted before work exposing them to lead commences or within 14 working days of commencement of such work. Subsequent to this, medical surveillance must be conducted at intervals no greater than 12 months, or shorter if specified by a doctor.

Biological monitoring for most employees must normally be conducted at intervals no greater than 6 months, unless air and blood monitoring indicates low levels of exposure (specified in the regulations) where the interval can be increased to once per year. Biological monitoring of young people or females of reproductive capacity is at intervals no greater than 3 months. The biological monitoring will establish blood-lead and urinary-lead levels in workers. The CLAW 2002 set out blood-lead and urinary action levels. The regulations also specify suspension levels at which employees must not continue work which could expose them to concentrations that could lead to lead poisoning. ***See also - 'Element B4.1 - Workplace exposure limits (WELs)'.***

The objectives of medical surveillance are to:

- Evaluate the effect of lead absorbed by employees and to advise them on their state of health.
- For those where there is continuing exposure to lead, assess the suitability to carry on working.
- Detect early signs of excessive lead absorption or early adverse health effects.
- Identify where there is a need for removal from working with lead.
- Prevent lead poisoning.

CONTROL OF ASBESTOS REGULATIONS

Regulation 22 of the Control of Asbestos Regulations (CAR) 2012 places a duty on employers to ensure that employees who may be exposed to asbestos are subject to regular medical surveillance by a relevant doctor (a registered medical practitioner appointed by the HSE's Employment Medical Advisory Service or an employment medical adviser). Employers are required to ensure that any employees who are exposed to asbestos are under adequate medical surveillance by a relevant doctor to include:

- A medical examination not more than 2 years before the beginning of exposure.
- Periodic medical examinations at intervals of not more than 2 years or a shorter time as the relevant doctor may require, the examination to include a specific examination of the chest.
- Maintain a health record for each employee who is exposed to asbestos.
- Provide the employee and employer with a certificate of examination (kept by the employer for at least 4 years).
- Ensure the health record is kept available, while such exposure continues, for at least 40 years from the date of the last entry or until the employee would have been 80 years old.

From 2015 employers must provide medical examinations every three years for employees engaged in notifiable non-licenced work (NNLW) and keep appropriate records of those employees.

The employer has a duty to ensure suitable facilities are made available to the doctor for the purposes of medical surveillance. The decision to perform a chest X-ray is the appointed doctor's. Due to the long latency period between exposure to asbestos and the potential radiological manifestation of pathology, a chest X-ray taken early during exposure will only reveal limited useful information. For epidemiological research into asbestos related diseases, the appointed doctor will record certain non-medical information about work practices and smoking habits on the medical surveillance form (FODMS75) during the course of each worker's medical examination.

IONISING RADIATION REGULATIONS

Regulation 24 of the Ionising Radiation Regulations (IRR) 1999 places a duty on employers to ensure that employees who may be exposed to ionising radiation are subject to regular medical surveillance by an appointed doctor (a registered medical practitioner appointed by the HSE's Employment Medical Advisory Service) to include:

- Medical surveillance to determine fitness for work with ionising radiation before first being designated as a classified person.
- Periodic reviews of the health of classified persons, at least once every year.
- Special medical surveillance of an employee who is not a classified person when a relevant dose limit has been exceeded.
- Determining whether specific conditions related to work with ionising radiation are necessary.

- A review of health after cessation of work where this is necessary to safeguard the health of the individual.
- Health records are established and kept until the person reaches the age of 75 and in any event 50 years from the date of the last entry.

Regulation 24 states:

"(1) This Regulation shall apply in relation to:

(a) *Classified persons and persons whom an employer intends to designate as classified persons.*

(b) *Employees who have received an overexposure and are not classified persons.*

(c) *Employees who are engaged in work with ionising radiation subject to conditions imposed by an appointed doctor or employment medical adviser under paragraph (6).*

(2) The employer shall ensure that each of his employees to whom this regulation relates is under adequate medical surveillance by an appointed doctor or employment medical adviser for the purpose of determining the fitness of each employee for the work with ionising radiation which he is to carry out."

Legal requirements for keeping health records and medical records

HEALTH RECORDS

A health record is different to a medical record and should not include confidential medical data. The Ionising Radiation Regulations (IRR) 1999 specifies the content of a health record in Schedule 7 to the regulations.

The following particulars shall be contained in a health record made for the purposes of Regulation 24(3):

(a) The employee's:

(i) Full name. (ii) Sex. (iii) Date of birth. (iv) Permanent address, and (v) National insurance number.

(b) The date of the employee's commencement as a classified person in present employment.

(c) The nature of the employee's employment.

(d) In the case of a female employee, a statement as to whether she is likely to receive in any consecutive period of three months an equivalent dose of ionising radiation for the abdomen exceeding 13 mSv.

(e) The date of last medical examination or health review carried out in respect of the employee.

(f) The type of the last medical examination or health review carried out in respect of the employee.

(g) A statement by the appointed doctor or employment medical adviser made as a result of the last medical examination or health review carried out in respect of the employee classifying the employee as fit, fit subject to conditions (which should be specified) or unfit.

(h) In the case of a female employee in respect of whom a statement has been made under paragraph (d) to the effect that she is likely to receive in any consecutive period of three months an equivalent dose of ionising radiation for the abdomen exceeding 13 mSv, a statement by the appointed doctor or employment medical adviser certifying whether in his professional opinion the employee should be subject to the additional dose limit specified in paragraphs 5 and 11 of Schedule 4.

(i) In relation to each medical examination and health review, the name and signature of the appointed doctor or employment medical adviser.

(j) The name and address of the approved dosimetry service with which arrangements have been made for maintaining the dose record in accordance with Regulation 21.

The IRR prescribes that the health record contain facts regarding what type of medical examination was conducted, by whom and when, but limits details to a statement of fitness for work - no medical details are included.

MEDICAL RECORD

Medical records are a confidential record. Clinical information used to make decisions on fitness to work must not be held on the health record. It is the responsibility of the occupational health service provider to keep the clinical data in confidence.

Health and medical records kept to meet a specific legal requirement should typically be kept for a minimum of 40 years because of the long latency period of some illnesses and they also provide useful occupational hygiene data. On cessation of the business, all such records should offered to the Employment Medical Advice Service (EMAS) or the HSE. Health records, if requested, must be made available to the HSE.

CONFIDENTIALITY AND ACCESS

An employee must be allowed access to their health records that are kept are a specific requirement of the Control of Lead at Work Regulations (CLAW) 2002 regulations, providing they give reasonable notice of their intention to the employer.

Generally, medical records may only be shared with others by an employee giving written consent for their details to be released, unless the statutory provision requires that they be made available to a third party, usually the HSE or their EMAS staff.

The Data Protection Act (DPA) 1998 provides a right of access to personal information, including health and medical records, held by public and private organisations, regardless of the form in which it is held - electronic or paper (structured files relating to the individual).

Individuals have the right to know:

- Whether the organisation, or someone else on its behalf, is processing personal information about them.
- What information is being processed, why it is being processed and who it may be disclosed to.

Under the DPA 1998 the people on whom information is held must be given access to their records and they have the right to have any inaccurate information corrected. The contents of individual records can only be disclosed in normal use in line with the purposes for which the information is kept.

Individuals have additional rights of access to their own health records. Under the DPA 1998 they are entitled to see all information relating to their physical or mental health that has been recorded by or on behalf of a 'health professional' in connection with their care. This applies not just too computerised data and structured files, but to 'unstructured' data as well. The right of access covers both NHS and private medical records, and information of any age, however long ago it was recorded.

The health professionals whose records can be seen are doctors, dentists, opticians, pharmacists, nurses, midwives, health visitors, clinical psychologists, child psychotherapists, osteopaths, chiropractors, chiropodists, dieticians, occupational therapists, physiotherapists, radiographers, speech therapists, music and art therapists, orthotists, prosthetists, medical laboratory technicians and scientists who head health service departments.

Access rights are more limited if:

- Information about the individual's health is held by someone who does not fall within the definition of a 'health professional', such as, records held by various kinds of psychotherapists (for example, counsellors) or alternative practitioners.
- Information is held by a health professional that is not, and never has been, responsible for the individual's care - such as a doctor conducting a fitness for work assessment.

In these cases, the individual is entitled to see computerised data and structured files, but not unstructured information.

However, the individual has additional rights to see medical reports supplied for employment or insurance purposes. Under the Access to Medical Reports Act (AMRA) 1988, if a doctor writes a report on the health of an individual for an employer or an insurance company, the individual has the right to see it before it is sent.

Only reports by doctors who are, or have been, involved in the individual's medical care are covered by the AMRA 1988. A report by an independent doctor, who has never treated the individual, and acts solely for an employer or insurer is not subject to the AMRA 1988, and records will only be accessible under the DPA 1998.

An employer or insurer cannot contact the individual's doctor unless they have their written consent and have informed them of their rights under the AMRA 1988. The individual must be invited to say whether they want to see the report before it is sent. If the individual says they would like to see it, the doctor should wait 21 days before sending it, to allow time to arrange to see it.

The individual may request the doctor to amend parts of the report believed to be inaccurate or misleading; if the doctor declines to make amendments the individual can insist on their views being attached to the report.

Biological monitoring

Biological monitoring is concerned with finding out if there is the presence of a hazardous substance (or its metabolites) in the human body or if any damage has been done to the body by hazardous substances.

While atmospheric monitoring is concerned with how much of a hazardous substance there is in the atmosphere that can be inhaled, biological monitoring finds out if the person has inhaled it and if it has done any harm.

There are several methods available: blood samples, urine samples, hair and nail samples, x-rays, lung function tests, breath samples, etc. Not all methods are available for choice, as some substances will only show up in the blood and some only by breath samples. Not all hazardous substances can be biologically monitored.

ADVANTAGES

- It allows for individual susceptibility. The atmosphere may contain a level below the WEL for the substance in question, but certain individuals may still show an uptake.
- It can show damage caused by an uptake from other routes of entry. The atmospheric measurement may be below the WEL, but some substances can go through intact skin. In this case, the person may be inhaling very little, but may be absorbing it through the skin. It will cause damage in the body just the same.
- It can show changes in the body that can be reversed, rather than wait for damage to be obvious, which may be irreversible.
- It can check that the controls used are in fact working.
- It may show an uptake from more than just workplace exposure, which also could be seen as a disadvantage.

DISADVANTAGES

- It can be invasive, such as taking blood samples.
- Some people worry that the samples will be used to find out more than just the presence of a workplace substance, for example, drugs, HIV, HepB, and then the information may be used against them.
- These methods require a medically qualified person to carry them out and analyse the results.
- There are not that many methods to choose from.
- Not all hazardous substances can be measured in the body.
- A problem can show and may result in the employer spending a lot of resources finding the cause, when it is non-workplace exposure that is causing the measurable presence of the substance.

THE ROLE OF BIOLOGICAL MONITORING GUIDANCE VALUES

Biological monitoring can be a very useful complementary technique to air monitoring when air sampling techniques alone may not give a reliable indication of exposure. The technique involves the measurement and assessment of hazardous substances or their metabolites in tissues; secretions excreta or expired air or any combination of these, in exposed workers.

Measurements reflect absorption of a substance by all routes, including skin or gastrointestinal tract uptake following ingestion.

The method is used where there is a reasonably well defined relationship between biological monitoring and the effect of a substance; or where it gives information on accumulated dose and target organ burden, which is related to toxicity.

There are two types of biological monitoring guidance values:

1) **Health Guidance Values** is a health based guidance value that is set at a level at which there is no indication from the available scientific evidence that the substance is likely to be injurious to health. It is set where a clear relationship can be established between biological concentrations and health effects.

2) **Benchmark Guidance Value** is a hygiene-guidance value set at around the 90th percentile of available validated data, provided by a representative cross sectional study of workplaces with good occupational hygiene practices; it can therefore be achieved by the great majority of industries that employ good workplace practice.

Biological monitoring guidance values (BMGVs) are non-statutory and any biological monitoring undertaken in association with a guidance value needs to be conducted on a voluntary basis (with the informed consent of all concerned).

BMGVs are intended to assist employers to ensure adequate control under COSHH. If the values are exceeded the employer should review current control measures since some workers may experience headaches and other symptoms at the benchmark value.

BMGVs are not an alternative to or replacement for airborne occupational exposure limits.

Some examples of BMGVs from EH40 are listed in the following table:

Substance	Biological monitoring guidance values			
	Health guidance values	Sampling time	Benchmark guidance values	Sampling time
Carbon monoxide.	30 ppm in end - tidal breath.	Post shift.		
Mercury.	20 µmol/mol creatinine in urine.	Random.		
Glycerol trinitrate (nitroglycerine).			15 µmol/mol creatinine in urine.	At the end of period of exposure. This may be mid shift or at the end of a shift.
Xylene, o-, m-, p- or mixed isomers.	650 mmol methyl hippuric acid/mol creatinine in urine.	Post shift.		

Figure B4-37: Examples of BMGVs.   Source: HSE, EH40.

STATUTORY BIOLOGICAL MONITORING

Statutory biological monitoring of employees is a requirement of the Control of Lead at Work Regulations (CLAW) 2002, Regulation 10 together with on-going medical surveillance *(see 'B4.1 - Workplace exposure limits (WELs)' earlier in this Element)*. CLAW establishes statutory biological limit values for blood-lead and urinary-lead concentrations. The limits set action limits, for the employer to control exposure and suspension limits, which require that any worker that reaches the suspension limit must be removed from work exposing them to lead.

BLOOD-LEAD CONCENTRATION AND URINARY-LEAD CONCENTRATION ACTION LEVELS AND SUSPENSION LEVELS FROM WORK

See figure ref B4-38 for a table showing the current blood-lead action levels, urinary-lead action levels and suspension levels:

Blood - lead level concentration

Category	Blood action level	Suspension (from work) level
General employees	50 µg/dl	60 µg/dl
Women capable of having children	25 µg/dl	30 µg/dl
Young people under 18 (other than at general employees)	40 µg/dl	50 µg/dl

Urinary - lead level concentration

Category	Urinary action level	Suspension (from work) level
General employees	50 µg/dl	110 µ Pb/g creatinine
Women capable of having children	25 µg/dl	25 µg Pb/g creatinine
Young people under 18 (other than at general employees)	40 µg/dl	No figure quoted

Figure B4-38: Blood-lead action level and urinary-lead action levels and suspension levels.   Source: HSE, CLAW 2002, ACOP.

When investigating why the action level has been breached, the employer's review should include the following:

- Work practices are being followed.
- Check effectiveness of all control measures.

When the initial result relates to a young person or a woman of reproductive capacity, or it reaches or exceeds 60 µg/dl for all other employees, the employer should consult the relevant doctor and consider whether to remove the employee(s) concerned from work (suspension) involving exposure to lead until the result of the repeat test becomes available.

If the result of the repeat test is also equal to or greater than the appropriate suspension level, the doctor should certify that the employee be taken off work which further exposes the employee to lead.

Some employees, excluding women of reproductive capacity, who have worked for many years in the lead industry, may have built up a high body burden of lead, which could take a long time to fall below their suspension level of 60 µg/dl.

These employees are those who either:

- Have been exposed to lead for at least 20 years.
- Are aged 40 years and exposed for at least 10 years.

The CLAW 2002 Regulation 10 makes provision for employees, who were at work with lead before the Control of Lead at Work Regulations 2002 came into force, to continue in work which exposes them to lead as long as their blood-lead concentration is not more than 80µg/dl and the ZPP (zinc proto porphyrin level in blood determination by a portable haemato fluorimeter - a screening device for lead poisoning) level remains lower than 20 µg/g haemoglobin.

For these longer-term employees who continue in work under these circumstances, the employer should nevertheless make every effort to reduce the employee's blood-lead concentration to below 60 µg/dl, and the doctor should consider increasing the frequency of blood-lead and haemoglobin testing should be taken.

Biological agents

On completion of this element, candidates should be able to demonstrate understanding of the content through the application of knowledge to familiar and unfamiliar situations. In particular, they should be able to:

B5.1	Explain the types and properties of biological agents found at work.

B5.2	Explain the assessment and control of risk from deliberate and non-deliberate exposure to biological agents at work.

Content

Relevant statutory provisions

Control of Substances Hazardous to Health Regulations (COSHH) 2002 (as amended)

Control of Substances Hazardous to Health Regulations (Northern Ireland) 2003

Sources of reference

N/A

B5.1 - Types and properties of biological agents

Meaning of 'biological agent'

The definition of biological agents, as set out in the Biological Agents appendix (appendix 2) to the Approved Code of Practice (ACOP) that accompanies the Control of Substances Hazardous to Health Regulations (COSHH) 2002 (as amended), includes the general class of micro-organisms, and also cell cultures and human endoparasites, provided that they have one or more harmful properties that are specified in the definition.

> *"The definition of 'biological agent' includes:*
> *(a) Micro-organisms such as bacteria, viruses, fungi.*
> *(b) The agents that cause transmissible spongiform encephalopathies (TSEs).*
> *(c) Parasites (for example, malarial parasites, amoebae and trypanosomes).*
> *(d) The microscopic infectious forms of larger parasites (for example, the ova and infectious larval forms of helminths).*
> *(e) Cell cultures and human endoparasites, providing they have one or more of the harmful properties in the definition (cause any infection, allergy, toxicity or otherwise create a hazard to human health)."*

Figure B5-1: Definition of biological agents.   *Source: Biological Agents appendix (appendix 2) to the ACOP.*

Biological agents are classified into four hazard groups according to their ability to cause infection, the severity of the disease that may result, the risk that infection will spread to the community, and the availability of vaccines and effective treatment. These infection criteria are the only ones used for classification purposes, even though an infectious biological agent may have toxic, allergenic or other harmful properties, and some biological agents are not infectious at all. Although a non-infectious biological agent falls into group 1 (the lowest group of the four groups), substantial control measures may still be needed for it, depending on the harmful properties it has.

Types and sources of biological agent

Most biological agents are micro-organisms, of which there are five basic groups:

1) Bacteria.
2) Rickettsiae and Chlamydiae.
3) Viruses.
4) Fungi.
5) Protozoa.

The micro-organisms considered in this Element are fungi, bacteria, viruses and some parasites. Included in the term fungi are yeasts and moulds.

Micro-organisms can be:

- Saprophytic - they live freely on decaying matter.
- Parasitic - live in or on a living host (for example, skin flora).
- Commensals - live in harmony with a host.
- Symbiotic - live in harmony for mutual benefit.
- Pathogenic - produce disease in the host.

FUNGI

Fungus: any of a group of unicellular, multi-cellular or multinucleate non-photosynthetic organisms feeding on organic matter, which include yeast, moulds, mushrooms and toadstools. They are simple, parasitic life forms with more than 100,000 different species. Most are either harmless or positively beneficial to health. There are, however, a number of fungi that can cause sometimes fatal disease and illness in humans. For example, the cellulose fibres of cane-sugar after the sugar has been extracted, called bagasse, and used in the manufacture of fibreboard, was found to contain as much as 240 million fungal spores per gram.

Many fungi form minute bodies called spores, which are like seeds. The spores can be carried in air and, if they settle in a suitable location with nutrients available, they will grow.

Moulds: of the large group of fungi, they grow rapidly in moist conditions. Micropolyspora faeni is green-grey, dusty mould which grows on straw and hay. When handled, clouds of dusty spores are produced. The size of the spores is about 1 micron so when they are inhaled they penetrate to the alveoli.

Yeasts: are types of fungi which can cause infections of the skin or mucous membranes. The most important disease causing yeast is Candida Albicans, which causes candidiasis (thrush).

Sources and modes of transmission

Fungi can cause disease in a variety of ways. The fruiting bodies of some soil living fungi contain toxins that can produce direct poisoning if eaten. Some fungi that infect food crops produce toxins that cause a type of food poisoning if eaten, for example, a fungus that grows on peanuts produces aflatoxin. Chronic aflatoxin poisoning is suspected of causing liver cancer.

Inhaled spores of some fungi cause an allergic reaction in the lungs known as allergic alveolitis. One such fungi is a family of moulds called Aspergilla are ubiquitous and can be found in damp, warm, dark places.

Aspergilla mould is often present on hay when it is packed together and stored as winter feed for farm animals. The conditions in the centre of the hay roll are perfect for mould growth. When the hay roll is opened, millions of mould spores are released, which can then be inhaled by the farm worker.

Some fungal spores are responsible for causing asthma and allergic rhinitis (hay fever). Some fungi are able to invade and form colonies in the lungs, in the skin, beneath the skin or in various tissues throughout the body. They can be present on surgical instruments and introduced into the body during surgery, which can be fatal. Dependent on where the fungi or moulds colonise, conditions can occur ranging from mild skin irritation to severe, sometimes fatal, widespread infection and illness.

BACTERIA

These are a group of single-celled micro-organisms, some of which cause disease. They are commonly known as 'germs' and have been recognised as a cause of disease for over a century. Most bacteria are harmless to humans and some are beneficial. The bacteria that cause disease are known as pathogens.

Bacteria were discovered by Antonj van Leewenhoek in the 17th century following the development of the microscope. In the 19th century, the French scientist, Louis Pasteur, established beyond doubt that they were the causes of many diseases.

Pathogenic bacteria are classified into three main groups on the basis of shape: cocci (spherical); bacilli (rod-shaped); and spirochaetes or spirilla (spiral-shaped).

Sources and modes of transmission

Bacteria can be passed from:

- Human to human, for example, E. Coli.
- Animal to human, for example, Leptospira, which is carried by rats and cows.
- Environment to human, for example, Legionella Pneumophila, which may be present in water.

Bacteria that are passed from animal to human and cause disease in the human are called Zoonotic. The range of diseases are called Zoonoses.

Bacteria can enter the body through the lungs if droplets that are breathed, coughed or sneezed out by an infected person are inhaled. They can infect the digestive tract if contaminated food is eaten. Bacteria may be present in food at its primary source, brought to it by flies or from contamination on hands. Bacteria can enter the genito-urinary system. They can also penetrate the skin through various ways: through hair follicles, cuts and abrasions, 'sharps' injuries, bites and through deep wounds.

Bacteria cause disease because they produce toxins that are harmful to human cells. If they are present in sufficient quantity and the affected person is not immune, they will cause disease.

Some bacteria release endotoxins when the bacterial cells die, which cause fever, haemorrhage and shock. They cause a change in the blood capillary cell walls, which allows fluid to seep into the surrounding tissue and blood pressure to drop. This is life threatening and is known as endotoxic shock.

Other bacteria produce exotoxins, where the toxin is released from live bacteria, which account for the damage done in diseases such as diphtheria, tetanus and toxic-shock syndrome.

Some bacteria causes harm in the intestinal region of the body by releasing enterotoxins. These are usually found on food or bodily waste products. E. coli, the cholera bacterium, Salmonella and Staphylococci are bacteria of this type.

Outbreaks of severe sickness and diarrhoea have occurred amongst children who have petted farm animals. The animals probably had traces of animal excrement on their coats and the petting transferred the bacteria to the child's hands, which was later transferred to their mouth.

E. coli is a very fast replicating bacterium and is used in genetic manipulation. The bacterial cells are genetically modified to produce hormones the body needs, such as insulin. Working on this process puts the person at great risk of accidental injection of these fast growing hormone-producing bacteria.

VIRUSES

These are the smallest known types of infectious agent. They are about one half to one hundredth the size of the smallest bacteria. It is debatable whether they are truly living organisms or just collections of large molecules capable of self-replication under very specific, favourable conditions. Their sole activity is to invade the cells of other organisms, which they take over to make copies of themselves. Outside living cells, they are totally inert. They do not carry out activities that are typical of life, such as metabolism.

There are probably more types of virus than the number of types of all other organisms. They parasitise all recognised life forms, and while not all cause disease, many of them do.

Infections caused by viruses range from the common cold to serious diseases such as rabies, and can lead to the development of AIDS, and probably various cancers.

Sources and modes of transmission

Viruses gain entry to the body by all possible routes. They are inhaled in droplets; swallowed in food and liquids; passed through punctured skin on infected needles, in the saliva of feeding insects or rabid dogs; viruses are accepted directly by the mucous membranes of the genital tract and by the conjunctiva of the eye after accidental contamination.

Viruses can pass from:

- Human to human, for example, HIV, Hepatitis B, various strains of influenza and the common cold.
- Animal to human, for example, the rabies virus.
- Environment to human: there are vast numbers of viruses present in the environment, their source not always known.

Many viruses invade cells and multiply near their site of entry. Some enter the lymphatic vessels and spread to the lymph nodes. Some, such as the HIV virus, invade the lymphocytes (a type of white cell). Many pass from the lymphatics to the blood and are quickly carried to every part of the body. They may invade specific target organs, such as the liver, lungs, brain, and start to multiply. Some viruses travel along the nerve fibres to their target organs.

Viruses cause disease in a variety of ways:

- They can destroy or disrupt the cells they invade, causing serious illness if it is a vital organ.
- The response of the body's immune system may lead to fever and fatigue.
- Antibodies produced by the immune system may attach to viral particles and be circulated in the blood. The antibodies may then be deposited in various parts of the body causing inflammation and tissue damage.
- They may interact with the chromosomes of the host cells causing cancer.
- They may weaken the activity of the t-lymphocytes and so interfere with the immune system. The body's normal defences against a whole range of infections can be lost. This is how HIV works.

PROTOZOA

Protozoans are living, motile, single-celled organisms that can live inside or outside host cells or organisms. They can only divide within a host organism. Most antibiotics, such as penicillin, don't work on protozoan diseases.

One of the basic requirements of all protozoans is the presence of water, but within this limitation, they may live in the sea, in rivers, lakes, stagnant ponds of freshwater, soil, and in some decaying matters. Many protozoa cause diseases in animals and humans. Some, like Plasmodium, which causes malaria, can be devastating to people worldwide. This parasite is transferred from a malarial patient to a healthy person by the bite of female mosquitoes.

While malaria is one of the best known diseases known to be caused by protozoans, a wide range of other equally devastating ailments are also caused by protozoan infections, for example, Amoebic dysentery and African sleeping sickness which is spread by the bite of the tsetse fly.

Special properties of biological agents

RAPID MUTATION

The body's immune system protects a person against most biological agents. The white blood cells in the immune system recognise the biological agent as a pathogen, if exposed to it more than once. They are then able to deal with it and prevent it infecting the body. If human beings did not have this well-developed immune system, we would not have survived as a species. However, the sheer number of biological agents, for example bacteria or viruses, may overwhelm a person's immune system. The person may also have a weakened immune system due to illness, age, medical treatment or drug use, which leaves them unable to sufficiently respond to the biological agents. The biological agent may be new to the person and they have therefore not built up immunity.

In many cases the biological agent has mutated since the initial exposure and the immune system does not recognise it. This happens each year when a new strain of influenza develops through mutation. It is difficult to build up immunity to a virus that can rapidly mutate. HIV is proving difficult to deal with because each time the researchers come close to developing a vaccine the virus mutates and is able to bypass the immune system. Bacteria can also rapidly mutate. It is thought that the bacterium discovered to have caused Legionnaires' disease was not a new bacterium as such, but a mutation of one that had already been cited as a pathogen.

INCUBATION PERIOD

Incubation is the time elapsed from exposure to an infectious biological agent and the appearance of the first signs and symptoms. This period may vary considerably, for example, the incubation period for influenza is 1-2 days and for chickenpox it is 14-16 days. The infectious biological agent multiplies and spreads through the body during this period. The person's blood would show the presence of high levels of white blood cells during this time, as the immune system battles against the infection.

INFECTIOUS

Some biological agents cause illnesses that are known as infections, because they can be spread from person to person, i.e. they are infectious or contagious. The biological agent can be passed in the air, by one person coughing and sneezing and another person inhales it. Actual contact between persons may not be necessary, but some infections require contact with the person or infected clothing, for example, bacterial skin infections such as impetigo or scabies from mites. The infectious nature of biological agents can vary from not infectious to extremely infectious/contagious. For example, you can only get Legionnaires' disease from inhaling the Legionella bacterium and not from an infected person, whereas many viral infections, such as various types of Influenza, are extremely infectious/contagious.

RAPID MULTIPLICATION

Biological agents are different from other hazardous agents in that they are capable, in the right conditions, of multiplying; billions can grow from one in a few hours. Biological agents are said to be 'living' organisms. Moulds and fungi, and to a certain extent bacteria, fulfil the criteria of living organisms.

Viruses are a little more difficult to match to the criteria, but once inside a human cell, for example, they are capable of rapid multiplication. This means that a small amount of a biological agent can cause severe disease, which may be fatal in some cases. There can be no safe exposure limit.

Fungi are present all the time in the body and are prevented from multiplying through competition from bacteria. Fungal infections become a problem when people are taking antibiotic drugs, which destroy the bacterial competition and allow the fungi to rapidly multiply. For example, Candida Albicans can cause thrush following a course of antibiotics.

Zoonotic diseases

ANIMAL INFLUENZA

Influenza is a respiratory disease common to man and a limited number of including ducks, chickens, pigs, whales, horses and seals. There are three types of influenza viruses A, B and C:

- Types B and C are human viruses mainly affecting young children and causing a mild disease.
- Type A virus is the important type as far as cross-species infections are concerned.

Influenza type A is distributed worldwide and usually causes a mild respiratory disease in humans and animals. Human influenza epidemics due to new epidemic strains occur at regular intervals of 2 to 3 years and affect mainly elderly people. However, influenza is a potentially devastating disease in both humans and animals thereby very important for both human and veterinary medicine.

The World Health Organization (WHO, 2013) defines Influenza as a: 'Disease common to man and a limited number of lower animal species mainly horses, pigs, domestic and wild birds, wild aquatic mammals such as seals and whales, minks and farmed carnivores'.

Some animals show signs of upper respiratory infection and others may not show any illness but be a source of the virus. Prevention requires hygienic management practices and sick animals should be isolated and healthy animals vaccinated.

Pandemics are major epidemics characterized by the rapid spread of a novel type of virus to all areas of the world resulting in an unusually high number of illnesses and deaths in humans in most age groups.

Wild birds represent the primary natural reservoir for all subtypes of influenza A viruses and are considered as a source of influenza A viruses in all other animals. In addition to swine influenza viruses, pigs can be infected with both human and avian influenza viruses.

CRYPTOSPORIDIOSIS

A zoonotic disease caused by a parasite (protozoa). *Cryptosporidium* is now recognised as an important opportunistic infection, especially in immuno-compromised hosts. Members of the genus cryptosporidium are parasites of the intestinal tracts of fishes, reptiles, birds, and mammals.

Cryptosporidium isolated from humans is now referred to as *C. parvum*. Cryptosporidium infections have been reported from a variety of wild and domesticated animals, and in the last six or seven years literally hundreds of human infections have been reported, including epidemics in several major urban areas in the United States.

Cryptosporidium is a small parasite, measuring about 3-5 μm. It lives on (or just under) the surface of the cells lining the small intestine, reproduces asexually, and oocysts are passed in the faeces.

Transmission of the infection occurs via the oocysts. Many human infections have been traced to the contamination of drinking water with oocysts from agricultural 'run-off' (i.e., drainage of faeces from pastures), so it is considered zoonotic. Cryptosporidium survives for a long time in the environment and is resistant to chlorine and other disinfectants used to treat water, but are effectively removed by other physical treatment barriers.

The occupational context of exposure to this parasite includes water treatment workers, farm workers, workers involved with animals for leisure (zoos, equestrian, and show farms) and those involved with land management.

In most patients infected with **cryptosporidiosis** the infection causes a short term, mild diarrhoea, and often individuals may not seek medical treatment, and the infection may subside on its own. In persons with compromised immune systems, this parasite can cause pronounced chronic diarrhoea; in severe cases the infected individual may produce up to 15 litres/day of stools, and this may go on for weeks or months. Whilst, such an infection may not be fatal, it can exacerbate other infections common in immuno-compromised hosts.

MALARIA

Malaria is a mosquito-borne infectious disease of humans and other animals caused by protists (a type of microorganism) of the genus Plasmodium. The protists first infect the liver then act as parasites within red blood cells, causing symptoms that typically include fever and headache, in severe cases progressing to coma or death. The disease is widespread in tropical and subtropical regions in a broad band around the equator, including much of Sub-Saharan Africa, Asia, and the Americas.

Malaria is prevalent in tropical regions because the significant amounts of rainfall, consistently high temperatures and high humidity, along with stagnant waters, in which mosquito larvae readily mature, provide them with the environment they need for continuous breeding. Disease transmission can be reduced by preventing mosquito bites, by distribution of mosquito nets and insect repellents, or with mosquito-control measures, such as spraying insecticides and draining standing water.

PSITTACOSIS

This is one of the diseases called Avian Chlamydiosis and is a type of allergy similar to Farmer's Lung. The disease results from infection by chlamydophila psittaci bacteria. It occurs in those who work with turkeys, ducks, chickens, pigeons and exotic birds. It is caused by the inhalation of dried bird droppings or feathers. The health effects are respiratory problems and a type of pneumonia. Its common name is Pigeon Fancier's Lung as it was found amongst those who kept racing pigeons. Scraping the bird droppings to remove them during cleaning made the dust airborne and liable to be inhaled.

Those at risk are poultry workers, pet shop owners, those who remove bird droppings from buildings and other contaminated areas, vets, etc.

Controls include good ventilation, damping the area to prevent the dust becoming airborne and respiratory protective equipment.

Biological agents

BLOOD-BORNE VIRUSES: HEPATITIS

Hepatitis

A disease caused by a virus. Hepatitis is inflammation of the liver and is passed from human to human. There are five main hepatitis viruses, referred to as types A, B, C, D and E and the B variety being the most serious.

1) **Hepatitis A virus (HAV)** is present in the faeces of infected persons and is most often transmitted through consumption of contaminated water or food. Certain sex practices can also spread HAV. Infections are in many cases mild, with most people making a full recovery and remaining immune from further HAV infections. However, HAV infections can also be severe and life threatening. Most people in areas of the world with poor sanitation have been infected with this virus. Safe and effective vaccines are available to prevent HAV.

2) **Hepatitis B virus (HBV)** is transmitted through exposure to infective blood, semen, and other body fluids. HBV can be transmitted from infected mothers to infants at the time of birth or from family member to infant in early childhood. Transmission may also occur through transfusions of HBV-contaminated blood and blood products, contaminated injections during medical procedures, and through injection drug use. HBV also poses a risk to healthcare workers who sustain accidental needle stick injuries while caring for infected-HBV patients. Safe and effective vaccines are available to prevent HBV.

3) **Hepatitis C virus (HCV)** is mostly transmitted through exposure to infective blood. This may happen through transfusions of HCV-contaminated blood and blood products, contaminated injections during medical procedures, and through injection drug use. Sexual transmission is also possible, but is much less common. There is no vaccine for HCV.

4) **Hepatitis D virus (HDV)** infections occur only in those who are infected with HBV. The dual infection of HDV and HBV can result in a more serious disease and worse outcome. Hepatitis B vaccines provide protection from HDV infection.

5) **Hepatitis E virus (HEV)** is mostly transmitted through consumption of contaminated water or food. HEV is a common cause of hepatitis outbreaks in developing parts of the world and is increasingly recognized as an important cause of disease in developed countries. Safe and effective vaccines to prevent HEV infection have been developed but are not widely available.

BLOOD BORNE VIRUSES: HIV/AIDS

A viral infection. HIV (Human Immuno-deficiency Virus) is a virus which attacks the body's immune defence mechanism. The virus is not recognised by the immune system because its structure mimics a blood sugar which is not recognised as invasive. It is a delicate virus that does not live for very long outside the body. It is unlikely to survive, for example in old discarded needles. Contamination is more likely from direct contact with blood or other bodily fluids, or from blood samples freshly taken.

HIV is the infection which, through progressive destruction of specific immune cells (CD4 cells) leads to AIDS. Opportunistic infections, specific malignancies, HIV wasting or HIV encephalopthy are part of a complex case definition which comprise the Acquired Immuno Deficiency Syndrome. HIV is a sexually transmitted and bloodborne virus (BBV). This means it can be transmitted by unprotected sexual intercourse or by routes similar to other BBVs, i.e. shared needle use by injecting drug users, or mother to child transmission before, during or after (via breast milk form an infected mother to her child) the birth of the child. In countries that can afford anti-retroviral therapies the progression to AIDS is not inevitable and the patterns of survival have been fundamentally changed by these drugs. However, there is currently no cure and patients need to continue on therapy. There is currently no effective vaccine although much research is going on to seek to develop one.

Occupational transmission could be needlestick injuries during the taking of blood samples and other sharps injuries when dealing with the samples. Dentists, doctors, surgeons, nurses, other care workers and professions involved with drug users such as prison warders, police, social workers, are just some of the occupations at risk.

As mentioned previously, Acquired Immune Deficiency Syndrome (AIDS) is caused by the HIV, which attacks the body's immune defence mechanism. The virus can be found in most infected body fluids and is passed from human to human. It is delicate and easy to kill by heat and chemicals. It lives for only a short time outside its host.

Some people who contract HIV experience very strong symptoms, but others experience none at all. Those who do have symptoms generally experience fever, fatigue, and, often, rash. Other common symptoms can include headache, swollen lymph nodes, and sore throat.

These symptoms can occur within days or weeks of the initial exposure to the virus during a period called *primary or acute HIV infection.* Because of the non-specific symptoms associated with primary or acute HIV infection, symptoms are not a reliable way to diagnose HIV infection.

Testing for HIV antibodies is the only way to identify infection. The HIV antibody test only works after the infected person's immune system develops antibodies to HIV.

LEGIONELLA

An infection caused by a bacterium found in water. This disease is a type of pneumonia caused by Legionella pneumophila (of which there are over 20 species and 120 serogroups).

It is an occupational disease for those who maintain water systems, though not for the passer-by infected from a cooling tower. The organism is ubiquitous in water and frequently present in water cooling systems and domestic hot water systems. Large workplace buildings are therefore susceptible to infected water systems, especially hotels, hospitals and leisure centres. It is a particular hazard in hospitals, where there are many people with naturally reduced defences.

The organism is widespread in the environment, and is readily found in rivers, lakes, streams, ponds and soil. It needs certain conditions to multiply, for example, presence of sludge, scale, algae, rust and organic material plus a temperature of 20-50°C.

Victims become infected by inhaling the organism in contaminated water aerosols. Smoking, age and alcohol may increase susceptibility. There is no risk of person-to-person spread. Incubation of the infection is typically 2-10 days after exposure.

The symptoms are:

- Aching muscles, headaches, fever followed by cough.
- Confusion, emotional disturbance and delirium may follow the acute phase.
- Fatality rate about 12% in the UK.

There are no distinguishing clinical or radiological features. Diagnosis is based on identifying specific antibodies.

The greatest risks are from:

- Showers.
- Air conditioning sprays.
- Water cooling towers.
- Recirculating water cooling systems.

Workers involved with the maintenance of any of the previous points are at risk, as are those who work in buildings where showers and air conditioning are provide. Additionally, the public may be at risk from contaminated water from cooling towers or showers in places like leisure centres. A number of industrial processes create or use water in droplet form, these processes can also put workers at risk.

Control is established by the careful commissioning, adequate maintenance, routine cleaning of systems and use of biocides are requirements for all installations *(see also – HSE 'Control of Legionella Bacteria in Water Systems ACOP L8).*

The bacterium can be controlled by:

- Proper design of water systems (being aware of features that may contribute to an outbreak).
- Chlorination of water to concentration 2-4 ppm.
- Heating water to 55-60°C and above.

LEPTOSPIRA

A zoonotic disease caused by a bacterium. Caused by bacteria Leptospira, spiral shaped bacteria of which there are various species (serotypes). Two types of leptospirosis infection are the main ones that can affect workers in the UK:

1) Weil's disease - this transmitted to humans by contact with urine from infected rats.
2) The Hardjo form of leptospirosis - this is transmitted from cattle to humans.

Survival of the bacterium is encouraged by warm surroundings; therefore cases in the UK are mainly observed during the summer. Its survival depends on protection from direct sunlight, so it survives well in water courses and ditches protected by vegetation. It can survive in soil for 15 days or more. However, its survival is poor in badly polluted water or water with pH below 7 and does not survive in salt water, which kills it.

Incubation of the infection is 4-19 days, usually around 10 days, after exposure. Symptoms vary according to serotype but start with flu-like illness with a persistent and severe headache, which can lead to vomiting and muscle pains and ultimately to jaundice, meningitis and kidney failure. Mortality is low, but the age of the person increases risk, the young and older people are most vulnerable. Illness usually lasts between a few days and 3 weeks. Following illness, immunity lasts for years. Rodents represent the most important source of infections, especially rats (also gerbils, voles, field mice). Other sources of infection are dogs, hedgehogs, foxes, pigs, cattle. These animals are not necessarily ill, but carry leptospires in kidneys and excrete it in urine. Man is not a natural host. Transmission from person to person is rare (though possible).

Infection can be transmitted:

- Directly via direct contact with blood, tissues, organs or urine.
- Indirectly by contaminated environment, for example, contaminated water, such as in sewers, ditches, ponds and slow-flowing rivers.

Infection enters through:

- Broken skin or mucous membrane.
- By inoculation, for example, animal bites or accidental laboratory infection.
- By handling infected animals.
- By inhalation of aerosolised leptospires, for example, following urination of a cow in a dairy.

Workers in the following occupations are most at risk:

- Water and sewage work.
- Workers involved in maintenance of canals and rivers.
- Farming, for example, rat contamination of feed stuffs and infestation of farm buildings during winter, inhalation of leptospira when cows urinate when being milked - there is probably a large under-diagnosis in this group where illness is either less severe (so the patient does not call the doctor) or fatal (but not recognised as the cause of death).
- Water sports and leisure activities - activities are more likely to take place in the summer when water is more contaminated by the bacterium.

This disease can be controlled by:

- Effective and continued rodent control.
- Wearing protective clothing for example, rat-catcher, sewer worker.
- Sensible, basic health and hygiene precautions, for example, protection of cuts and removal of contamination by washing with soap and water.
- Prompt reporting of suspected infection.
- Training to ensure GPs are advised of a person's occupation or possible exposure, to assist early diagnosis.
- Carrying Leptospirosis information card.

NOROVIRUS

This virus is named after a town in Ohio (Norwalk) where there was an outbreak in 1968. Norovirus causes a form of gastroenteritis, the symptoms being sickness, diarrhoea, stomach pains, raised temperature and tiredness. The incubation period is 12-72 hours and the illness is of short duration, with recovery normally within 2 days. It is highly contagious and the main effect of the virus is found in institutional settings such as hospitals, schools and prisons. It has caused many wards to be closed in hospitals, due to the high infection rate, as many patients will be more vulnerable due to their reduced ability to fight infections. Good personal hygiene and deep cleaning is required to prevent and control norovirus as it is transmitted by the faecal - oral route (for example, using the toilet and not washing the hands properly). Where institutions have been closed, for example, hospital wards, schools, there is likely to be high absence levels amongst health care and teaching staff that may also contract the illness, typically through hand to infected hand or surfaces.

B5.2 - Assessment and control of risk from exposure to deliberate and non-deliberate biological agents

Distinction between deliberate vs non-deliberate infection

DELIBERATE WORK

Where the work intentionally involves biological agents, for example, with the development of vaccines or purposely infecting animals for research, the employer must ensure an assessment of the risks to health is carried out. This will involve knowing the group that the agent belongs to and ensuring the controls are appropriate to that group. If the agent is modified, so it is more hazardous than the named agent on the Approved List, it should be classified as though it were in a higher group. It should be easier to control biological agents in intentional work, as they are obviously known to be present.

NON-DELIBERATE INFECTION

Infection may develop from work, where the biological agent is present by chance or is present because that is the environment in which it exists. For example, a worker does not have to work with the Leptospira bacterium to become infected. It may be present in the working environment because that is the habitat of the rats that carry the bacteria. The possible presence of biological agents that can cause infection must be part of the risk assessment, for example, exposure to Legionella contaminated airborne water droplets. Opportunist infection may occur in vulnerable people, i.e. people who are taking immuno-suppressant drugs. Where the immune system is suppressed by drugs such as steroids, the person cannot fight off the bacteria and viruses that attack.

Also, antibiotics taken for certain infections will kill off beneficial bacteria and allow other pathogens to attack. For example, following a course of antibiotics, the depletion of beneficial bacteria will allow the fungus Candida albicans to attack and cause infection. Candida albicans is naturally occurring but beneficial bacteria inhibit its development. These are issues that need to be considered in the biological agents risk assessment for completeness.

Purpose of approved list of biological agents

Biological agents are classified in an Approved List in the document 'Advisory Committee on Dangerous Pathogens, The Approved List of Biological Agents'. The Control of Substances Hazardous to Health Regulations (COSHH) 2002 (as amended) refers to this list, which makes any requirements relating to it legally binding in much the same way as references in COSHH make requirements with the list of WELs in EH40 legally binding. The Approved List of Biological Agents implements the European Community classification of biological agents set out in European Community Directive 2000/54/EC.

The Approved List of Biological Agents should be used in conjunction with COSHH 2002, in particular COSHH 2002 schedule 3 - 'Additional provisions relating to work with biological agents'.

Biological agents are classified in the Approved List of Biological Agents according to their ability to cause disease by infection. The classification comprises Groups 1, 2, 3 and 4 *(see the definitions of the groups under 'Factors to be taken into account in risk assessment')*. Group 1 biological agents are unlikely to cause human disease and therefore are not listed. Any agent not listed should not be implicitly classified as Group 1; further information should be sought.

In the Approved List of Biological Agents bacteria, viruses, parasites and fungi are listed in sections in that order with the classification Group 2, 3 or 4 alongside. This classification is necessary for health risk assessments under COSHH 2002. It aids the decision on the level of risk that may be present and the control measures necessary.

The Approved List of Biological Agents gives further information on the listed biological agents according to their capability of causing allergic or toxic reactions or whether there is an effective vaccine available.

The list is under constant review and with further evidence biological agents may be placed in different groups, removed from the list altogether or new agents added.

Reportable diseases caused by biological agents

Any disease attributed to an occupational exposure to a biological agent must be reported under the Reporting of Injuries, Diseases and Dangerous Occurrences Regulations (RIDDOR) 2013 Regulation 9. Shown in *figure ref B5-2* are some of the diseases that were listed in Schedule 3 of RIDDOR 1995:

Infections due to biological agents	
Infection	**Work activity**
Anthrax	(a) Work involving handling infected animals, their products or packaging containing infected material.
	(b) Work on infected sites.

Brucellosis	Work involving contact with: (a) Animals or their carcasses (including any parts thereof) infected by brucella or the untreated products of same. (b) Laboratory specimens or vaccines of or containing brucella.
Avian chlamydiosis	Work involving contact with birds infected with chlamydia psittaci, or the remains or untreated products of such birds.
Ovine chlamydiosis	Work involving contact with sheep infected with chlamydia psittaci or the remains or untreated products of such sheep.
Hepatitis	Work involving contact with: (a) Human blood or human blood products. (b) Any source of viral hepatitis.
Legionellosis	Work on or near cooling systems which are located in the workplace and use water; or work on hot water service systems located in the workplace which are likely to be a source of contamination.
Infection	*Work activity*
Leptospirosis	(a) Work in places which are or are liable to be infested by rats, field mice, voles or other small mammals. (b) Work at dog kennels or involving the care or handling of dogs. (c) Work involving contact with bovine animals or their meat products or pigs or their meat products.
Lyme disease	Work involving exposure to ticks (including in particular work by forestry workers, rangers, dairy farmers, game keepers and other persons engaged in countryside management).
Q fever	Work involving contact with animals, their remains or their untreated products.
Rabies	Work involving handling or contact with infected animals.
Streptococcus suis	Work involving contact with pigs infected with streptococcus suis, or with the carcasses, products or residues of pigs so affected.
Tetanus	Work involving contact with soil likely to be contaminated by animals.
Tuberculosis	Work with persons, animals, human or animal remains or any other material which might be a source of infection.
Any infection reliably attributable to the performance of the work specified in the entry opposite hereto	Work with micro-organisms; work with live or dead human beings in the course of providing any treatment or service or in conducting any investigation involving exposure to blood or body fluids; work with animals or any potentially infected material derived from any of the above.

Figure B5-2: Examples of reportable diseases.   *Source: RIDDOR 1995.*

There should be a procedure for internal reporting to take account of any accidental exposure or possible exposure to a biological agent. The person should not wait until there are symptoms of infection before making it known that they may have been exposed to a biological agent.

However, it may be that symptoms are the first sign that there could have been an exposure, for example, a person having worked in a sewer develops flu-like symptoms of the type of someone suffering from Weil's disease. Exposure may be obvious if the work with biological agents is intentional, for example, research projects; but a little more difficult if the person does not know if a biological agent is present, for example, it is not obvious working in a warehouse that Leptospira is present. This is where risk assessments and the communication of their findings are extremely valuable.

Internally reported possible exposures could include circumstances that could introduce a virus (Hepatitis B) or a bacterium (Leptospira) directly into the blood:

■ Needle-stick injuries.
■ Any 'sharps' injuries, such as broken glass, rusty nails.
■ Animal or human bites and scratches.

COSHH 2002, schedule 3 requires the employer to keep a list of employees exposed to biological agents in Groups 3 and 4 for at least 40 years after the last exposure. This also applies to one Group 2 agent: Human herpes virus type 8.

Factors to take into account in risk assessment

Factors to take into account in risk assessment, are outlined in the following points:

- Hazard category.
- The criteria for categorisation.
- The pathogenicity of the agent and infectious dose.
- The activities and people at risk.
- The likelihood and nature of resultant disease.
- The modes of transmission with examples.
- The stability of the agent in the environment.
- The concentration and amounts.
- The presence of a suitable host (human or animal).
- Data available.
- The nature of activity.
- The local availability of prophylaxis/treatment.

HAZARD CATEGORY

Biological agents are classified according to the level of risk of infection:

Group 1 - unlikely to cause human disease.

Group 2 - can cause human disease and may be a hazard to employees; it is unlikely to spread to the community and there is usually effective prophylaxis or treatment available.

Group 3 - can cause severe human disease and may be a serious hazard to employees; it may spread to the community, but there is usually effective prophylaxis or treatment available.

Group 4 - causes severe human disease and is a serious hazard to employees; it is likely to spread to the community and there is usually no effective prophylaxis or treatment available.

Risk assessment must be carried out according to the general requirement of Regulation 6 of the COSHH Regulations 2002. If the work to be carried out exposes employees to any biological agent, the employer must take into account the group into which that agent is classified. *See also - Relevant statutory provisions - COSHH 2002, Schedule 3 (Regulation 7) Biological agents.*

An assessment should include consideration of:

- The biological agents that may be present.
- What hazard groups they belong to.
- What form they are in, for example, may form spores or cysts that are resistant to disinfection, or in the development cycle a form may be dependent on an intermediate host.
- The diseases they may cause.
- How and where they are present and how they are transmitted.
- The likelihood of exposure and consequent disease (including the identification of susceptible workers, for example, Immuno-compromised).
- Whether there can be substitution by a less hazardous agent.
- Control measures and minimisation of numbers exposed.
- Monitoring procedures.
- Health surveillance.

The selection of control measures for biological agents should take into account that there are no exposure limits for them. Exposure may have to be reduced to levels that are at the limit of detection.

If exposure cannot be prevented, then it should be controlled by the following measures, which can be applied according to the results of the assessment:

- Keep as low as possible the number of people exposed or likely to be exposed.
- Design work processes and engineering controls to prevent or minimise the release of biological agents into the workplace.
- Display the biohazard sign and any other relevant warning signs.
- Draw up plans to deal with accidents involving biological agents.
- Specify appropriate decontamination and disinfection procedures.
- Arrange the means for safe collection, storage and disposal of contaminated waste, including safe and identifiable containers, after suitable treatment where necessary.
- Arrange for the safe movement of biological agents within the workplace.
- State procedures for taking, handling and processing samples.
- Provide collective protection where possible and individual protection otherwise, for example, PPE.
- Provide vaccines for those not immune.
- Provide washing and toilet facilities.
- Prohibit eating, drinking, smoking and the application of cosmetics in the workplace.

There are special control measures for health and veterinary care facilities and for laboratories, animal rooms and industrial processes. *(See the end of this Element for an extract taken from General COSHH*

ACOP/Carcinogens ACOP/Biological Agents ACOP - Control of Substances Hazardous to Health (COSHH) 2002).

Controls for intentional work should be chosen from these lists, with the minimum containment level being: level 2 for handling Group 2 biological agents, level 3 for Group 3 and level 4 for Group 4.

Level 2 is the minimum for laboratories which do not intentionally work with biological agents, but handle materials where there exist uncertainties about the presence of a Group 2, 3, or 4 biological agents. Level 3 or 4, where appropriate, should be used in unintentional work if the employer knows or suspects such a containment level is necessary.

THE CRITERIA FOR CATEGORISATION

Biological agents are classified according to the level of risk of infection:

Group 1 - unlikely to cause human disease.

Group 2 - can cause human disease and may be a hazard to workers; it is unlikely to spread to the community and there is usually an effective prophylaxis (vaccination) or treatment available.

Group 3 - can cause severe human disease and may be a serious hazard to workers; it may spread to the community, but there is usually effective prophylaxis or treatment available.

Group 4 - causes severe human disease and is a serious hazard to workers; it is likely to spread to the community and there is usually no effective prophylaxis or treatment available.

The World Health Organization (WHO) classifies the biological agents in a country by risk group based on pathogenicity of the organism, modes of transmission and host range of the organism. These may be influenced by existing levels of immunity, density and movement of host population presence of appropriate vectors and standards of environmental hygiene.

Availability of effective preventive measures

An assessment should include consideration of:

- The biological agents that may be present.
- What hazard groups they belong to.
- What form they are in, for example, the biological agent may form spores or cysts that are resistant to disinfection. In the development cycle the form may be dependent on an intermediate host.
- The diseases they may cause.
- How and where they are present and how they are transmitted.
- The likelihood of exposure and consequent disease (including the identification of susceptible workers, for example, immuno-compromised).
- Whether there can be substitution by a less hazardous agent.
- Control measures and minimisation of numbers exposed.
- Monitoring procedures.
- Health surveillance.

The selection of control measures for biological agents should take into account that there are no exposure limits for them.

Exposure may have to be reduced to levels that are at the limit of detection.

If exposure cannot be prevented, then it should be controlled by the following measures, which can be applied according to the results of the assessment:

- Keep as low as possible the number of people exposed or likely to be exposed.
- Design work processes and engineering controls to prevent or minimise the release of biological agents into the workplace.
- Display the biohazard sign and any other relevant warning signs.
- Draw up plans to deal with accidents involving biological agents.
- Specify appropriate decontamination and disinfection procedures.
- Arrange the means for safe collection, storage and disposal of contaminated waste, including safe and identifiable containers, after suitable treatment where necessary.
- Arrange for the safe movement of biological agents within the workplace.
- State procedures for taking, handling and processing samples.
- Provide collective protection where possible and individual protection otherwise, for example, PPE.
- Provide vaccines for those not immune.
- Provide washing and toilet facilities.
- Prohibit eating, drinking, smoking and the application of cosmetics in the workplace.

The biohazard symbol, which is used internationally, was developed in 1966 by Charles Baldwin, an environmental-health engineer working for the Dow Chemical Company on containment systems. He noticed a lot of different warning symbols in use and along with the Dow marketing people developed the current symbol. There are special control measures for health and veterinary care facilities and for laboratories, animal rooms and industrial processes.

Figure B5-3: International symbol. *Source: Rivington Designs.*

Figure B5-4: UK symbol. *Source: Rivington Designs.*

See the end of this Element for an extract taken from General COSHH ACOP/Carcinogens ACOP/Biological Agents ACOP - Control of Substances Hazardous to Health (COSHH) 2002.

Availability of effective treatment

Effective treatment includes passive immunisation and post-exposure vaccination, antibiotics, and chemotherapeutic agents, taking into consideration the possibility of the emergence of resistant strains. It is important to take prevailing conditions in the geographical area in which the micro-organisms are handled into account.

PATHOGENICITY OF THE AGENT AND INFECTIOUS DOSE

Some biological agents cause illnesses that are known as infections, because they can be spread from person to person, i.e. they are infectious or contagious. The biological agent can be passed in the air, by one person coughing and sneezing and another person inhaling it. Actual contact between persons may not be necessary, but some infections do require contact with the person or infected clothing, for example, bacterial skin infections such as impetigo or scabies from mites. The infectious nature of biological agents can vary from not infectious to extremely infectious/contagious. For example, you can only get Legionnaires' disease from inhaling the Legionella bacterium and not from an infected person, whereas many viral infections, such as various types of Influenza, are extremely infectious/contagious.

The infectious dose is the quantity of a pathogen (measured in number of organisms) that is necessary to cause infection in a susceptible host.

ACTIVITIES AND PEOPLE AT RISK

Occupational groups at risk of illness through biological agents are many and varied:

- Health care workers, dentists, laboratory workers, custodial and emergency service workers, police, fire ambulance, paramedics, street cleaners who may come into contact with contaminated needles and other sharps that could be contaminated with the Hepatitis B virus and/or HIV.
- Farm workers and others who may be involved with feeding animals may come into contact with aspergilla.
- Anyone who handles animal hides may come into contact with anthracis, i.e. dock workers, leather tanning workers, carpet makers.
- Workers in contact with cooling towers, showers or other areas where water may form an aerosol may inhale droplets contaminated with Legionella.
- Anyone who works with animals is at risk from a zoonotic disease: vets, veterinary nurses, abattoir workers, grooms.
- Laboratory workers who work with any of the biological agents.
- Workers who come into contact with rats or areas in which rats have been: sewer workers, warehouse workers, farmers, operatives in the food industries, and cleaners who may come into contact with Leptospira.
- Farm workers who drink unpasteurised milk may be at risk from Brucella.

LIKELIHOOD AND NATURE OF RESULTANT DISEASES

Because biological agents can multiply and infect, there is not a standard dose-response relationship. This means that the severity of any subsequent disease is not easily predicted. It does not depend on the level of exposure. The severity of any resultant disease will depend on the type and virility of the biological agent, and the vulnerability of the exposed person. The exposed person may be someone who already has a health problem, is on medication, elderly or very young, has no immunity, or is pregnant.

For biological agents that are known pathogens, those agents capable of causing disease, the resultant disease will be known and documented, but the level of severity is dependent on other factors. It is not always known that the agent is present; therefore the risk assessment should take account of these uncertainties. For example, the likelihood of Leptospira exposure is greater for those working where rats may be or have been (sewers) and the likelihood of disease from the exposure is greater for those who have skin cuts and abrasions and no hand washing facilities.

The resultant disease depends on the specific biological agent and what sort of infection it causes. Many infections start with flu-like symptoms as the body's immune system tries to fight it off. This can be mistaken for flu and therefore it is important that that risk assessment makes it clear that these symptoms could mean something else. For example, failure to recognise Leptospirosis (Weil's disease) and treating the symptoms as flu could lead to severe illness or death. As Leptospira are bacteria, Leptospirosis can be treated with antibiotics, while flu is a response to the flu virus and cannot.

MODES OF TRANSMISSION WITH EXAMPLES

The transmission of biological agents must be known in order to assess the likelihood of infection. They may be transmitted by a number of modes.

Human to human:

- HIV and Hepatitis B viruses may be transmitted by human direct contact with another person's body fluids, particularly blood.

Animal to human:

- E. coli bacteria, which may be transmitted between animals and humans via contact with their faeces during farm handling or in food preparation.
- Leptospira bacteria are carried by an animal, such as rats, but when released in urine can survive outside the animal for a period of time and be carried in water or reside on the land. It may then be transmitted to humans by direct contact with the bacteria.
- The rabies virus, which may be transmitted by animal bites.

Environment to human:

- Fungi may be transmitted by direct contact, including being eaten by humans or by their spores carried in the air, for example in the case of the moulds called Aspergilla whose spores are released into the air when infected hay is handled and then breathed in by humans.
- Legionella pneumophila bacteria, which may be present in water and transmitted to humans when the water carrying it becomes airborne in breathable droplets.
- Viruses can easily become airborne or be transmitted from contact with contaminated soil.
- Cryptosporidia parasite, which is found in the faeces of animals and humans and may be in water that is untreated. This can be transferred to humans when deliberately or accidentally taken into their mouth.

STABILITY OF THE AGENT IN THE ENVIRONMENT

Bacteria

Bacteria can be passed from:

- Human to human, for example, E. coli.
- Animal to human, for example, Leptospira, which is carried by rats and cows.
- Environment to human, for example, Legionella Pneumophila, which may be present in water.

Bacteria that are passed from animal to human and cause disease in the human are called Zoonotic. The range of diseases is called Zoonoses. Bacteria can enter the body through the lungs if droplets, that are breathed, coughed or sneezed out by an infected person, are inhaled. They can infect the digestive tract if contaminated food is eaten. Bacteria may be present in food at its primary source, brought to it by flies or from contamination on hands. Bacteria can enter the genito-urinary system. They can also penetrate the skin in various ways: through hair follicles, eyes, cuts and abrasions, 'sharps' injuries, bites and through deep wounds. Bacteria cause disease because they produce toxins that are harmful to human cells. If they are present in sufficient quantity and the affected person is not immune (resistant) they will cause disease.

Some bacteria release endotoxins when the bacterial cells die, which cause fever, haemorrhage and shock. They cause a change in the blood capillary cell walls, which allows fluid to seep into the surrounding tissue and blood pressure to drop. This is life threatening and is known as endotoxic shock.

Other bacteria produce exotoxins, where the toxin is released from live bacteria, for example, in diseases such as diphtheria, tetanus and those diagnosed with toxic-shock syndrome.

Some bacteria causes harm in the intestinal region of the body by releasing enterotoxins. These are usually found on food or bodily waste products. E. coli, the cholera bacterium, Salmonella and Staphylococci are bacteria of this type.

Outbreaks of severe sickness and diarrhoea have occurred amongst children who have touched farm animals. The animals probably had traces of animal excrement on their coats and on being touched the bacteria was transferred to the child's hands and later transferred to their mouth.

Viruses

Viruses gain entry to the body by all possible routes. They are inhaled in droplets; swallowed in food and liquids; passed through punctured skin on infected needles, in the saliva of feeding insects or rabid dogs;

viruses are accepted directly by the mucous membranes of the genital tract and by the conjunctiva of the eye after accidental contamination.

Viruses can pass from:

- Human to human, for example, Human Immunodeficiency Virus (HIV), Hepatitis B, various strains of influenza and the common cold.
- Animal to human, for example, the rabies virus.
- Environment to human: there are vast numbers of viruses present in the environment, such as Leptospirosis, a bacterial disease spread by coming into contact with water contaminated with infected animal urine.
- Although rats, mice, and moles are important primary hosts, a wide range of other mammals including dogs, deer, rabbits, hedgehogs, cows, sheep, raccoons, opossums, skunks, and certain marine mammals are able to carry and transmit the disease as secondary hosts. In Africa, the banded mongoose has been identified as a carrier of the pathogen.

Many viruses invade cells and multiply near their site of entry. Some enter the lymphatic vessels and spread to the lymph nodes. Some, such as HIV, invade the lymphocytes (a type of white cell). Many pass from the lymphatics to the blood and are quickly carried to every part of the body. They may invade specific target organs, such as the liver, lungs, brain, and start to multiply. Some viruses travel along the nerve fibres to their target organs.

Viruses cause disease in a variety of ways:

- They can destroy or disrupt the cells they invade, causing serious illness if it is a vital organ.
- The response of the body's immune system may lead to fever and fatigue.
- Antibodies produced by the immune system may attach to viral particles and be circulated in the blood.
- The antibodies may then be deposited in various parts of the body causing inflammation and tissue damage.
- They may interact with the chromosomes of the host cells causing cancer.
- They may weaken the activity of the t-lymphocytes and so interfere with the immune system. The body's normal defences against a whole range of infections can be lost. This is how HIV works.

CONCENTRATION AND AMOUNTS

Laboratory environments must demonstrate they have strict controls. A factor in the laboratory risk assessment is, therefore, knowing the quantities of an agent likely to be present. Usually the quantities, though measured in millions of bacteria, will be very small. Outside the laboratory environment such control is not possible.

PRESENCE OF A SUITABLE HOST (HUMAN OR ANIMAL)

The suitable host may be a human or an animal. Some biological agents can survive outside the host (Legionella bacteria) while others require a host to grow and survive.

DATA AVAILABLE

The laboratory risk assessment should also consider the data available (for example, from animal studies) on the biological agent, the type of work where exposure is possible and the availability of a vaccination or treatment.

NATURE OF ACTIVITY

E. coli is a very fast replicating bacterium and is used in genetic manipulation. The bacterial cells are genetically modified to produce hormones the body needs, such as insulin. Workers involved with such processes must take extra care to avoid the risk of accidental injection of these hormone-producing bacteria when handling them.

LOCAL AVAILABILITY OF PROPHYLAXIS/TREATMENT

Such measures may include: prophylaxis (prevention) by vaccination or antisera (passive antibody transfusion from a previous human survivor); sanitary measures, for example, food and water hygiene; the control of animal reservoirs or arthropod vectors; the movement of people or animals; and the importation of infected animals or animal products.

Special control measures required when working with biological agents

In the control of occupational health hazards many approaches are available, the use of which depends upon the severity and nature of the hazard. The principal control strategies are outlined in the following sections:

ERADICATION/ELIMINATION

This involves the eradication or elimination of a hazard by design or specification and consists of hazard prevention and control features.

This represents an extreme form of control and is appropriately used where high risk is present. It is usually achieved through the prohibition of use of these substances or not carrying out the work that involves their use. However, if the work with the micro-organisms is intentional, it would not make sense to eliminate them. Care must be taken to ensure that discarded stock is safely disposed of and that controls are in place to prevent their re-entry, even as a sample or for research. If this level of control is not achievable then another must be selected.

REDUCED VIRULENCE

During research, pathogens may be genetically modified to be less virulent and still allow the research to continue. For example, Salmonella typhimurium (which causes gastro-enteritis) could be made less virulent, but still have the necessary properties to be used in the Ames test for signs of a substance having mutagenic properties. However, if the virulent properties are the purpose of the research, then this control is not suitable.

CHANGE OF WORK METHOD TO MINIMISE OR SUPPRESS GENERATION OF AEROSOLS

Aerosols or finely suspended droplets containing pathogens may become airborne. The best way to suppress or minimise aerosols is to change the work method so aerosols are not produced or if they are, they are contained. Aerosols carrying micro-organisms can also be scrubbed (passed through water and disinfectant or biocide curtains to kill the pathogens). Local exhaust ventilation can be used to remove the aerosol at source. Processes such as aeration are used in effluent treatment and it is usual to have remote positioning for this treatment.

ISOLATION AND SEGREGATION

If a biological agent cannot be eliminated or replaced, another option is to enclose it completely to prevent harm to workers, for example, handling substances in a glove box. A dedicated room away from the general work area with its own air movement and ventilation system, completely isolated from all other systems could also be used. Any accidental loss of containment within the room would not then have an impact on the general working area. A relatively simple process is to restrict the numbers exposed to the hazard. This technique is used when dealing with highly infectious diseases either in patient treatment or for purposes of research. The fewer the people who come into contact, the better. The ones who are allowed to work with the infectious patient may also be protected by vaccine, where one exists.

CONTAINMENT/ENCLOSURE APPROACH

This strategy is based on the containment or enclosure of an offending agent to prevent its free movement in the working environment. It may take a number of forms, for example, pipelines, closed conveyors, laboratory fume cupboards. The containment measures taken from COSHH 2002, schedule 3, Part II and Part III are set out in *figure refs B5-5 and B5-6.*

CONTROL FOR SPECIFIC EXAMPLES

Hospitals

There are two main issues that hospitals and clinics need to manage effectively:
1) Patient infection.
2) Patient staff-cross infection.

Hygiene standards should be such that the cross infection of patients and staff is prevented or kept to a minimum. Patients will often be at risk from everyday illness such as the common cold, which will be introduced on a daily basis from both staff and visiting members of the public. It is very difficult to protect against the introduction of such common infections for the general patient population. High-risk patients are therefore segregated from the general population into isolation wards. Here precautions are taken to ensure both patient and staffs are segregated to prevent cross infection.

A number of procedures are used depending on the risk-factors from basic hygiene, washing of hands, equipment, such as stethoscope and thermometers issued on personal patient basis, when dealing with the medium risk level. Where there is potential for significant risk additional safeguards may be required in addition to hygiene or equipment issue. With highly virulent infections, it may be necessary to have sealed wards, with airlock access and negative pressure air supply. Staff may be required to wear bio-bodysuits (staff wear total protection with a controlled air feed line). Provision of suitable (stainless steel work surfaces, tiled floors, walls etc.) cleanable surfaces, rigorous sterilisation programme and swab testing following cleaning.

Laboratory work

There are many factors to consider when scaling up production from the laboratory to commercial quantities. A bacteriological process, such as yeasts, moulds and fungi used in a fermentation process, may not be suitable for scaling up if it presents a major health risk to workers and the general public. While these may be controllable on the small scale, the problems associated with their large scale use can mean such a substantial investment in containment and other safety equipment that the process would be uneconomic.

An additional factor to take into account is the plant may be expected to run for long periods of time, which does not usually happen in the laboratory. Most experiments are carried out using glass (in vitro) and are usually carefully monitored and controlled.

Many reaction vessels are made from materials such as mild or stainless steel, which can corrode and become difficult to decontaminate. This can be mitigated by the use of steel clad with titanium or glass-lined metal reaction vessels. These provide a corrosion resistant surface, which is relatively easy to clean when decontamination is necessary. Ease of decontamination can be critical where only small amounts of product are required; the same vessels may be used for several processes thereby necessitating frequent cleaning. While laboratory experiments are usually closely monitored, this is more difficult in large-scale production.

Animal houses

Containment measures as with hospitals will be dependent on the risk; good guidance is provided in the 'COSHH ACOP Part II, *see figure ref B5-5* for containment measures of health and veterinary care facilities, laboratories and animal rooms'.

Part II - Containment measures for health and veterinary care facilities, laboratories and animal rooms

Containment measures	Containment levels		
	2	3	4
The workplace is to be separated from any other activities in the same building.	No	Yes	Yes
Input air and extract air to the workplace are to be filtered using HEPA or equivalent.	No	Yes, on extract air	Yes, on input air and double on extract air
Access is to be restricted to authorised persons only.	Yes	Yes	Yes, via air-lock key procedure
The workplace is to be sealable to permit disinfection.	No	Yes	Yes
Specified disinfection procedures.	Yes	Yes	Yes
The workplace is to be maintained at an air pressure negative to atmosphere.	No, unless mechanically ventilated	Yes	Yes
Efficient vector control, for example, rodents and insects.	Yes, for animal containment	Yes, for animal containment	Yes
Surfaces impervious to water and easy to clean.	Yes, for bench	Yes, for bench and floor (and walls for animal containment)	Yes, for bench, floor, walls and ceiling
Surfaces resistant to acids, alkalis, solvents, disinfectants.	Yes, for bench	Yes, for bench and floor (and walls for animal containment)	Yes, for bench, floor, walls and ceiling
Safe storage of biological agents.	Yes	Yes	Yes, secure storage
An observation window, or alternative, is to be present, so that occupants can be seen.	No	Yes	Yes
A laboratory containing its own equipment.	No	Yes, so far as is reasonably practicable	Yes
Infected material, including any animal, is to be handled in a safety cabinet or isolator or other suitable containment.	Yes, where aerosol produced	Yes, where aerosol produced	Yes (Class III cabinet)
Incinerator for disposal of animal carcases.	Accessible	Accessible	Yes, on site

Figure B5-5: Biological agent containment measures - veterinary, laboratory and animal rooms. *Source: COSHH, Schedule 3.*

COSHH Schedule 3 states that in this Part, 'Class III cabinet' means safety cabinet defined as such in British Standard 5726: Part I: 1992, or unit offering an equivalent level of operator protection as defined in British Standard 5726: Part I: 1992.

British Standard 5726 has been superseded by BS EN 12469:2000 'Biotechnology: performance criteria for microbiological safety cabinets'.

Part III - Containment measures for industrial processes

	Containment measures	Containment levels		
		2	**3**	**4**
1.	Viable micro-organisms should be contained in a system which physically separates the process from the environment (closed system).	Yes	Yes	Yes
2.	Exhaust gases from the closed system should be treated so as to -	Minimise release	Prevent release	Prevent release
3.	Sample collection, addition of materials to a closed system and transfer of viable micro-organisms to another closed system, should be performed as to -	Minimise release	Prevent release	Prevent release
4.	Bulk culture fluids should not be removed from the closed system unless the viable micro-organisms have been -	Inactivated by validated means	Inactivated by validated chemical or physical means	Inactivated by validated chemical or physical means
5.	Seals should be designed so as to -	Minimise release	Prevent release	Prevent release
6.	Closed systems should be located within a controlled area -	Optional	Optional	Yes, and purpose-built
	Biohazard signs should be posted.	Optional	Yes	Yes
	Access should be restricted to nominated personnel only.	Optional	Yes	Yes, via air-lock
	Personnel should wear protective clothing.	Yes, work clothing	Yes	Yes, a complete change
	Decontamination and washing facilities should be provided for personnel.	Yes	Yes	Yes
	Personnel should shower before leaving the controlled area.	No	Optional	Yes
	Effluent from sinks and showers should be collected and inactivated before release.	No	Optional	Yes
	The controlled area should be adequately ventilated to minimise air contamination.	Optional	Optional	Yes
	The controlled area should be maintained at an air pressure negative to atmosphere.	No	Optional	Yes
	Input and extract air to the controlled area should be HEPA filtered.	No	Optional	Yes
	The controlled area should be designed to contain spillage of the entire contents of closed system.	Optional	Yes	Yes
	The controlled area should be sealable to permit fumigation.	No	Optional	Yes
7.	Effluent treatment before final discharge.	Inactivated by validated means	Inactivated by validated chemical or physical means	Inactivated by validated physical means

Figure B5-6: Biological agent containment measures - industrial processes. *Source: COSHH, Schedule 3.*

SHARPS CONTROL

Biological agents can be passed directly into the body by injection. They can be injected by needles or other sharp objects, which form part of the work equipment. Injection can also occur when glass gets broken and is handled. Careful disposal of sharps is necessary to prevent accidental injection. Sharps containers should be provided and used so needles and syringes are not left lying about.

Care must be taken when dealing with the possibility that someone may come into contact with discarded needles. Needle stick injuries can give rise to infections by agents such as hepatitis B, hepatitis C, HIV, malaria and syphilis. A recent study has identified 22 biological agents that could be transmitted by means of sharps.

Sharps should be handled as little as possible and disposed of in a special container, which should be lidded.

IMMUNISATION

How vaccines work

Disease-causing organisms have two distinct effects on the body. The first effect is very obvious; we feel ill, exhibiting a variety of symptoms such as fever, vomiting and diarrhoea. The disease-causing organism also induces an immune response in the infected person. As the response increases in strength over time, the infectious organisms are slowly reduced in number until symptoms disappear and recovery is complete.

The host recovery occurs because the disease-causing organisms contain proteins called 'antigens' which stimulate the body's immune response. The main effect of antigens is to cause the body to produce proteins called 'antibodies.' The proteins bind to the disease causing organisms and cause dysfunction and eventual destruction. At the same time, 'memory cells' are produced in an immune response. Memory cells are cells which remain in the blood stream, sometimes for the life span of the host, ready to initiate a quick protective immune response against subsequent infections with the particular disease causing organism that induced their production. This response is often so rapid that infection does not develop and the person is unaware.

Vaccines exist for protection against some of the occupational diseases caused by biological agents, but not all. There is a vaccine available that offers some immunisation protection against Hepatitis B and one for Leptospira, but it is not always available. There are also a number of vaccines available that can provide immunisation for those who intentionally work with certain bacterium and viruses, such as the TB vaccine and protection against malaria.

DECONTAMINATION AND DISINFECTION

Decontamination and disinfection should be carried out according to COSHH Regulations 2002. For example, decontamination and washing facilities should be provided and disinfection should be carried out for all containment levels, i.e. 2, 3 and 4.

Protective clothing and personal protective equipment may need to be decontaminated and disinfected, as will equipment that is used. All surfaces, including walls, floors, ceilings, tables, ventilation ducts, may also need to be dealt with. Changing areas and showering facilities will need decontamination and disinfection, and regular swabs taken to ensure the cleaning is working.

EFFLUENT AND WASTE COLLECTION

Effluent and waste collection should be handled according to the containment measures for health and veterinary care facilities, laboratories and animal rooms and the containment measures for industrial processes listed in COSHH Regulations 2002. For example, all effluent from Groups 2, 3 and 4 must be treated and inactivated. Animal carcasses must be incinerated and for Group 4 agents, they must be incinerated on site. Other effluent and waste collection methods can be seen in *figure ref B5-5 and B5-6*.

PERSONAL HYGIENE MEASURES

Personal hygiene to prevent spread of disease will require segregation of domestic and work clothing and, as a minimum, regular hand washing (using a prescribed technique for a minimum of 30 seconds). Normally workers will use an apron, which is impervious and disposable. Where it is necessary to take increased precautions, it may be necessary to shower before and at the end of a shift or exposure period. Eating and drinking should not be allowed in exposed areas. A smoking prohibition already exists for workplaces in specific legislation, which should be rigorously enforced.

PERSONAL PROTECTIVE EQUIPMENT

Where personal protective equipment (including protective clothing) is provided to meet the requirements of COSHH 2002 relating to biological agents, the employer must ensure that it is:

- Properly stored in a well-defined place.
- Checked and cleaned at suitable intervals.
- Repaired or replaced before it is used again if it is found to be defective.

If there is a possibility that PPE may be contaminated by biological agents, it should be removed on leaving the work area and kept apart from uncontaminated clothing and equipment. Contaminated equipment should be decontaminated and cleaned or destroyed.

BIOHAZARD SIGNS

COSHH 2002 (as amended) Part IV Biohazard sign specifies:

"The biohazard sign required by Regulation 7(6) (a) shall be in the form." **See figure ref B5-7**.

BASELINE TESTING AND HEALTH SURVEILLANCE

This is necessary to assess employees' immunity before or after vaccination. If there is an indication that infection has occurred, then it may be appropriate to take specimens in order to attempt to isolate infectious agents.

If someone shows signs of illness that could be due to exposure at work, then others who may have been exposed should be placed under surveillance.

Figure B5-7: Biohazard sign. *Source: Stocksigns.*

People may show a susceptibility to microbial allergens and immunological testing may show which agents are responsible. All workers exposed to respiratory sensitisers of biological origin should be under surveillance. The level of surveillance should be related to the risk identified in the risk assessment.

COSHH 2002 (as amended) Regulation 11(9) states:

"Where, as a result of health surveillance, an employee is found to have an identifiable disease or adverse health effect which is considered by a relevant doctor or other occupational health professional to be the result of exposure to a substance hazardous to health the employer of that employee shall:

(a) Ensure that a suitably qualified person informs the employee accordingly and provides the employee with information and advice regarding further health surveillance.

(b) Review the risk assessment.

(c) Review any measure taken to comply with Regulation 7, taking into account any advice given by a relevant doctor, occupational health professional or by the Executive.

(d) Consider assigning the employee to alternative work where there is no risk of further exposure to that substance, taking into account any advice given by a relevant doctor or occupational health professional.

(e) Provide for a review of the health of any other employee who has been similarly exposed, including a medical examination where such an examination is recommended by a relevant doctor, occupational health professional or by the Executive."

Health surveillance related to biological agents may include:

- **Biological monitoring** is the measurement and assessment of the quantity of biological agents, or the cells formed when the body responds to the biological agent, in exposed workers. This may involve samples of skin, saliva, urine or blood.
- **Biological effect monitoring** is the measurement and assessment of early biological effects in exposed workers caused by biological agents.
- **Medical surveillance** is both health surveillance under the supervision of a medical inspector of the HSE's Employment Medical Advisory Service, or an appointed doctor for the purpose of regulation 11(5) and under the supervision of a registered medical practitioner. It may include clinical examinations and measurements of physiological effects of exposure to biological agents, for example, lung function testing.
- **Enquiries about symptoms**, inspection or examination by a suitably qualified person, for example, an occupational health nurse.
- **Inspection** by a responsible person such as a supervisor or manager, for example, for effects of anthrax on the skin.
- **Review** of records and occupational history during and after exposure.

The Approved Code of Practice to COSHH 2002 (as amended) requires the employer to:

"Keep individual exposure and health records for employees where:

(a) Personal exposure monitoring is carried out for an identified employee.

(b) The employee concerned is also under health surveillance in accordance with Regulation 11."

This page is intentionally blank

ELEMENT B5 - BIOLOGICAL AGENTS

This page is intentionally blank

Noise and vibration

Learning outcomes

On completion of this element, candidates should be able to demonstrate understanding of the content through the application of knowledge to familiar and unfamiliar situations. In particular, they should be able to:

B6.1   Explain the basic physical concepts relevant to noise.

B6.2   Explain the effects of noise on the individual and the use of audiometry.

B6.3   Explain the measurement and assessment of noise exposure.

B6.4   Explain the principles and methods of controlling noise and noise exposure.

B6.5   Explain the basic physical concepts relevant to vibration.

B6.6   Explain the effects of vibration on the individual.

B6.7   Explain the measurement and assessment of vibration exposure.

B6.8   Explain the principles and methods of controlling vibration and vibration exposure.

Content

Relevant statutory provisions

Control of Noise at Work Regulations (CNWR) 2005

Control of Noise at Work Regulations (Northern Ireland) 2006

Control of Vibration at Work Regulations (CVWR) 2005

Control of Vibration at Work Regulations (Northern Ireland) 2005

Personal Protective Equipment at Work Regulations (PPER) 1992

Regulation (EU) 2016/425 on personal protective equipment at work

Sources of reference

Reference information provided, in particular web links, was correct at time of publication, but may have changed.

Controlling Noise at Work, The Control of Noise at Work Regulations, Guidance on Regulations, L108, HSE Books, ISBN: 978-0-7176-6164-4, http://www.hse.gov.uk/pUbns/priced/l108.pdf

Hand-arm vibration, Control of Vibration at Work Regulations 2005, Guidance on Regulations, L140, HSE Books, ISBN: 978-0-7176-6125-1, http://www.hse.gov.uk/pubns/books/l140.htm

Hand-arm vibration exposure calculator, HSE, http://www.hse.gov.uk/vibration/hav/vibrationcalc.htm

Noise calculators, HSE, http://www.hse.gov.uk/noise/calculator.htm

Personal protective equipment at work, Guidance on Regulations, L25, HSE Books, ISBN: 978-0-7176-6597-6, http://www.hse.gov.uk/pubns/priced/l25.pdf

Personal protective equipment (PPE) at work; A brief guide, INDG174, HSE Books, http://www.hse.gov.uk/pubns/indg174.pdf

The health and safety toolbox, How to control risks at work, HSG268, HSE Books, ISBN: 978-0-7176-6587-7, http://www.hse.gov.uk/pubns/books/hsg268.htm

Whole body vibration calculator, HSE, http://www.hse.gov.uk/vibration/wbv/calculator.htm

Whole-body vibration; The Control of Vibration at Work Regulations 2005, Guidance on Regulations, L141, HSE Books, ISBN: 978-0-7176-6126-8, http://www.hse.gov.uk/pUbns/priced/l141.pdf

Work Related Upper Limb Disorders - A Guide, HSG60, HSE Books, ISBN: 978-0-7176-1978-8, http://www.hse.gov.uk/pubns/books/hsg60.htm

The above web links along with additional sources of reference, which are additional to the NEBOSH syllabus, are provided on the RMS Publishing website for ease of use - www.rmspublishing.co.uk.

B6.1 - Basic physical concepts relevant to noise

Definition of noise

Noise may be defined as any signal that does not convey useful information. Noise is also defined as unexpected, unpleasant or undesired. Some sounds can cause annoyance or stress and loud noise can cause damage to the ear.

The meaning of noise under the Control of Noise at Work Regulations 2005

The Control of Noise at Work Regulations (CNWR) 2005 implements the European Union Directives to protect workers from the health risks caused by noise.

They do not apply to members of the public exposed to noise from their non-work activities, or when they make an informed choice to go to noisy places or from nuisance noise.

"Any audible sound."
"Noise is a reference to the exposure of that employee to noise which arises while he is at work, or arises out of or in connection with his work."

Figure B6-1: Definition of noise.   *Source: CNWR 2005 Regulation 2.*

WORKPLACE EXAMPLES OF NOISE

Workplaces where high levels of noise may be found include those where the following are present:

- Metal processing and fabrication.
- Packaging into glass or metal containers.
- Veterinary clinics and charity animal sanctuaries.
- Wood cutting and finishing.
- Grinding equipment.
- Highway repair work.
- Heavy machinery, including transport, construction and mining.
- Orchestral music.

See figure ref B6-2, which gives a range of activities with their corresponding noise levels.

Figure B6-2: Noise levels. *Source: Noise at work: HSE Guidance for employers.*

The basic concepts of sound

The ear senses sound, which is transmitted in the form of longitudinal waves travelling through a medium/substance, for example, air, water, metals. Any audible sound is noise. There is a lot of terminology with regards to sound. Key concepts of sound have been briefly defined as follows.

WAVELENGTH

Wavelength (λ) - is the distance covered during one complete cycle (i.e. the distance between wave peaks).

The relationship between wavelength and frequency is described by the formula:

$$\text{Wavelength} = \frac{\text{speed of sound}}{\text{frequency}}$$

Thus, as the speed of sound in air at normal temperature is 344 metres per second. A frequency of 20 Hz = a wavelength of 17 m and 20 kHz = 0.017 m.

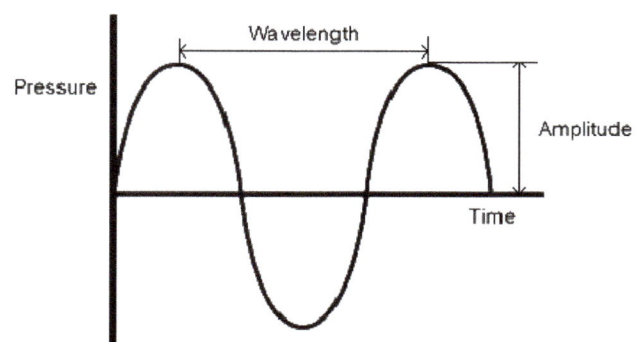

Figure B6-3: Features of a sound wave.   *Source: RMS.*

AMPLITUDE

Amplitude is important in the description of a wave phenomenon such as light or sound. In general, the greater the amplitude of the wave, the more energy it transmits (for example, a brighter light or a louder sound).

FREQUENCY

Frequency is the number of cycles (completed wavelengths) that the wave makes per second, expressed in the unit Hertz (Hz). Sound may be of a single frequency (i.e. a pure tone such as a tuning fork) but is usually a complex mixture of frequencies. The normal range of human hearing is 20 - 20,000 Hz in a young healthy adult. *Broad band noise* is the term often used to describe occupational noise, because it contains a wide mixture of frequencies. The range of frequencies that we encounter is often divided into Octave Bands. A noise can be measured in each octave band and these levels can be used when assessing the attenuation of hearing protectors, or when diagnosing noise problems.

PITCH

The frequency of the sound is related to *pitch*. High frequency sound, such as that caused by steam escaping from a valve, would be described as high pitch whereas low frequency sound, like that emitted from a passing large goods vehicle or bus would have a low pitch.

SOUND INTENSITY

Intensity refers to the strength of a sound source and, separately, the magnitude of a sound field. A sound source is characterised by the sound power it produces and this is expressed in decibels relative to 1 pico watt (pW) as a sound *power* level. The magnitude of a sound field is measured in decibels relative to 20 micropascals (μPa) as a sound *pressure* level.

The fact that both power and pressure are expressed in decibels can be confusing. Though the power intensity of the sound source in decibels may indicate possible levels of pressure intensity, in the sound field they do not always correlate. This is because many factors may influence the sound pressure level of the sound field, causing absorption or magnification. When expressing sound intensity it is important to clarify the unit used when establishing the relative decibel level; sometimes this is done by stating that it is sound power level or sound pressure level.

The highest sound pressure level that does not produce a sensation of pain in humans is approximately 20 Pa - called 'the threshold of pain'. The lowest sound intensity, sound pressure level, that the ear can perceive, at a frequency of 1,000 Hz, is approximately 20 μPa - called 'the threshold of hearing'.

Sound pressure levels are expressed in decibels relative to the threshold of hearing, 20 μPa. Thus, the usual range of sound field intensity affecting humans is a sound pressure range of 20 μPa to 20 Pa.

Because sound pressure varies as the sound wave goes through its pressure cycle it is necessary to average out this pressure to a value equivalent to a pressure at a steady level; sound pressure level (SPL) meters do this automatically.

This averaged value of sound pressure is the root mean square or rms value.

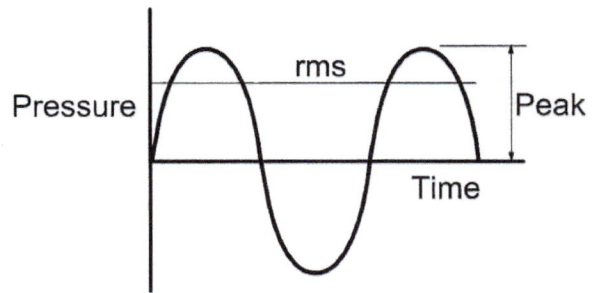

Figure B6-4: Rms and peak levels of a sound wave. *Source: RMS.*

THE DECIBEL (DB)

The decibel is the unit used to measure the intensity of sound. The Bel was named after A. G. Bell; however this was considered to be a large unit and is rarely used. The decibel is one tenth of a Bel and is a more convenient unit for general use. The decibel is a measure of sound intensity relative to a reference level.

This is calculated for sound *power* levels by the formula $dB = 10 \log_{10} (P_1/P_2)$, where P_1 and P_2 are the power levels of the two sounds being compared. The larger sound power level is divided by the smaller. Because the decibel is a comparative unit the units of power (W) cancel each other out in the calculation.

Similarly, a decibel level for a sound *pressure* level is calculated by the formula $dB = 20 \log_{10} (P_1/P_2)$, where P_1 and P_2 are the pressure levels of the two sounds. The larger sound pressure level is divided by the smaller and because the decibel is a comparative unit the units of pressure (μPa) cancel each other out in the calculation. For most measurements of the sound pressure level of a sound field that are taken for occupational health and safety reasons P_2 will be the threshold of hearing, 20 μPa.

A-WEIGHTING (dB (A))

The human ear can hear sound over a range of frequencies, typically from 20 Hz to approximately 20,000 Hz (20 kHz). However, the ear does not hear the same at all frequencies; it naturally reduces (attenuates) low frequencies and very high frequencies.

To take account of the response of the ear to a range of frequencies measured sound pressure levels are usually 'A weighted'. *'A' weighted* is a weighting added to a sound pressure level meter reading (by electronic filters) to represent the way the ear hears sound of different frequencies; the measured noise levels are given the abbreviation dB (A).

The majority of sound pressure level measurements are made in terms of dB (A), although there are other weightings that are used in some circumstances. One of these is the C weighting, which is used to assess the acoustic emissions of machines, in the selection of hearing protectors and in the analysis of environmental noise. The C weighting is also used to determine peak sound pressure levels and is particularly useful for impact or explosive noises.

It has a broader spectrum than that of the A weighting and is more accurate at higher sound levels. Measurements made on this scale are expressed as dB(C). There is also a (rarely used) B weighting scale, intermediate between A and C.

WEIGHTING CURVES

dB(A)	The A weighting curve shown in **figure ref B6-5** mimics the sensitivity of the human ear. The ear is less sensitive to low frequency sound and is non-linear in its assessment of loudness. The ear is most sensitive to sounds around 1-4,000 Hz. The measured noise levels are given the abbreviation **dB (A)**.
dB(B)	Not so much weighting (attenuation) at low frequencies as dB (A), sometimes used to measure music noise with a 'bass' beat.
dB(C)	Weighting only at very low and high frequencies, used to measure 'peak' noise. This should not be used for measuring peak noise levels which have energy components at both the high and low frequency ends of the spectrum (0-20 Hz and greater than 15 kHz).
dB(Lin)	No weighting at all, also used to measure 'peak' noise.

Note that the 'D' scale was previously used in aircraft noise measurements.

It is important to note that the use of the term dB, alone, indicates an un-weighted linear measurement which is sometimes denoted by the term dB (Lin); however, this is not strictly necessary.

The expression 'dB' should not be substituted for dB (A).

Noise rating curves

There are a standard series of Noise Rating (NR) Curves which are stylised forms of the loudness response curves.

These NR curves are often used as a criterion for noise control, and as such are internationally accepted.

Other criteria may also be encountered such as NC curves (Noise Criteria used in the USA).

Sound pressure levels measured in octave bands are compared with the curves from which a noise rating (NR) is obtained.

The higher frequencies, where the ear is more sensitive, are given heavier noise ratings than the lower frequencies.

For example, 30 d(B) on the 4,000 Hz frequency has an NR of 35, while 30 d(B) on the 250 Hz band has an NR of 20.

There are European agreed Acceptable NRs, which should not be exceeded and are read off the 1,000 Hz band:

- 45 General offices.
- 50 Office with business machines.
- 60 Light engineering works.
- 70 Heavy engineering works.

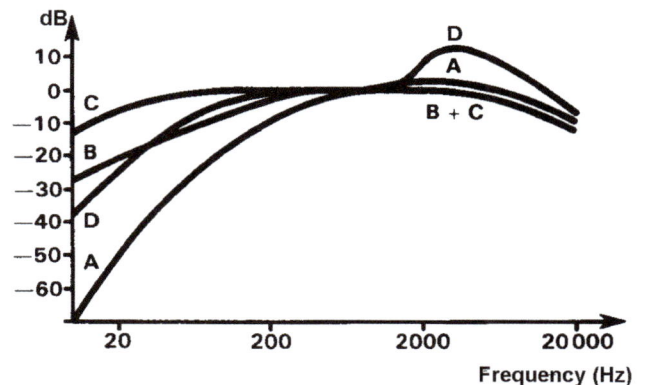

Figure B6-5: 'A', 'B', 'C' and 'D' weighting curves. *Source: Ambiguous.*

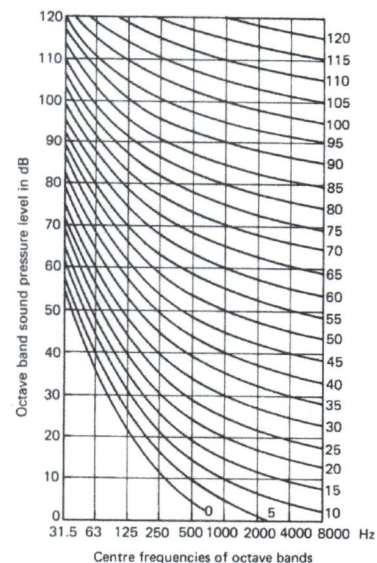

Figure B6-6: Noise rating curves.
Source: Safety at Work, Fourth Edition, J Ridley.

The significance of logarithmic scales in relation to dB and concepts of addition of combined sounds

THE SIGNIFICANCE OF LOGARITHMIC SCALES IN RELATION TO DB

The range of sound pressures that the human ear can process is quite wide, 20 µPa to 20 Pa. But as a measurement scale this was found to be quite cumbersome.

In order to make the scale more manageable, the logarithmic **Bel** (B) scale was devised. 1 Bel represents a tenfold change in sound pressure. Thus logarithms to the base 10 ($\log_{10}$) are used to establish a given Bel level, which is denoted by the term log in the following text.

If the threshold of hearing is taken as the basis for the scale (i.e. as a reference value) then the sound pressure level (SPL) in Bels will equate to:

- $SPL = \log (P^2_{rms} / P^2_{ref})$ B.

Where P_{rms} is the measured intensity:

■ P_{ref} is the reference intensity (20 μPa or 2 x 10^{-5} Pa).

The Bel forms too narrow a scale, i.e. it would only be 0, 1, 2, 3, 12 for the range of human hearing. For a more useful range the Bel is divided into 10 parts to give the decibel (dB). This makes the range 0-120. In order to convert the above equation to dB, it becomes:

■ SPL = 10 log (P^2_{rms} / P^2_{ref}) dB under the rules of logarithms, taking out the square terms this becomes:

■ SPL = 20 log (P_{rms} / P_{ref}) dB.

Thus the threshold of hearing (20 μPa or 2 x 10^{-5} Pa):

■ SPL = 20 log (2 x 10^{-5} / 2 x 10^{-5}) dB = 20 log (1)   =   0dB as the log of 1 = 0.

The threshold of pain (about 20 Pa):

■ SPL = 20 log (20 / 2 x 10^{-5}) dB   =   120 dB.

Because the scale is not linear, but is logarithmic, an *increase of three decibels* anywhere along the scale denotes a *doubling* of intensity and a doubling of the potential for harm. Similarly, for each 10 dB added the sound pressure level (in Pa), and therefore the intensity the ear receives, becomes 10 times higher. So, 80 dB is quite loud, but 90 dB is ten times louder, and 100 dB is 100 times louder than 80 dB, etc.

The human ear is, however, a poor sound pressure level meter. It might be expected that upon doubling the sound pressure it would double the loudness (intensity) a person hears. However, this is not the case. To obtain a subjective doubling of the loudness a person hears, the sound pressure level needs to be increased by about 10 dB, which is 10 times the original intensity. This can lead to workers believing that little harm is done to their hearing when they move into a noisier environment.

ADDITION OF COMBINED SOUNDS, EQUAL AND UNEQUAL

Equal sounds

Calculating the effects of combined noise levels can be achieved by using the rules of logarithms. As previously stated, an increase of three decibels represents a doubling in intensity. Thus if *two* sound sources, each producing 84 dB (A), are switched on then the expected noise level for a worker would be:

■ 84 dB (A) + 84 dB (A) = 87 dB (A).

If *four* of these sound sources are switched on then the intensity will increase to:

■ 87 dB (A) + 87 dB (A) = 90 dB (A).

Although this technique can be used for direct multiples (i.e. 2, 4, 8, 16 etc.), it cannot be used to calculate the intensity of for example, 3, 5, or 6 machines. This can be calculated by first converting to Bels - (for example, 84 dB (A) = 8.4 Bels), and then using the formula:

■ SPL   =   Log (10^s x n) B (A).

Where 's' is the sound pressure level in Bels and 'n' is the number of equal intensity level sound sources.

The noise level for five of these sound sources (84 dB (A) or 8.4 Bels each) can be calculated by:

■ SPL   =   Log ($10^{8.4}$ x 5) B (A).

              =   9.099 B (A).

              =   91 dB (A).

Unequal sounds

Where there are sound sources of differing levels, then this can be calculated using the formula:

■ SPL = Log (10^{s1} + 10^{s2} + 10^{s3} +....10^{sn}) B (A).

Where 'sn' is the sound pressure level of individual sound sources in Bels.

Thus, if four sound sources produce sound pressure levels at a worker's position of 91 dB(A), 93 dB(A), 89 dB(A) and 86 dB(A) respectively, when each is operated in turn with the other three switched off, then the calculated noise level would be:

■ SPL   =   Log ($10^{9.1}$ + $10^{9.3}$ + $10^{8.9}$ + $10^{8.6}$) B (A).

              =   9.648 B (A).

              =   96.48 dB (A).

              =   96.5 dB (A) rounded to one decimal.

CONCEPT OF NOISE DOSE

Dose is derived from exposure to an amount of 'something' over a period of time. In the concept of noise dose there is an acceptance that below the nominated lower action value the risk of harm to the health of the average person is unlikely.

Noise dose is often referenced over an eight hour period, for example, the Control of Noise at Work Regulations (CNWR) 2005 specifies a lower personal noise exposure action value of 80 dB (A) over an eight hour period. Therefore action is required if this level is exceeded. The general duty to ensure the dose is as far below the action value as is reasonably practicable remains, even though harm is unlikely.

The concept of noise dose uses the principle of equal sound energy, such that a sound of a given sound pressure level for a given time is equally hazardous to a person's hearing as one of half the sound pressure level for twice the time.

Thus, where a worker is exposed to a noise for a shorter period than eight hours the equal energy concept of noise dose allows an increase of sound pressure level of 3 dB for every halving of time of exposure. For example, compared with a noise dose of 80 dB (A) over an eight hour period a worker would be allowed 83 dB (A) for a four hour period, 86 dB (A) for a 2 hour period.

The concept of noise dose may also be applied in the other way, in that if a worker is exposed to a high sound pressure level noise they only need to be exposed for a short period to be at risk of harm to their hearing.

For example, a worker that moves into an area with a sound pressure level of 95 dB (A) for a period of 15 minutes, without personal protection, is at risk of harm to their hearing.

dB (A)	80	83	86	89	92	95	98	101	104
Time	8 hr	4 hr	2 hr	1 hr	30 min	15 min	7.5 min	3.75 min	1.875 min

Figure B6-7: Noise dose.   *Source: RMS.*

Where the amount of exposure varies over a period of exposure the dose over that period is calculated using the equal energy concept and a dB L$_{Aeq}$ is calculated, which is the average sound pressure level over the period or level of constant noise of equivalent energy.

B6.2 - Effects of noise on the individual

The physiology of the ear in relation to the mechanism of hearing

Sound is received by the external ear and transmitted along the ear canal, setting the tympanic membrane (the ear drum) in motion. This motion is transmitted via the middle ear bones (the ossicles) to the inner ear, a liquid-filled cavity of complex shape lying within the bony structure of the skull.

The ossicles consist of three bones known as the malleus ('hammer'), incus ('anvil') and stapes ('stirrup') which provide a threefold mechanical amplification of the sound vibrations. This causes the fluid in a portion of the inner ear, known as the cochlea, to move in response to the pressure wave created by the vibrations moving the membrane of the oval window.

The pressure wave passes through the fluids of the two canals, the scala vestibule and the scala tympani, causing movement of the fluids at the same frequency of the original sound waves in contact with the ear drum.

This pressure wave is transferred to the fluid in the scala media and via the basilar membrane to the organ of corti. The organ of corti contains the hair cells that sense the pressure wave.

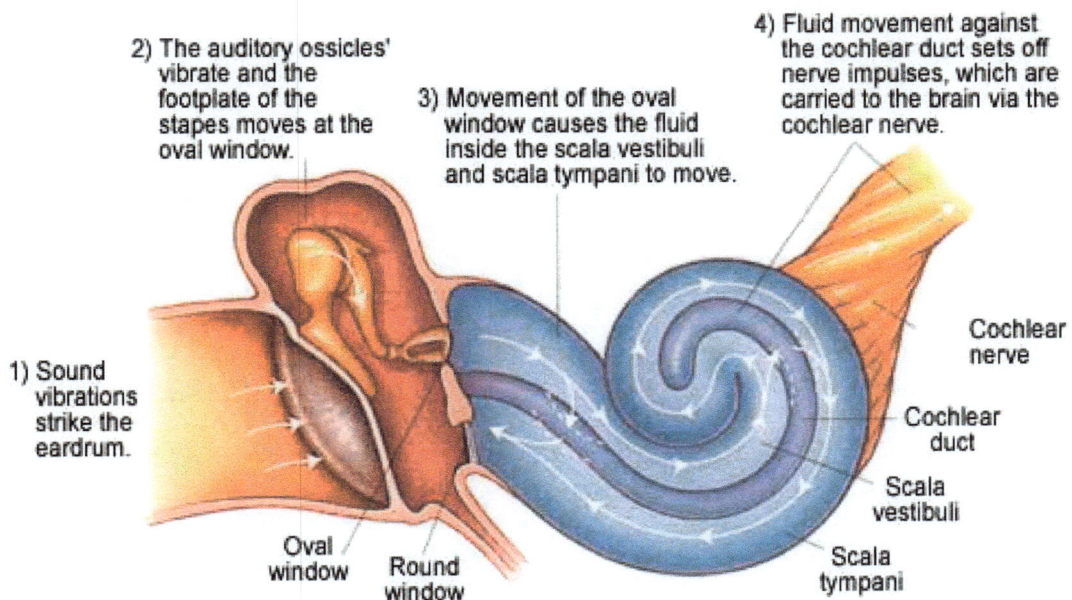

Figure B6-8: Mechanism of hearing.   *Source: Hearing Central LLC.*

The hair cells attached to the basilar membrane of the cochlear duct only move when they recognise a specific frequency. When this occurs they push against a specific part of the basilar membrane. This in turn, stimulates the sensor neurons that carry the impulses to the auditory centres of the brain, where they are decoded into what we recognise as the sensation of sound.

The basilar membrane is very narrow at the beginning (base) of the cochlea and becomes three to four times wider at the helicotrema (apex).

A sound pressure with a high frequency will only affect the initial part of the basilar membrane, which is stiffer than the final segment that responds to low frequencies.

The normal, young, healthy human ear can detect sound in a frequency range of between 20 Hz to 20,000 Hz.

The ear is most sensitive to frequencies of between 1,000-4,000 Hz. Above and below these frequency limits the sound intensity of the threshold of hearing changes.

The threshold of hearing at 100 Hz is 100 times greater, i.e. 20 dB more than the threshold at 1,000 Hz. If a low frequency sound (for example, 100 Hz) and a middle frequency sound (for example, 1,000 Hz) of equal intensity were played alternately the low frequency would sound much quieter.

The higher frequency waves are sensed at the base of the cochlea opening and the lower frequency waves are sensed towards the apex.

It should be realised that the cochlea can respond to both air conduction (the mechanism of hearing described above) and to bone conduction, where the sound directly vibrates the skull or the ear canal walls, which in turn stimulates the cochlea.

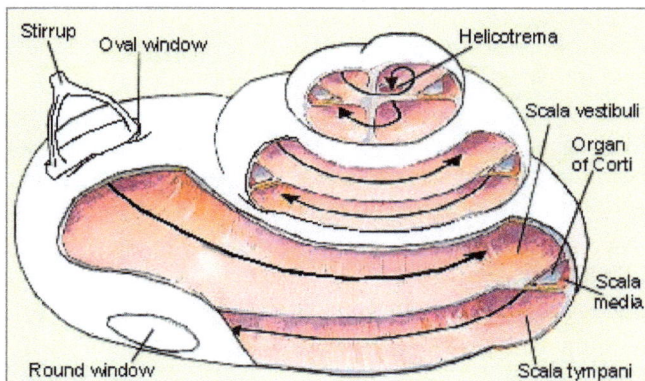

Figure B6-9: The cochlea.   Source: Hearing Central LLC.

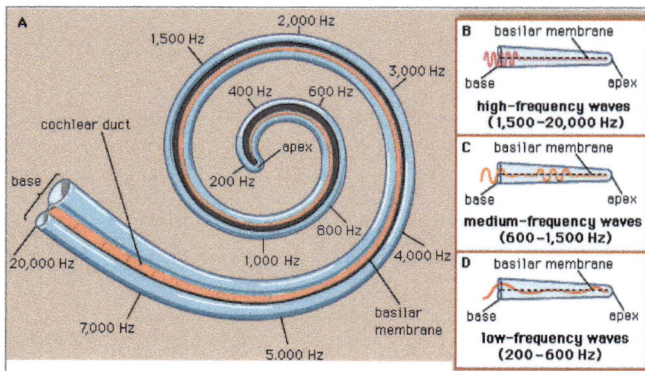

Figure B6-10: Detection of frequencies by the cochlea.
Source: Hearing Central LLC.

For a person with normal hearing, bone conduction sensitivity is much lower than its corresponding air conduction sensitivity, although both paths make a contribution to the overall sound. In practice, this means that even if the auditory canal was completely blocked the ear could still detect some sound.

The physical and psychological effects on the individual

PHYSIOLOGICAL EFFECTS OF EXPOSURE TO HIGH NOISE LEVELS

Types of hearing loss (acute and chronic physiological effects)

Causes of hearing loss include:

1) *Sensorineural loss* caused by such factors as:
- *Congenital deafness,* for example, associated with the mother's exposure to rubella, flu and some drugs during pregnancy.
- *Ototoxic drugs,* for example, some antibiotics such as streptomycin, anti-rheumatic drugs and diuretics.
- *Presbycusis,* for example, the normal decline of hearing sensitivity at the higher frequencies as a person ages.

2) *Conductive loss* such as ear wax blocked eustachian tube and ruptured ear drum.

3) *Central deafness* whereby the hearing mechanism is in good condition but the brain does not recognise sounds.

4) *Acoustic trauma* is the acute effect of exposure to very high noise levels is acoustic trauma. This involves sudden damage caused by exposure to a burst of very high energy noise (for example, bomb blasts) which can cause physical rupture of the eardrum and displacement of the ossicles.

5) *Occupation deafness* can be categorised into two main types, which are based on their acute and chronic effects respectively:
- *Temporary threshold shift (acute).* Short periods of exposure to excessive noise levels produce varying degrees of inner ear damage that is initially reversible. This auditory fatigue is known as *'temporary threshold shift' (TTS).* As its name implies, this type of exposure produces an elevation of the hearing threshold which progressively reduces with time after leaving the excessively noisy environment. The time taken to   recover from the temporary threshold shift may be anything from a few minutes to days, depending upon the degree of exposure.
- *Permanent threshold shift (chronic).* Permanent damage, known as *'noise induced hearing loss' (NIHL)*, occurs when exposure to excessive noise continues over a long period of time.

The full relationship between temporary threshold shift and noise induced hearing loss is not fully understood. It is possible for a person with noise induced hearing loss to be affected by temporary threshold shift due to exposure to noise, but the degree of temporary threshold shift is reduced by the extent of the noise induced hearing loss.

Both the temporary threshold shift and the permanent noise induced hearing loss are frequency dependent insofar as the greatest loss generally occurs at frequencies about one half to one octave higher than the frequency of the noise source.

However, noise induced hearing loss generally occurs first in the 4,000 Hz octave band. This is commonly known as the *'4k dip'*. If exposure continues over a number of years, the hearing loss at 4,000 Hz increases along with losses in lower octave bands. Not everyone develops the same hearing loss when exposed to the same noise. Similarly, not everyone has the same level of threshold - anything up to 20 dB deviation from the accepted standard has been noted.

There is generally a long latency period before the effects of noise induced hearing loss may be identified. For example, the HSE determined that continuous occupational exposure to noise at 90 dB (A) would result in about 5% of the population sustaining a 30 dB hearing loss (considered moderate disability) within 10 years, but would be likely to rise to approximately 50% over a working lifetime of exposure, though some of this hearing loss would be the result of the normal ageing process.

6) **Tinnitus** is a condition associated with excessive noise. When the cochlea becomes damaged, as a consequence of exposure to excessive noise or an infection, it stops sending information to the brain. The brain then effectively seeks out signals from the parts of the cochlea that are still working. These signals become over represented in the brain and cause the sounds associated with tinnitus. Sufferers of NIHL often complain of a high-pitched tinnitus, usually intermittent at first, becoming continuous in up to 20% of cases. Tinnitus produces 'phantom' noise in the ear of varying degrees of seriousness and can lead to disturbed sleep.

7) **Loudness recruitment** is also associated with NIHL. When damage to the cochlea or the brain has occurred, the perception of loudness is altered. The person cannot hear sounds below a certain level and, when sounds rise above this level, 'normal' hearing is suddenly restored and the listener turns the volume down or asks the speaker not to shout.

Research supporting the regulatory impact of the introduction of the Control of Noise at Work Regulations (CNWR) 2005 determined that over 1.1 million people are exposed to noise levels above 80 dB(A) at work, with an estimated 170,000 suffering deafness, tinnitus or other ear conditions as a result.

Association of British Insurers (ABI) figures show that deafness accounts for approximately 80% of occupational disease claims up to 1997. The number of cases has fallen since then, which is likely to be due to the decline in heavy industry in the UK.

Noise level	80-85 dB(A)	85-90 dB(A)	90-95 dB(A)	95-100 dB(A)	100-110 dB(A)	>110 dB(A)
People exposed	1,097,000	696,800	273,000	124,000	37,100	4,200

Figure B6-11: Number of people exposed to noise in the workplace. Source: HSE.

The 2008/09 Labour Force Survey (LFS) shows an estimated 17,000 individuals who worked in the previous 12 months, who believed their hearing problems were the most serious of their work related illnesses.

Figure B6-12: Significance of hearing loss in the workplace. Source: HSE.

PSYCHOLOGICAL EFFECTS ON THE INDIVIDUAL

Noise can also cause or contribute towards other work-related stress, causing a loss of concentration, fatigue, and tension, and increasing the risks of ill-health associated with stress.

Damage to the hearing can result in reduced social interaction and a feeling of isolation by the person affected. NIHL particularly affects the frequencies that relate to speech, causing the sufferer to have difficulty hearing sounds like 't', 'd' and 's'.

This means that they often have difficulty following and contributing to conversations or listening to speech on the radio or television. This can lead to strong psychological effects that can include lack of confidence, anxiety about situations where speech is involved and depression.

Medical science cannot currently replace the full sound sensing features of a cochlea, and technology such as surgical implants and hearing aids are extremely poor substitutes for normal hearing as they merely magnify sound without correcting distortion.

In essence, even relatively moderate hearing damage can lead to a significant drop in the sufferer's quality of life and all reasonable steps should be taken to prevent impairment.

Health surveillance

THE LEGAL REQUIREMENTS FOR AUDIOMETRY

The aim of the Control of Noise at Work Regulations (CNWR) 2005 is to ensure that workers' hearing is protected from excessive noise at their place of work that could cause harm to their hearing.

Regulation 5 requires an employer who carries out work which is liable to expose any employees to noise at or above a lower exposure action value, to make a suitable and sufficient assessment of the risk from that noise to the health and safety of those employees. If the noise assessment indicates that there is a risk to the health of employees, the employer must ensure that the employees are placed under suitable health surveillance, which must include testing of their hearing. Employees, when required, must attend audiometric health surveillance, at the cost of the employer and during normal working hours.

Regulation 9 requires the employer to:

- Ensure a record of the results of health surveillance is kept available in a suitable form.
- Allow an employee access to their personal health record.
- Provide the enforcing authority with copies of such health records as it may require.
- If an employee is found to have hearing damage, ensure that the employee is examined by a doctor.

If the doctor or any specialist to whom the doctor considers it necessary to refer the employee considers that the damage is likely to be the result of exposure to noise, the employer must:

- Ensure that a suitably qualified person informs the employee accordingly.
- Review the risk assessment.
- Review any measure taken to comply with CNWR Regulations 6 - Control of Noise, 7 - Hearing Protection and 8 - Maintenance and Use of Equipment.
- Consider assigning the employee to alternative work where there is no risk from further exposure to noise.
- Ensure continued health surveillance and provide for a review of the health of any other employee who has been similarly exposed.

THE USE OF AUDIOMETRY TO MEASURE HEARING AND HEARING LOSS

Audiometry is a technique for evaluating the degree of hearing loss or impairment over the range of frequencies most necessary for normal conversation (4-6 kHz). For occupational purposes audiometry can be used for the early detection and the assessment of the degree of noise induced hearing loss. Perhaps a less obvious advantage to audiometry is that it can identify those whose individual differences mean that they are not adequately protected even below the action values.

Employers should consult their employees or their health and safety representatives before introducing audiometric health surveillance. It is important that everyone understands that the aim of audiometric health surveillance is to protect the worker's hearing. Understanding and co-operation are essential if the audiometric health surveillance is to be effective.

METHOD

The subject is placed in a booth which is soundproof in order to mask ambient noise. Headphones are fitted and tones, which are generated by an audiometer, are played in sequence to each ear in turn. The audiometer generates pure tones at 0.5, 1, 2, 3, 4, 6 and 8 kHz at intensities which are increased in 5 dB steps until the subject responds by pushing a button.

The reaction is recorded and a graph generated. Hearing levels from -10 dB to over 90 dB can be recorded. For obvious reasons it is necessary to ensure that readings are not affected by factors such as earwax and other obstructions, a heavy cold or from temporary threshold shift due to having been exposed to high levels of noise.

INTERPRETATION

The 'normal' subject will show an almost horizontal line high up on the chart with a slight dip at high frequencies depending on the subject's age (due to presbycusis).

A flat audiogram curve lower down on the chart indicates a similar hearing loss at all frequencies (indicative of conductive loss).

The classic pattern of occupational noise induced hearing loss is the *'four-K-dip'*.

This is the dip or notch centred on the 4 or 6 kHz region, which becomes larger as the years of exposure progress.

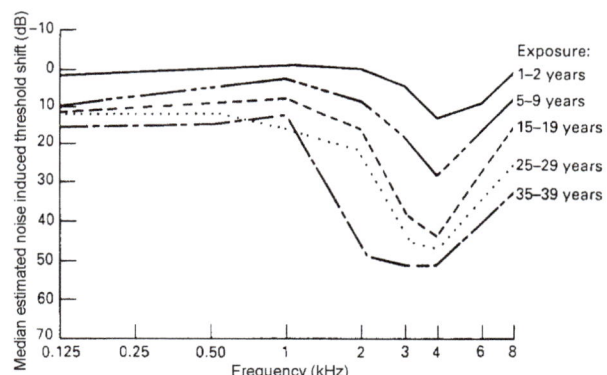

Figure B6-13: Audiogram.   *Source: Ambiguous.*

USE OF RESULTS

Audiometric health surveillance can be part of pre-employment health screening, where it is used to establish a reference level of hearing ability and identify those people who have a hearing loss that will require particular arrangements to be made to protect their remaining hearing. Subsequent to pre-employment screening, audiometric health surveillance is used to identify early signs of noise induced hearing loss in order to establish interventions to protect hearing and to confirm the degree to which noise controls are effective.

For many individuals the audiometry will show that there is no significant hearing problem, but some may find that their hearing is in the early stages of deterioration. Workers should be informed that whatever the outcome, the results should be viewed as an opportunity to identify any deterioration at an early stage and to ensure that appropriate measures are taken to prevent any further harm.

ADVANTAGES AND DISADVANTAGES OF AUDIOMETRY PROGRAMMES

Advantages:

- Pre-employment audiometric health screening following a job offer will ensure a reference level hearing ability is determined, against which subsequent measurements can be compared in the future.
- Reduced claims for noise induced hearing loss.
- Reduced insurance premiums.
- Identifies high risk susceptibility of new workers if carried out three months from work commencement.
- Improves worker moral, demonstrates management commitment to ensure healthy workforce.
- Enables the effectiveness of controls to limit noise exposure to be evaluated.
- Ensures best equipment (low noise footprint) is sourced and provided for the work to be done.

Disadvantages:

- Increased costs to the business for health screening and medical referral.
- Increased costs to remedy current controls and systems of work if a group of workers is identified with hearing loss.
- It is not uncommon for some that already suspect that their hearing is deteriorating to fear being transferred into a lower paid job or dismissal when the results confirm their concerns.

CIVIL LAW IMPLICATIONS OF AUDIOMETRY

The Control of Noise at Work Regulations (CNWR) 2005, Regulation 9 requires the employer to inform employees of the results of audiometric testing if the audiogram indicates that damage to the employee's hearing is likely to be the result of exposure to noise.

This will potentially expose the employer to civil action for the loss. If a reference audiogram has been taken and a subsequent one shows hearing loss that was not evident from the first audiogram, this may provide significant evidence to support a common law claim of negligence and breach of statutory duty in failing to control noise in accordance with the Control of Noise at Work Regulations (CNWR) 2005.

However, a programme of health screening will help to identify early signs of the effects of noise and may prevent claims or provide part of a defence in negligence, providing some evidence that reasonable care was taken. In some cases, where audiometry is conducted prior to employment it may be possible to see evidence that hearing loss experienced by a worker was in existence before they were taken into employment.

It is therefore important to conduct audiometry to screen all new employees likely to be exposed to noise at work and establish a reference level of hearing (base line hearing profile) before they are put at risk.

It is also good practice to measure hearing at a three month period after work commences to identify if the individual is unusually susceptible to excessive noise risk.

B6.3 - Measurement and assessment of noise exposure

Noise risk assessment

Employers have a duty under the Control of Noise at Work Regulations (CNWR) 2005 to reduce the risk of hearing damage to their employees by controlling exposure to noise. There is a further requirement that the employer obtain an adequate noise assessment, which will enable compliance with duties to control noise exposure.

The employer should make a suitable and sufficient assessment of the risk created by work that is liable to expose workers to risk from noise to assist in meeting this responsibility. An assessment will establish which workers are at risk of hearing damage, and the level of risk. It will also identify sources of noise that particularly contribute to the noise level workers are exposed to, for example, equipment and specific activities.

This will enable analysis of the options to control the noise at source or by other means. The assessment will also help the employer when providing suitable hearing protection, marking out ear protection zones and giving information, instruction and training to workers. The record of the assessment of the risks from noise and controls required or implemented provides evidence of a structured risk assessment process for noise exposure in the workplace.

When conducting the risk assessment, consideration should be given to workers at particular risk. The assessment will need to identify the type, duration, effects of exposure including any additional exposure at the workplace (for example, in rest facilities) to determine whether the exposure limit/action values have been exceeded. Information from health surveillance records and manufacturer's information on noise levels should also be reviewed.

The level of noise workers are exposed to should be assessed by:

- Observation.
- Reference to information on expected levels for work conditions and equipment.
- If necessary by measurement of the level of noise to which their workers may be exposed.

It is not always necessary to carry out measurements of noise exposure as part of the assessment; an estimate of noise levels may be enough to decide that controls are required. Estimation by observation may be sufficient to indicate there is a noise problem. This is where the assessor determines how easy it is for two people to hold a conversation at a distance of one and two metres from each other.

Test	Probable noise level	A risk assessment will be needed if the noise is at the level for more than:
The noise is intrusive but normal conversation is possible	80 dB	6 hours
You have to shout to talk to someone two metres away	85 dB	2 hours
You have to shout to talk to someone one metre away	90 dB	45 minutes

Figure B6-14: Simple tests to get a rough estimate on whether a risk assessment is required.   Source: RMS.

- **Two metre rule:** If the conversation is difficult (need to raise the voice or repeat words) at a distance of two metres apart the noise level is likely to be above 85dB.
- **One metre rule:** If the conversation is difficult (need to raise the voice or repeat words) at a distance of one metre apart the noise level is likely to be above 90dB.

Similarly, it may be possible to determine if there is a potential noise problem by considering expected levels of noise for work conditions and equipment.

This may be carried out by considering noise data provided by the manufacturer or other industry based data obtained in similar work situations, this can help to indicate expected noise levels. If it is necessary to be more certain whether noise exposure levels exceed acceptable values, *measurements* of actual noise levels may be preferred.

A detailed noise risk assessment should include consideration of:

- The risk of hearing impairment, impairment of communications, fatigue.
- The identification of sources, tasks.
- The expected noise emission levels from equipment.
- The expected time of exposure.
- Planning (who, how, where, how often).
- The types of instrumentation.
- The importance of calibration.
- The types of measurements to be taken.
- The use of specialist noise consultants.
- The interpretation and evaluation of results.
- The use of noise calculators to determine mixed exposures.
- Comparison with legal limits to make control decisions.

Instrumentation used for the measurement of noise

TYPES

There are 4 basic types of sound pressure level (SPL) meter:

Type 0	Very accurate meter designed for use as a laboratory reference standard. Has a high standard of tolerance over a wide frequency range. Not generally used in the 'field'.
Type 1	Used in laboratories or in the field where high precision is required.
Type 2	General field meter.
Type 3	Basic level indicator which could be used to establish if noise limits were being exceeded. The least accurate type of noise survey meter.

The instrument used for noise surveys should be at least a type 2 sound pressure level meter and should meet at least type 2 of BS EN ISO 61672-1:2003.

METHODOLOGY

Noise is measured using a sound pressure level meter which works in simple terms, by converting pressure variations into an electric signal. This is achieved by capturing the sound with a microphone, pre-amplification of the resultant voltage signal and then processing the signal into the information required, dependent on the type of meter (for example, 'A', 'B', 'C', or 'D' weighting, integrating levels, fast or slow response). The microphone is the most important component within the meter as its sensitivity and accuracy will determine the accuracy of the final reading. Meters can be set to fast or slow response depending on the characteristics of the noise level. Where levels are rapidly fluctuating, rapid measurements are required and the meter should be set to fast time weighting.

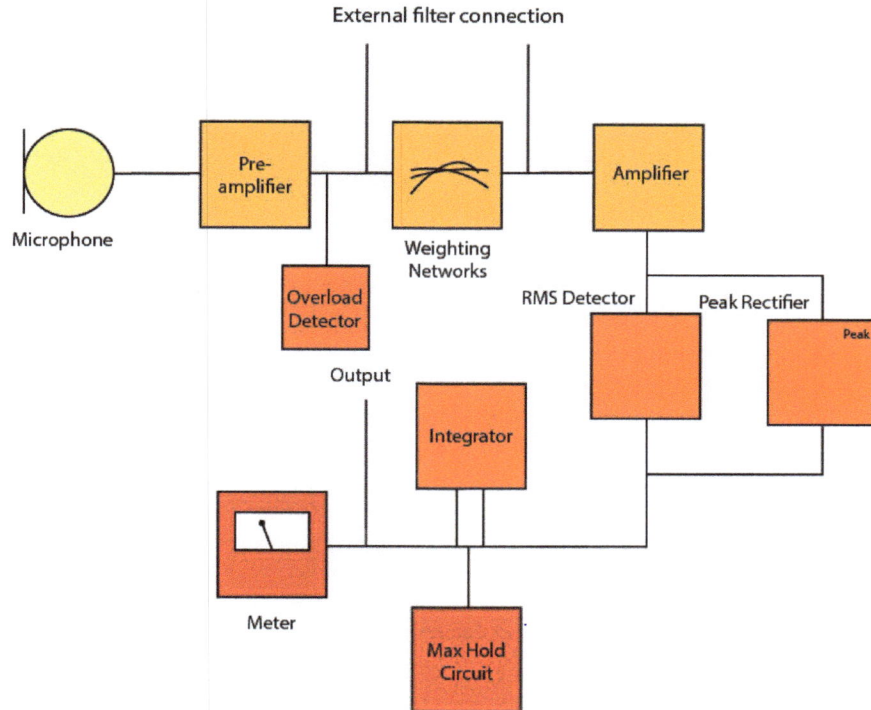

Figure B6-15: The basic integrating sound level meter. *Source: RMS.*

Meters are used to measure the:

- *Sound pressure level (L_A)* - the A weighted intensity of sound at a given moment in time at a given position (i.e. the instantaneous level) or an un-weighted linear measurement dB (Lin).
- *Equivalent continuous sound pressure level (L_{Aeq})* - an average measure of intensity of sound over a reference period, usually the period of time over which the measurement was taken. Measured in dB (A). As the intensity and spacing of the noise levels usually vary with time, an integrating meter is used. This meter automatically calculates the LAeq by summing or integrating the sound pressure level over the measurement period.
- *Peak pressure level* - under the Control of Noise at Work Regulations (CNWR) 2005 the exposure limit value for a peak sound pressure is 140 dB (C-weighted). Some sound pressure level meters produce peak pressure values for impulsive noise or where a fast time weighting reading exceeds 125 dB (A).

Figure B6-16: Sound level meter. *Source: Cirrus Research plc.*

An instrument used for noise surveys should be at least a type 2 integrating sound pressure level meter, capable of taking readings at:

- A weighted L_{eq}.
- C-weighted maximum peak sound pressure to above 140 dB (C).

An octave band analysis feature together with C-weighted L_{eq} would be needed to assess ear protection and noise control measures. *See 'Frequency analysis' later in this Element.*

Ideally, sound pressure level meter readings should be taken at the worker's ear, without any extraneous reflections from the operator, the meter itself or the subject's body. The meter should be either held at arm length or, ideally, mounted on a tripod. In practice, mounting on a tripod is rarely achievable in a typical workplace where the worker is moving around.

The person using the meter must therefore try to minimise this by careful positioning of the microphone to ensure a clear path between the (appropriate) worker's ear and the source of the noise(s). Due to the physical layout of the workplace, or the hazards present, this 'ideal' positioning may not be possible. In these cases personal dosimetry may have to be considered. This involves the use of a portable device that is fitted to the worker and obtains a noise dose by recording all the noise the worker is exposed to over a period of time.

CALIBRATION

When undertaking measurements of noise it is essential that the measurement equipment provides valid data. The regular calibration of such equipment is therefore an important requirement. This takes place on two levels: field calibration at time of use and verification calibration every two years by a competent specialist, using a process that meets the BS 7580 'Specification for the verification of sound level meters', comprehensive procedure or shortened procedure for type 2 sound level meters.

Calibration at time of use utilises a portable acoustic calibrator in the form of a signal generator placed over the microphone that produces a specific sound output at a specific frequency, for example, 94dB at 1kHz. This is used to check the sound pressure level meter is reading accurately or whether it requires adjustment, either by the user where this is available or by a specialist technician.

A calibration check of this kind is conducted before and after a noise assessment session. It is important that the calibration device is at least the same type as the instrument it is calibrating, for most circumstances this will be type 2. The calibration device much match the specified microphone design used on the sound pressure level meter; it may have to be used with an adapter so that it matches properly.

Some sound pressure level meters have an internal electronic calibration; although this only checks the meter's electronics and does not check the microphone, it may be a useful cross-check. Details of the calibration device used to calibrate equipment must be noted as part of the record of the assessment. Calibration devices are also subject to verification of accuracy at intervals, in the same way that a sound pressure level meter would be. For a sound pressure level meter reliant on batteries, a check of their status is included as part of the calibration process; the instrument usually provides a facility to check this as a direct reading.

BS 7580 specifies a range of tests to verify the accuracy of the measurement instrumentation. As a general requirement the standard recommends that verification shall be performed at least every two years. The majority of the test procedures are carried out with an electrical signal substituted for the measurement microphone with a series of test signals being used to assess the performance of the sound level meter in respect of noise, linearity, frequency weightings, time weightings, peak response, RMS accuracy, time averaging, pulse range, sound exposure level and overload indication.

Many of the test signals used to assess performance are complex and are not usually available to the majority of people involved with the measurement of sound on a day-to-day basis. The test for RMS accuracy, for example, requires that a reading obtained from a continuous 2 kHz sinusoidal signal be compared with that obtained from a sequence of tone bursts consisting of 11 cycles of a 2 kHz sine wave repeated 40 times a second and with amplitude 6.6 dB higher than the continuous signal.

The practical use of instrumentation

TERMINOLOGY

$L_{EP,d}$	Daily personal noise exposure level.
LA_{eq}	Equivalent 'A' weighted continuous sound level.
L_A	'A' weighted sound pressure level, dB (A).
L_C	'C' weighted sound pressure level, dB(C).
L_{WA}	'A' weighted sound power level, dB (A).
Octave band	a band of frequencies, Hz.

CALCULATION OF LEP,D

The daily personal noise exposure is a measure of the noise energy a person receives over the working day (8 hours); in other words their noise dose. $L_{EP,d}$ is numerically the same as the equivalent continuous sound level (Leq) measured over a full shift and normalised to 8 hours.

It is the measure of the true average level over 8 hours calculated from the energy dose received over a full working day. The person carrying out the assessment should take representative readings that reflect the worker's working day, including the various positions and tasks that they will undertake (i.e. 'A' weighted Leq readings).

Once these fractional exposures have been taken, then the L$_{EP,d}$ can be determined by either calculation, nomograms or HSE's noise exposure ready reckoner.

Calculating fractional exposures:

Step 1	Determine the fractional noise exposure values (f) for each noise exposure, using the formula:
	$$f = \frac{t}{8} \, anti\log[0.1(L - 90)]$$ Where t = exposure time in hours L = 'A' weighted L$_{eq}$
Step 2	Add together the fractional exposure values to give a total fractional exposure: f_T
Step 3	Determine the daily noise exposure given by: LEP,d = 90 + 10 log f_T

Example: A person works at a machine for 2 hrs 15 minutes per day. The 'A' weighted L$_{eq}$ is measured at 102 dB (A). The remainder of the day is spent in an area which is below 70 dB (A) - this is low enough to be ignored.

Calculation	1	Convert dB (A) to B (A) =	10.2 B (A)
	2	Convert to antilog =	1.58 x 10^{10}
	3	Multiply by t (duration in hours) =	3.57 x 10^{10}
	4	Divide by 8 =	4.46 x 10^9
	5	Convert to log =	9.65
	6	Multiply by 10 =	96.5
		LEP,d =	96.5 dB (A)
		Note: where there are multiple levels, sum the values for each exposure at step 3.	

For exposures at different levels: the HSE ready-reckoner (calculation table) provides a simple way of working out the daily personal noise exposure of employees, based on the level of noise and duration of exposure.

Noise exposure points can be used to help prioritise the noise control programme by showing which tasks contribute the most to the overall noise exposure. Dealing with these first will have the greatest effect on reducing the personal noise exposure.

The following ready-reckoner (calculation table) and examples have been taken from HSE Guidance on Regulations Controlling Noise at Work: L108.

An employee has the following typical work pattern: 5 hours where a 'listening check' suggests the noise level is around 80 dB; 2 hours at a machine for which the manufacturer has declared 86 dB at the operator position (a listening check suggests this is about right); 45 minutes on a task where noise measurements have shown 95 dB to be typical.

	Noise level	*Duration*	*Notes*	*Exposure points*
O	80	5 hours	No column for 5 hours, so add together values from 4 and 1 hour columns in row corresponding to 80dB.	16 + 4 = 20
O	86	2 hours	Directly from table.	32
O	95	45 minutes	No column for 45 minutes, so add together values from 30 and 15 minute columns in row corresponding to 95dB.	65 + 32 = 97
Total noise exposure points				149
L$_{EP,d}$				86 to 87 dB

Figure B6-17: Ready-reckoner (calculation table).   Source: HSE, Reducing Noise at Work, L108.

Sound pressure level, L_{Aeq} (dB)	Duration of exposure (hours)							
	1/4	1/2	1	2	4	8	10	12
95	32	65	125	250	500	1000		
94	25	50	100	200	400	800		
93	20	40	80	160	320	630		
92	16	32	65	125	250	500	625	
91	12	25	50	100	200	400	500	600
90	10	20	40	80	160	320	400	470
89	8	16	32	65	130	250	310	380
88	6	12	25	50	100	200	250	300
87	5	10	20	40	80	160	200	240
86	4	8	16	32	65	130	160	190
85		6	12	25	50	100	125	150
84		5	10	20	40	80	100	120
83		4	8	16	32	65	80	95
82			6	12	25	50	65	75
81			5	10	20	40	50	60
80			4	8	16	32	40	48
79				6	13	25	32	38
78				5	10	20	25	30
75					5	10	13	15

Total exposure points	Noise exposure $L_{EP,d}$ (dB)
800	94
630	93
500	92
400	91
320	90
250	89
200	88
160	87
130	86
100	85
80	84
65	83
50	82
40	81
32	80
25	79
20	78
16	77

Figure B6-18: Noise exposure points - worked example of daily exposure.   *Source: HSE, Reducing Noise at Work, L108.*

The pattern of noise exposure gives an $L_{EP,d}$ of between 86 and 87 dB. The priority for noise control or risk reduction is the task involving exposure to 95 dB for 45 minutes, since this gives the highest individual noise exposure points.

Weekly noise exposure ready-reckoner

The weekly noise exposure level ($L_{EP,w}$) takes account of the daily personal exposures for the number of days worked in a week (up to a maximum of seven days).

It may be calculated using the formula given in Schedule 1 Part 2 of the Regulations (CNWR).

	$L_{EP,d}$	Day	Notes	Exposure points
○	80	1, 2, 5	Work in general area away from major noise sources.	32 + 32 + 32 = 96
○	86	3	Work alongside others, including work adjacent to a major noise source.	130
○	92	4	Work on equipment that constitutes a major noise source.	500
Total noise exposure points				726
$L_{EP,w}$				86 to 87 dB

Figure B6-19: Weekly noise exposure ready-reckoner (calculation table).   *Source: HSE, Reducing Noise at Work, L108.*

In this example the worker's work pattern varies significantly during the week and although the worker is exposed to significant noise levels on one of the days when considered as a weekly exposure this is a lot lower.

Again the priority for noise control or risk reduction is the tasks on day 4 involving exposure that leads to an LEP,d of 92, since this gives the highest individual noise exposure points.

Daily noise exposure, $L_{EP,d}$	Points							Total exposure points	Weekly noise exposure, $L_{EP,w}$
	Day 1	Day 2	Day 3	Day 4	Day 5	Day 6	Day 7		
95	1000	1000	1000	1000	1000	1000	1000	5000	95
94	800	800	800	800	800	800	800	4000	94
93	630	630	630	630	630	630	630	3200	93
92	500	500	500	500	500	500	500	2500	92
91	400	400	400	400	400	400	400	2000	91
90	320	320	320	320	320	320	320	1600	90
89	250	250	250	250	250	250	250	1300	89
88	200	200	200	200	200	200	200	1000	88
87	160	160	160	160	160	160	160	800	87
86	130	130	130	130	130	130	130	630	86
85	100	100	100	100	100	100	100	500	85
84	80	80	80	80	80	80	80	400	84
83	65	65	65	65	65	65	65	320	83
82	50	50	50	50	50	50	50	250	82
81	40	40	40	40	40	40	40	200	81
80	32	32	32	32	32	32	32	160	80
79	25	25	25	25	25	25	25	130	79
78	20	20	20	20	20	20	20	100	78

Figure B6-20: Weekly noise exposure - worked example based on HSE table. *Source: HSE, Reducing Noise at Work, L108 and RMS.*

CALCULATION OF L$_{AEQ}$

Used to describe noise whose level varies with time; L_{Aeq} is defined as the A-weighted energy average of the noise level, averaged over the measurement period.

It can be considered as the notional continuous steady noise level which would have the same total A-weighted acoustic energy as the real fluctuating noise measured over the same period of time. Thus, to maintain the L_{Aeq} when the sound pressure level is doubled (increased by three decibels) exposure must be halved.

Noise levels vs. exposure times for an equivalent noise level (L_{Aeq}) of 90 dB(A)	
87	16 hours
85	8 hours
93	4 hours
96	2 hours
99	1 hours
102	0.5 hours
105	0.25 hours

Figure B6-21: Calculation of Leq. *Source: RMS.*

FREQUENCY ANALYSIS

Complex noise can be analysed by examination of its frequency spectrum. This consists of a value of sound pressure level for each frequency or frequency band (i.e. a section of the frequency spectrum).

For the purposes of examining occupational noise these bands have a bandwidth of one octave. An octave band is a band where the highest frequency is twice the lowest frequency. The band is denoted by its centre frequency.

For example the 1 kHz octave band includes all frequencies between the ranges 707-1,414 Hz:

- Lowest frequency (Hz): 707.
- Centre frequency (Hz): 1000.
- Highest frequency (Hz): 1414.

When measuring a noise source, such as a machine, the level of background noise can have an influence. Obviously the background noise must not be greater (or drown out) than the noise level of interest. For practical purposes the sound pressure level of the background noise must be at least 3 dB less than the sound pressure level of the noise source of interest. A correction level may, however, be required to ensure accuracy.

This can be achieved by:

1) Measuring total noise with the noise source running (for example, a machine).
2) Measuring the background noise with the noise source switched off.
3) Calculating the difference between 1 and 2 above. If this is less than 3 dB then the background noise is too high for accurate measurement. If between 3 and 10 dB then a correction is necessary. No correction is necessary if the difference is greater than 10 dB.
4) Using the chart shown to find the correction factor. The x-axis is the difference value found in step 3. The y-axis gives the correction value.
5) Subtracting the correction value from the original noise source reading taken at step 1. This gives the noise level of the source.

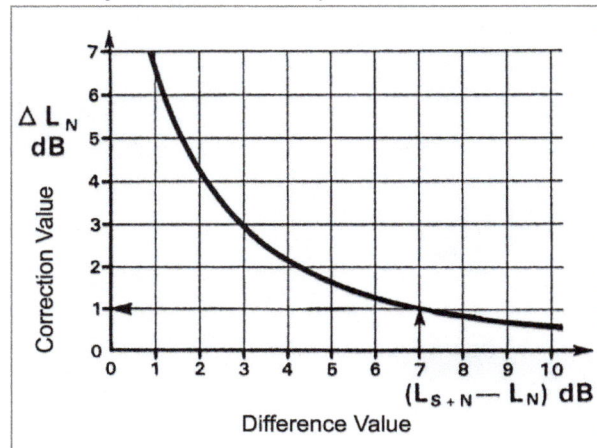

Figure B6-22: Chart for measuring the sound level under conditions of high background noise. *Source: Measuring sound, Brüel and Kjaer.*

Methodology of undertaking workplace noise surveys and personal noise exposure assessments

LEGAL REQUIREMENTS

The purpose of a noise assessment, as required by Regulation 5 of the Control of Noise at Work Regulations (CNWR) 2005, is to:

- Identify whether people are at risk from noise exposure causing hearing damage.
- Determine the level of risk, by quantifying daily or weekly personal noise exposure level.
- Identify priorities for action in noise reduction or protection.

This demonstrates that an employer has considered:

- All the factors relating to the risks from noise exposure.
- The steps which need to be taken to achieve and maintain adequate control of the risks.
- The need for health surveillance.
- How any further action plan for noise reduction or protection will be implemented.

An employer is required to conduct a noise assessment when any employee is likely to be exposed to noise at or above the lower exposure action value identified in the Control of Noise at Work Regulations 2005.

Factors to be considered include:

- Level, type and duration of exposure.
- Employees or groups of employees whose health is at particular risk, such as those with pre-existing conditions, young people and pregnant women.
- Any effects from the interaction between noise and the use of toxic substances at work.
- Any effects from the interaction between noise and vibration.
- Indirect effects resulting from the interaction between noise and audible warning signals, such as the masking of warning signals by the noise environment or any hearing protection worn.
- Information supplied by the manufacturers of work equipment.
- Availability of alternative equipment with lower noise emission.
- Extension of exposure to noise at the workplace beyond normal working hours, including overtime, rest or lunch breaks.
- Appropriate information obtained following health surveillance, indicating if noise induced hearing loss is developing.
- Availability of personal hearing protectors with adequate attenuation characteristics.

The assessment may take two forms, workplace noise surveys and personal noise exposure assessments.

PLANNING

A noise assessment, workplace noise survey or personal noise exposure assessment, conducted for the purposes of the Control of Noise at Work Regulations (CNWR) 2005 must be conducted by a competent person. **Workplace noise surveys** are concerned with measuring noise levels in an area where employees are likely to be exposed to noise at or above the lower exposure action value, to determine if further action is required to protect exposed workers.

It will assist with the identification of noise sources and provide information of the noise level in areas of the workplace. This is necessary in order to identify areas that exceed the lower and upper exposure action values for the purposes of taking further action, including providing information and signs to warn workers. A workplace noise survey will involve taking a number of noise measurements throughout the workplace in locations where it is believed levels might exceed the lower exposure action value. The readings need to be taken in such a way that they will represent typical levels of noise present in the workplace. This may involve taking a number of readings to identify what the sound pressure level is while different tasks or activities are taking place. This will involve an amount of planning in order to determine the factors that can influence the noise level in the workplace.

Personal noise exposure assessments provide information on exposure of individual workers. They may be necessary where the worker does not receive a steady exposure to noise due to the changing noise level caused by activities or if they move between a number of locations where steady noise exists. Personal noise exposure assessments may provide a more accurate measurement of an individual's exposure as readings are be taken at the position of the worker's head.

Personal noise exposure assessments may also be necessary in circumstances where a specific individual in an affected workgroup has experienced hearing loss identified by an audiogram. The noise assessment, whichever method is used, may involve taking a number of readings over a period of time if processes and activities vary. In order to conduct the noise survey and identify noise from specific sources it may be necessary to plan to operate equipment individually and together to identify individual and combined noise levels.

Factors to consider at time of planning can include:

- Who will conduct the assessment, in particular their competence.
- How the assessment will be done, use of a workplace noise survey and/or a personal noise exposure assessment.
- Where the assessment will be made, including points and timing of measurements as well as workers involved in the personal noise exposure assessment.
- How often measurements will be required to get a representative measurement.
- The identity of workers employed in the area and groups with the same exposure.
- Shift patterns and lengths.
- Working positions.
- Time in positions over the working day/shift.
- Characteristics of activities that influence noise levels.
- Overall noise levels.
- When the assessment will be repeated.

The noise assessments should be repeated whenever:

- Audiometric results indicate that controls are not working.
- New machinery is introduced.
- Workplace layout and/or processes are redesigned.
- Changes in working patterns occur.
- Modifications to machinery are made.
- Equipment becomes worn and noisier.

CHOICE OF INSTRUMENTATION

Observations made at the planning stage will allow the assessor to decide on the type of sound pressure level meter required. For example, whether a static or a hand held sound pressure level meter is required.

A static sound pressure level meter will provide a cumulative reading over a period of time whereas the hand held meter will provide a reading for a short duration. This can mean that several readings need to be taken with a *hand held sound pressure level meter* over a period of time to determine representative exposure levels.

A hand held sound pressure level meter can provide measurements at the worker's head position and can therefore provide a personal noise exposure assessment. Hand held sound pressure level meters have the advantage of being highly portable and because measurements are taken under the direct control of the competent person they are less prone to interference, provide prompt readings and a poor reading may be promptly repeated.

Figure B6-23: Sound level meter with truck. *Source: Cirrus Research plc.*

This instrument is adequate for circumstances where the worker stays in a fixed point for a significant part of their work time or if there is a simple routine where a number of specific readings can be taken to represent the work pattern and noise dose of the worker.

Alternatively a personal sound pressure level meter, sometimes called a dosemeter, may be fitted to the worker to establish an exact profile of the worker's exposure over the work period.

The **dosemeter** provides an alternative means of measuring noise, which is especially useful where a person is highly mobile or working in places where access for measurement is difficult.

Dosemeters should therefore only be used in one of two preferred positions (shoulder or head mounted), worn or clipped to the clothing.

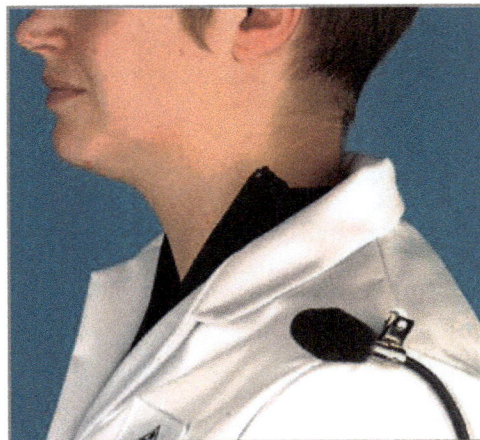

Figure B6-24: Dosemeter microphone.
Source: Controlling noise at work L108 HSE.

Workers selected to wear dosimeters must be given appropriate instruction to ensure results are valid, which includes not interfering with the microphone or instrument during the course of measurements, and limiting their own speech as much as possible as an individual's own voice should not be included in an assessment of their daily personal noise exposure. Dosemeters do not meet the design specifications for Type 1 sound level meters because of the position of the dosemeter. This affects the accuracy of the reading, which will be reduced by reflections from the body. The degree of error will depend on the position of the microphone and the nature and direction of the sound. Dosemeters should be fitted with an overload indicator which is triggered when the sound has exceeded the range of the meter. When this happens the level is not integrated and consequently the dose reading will be inaccurate.

Readings taken when the overload indicator is triggered should be discarded. Unfortunately the overload indicator can be tripped by factors other than excessive workplace noise.

For example:

- Microphone impact including blowing into or shouting into the microphone.
- Momentary tampering (for example, sudden removal of the microphone cable from the body of the instrument).
- Close proximity of an airline to the microphone.
- Close proximity to a radio transmitter.

Dosemeters are used to measure the total noise dose received over the measurement period, which can then be used to calculate the daily personal exposure level.

However, for reliability the measurement period is normally several hours, preferably an entire shift to reduce inaccuracies.

If a shorter period is sampled than care must be taken to ensure that the result is representative of the full shift exposure. This will require an understanding of the tasks performed and the cycle of those tasks.

One limitation of dosimetry is that while the instrument records the accumulated dose over the measurement period it will not give any information about the causes (sources) of the noise.

This will require a separate noise survey investigation. In addition, any result obtained will only be representative of the role conducted by that individual. Identification of a number of individuals to create a representative sample across all employees affected by exposure to noise will be required.

Finally, dosimetry relies on the co-operation of both the subject and co-workers. If the work is not carried out normally or tampering occurs then the readings may be of little value.

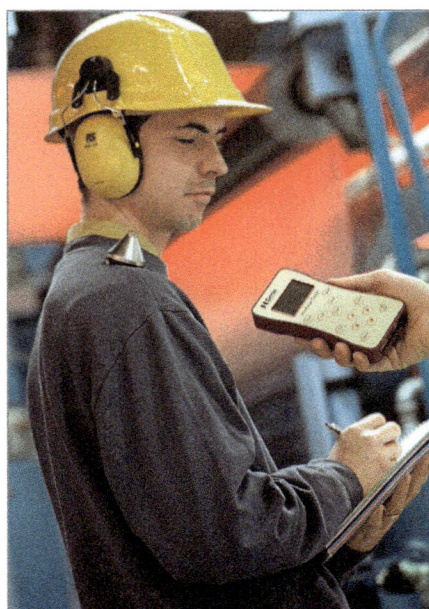

Figure B6-25: Dose badge - personal noise dosemeter.
Source: Cirrus Research plc.

Training and supervision is needed to minimise this 'novelty' effect. Even so supervision could be so time consuming as to become impracticable. The expense and time consuming nature of personal dosimetry taken together with the limitations outlined above means that they should only be used when other techniques are unsuitable.

MEASUREMENTS TO BE TAKEN

Representative measurements should then be taken at each of the identified working positions and at points where workers may pass or rest.

When measuring to estimate the exposure of a specific worker to noise the (calibrated) meter should be positioned as close to the worker's ear as possible, with the microphone of the sound pressure level meter directed towards the source, avoiding any shielding effect or reflections.

The results should not usually be affected if the microphone is kept at least 15cm away from the operator. The aim is to measure the individual's exposure, therefore the meter should follow the worker's head as they move and be positioned on the side most affected by noise. Enough measurements need to be taken in order to accurately estimate the worker's personal daily exposure.

Figure B6-26: Measuring noise with a hand held meter.
Source: Controlling noise at work L108 HSE.

In order to avoid taking a lot of readings of operations with varying noise levels it may be decided to identify the highest noise level and assume the 'worst case' applies at all times. If more detailed measurements are to be taken the measurement must represent the pattern of noise; this may mean taking a reading for sufficient time that it represents the pattern or taking multiple measurements.

The duration of the measurement will depend on the type of noise. A steady noise can be measured in a short period of time as it will quickly provide a stable reading, a short sample L_{Aeq} may be enough. If the noise is changing it will be necessary to continue the reading until it settles to a L_{Aeq} that does not vary more than 1 dB. In more complex situations the reading may have to be taken over the duration of a whole cycle of an operation.

Full and accurate details should be noted as measurements are taken. Observations, methods and results, together with details of the instruments used, should be recorded. If the measurement environment changes during measurement the change in status should be recorded, the time it occurred and other related information on the assessment record. For example, if a machine stops or someone passes in front of the sound pressure level meter, this should be recorded.

Differentiate recording points by numbers or other means and mark them on the prepared documents beforehand. Also include the distance from the source, walls, etc. In order to verify the measurement point after measurement, take photographs of the site.

The records required would include:

- Details of equipment used.
- Noise sources.
- Measurement points and/or individuals concerned.
- Sound pressure level L_{Aeq}.
- Exposure times.
- Fractional exposures *(for calculating daily noise exposures see 'noise calculators').*
- Peak sound pressure level dB(C).

INTERPRETATION AND EVALUATION OF RESULTS

The sound level pressure meter used may give readings in the form of an $L_{EP,d}$, which is acceptable if it is a noise dose meter that has been worn for eight hours. However, if it has not it will be necessary to obtain a reading in the form of a percentage dose or L_{Aeq} in order to obtain an $L_{EP,d}$.

A noise assessment should convey clear, unambiguous, information to the intended recipient, who may not be conversant with the terms used.

Essential information should include:

- The date the assessment was made.
- Who made the assessment, including who took any measurements involved?
- The workplace, areas, jobs or people assessed, including details of the work taking place.
- Measurement locations and durations and any noise control measures being used at the time.
- Daily personal noise exposures ($L_{EP,d}$) and indication where they are above the action and limit values.
- Peak noise exposures and indication where they are above the action and limit values for peak sound pressure.
- The sources of noise.
- Any further information necessary to help comply with the reduction of noise exposure.

Additional information which may be added to a noise assessment report:

- Details of instruments used - including serial numbers, details of field calibration checks and the dates they were calibrated.
- Detailed field readings and the people, tasks and conditions in which readings were taken.
- A detailed plan of the survey area including indications of people and their workstations together with exposure times.
- Details of work patterns where workers move between areas.
- The assessment of and recommendations for existing or new ear protectors, derived from octave band analysis.
- Recommendations for action.

From the noise assessment report the employer should be able to interpret the significance of the findings and develop an action plan, including the immediate measures needed to control noise exposure and the introduction of noise reduction measures.

A strategy may include:

- Engineering control of existing noisy machines.
- Modifications to noise transmission paths.
- Changes in methods of working.
- The provision of ear protection.
- Information, instruction and training.
- Health surveillance.

The noise assessment must be reviewed and updated regularly as part of the employer's ongoing noise risk management and control programme. Reports should be kept readily available for as long as they are relevant to the premises and to the staff who work there - and at least until the next assessment has been completed.

An update will be required if there are any significant changes to the workplace, plant or processes, working hours and after the implementation of noise control measures. In any event, a regular review should be planned to ensure continued effectiveness.

Regulation 5(6) of CNWR sets out the following requirements regarding recording the results of a noise assessment:

"The employer shall record:
(a) The significant findings of the risk assessment as soon as is practicable after the risk assessment is made or changed.
(b) The measures which he has taken and which he intends to take to meet the requirements of regulations 6, 7 and 10."

The guidance to the CNWR emphasises the need to record which employees are at risk, the level of risk and exposure, and under what circumstances the risk occurs. Records should also include details of the action taken or intended, with timescales and allocation of responsibilities.

USE OF NOISE CALCULATORS TO DETERMINE MIXED EXPOSURE
Concepts of addition of combined sounds (equal and unequal)

The purpose of the measurements is to determine the personal exposure of workers over a full working period. The obtaining, through measurements taken, of a 'noise map' of sound levels relating to various sound sources is an early step in this process. Measurements obtained from the noise assessment may contain data on exposure from a mixture of noise sources over a mixture of periods of time.

The equivalent noise exposure can be calculated mathematically, as explained in section 6.1 of this element or by use of the HSE's noise exposure calculators and noise exposure ready-reckoners. The noise exposure calculators enable the daily and weekly noise exposures to be estimated using Excel spreadsheets and the noise ready-reckoners enable similar exposures to be estimated using ready-reckoner sheets. *See calculation of Lep,d earlier in section 6.3.*

When details of personal exposure have been obtained it is then possible to compare them with action and limit values set out in legislation.

COMPARISON WITH LEGAL LIMITS

The results of the noise assessment should be compared with the Control of Noise at Work Regulations (CNWR) 2005, Regulation 4, exposure limit values and action values, *see figure ref B6-27*.

The lower exposure action values are:

- A daily or weekly personal noise exposure of 80 dB (A-weighted) and a peak sound pressure of 135 dB (C-weighted).

The upper exposure action values are:

- A daily or weekly personal noise exposure of 85 dB (A-weighted) and a peak sound pressure of 137 dB (C-weighted).

The exposure limit values are:

- A daily or weekly personal noise exposure of 87 dB (A-weighted) and a peak sound pressure of 140 dB (C-weighted).

This will enable the person conducting the assessment to identify areas and tasks creating the highest risk and individuals that are most at risk. The findings of the assessment should clearly identify where readings exceed these action and limits values.

B6.4 - Controlling noise and noise exposure

Legal requirements and duties to manage exposure to noise

The duties to manage exposure to noise are expressed in regulation 6, of the control of Noise at Work Regulations (CNWR) 2005. This requires the employer to ensure that risk from noise exposure to his employees is either eliminated at source or, where this is not reasonably practicable, reduced to as low a level as is reasonably practicable.

If any employee is likely to be exposed to noise at or above an *upper exposure action value*, the employer must reduce exposure to as low a level as is reasonably practicable by establishing and implementing a programme of organisational and technical measures, excluding the provision of personal hearing protectors, which is appropriate to the activity.

Regulation 6 (3) states that the above actions taken by an employer must be based on the general principles of prevention set out in Schedule 1 to the Management of Health and Safety Regulations 1999 and consideration should be given to:

- Other working methods which reduce exposure to noise.
- Choice of appropriate work equipment emitting the least possible noise, taking account of the work to be done.
- The design and layout of workplaces, work stations and rest facilities.
- Suitable and sufficient information and training for employees, such that work equipment may be used correctly, in order to minimise their exposure to noise.
- Reduction of noise by technical means.
- Appropriate maintenance programmes for work equipment, the workplace and workplace systems.
- Limitation of the duration and intensity of exposure to noise.
- Provision of appropriate work schedules with adequate rest periods.

The employer carries a duty not to expose employees to noise above an *exposure limit value* or if they do immediately reduce exposure to below it, find out why it happened and modify controls in place to prevent it happening again.

There is a specific legal duty on the employer to ensure that noise levels in rest facilities are reduced to a suitable level. Measures taken in order to comply need to be adapted to suit any individual or group at particular risk, for example, an individual that has already experienced hearing loss or a young person. The employer must ensure that employees or their representatives are consulted regarding measures taken.

Lower and upper exposure action values and exposure limit values

Regulation 4 of CNWR sets out the definition of 'daily average noise exposure' as the time weighted average of the levels of noise to which a worker is exposed over an 8 hour working day, taking account of levels of noise and duration of exposure and including impulsive noises.

The exposure action values are the levels of exposure to noise at which the employer is required to take certain actions.

	Lower exposure action values	*Upper exposure action values*	*Exposure limit values*
Daily or weekly personal noise exposure (A-weighted)	80 dB (A)	85 dB (A)	87 dB (A)
Peak sound pressure (C-weighted)	135 dB (C)	137 dB (C)	140 dB (C)

Figure B6-27: Noise exposure values.   *Source: CNWR 2005.*

Under regulation 4 of CNWR 2005 if the noise levels in the workplace vary greatly from day to day, then the employer can choose to use weekly noise exposure levels instead of daily noise exposure levels. This is only likely to be appropriate where daily noise exposure on one or two days is at least 5 dB (A) higher than other days, or the working week comprises three or fewer days of exposure.

Weekly noise exposure level means the average of daily noise exposure levels over a week and normalised to five working days.

When using weekly averaging, there must be no increase in risk to health. It is not acceptable to expose employees to very high noise levels for one day when using weekly averaging. Specific regard to the exposure limit values must also be made.

LOWER EXPOSURE ACTION VALUE

Where the daily or weekly exposure level of 80 dB (A) or a peak sound pressure level of 135 dB (C) is likely to be exceeded the employer must make hearing protection available to employees upon request and provide the employees and their representatives with suitable and sufficient information, instruction and training.

The information, instruction and training must include:

- Nature of risks from exposure to noise.
- Organisational and technical measures taken in order to comply.
- Exposure limit values and upper and lower exposure action values.
- Significant findings of the risk assessment, including any measurements taken, with an explanation of those findings.
- Availability and provision of personal hearing protectors and their correct use.
- Why and how to detect and report signs of hearing damage.
- Entitlement to health surveillance.
- Safe working practices to minimise exposure to noise.
- The collective results of any health surveillance compiled in a form in a format which prevents those results from being identified as relating to a particular person.

UPPER EXPOSURE ACTION VALUE

Where the daily or weekly exposure level of 85 dB (A) or a peak sound pressure level of 137 dB (C), the upper exposure action value, is likely to be exceeded, the employer must provide employees with hearing protection.

In addition the employer must:

- Ensure that the area is designated a hearing protection zone, fitted with mandatory hearing protection signs.
- Ensure access to the area is restricted where practicable.
- So far as reasonably practicable, ensure those employees entering the area wear hearing protection.

Figure B6-28: Ear protection zone sign.
Source: Controlling noise at work L108 HSE.

EXPOSURE LIMIT VALUE

The exposure limit values are the levels of noise exposure which must not be exceeded. They are expressed as a daily/weekly value and as a peak value. When assessing exposure, to ensure limit values are not exceeded, the effect of personal hearing protection provided to the employee can be taken into account. This will depend on the attenuation provided by the hearing protection, its appropriateness for the type of noise, practical aspects of use and maintenance.

The employer must ensure that employees are not exposed to levels above an exposure limit value or if one is exceeded immediately, reduce exposure below that level, investigate the reasons and modify measures to prevent a reoccurrence.

The hierarchy of noise control

ELIMINATE OR CONTROL AT SOURCE

Relocation

Noise reduction can be achieved by moving the noise source away from areas where people are working. Moving the noise source to an adjacent room will also create a reduction in sound. In other cases, when noise is not emitted equally in all directions, turning it round can achieve significant reduction. Similarly, noisy exhausts should be directed away from workers and discharged further away by the use of a flexible hose.

Redesign

Noise is generated by vibrating surfaces and fluids. If this vibration can be reduced or eliminated then so will the noise. Possible changes in design that will aid this process are:

- Cushioning impact with plastic or nylon surfaces.
- Replacing rigid pipework with flexible materials.
- Reduce the distance objects fall.
- Using large diameter fans that run at low speed. This will move the same amount of air as small fast running fans, but will do it quieter.
- Replacing metal gears with nylon or belts.
- Similarly, using large diameter, low pressure ducting.
- Stiffening structural parts.
- Streamlining ducting to avoid turbulence. This will have the added benefit of increasing efficiency.

Maintenance

Vibrated noise can be reduced by adopting a programme of regular maintenance. Regular maintenance will ensure that:

- Worn or badly fitting parts are replaced.
- Rotating and moving parts are balanced correctly.
- Loose parts are secured.
- Moving parts are well lubricated.

CONTROL ALONG TRANSMISSION PATH

Isolation

Positioning an elastic element (for example, rubber mount etc.) in the path of vibration can isolate a noise radiating area from a vibration input.

Barriers

In some situations noise levels, at a distance from the source, can be reduced by the use of noise barriers. Screens and barriers place an obstacle in the noise transmission path. Barriers work on the principle of increasing the effective distance between the noise source and the receiver and elimination of the line-of-sight sound path.

The effectiveness of the barrier will depend on the frequency of the sound. At low frequencies barriers will give little or no noise reduction. Barriers are also not effective in reverberant sound fields or for people working close to the source. These limitations mean that they are best used in conjunction with other reduction techniques.

Enclosure

Acoustic enclosures are widely used in industry, but are not suitable for machines that need to be accessed frequently. This will include most workshop machines.

Typically enclosures can be used with machines like pumps, compressors and conveyors. An enclosure is simply a sound attenuating cover fitted over a noise source. If an operator works within a noise enclosure they are likely to be exposed to an increased sound level. Enclosures can be full or partial depending on what it is enclosing.

For an enclosure to provide adequate attenuation the following steps should be taken:

- The enclosure should be built with a material such as brick, metal or wood.
- Doors and inspection hatches must be made to fit tightly.
- The inside of the enclosure should be lined with a sound absorbent material such as mineral, wool or foam rubber etc.
- If openings are required for ventilation or material input/output they should have some form of sound attenuation fitted.
- Gaps for services (cables, pipes etc.) should be sealed.
- To prevent the transmission of vibration, the machine should be vibration isolated from the enclosure and the floor.
- All services to the machine must have flexible connections.

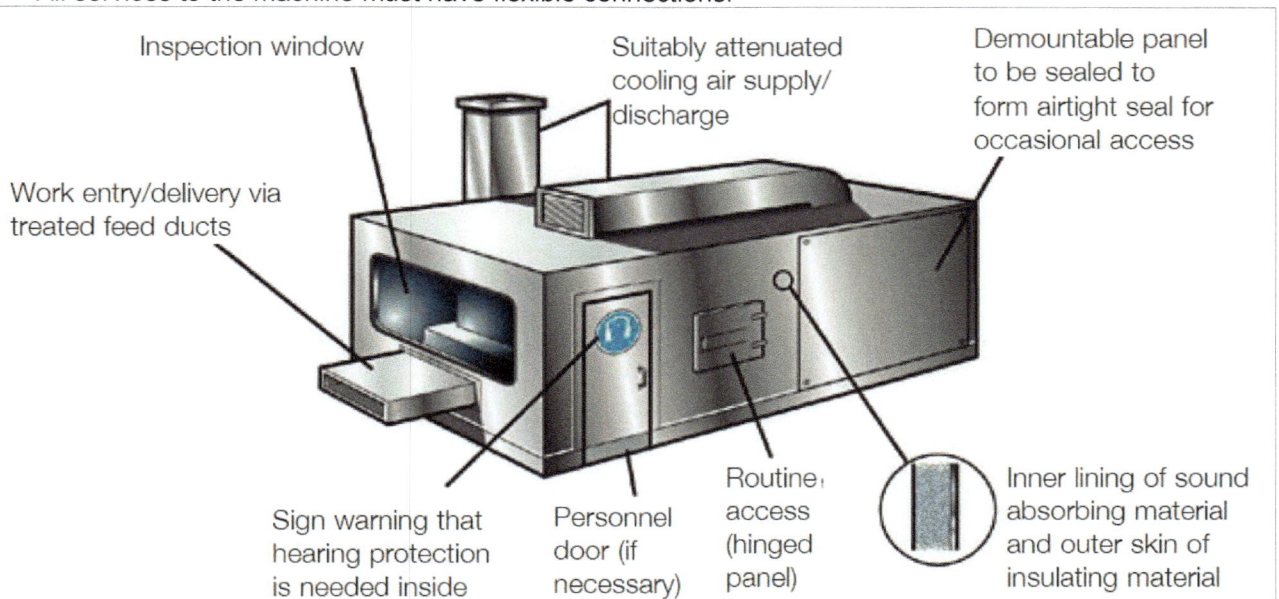

Figure B6-29: Noise enclosure.

Source: HSE, Controlling Noise at Work, L108.

Distance

A simple method of reducing noise levels is to increase the distance between the noise source and the worker/s. Doubling this distance can reduce the effect by 6 dB. This is because noise (for example, from an angle grinder) is dispersed in a pattern approximating a sphere. Therefore, noise follows the inverse square law.

$$\text{Intensity} \quad \propto \quad \frac{1}{\text{distance}^2}$$

So doubling the distance (x 2) reduces the intensity by 4 ($2^2 = 4$). This is only an approximation, however, since other environmental factors and reflections will need to be taken into consideration.

Active noise control

Active noise control/cancelation (ANC), or active noise reduction (ANR), is a method for reducing unwanted sound by the addition of a second sound specifically designed to cancel the first. This is a complex process; one common use is in noise-cancelling headphones. A miniature microphone in the earpiece picks up surrounding noise; the ear piece creates a noise-cancelling wave that is 180° out of phase with the background noise.

CONTROL EXPOSURE AT THE RECEIVER

Enclosures

Enclosures for workers include full enclosure in a noise proof room, for example, a control room for a remotely operated process where people can work without the need for hearing protection.

Enclosures may also be used to give a noise free environment for workers, enabling them to take a break away from a noisy workplace.

The design should take account of:

- Dense acoustically absorbent construction material.
- Acoustically double-glazed windows.
- Isolation from the floor.
- Adequate ventilation.
- Good door, window and ventilation ducting seals.
- Self-closing doors.

Acoustic havens

Where workers need to enter a noisy work area and enclosing the whole machine is difficult, they may be provided with an acoustic haven, which allows them to spend an amount of time looking after the process in a reduced noise level area. This may provide acoustic reduction all around the worker or in the main direction that the noise comes from. The refuge should be designed with consideration to reflected noise that may come to the worker from behind or above.

Hearing protection zones

Identification of hearing protection zones enables workers to be aware of the noise risk and to take action to limit their exposure by avoiding the zone, shutting down noisy activities while they are in the zone or using suitable personal hearing protection.

Personal protective equipment (PPE)

Hearing protection should not be the first control used to limit exposure, but after other controls have been applied it may help to reduce noise to an acceptable level for the remaining situations. This might be where maintenance work has to be conducted, where noisy equipment cannot be shut down or in situations where other reasonable practicable solutions are not available.

Personal hearing protection is available in a range of forms including ear muffs, re-usable ear plugs and disposable ear plugs. They have a range of effectiveness in reducing noise and some are better at different noise frequencies. It is therefore important to carefully select the right personal hearing protection for the noise and work activity. It is also important that it is used properly, for sufficient time, to provide the desired protection and it is maintained in good condition.

Active noise reduction (ANR) personal hearing protectors use an electronic sound cancelling system to reduce noise. This is particularly effective at low frequencies (50-500 Hz) where normal *(passive)* protectors are least effective.

Electronic hearing protection can be divided into two main categories: products intended to improve short range communications in noisy environments and products intended to provide entertainment during work or play.

Terms like 'noise cancelling ear muffs and/or active ear muffs' are sometimes used to describe this type of product. However, the technology used is not noise cancellation but rather different compression and filtration techniques.

Limiting exposure time

Where there is a requirement for continuous work in high noise environments, limiting individual worker's personal noise exposure should be achieved by other means than by the use of PPE. One such system is the use of job rotation, where workers rotate between high and low noise activities. In addition to protecting hearing, this will often improve workforce morale and productivity.

Surveillance (audiometry)

Regulation 9 of the CNWR 2005 states that if a risk assessment indicates a risk to the health and safety of employees who are, or are liable to be, exposed to noise, and then they must be put under suitable health surveillance (including testing of their hearing). The employer must keep and maintain a suitable health record. The employer will, providing reasonable notice is provided, allow the employees access to their health record.

Where, as a result of health surveillance, an employee is found to have identifiable hearing damage the employer must ensure that the employee is examined by a doctor.

If the doctor or any specialist to whom the doctor considers it necessary to refer the employee considers that the damage is likely to be the result of exposure to noise, the employer must:

- Ensure that a suitably qualified person informs the employee accordingly.
- Review the risk assessment.
- Review any measure taken to comply with the regulations.
- Consider assigning the employee to alternative work.
- Ensure continued health surveillance.
- Provide for a review of the health of any other employee who has been similarly exposed.

Employees must, when required by the employer and at the cost of the employer, present themselves during working hours for health surveillance procedures.

The results of health surveillance will be expressed as audiograms, the principles of which are outlined in an earlier section.

Audiometry is a supporting process to other controls and of itself it provides no protection, but may confirm that other controls are working or that they are not. Audiometry may provide an opportunity to identify a worker experiencing the first stages of hearing loss and enable action to be taken to prevent further loss.

Techniques to control noise generation, transmission and exposure

GENERATION

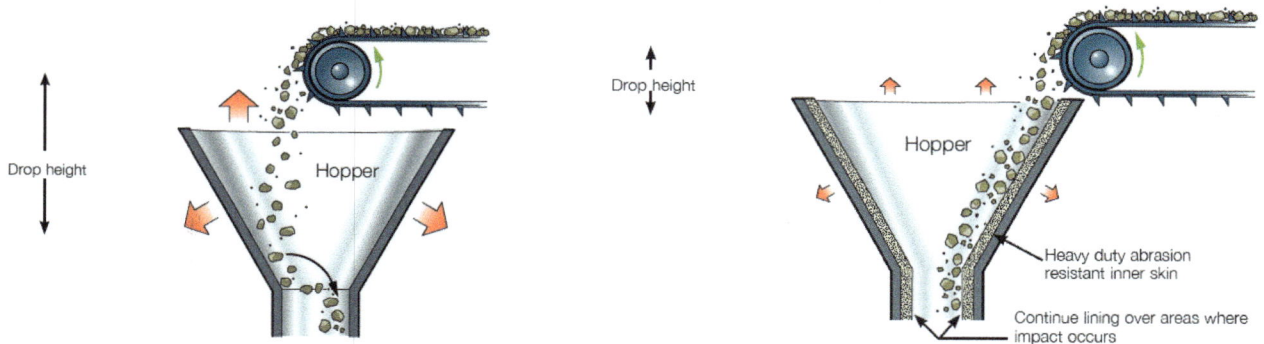

Figure B6-30: Reducing impact noise at source of generation.

Source: HSE, Controlling Noise at Work, L108.

Techniques to control noise generation frequently involve consideration of alternative materials, air turbulence, avoiding impacts and stiffening of parts that vibrate.

This will involve substitution of quieter, plastic, materials for metal, such as metal gears or actuating levers.

Reduced air turbulence might be achieved by replacing one large high pressure port with an increased number of ports operating at reduced pressure. Impacts may be avoided or reduced by reducing the 'drop height' that material has to fall.

Large panels may easily vibrate and this can be controlled by making the panel stiffer by adding stiffening bars.

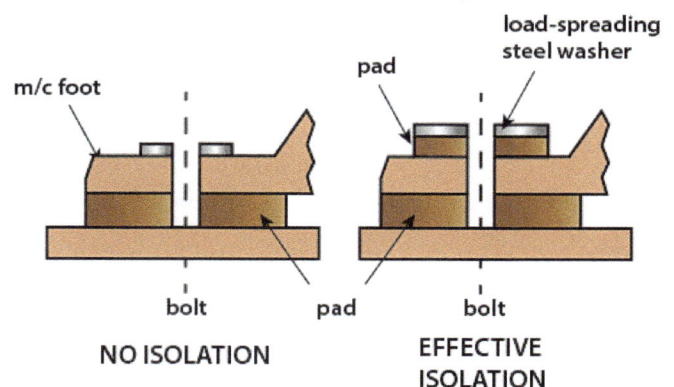

Figure B6-31: Noise control. Source: www.hse.gov.uk/publications/10 top noise control techniques.

TRANSMISSION

Mounting motors, pumps, gearboxes and other items of plant on rubber bonded cork (or similar) pads can be a very effective way of reducing transmission of vibration and therefore noise radiated by the rest of the structure. This is particularly the case where vibrating units are bolted to steel supports or floors.

However, a common error with the use of these pads is for the bolt to 'short-circuit' the pad, resulting in no isolation. Additional pads must be fitted under the bolt heads as shown in *figure ref B6-31.* There are many types of off-the-shelf anti-vibration mounts available, for instance rubber/neoprene or spring types. The type of isolator that is most appropriate will depend on, among other factors, the mass of the plant and the frequency of vibration to be isolated.

REFLECTION

Noise may be reflected back into a noisy area to limit the exposure of people outside this area. This may be done by using barriers of hard material that reflects the sound back to the source. A hard surface will reflect most of the energy at an angle equal to the angle of incidence. This can be used where the material totally encloses the sound source or as a single flat barrier sheet. This may control noise exposure beyond the barrier, but reflecting the sound back to the source could increase the sound pressure level inside the barrier.

ABSORPTION

The principle of sound absorption is to reduce the amount of reflected energy by transforming it into some form of energy other than vibration energy. Typically, friction rubbing within the material converts the sound energy into heat. Porous materials through which air can pass are often good sound absorbers. Thin layers of material will only absorb high frequencies, whereas thicker layers will absorb over a much larger frequency range. Absorbent materials are not usually effective for frequencies below 100 Hz. Porous materials, that are good thermal insulators, are usually good absorbers; examples include visco-elastics, mineral wool and foam.

In rooms with hard materials on the floor, walls and ceiling, any noise from a noise source that reaches these surfaces will be reflected back into the room. (A reverberant sound field will be created). This effect can be reduced by the use of absorbent material. Covering the ceiling with sound absorbent material will give a reduction of up to 8 dB at distances away from the source. Near the source levels will not be changed.

Covering the walls and ceiling may give up to a 5 dB reduction. Absorbent materials on the walls and ceiling near the operator will reduce local reflections and may give a noise reduction of a few dB for people near the source.

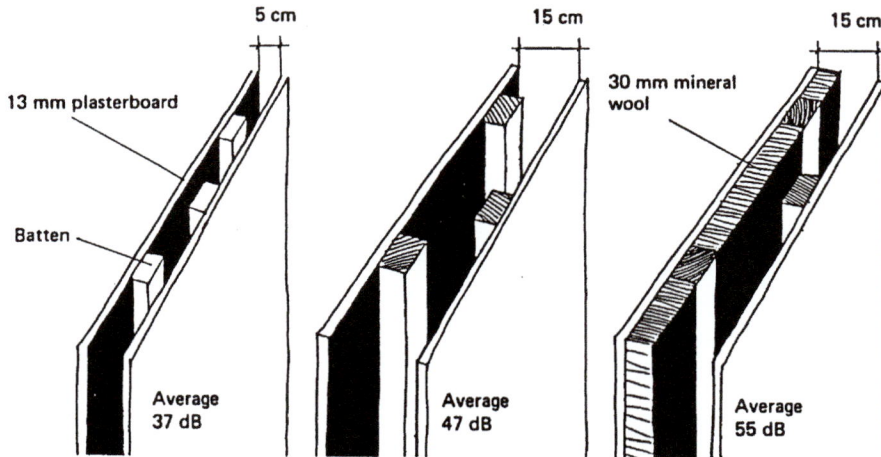

Figure B6-32: Sound absorption of partitions.   *Source: Ambiguous.*

DAMPING

There are 2 basic techniques:

1) Unconstrained layer damping where a layer of bitumastic (or similar) high damping material is stuck to the surface.

2) Constrained layer damping where a laminate is constructed.

Constrained layer damping is more rugged and generally more effective. This technique is suitable for chutes, hoppers, machine guards, panels, conveyors, tanks. Damping may be achieved by changing panels or other components for commercially available sound deadened steel or by the addition of self-adhesive sound deadened steel sheet.

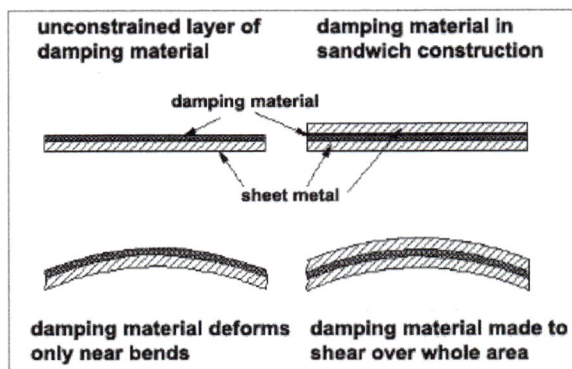

Figure B6-33: Noise damping.   *Source: www.hse.co.uk.*

The latter can simply be stuck on to existing components (inside or outside) covering about 80% of the flat surface area to give a 5 - 25 dB reduction in the noise radiated (use a thickness that is 40% to 100% of the thickness of the panel to be treated).

Limitations: the efficiency of damping diminishes for thicker sheets. Above about 3mm sheet thickness it becomes increasingly difficult to achieve a substantial noise reduction.

DIFFUSION

Diffusers are used to either complement sound adsorption or as an alternative. Unlike sound absorption materials they do not remove sound energy, but can be used to effectively reduce distinct echoes and reflections. A hard surface will reflect most of the energy at an angle equal to the angle of incidence whereas a diffuser will cause the sound energy to be radiated in many directions.

SOUND REDUCTION INDICES

A sound reduction indices (SRI) is defined as 'A set of values measured by a specific test method to establish the actual amount of sound that will be stopped by the material, partition or panel when located between two rooms'.

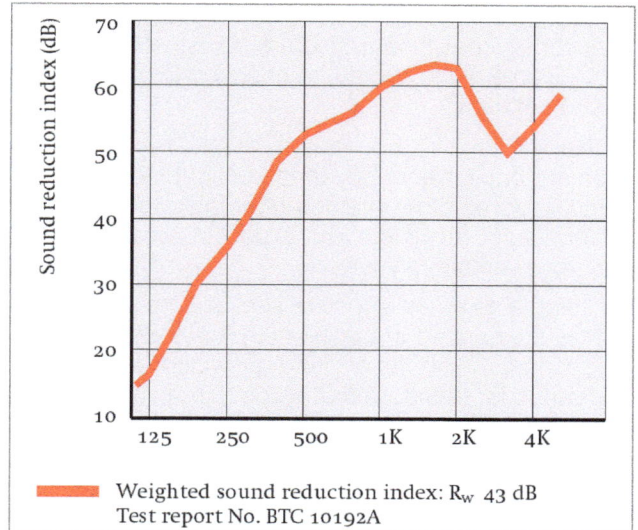

Figure B6-34: Sound reduction indicies.   *Source: www.rockwool.co.uk.*

The SRI is expressed as an amount of sound pressure reduction in dB. The performance of the material may vary over a range of frequencies and its performance may be expressed against this range and as a weighted sound reduction index rw that 'averages' the performance of the material over the range of frequencies.

SOUND ABSORPTION COEFFICIENTS

A material's sound absorbing properties are expressed by the sound absorption coefficient, α, (alpha), as a function of the frequency. α ranges from 0 (total reflection) to 1.00 (total absorption). Sound absorption coefficients are often expressed for a given material over a range of frequencies.

For example, one manufacturer of mineral fibre acoustic slab tested it and found 47mm of acoustic slab provided the following:

Frequency (Hz)	125	250	500	1K	2K	4K
Absorption co-efficient	0.20	0.50	0.85	1.00	1.00	1.00

Figure B6-35: Sound absorption coefficients.   *Source: www.rockwool.co.uk.*

The selection, maintenance and use of appropriate hearing protection

TYPES OF HEARING PROTECTION

Ear plugs

These are designed to fit into the outer portion of the ear canal and remain in position without any external fixing device. They may be made of rubber, plastic or similar resilient materials, in a variety of designs, and generally in a number of sizes to help obtain a good fit.

They can be corded or be fitted with a neckband to prevent loss and to aid monitoring of use. Designs available are suitable for permanent, reusable (i.e. used a few times) or one-use disposable types.

Where applicable, the correct size required should be determined by a competent person. Custom moulded plugs, which can offer high levels of comfort, can be individually moulded to a person's ear. Semi-insert (using a head-band) ear protectors rely on pressure of the band to maintain a seal. This can be useful for those spending short times in ear-protection zones.

Correctly selected earplugs have good attenuation characteristics at both high and low frequencies, and do not hinder the use of head protection or eye and face protection. It is, however, more difficult to see if earplugs are being worn, which increases the need for observant supervision to ensure their use. Earplugs are more effective in attenuating high frequencies than low.

Prolonged contact with the ear canal may cause a skin reaction, particularly when plugs are dirty. Accordingly, care must be taken to clean them after use. People suffering from ear infections, discharges or irritation should be referred to a doctor.

The rules for effective use of ear plugs are summarised as follows:

- Ear plugs should not be handled unless hands are clean.
- Reusable, washable types should be cleaned in accordance with manufacturer's instructions.
- Reusable plugs should be checked for suppleness and softness.
- Ensure ear plug is correctly inserted into ear canal.
- Ear plugs should be used only by the person they are issued to.
- Adequate supplies of disposable/replacement plugs must be kept available.

Ear muffs

These are cups, frequently made of light metal or plastic and filled with sound absorbent material, connected by a band or some other method of holding them in place over the ears. To ensure a light comfortable fit around the ear, muffs are lined with a pad of elastomeric material or a 'sausage like' roll filled with a high viscosity liquid, for example, glycerol.

This lining acts as an effective seal and helps damp vibration on the muffs. This type of 'over the ear' protector is either suspended in caps which are sized for proper fit or supported by a headband similar to that used for ear phones. The latter type provides a universal fit, and is particularly suitable for intermittent wear. Headbands for standard muffs come in a variety of designs. Usually these are worn over the head or under the chin.

Many safety helmets incorporate a fitting for attaching ear muffs, and this hearing protection will have different characteristics to a standard ear muff. Ear muffs give better attenuation at high frequencies than low and average attenuation of frequencies less than 1,000 Hz, which is usually lower than with plugs. Ear muffs can be uncomfortable in hot conditions.

This can be affected by:

- Weight of muffs.
- Headband pressure.
- The size of the muff cup.

In addition, they may be heavy and cumbersome and interfere with the use of head, eye and face protection. When using ear muffs people with long hair, wearers of spectacles and jewellery will have difficulty in ensuring a good seal around the ear, which will reduce their effectiveness.

Muffs are more expensive than plugs. One advantage with muffs is that they are highly visible and therefore easily monitored. In addition, communication equipment can be fitted - however this must not present a new noise hazard.

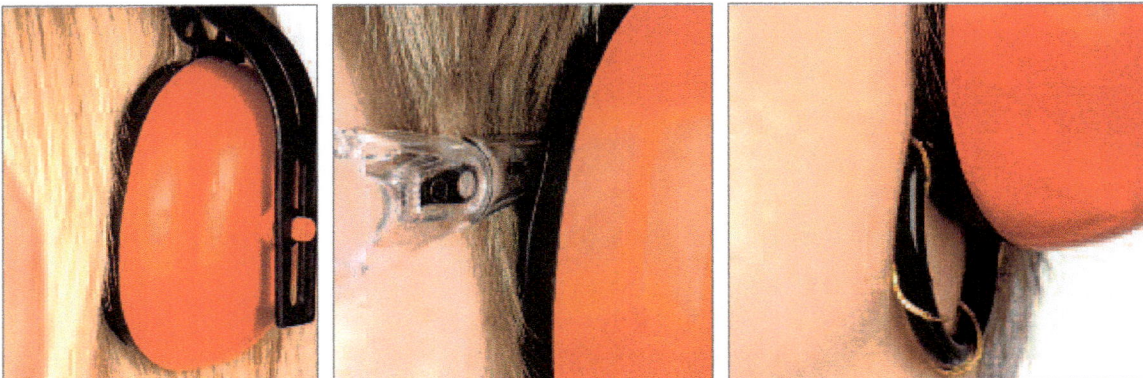

Figure B6-36: Effects of hair, spectacles and jewellery on ear muffs.   *Source: HSE, indg362.*

Other types of hearing protection

Level dependant, or amplitude sensitive, protectors are available. These come in two types - electronic and mechanical (by the use of carefully designed air ducts) designs. Both vary the level of protection given with varying levels of sound pressure. These are useful because they allow good communication in quieter areas.

Flat frequency response protectors are available where the ability to hear high frequency sounds is important (for example, musicians). These give similar attenuation across the whole spectrum whereas most protectors provide greater attenuation at high frequencies than they do at low.

Dual protection

At extreme noise levels a combination of both muffs and ear plugs may be required to provide effective attenuation. This is likely at an $L_{EP,d}$ of 115 dB(A) or where peak pressure levels exceed 160 dB, particularly if there is substantial noise at low frequencies under ~500 Hz.

Where this is the case, manufacturer's data must be obtained for the muff and plug combination. It cannot be assumed that the attenuation provided by the combination is simply the sum of the components.

USE OF HEARING PROTECTION
Plugs

Plugs may be unsuitable for use where they have to be removed frequently, particularly in dusty and dirty environments. Reusable plugs should be issued to an individual and not shared. It is important to ensure adequate supplies of disposable ear plugs are available.

Workers should ensure their hands are clean when fitting ear plugs. The manufacturer's instructions should be followed to ensure correct fitting in ear.

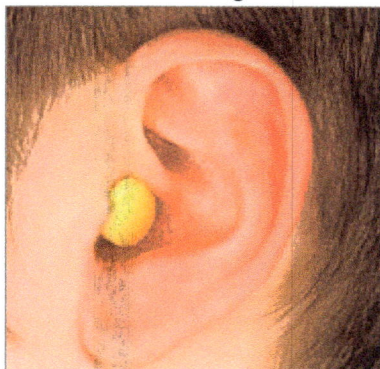

Figure B6-37: Correct use of ear plug. *Source: HSE, L108.*

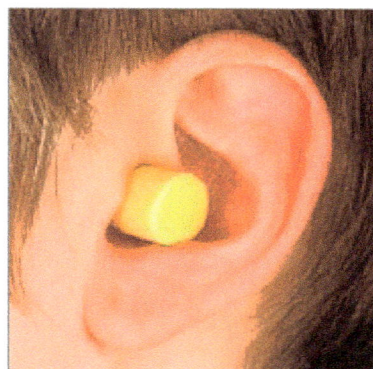

Figure B6-38: Incorrect use of ear plug. *Source: HSE, L108.*

Ear plugs can work loose over time and during movement and should be re-fitted periodically, perhaps hourly. This should preferably be done in an area outside any hearing protection zone and in a clean place.

Muffs

The headband on a standard ear muff can prevent use of a hard hat, though the headband can be worn behind the neck or under the chin if an under-hat support strap is provided. Wearing it in any position other than with the headband over the head will reduce the protection provided by the ear muff. Long hair, beards and jewellery may interfere with the seals of the muff and reduce protection.

They may not be appropriate for use with safety glasses and other forms of personal protective equipment. An alternative form of safety glasses or eye protection may be necessary, and this compatibility should be checked.

Ear muffs may be uncomfortable in warm conditions and lead to increased perspiration around the ears.

Personal hearing protection is only viable if:

- It is compatible with other PPE, for example, glasses, safety helmets.
- They are correctly fitted, for example, no interference with ear muff seals, ear plugs inserted and fitted correctly.
- They give sufficient protection with regard to the type and characteristics of the noise, i.e. they must offer protection to at least below the upper exposure action values and, in order to avoid giving limited protection, must give an assumed protection of at least 5dB(A).
- *'Real-world'* attenuation is considered. Practical work situations can result in about 4 dB less than predicted by manufacturer's average attenuation data. With ear plugs this can be up to 18 dB - mainly due to poor fitting.
- They are hygienic, comfortable to use and do not provoke a toxic reaction in the wearer.
- They are suitable for the working environment, for example, traceable types in the food industry.
- Hearing protection is used all the time in noisy areas, *see figure ref B6-39.*

Example: A worker is working in a noisy environment with a continuous sound pressure level of 113 dB (A) and wears hearing protection capable of 30 dB (A) reduction:

Percentage of time protector is worn	Time worn during an 8 hour day	Actual noise reduction/dB(A)
0%	Not worn	-
50%	4 hours	3
75%	6 hours	6
87%	7 hours	9
95%	7 hours 30 minutes	12
97%	7 hours 45 minutes	15
97.9%	7 hours.50 minutes	16.6
99%	7 hours 55 minutes	18.5

Percentage of time protector is worn	Time worn during an 8 hour day	Actual noise reduction/dB(A)
99.6%	7 hours 58 minutes	23
99.82%	7 hours 59 minutes	25.5
100%	8 hours (all day)	30

Figure B6-39: Effect of wearing hearing protection for par of required time.   *Source: RMS.*

Regulation 8 of the Control of Noise at Work Regulations (CNWR) 2005 requires that employees 'make full and proper use of personal hearing protectors provided' and if they discover 'any defect in any personal hearing protectors or other control measures as specified in sub-paragraph (a) report it to his employer as soon as is practicable'.

Information, instruction and training

Users must be instructed in the correct fitting and use of hearing protection, including:

- How to fit/wear their hearing protection, particularly in combination with other personal protection.
- How to avoid the potential interference of long hair, spectacles and earrings on the effectiveness of their hearing protection.
- The importance of wearing their hearing protection at all times in a noisy environment.
- Not to remove hearing protection when speaking to others in a noisy area and the need to speak up or move closer to the person they are talking to.
- The need to check correct fitting/wearing from time to time when in use.
- How to store their hearing protection correctly.
- How to check and care for their hearing protection at frequent intervals.
- Where to report damage to hearing protection and where to get replacements.

MAINTENANCE

Ear plugs

- Clean reusable plugs regularly and at the end of each work period.
- Ensure they are not damaged or degraded.
- Consider control of issue and maximum period of use before replacement.
- Return them to the container provided to keep them in when not in use.
- Ensure container does not become contaminated, keep closed where possible.
- Where a band is provided ensure it is effectively holding plugs in position, replace when not tight enough.
- Disposable earplugs should only be used once.
- Reusable plugs should be issued to an individual and not shared.
- Ensure adequate supplies of disposable earplugs.
- Hands should be clean when fitting earplugs.
- Follow manufacturer's instructions.

Ear muffs

- Check seals hardening, tearing and misshape, kits are available to replace seals periodically.
- Check cup condition for cracks, holes, damage and unofficial modifications.
- Avoid over bending or twisting the headband, which may degrade performance and its ability to hold the cup in place and create an effective seal. Check the tension of the headband; it is useful to compare it with a new earmuff.
- Ensure the seals of helmet mounted muffs do not sit on the side of the helmet for long periods as this can damage them and affect their performance leading to reduced tightness of seal around the ear.
- Clean seals periodically.
- Store in a clean environment.
- Consider control of issue and maximum period of use before replacement.
- Follow manufacturer's instructions for fitting, cleaning, maintenance and replacement.

Regulation 8 of the Control of Noise at Work Regulations (CNWR) 2005 requires that an employer ensure that anything provided in compliance with duties under CNWR 2005 is *"maintained in an efficient state, in efficient working order and in good repair."*

Regulation 8 of CNWR 2005 further requires that:

"(2) Every employee shall:

(a) Make full and proper use of personal hearing protectors provided to him by his employer in compliance with regulation 7(2) and of any other control measures provided by his employer in compliance with his duties under these Regulations.

(b) If he discovers any defect in any personal hearing protectors or other control measures as specified in sub-paragraph (a) report it to his employer as soon as is practicable."

PROBLEMS OF OVER-PROTECTION

Noise should always be reduced at source, using engineering controls.

Too much reliance on hearing protection has a number of disadvantages:

■ Loss of awareness of alarms, for example, fire exit alarms, vehicle warning signals and horns.
■ Personal feelings of isolation, which may lead the individual to not use or tamper with the hearing protection.
■ Personal hearing protection may be removed to wipe away perspiration, or to communicate with another in the workplace and this will remove protection for short periods during the working day. Over an 8 hour working day this may expose the worker to significant levels in excess of the exposure action values. For example, if the workplace noise level is 100 dB, a worker would exceed the upper exposure action value of 85db within 8 minutes of collective exposure resulting from intermittent removal of the ear muffs during the working day.

The HSE recommends that when selecting hearing protection over-protection should be avoided and that hearing protection that results in less than 70 dB (A) sound pressure level at the ear is unnecessary and results in over-protection.

See also - Relevant statutory provisions - for further information.

SELECTION OF HEARING PROTECTION

The following factors are likely to influence selection of hearing protection:

■ Characteristics of the noise workers are exposed to.
■ Types of hearing protector, and suitability for the work being carried out.
■ Noise reduction (attenuation) offered by the protector, assumed protection factor.
■ Pattern of the noise exposure.
■ Compatibility with other personal protective equipment.
■ The need to communicate and hear warning sounds.
■ Environmental factors such as heat, humidity, dust and dirt.
■ Comfort and user preference.
■ Medical disorders suffered by the wearer.
■ Cost of maintenance or replacement.

There are three principal methods of taking account of assumed protection factors provided by manufacturers and for estimating sound pressure levels at the ear when hearing protection is worn, as defined by BS EN 24869-2:1995. Octave band analysis, which as the name suggests requires A-weighted octave band analysis measurements, single number rating, which requires a single C-weighted measurement and HML method, which requires both a single A-weighted measurement and a single C-weighted measurement.

Whichever method is used to calculate the assumed sound pressure level at the ear when wearing the hearing protection, the Health and Safety Executive (HSE) recommends reducing the assumed protection provided by the hearing protection by 4 dB to take account of what they call 'real-world' factors, such as variability of fit and decline in performance of the hearing protection over time.

The HSE also recommend avoiding the provision of hearing protection that 'over-protects' and that protection that results in less than 70 dB at the ear should be avoided. The HSE website provides access to excel sheets that enable the easy calculation of the sound pressure level at the ear using each of the three methods.

USE OF OCTAVE BAND ANALYSIS

This is the most accurate prediction method of determining the effectiveness and suitability of hearing protection; however it requires the measurement of the octave band spectrum. It is the most complicated method of calculating the effective A-weighted sound pressure level (SPL) at the ear.

The purpose of carrying out such a frequency band analysis is to assess the suitability of hearing protection. There are many types and models of hearing protection available, giving a wide range of protection levels needed for the variety of industrial situations. The protection given by any hearing protection will depend on both the fit and the frequency of the noise. Manufacturers supply mean attenuation test data gained from measurements taken from a variety of subjects with differently shaped heads and a range of frequencies. If this means (average) test data were used then only 50% of the population would receive the protection assumed. For this reason the data is corrected by subtracting one standard deviation.

This gives an *assumed protection value (APV)* which takes greater account of testing inaccuracies, variations in manufacturing and the differences in the physiology of each individual ear. The aim when providing hearing protection is to ensure that the noise is attenuated sufficiently without 'over-protecting' the operator thereby impairing communication.

The assumed protection of any given hearing protection is calculated by first measuring the overall A-weighted sound pressure level and the frequency band pressure levels which contribute to the overall sound pressure level workers are exposed to. This is known as a frequency (or octave) band analysis. Levels are usually measured at 63, 125, 250, 500, 1k, 2k, 4k and 8k Hz centre frequencies for each band. Using the supplier's data for the hearing protection under consideration these measurements can be used to establish an assumed exposure level that workers receive when using the hearing protection. Some noise instruments/calculators programmed with the common models of hearing protection carry out this function automatically.

In essence the calculation involves:

- Calculating the A-weighted noise level.
- Calculating the octave band A-weighted noise level.
- Calculating the assumed attenuation provided by the hearing protection.
- Calculating the noise level at the ear when wearing the hearing protection.

An example calculation is as follows. A worker is exposed to 103.2 dB (A). Octave band analysis shows that this comprises 90, 92, 94, 94, 96, 98, 96, 94 dB respectively.

See figure ref B6-40 which shows the supplier's data for the ear protector:

Octave-band frequency	Hz	63	125	250	500	1k	2k	4k	8k
Exposure level before protection									
Band pressure level	dB	90	92	94	94	96	98	96	94
A-weighting correction	dB	-26.2	-16.1	-8.6	-3.2	0	+1.2	+1.0	-1.1
Weighted band noise level (1)	dB (A)	63.8	75.9	85.4	90.8	96	99.2	97	92.9
Level of protection provided by hearing protection being considered									
Mean attenuation	dB	7.4	10.0	14.4	19.6	22.8	29.6	38.8	34.1
Standard deviation	dB	3.3	3.6	3.6	4.6	4.0	6.2	7.4	5.2
Assumed protection value of hearing protection (2)	dB	4.1	6.4	10.8	15.0	18.8	23.4	31.4	28.9
Exposure level while using protection									
Assumed protection level lines (1)-(2)	dB(A)	59.7	69.5	74.6	75.8	77.2	75.8	65.6	64.0

The sum of the assumed protection levels (in bels) in the last row can be added:

$$= \text{Log} (10^{5.97} + 10^{6.95} + 10^{7.46} + 10^{7.58} + 10^{7.72} + 10^{7.58} + 10^{6.56} + 10^{6.40})$$
$$= 8.239 \text{ B (A)}$$
$$= 82 \text{ dB (A)}$$

Figure B6-40: Example calculation.   *Source: RMS.*

The 'A' weighted total noise level at the ear fitted with the hearing protection being considered is assumed to be 82 dB (A). Thus, the hearing protection gives a reduction of around 21 dB (A).

Obviously, this assumed protection can only be provided if the hearing protection fits properly and is correctly used and therefore 4 dB should be added to the example to take account of 'real-world' factors.

SNR (SINGLE NUMBER RATING)

The SNR value is used with a single measurement of the sound pressure level as a C-weighted average sound pressure level over a range of frequencies.

The SNR is subtracted from the C-weighted sound pressure level to give an assumed sound pressure level received at the ear when wearing the hearing protection.

The SNR provided by the manufacturer of the hearing protection already takes account of the mean and standard deviation values, but the 4 dB adjustment for 'real world' factors should still be made.

HML (HIGH, MEDIUM, LOW) METHODS

Three protection values high (H), medium (M) and low (L) are used to determine an assumed sound pressure level received at the ear when wearing the hearing protection.

The values for are supplied by the manufacturer of the hearing protection, for example, H31, M27, L24, and are used with two simple measurements of sound pressure level, one a A-weighted and one C-weighted average sound pressure levels.

This is the preferred method where the octave-band spectrum is not available. The sound pressure level at the ear is calculated using one of the following two equations depending on the difference between the A-weighted sound pressure level and the C-weighted sound pressure level.

If (L_C – L_A) >2 dB, then the predicted noise level reduction (PNR) is calculated by:

$$PNR = M - \left[\frac{(M - L)}{8} (L_C - L_A - 2)\right]$$

Otherwise the calculation uses the following formula:

$$PNR = M - \left[\frac{(M - L)}{4} (L_C - L_A - 2)\right]$$

The values for HML already take account of the mean and standard deviation values for the hearing protection. To obtain the assumed A-weighted sound pressure level at the ear the PNR is subtracted from the initial A-weighted measurement of noise exposure obtained from the workplace. The 4 dB adjustment should be added to the assumed sound pressure level at the ear to provide an effective sound pressure level at the ear, taking account of 'real-world' factors.

B6.5 - Basic physical concepts relevant to vibration

Vibration as defined in the Control of Vibration at Work Regulations 2005

Vibration may be defined as oscillating motion of a particle or body about a fixed point. The Control of Vibration at Work Regulations (CVWR) 2005 states that a reference to an employee being exposed to vibration is a reference to the exposure of that employee to mechanical vibration arising out of or in connection with their work.

- 'Hand-arm vibration' means mechanical vibration which is transmitted into the hands and arms during a work activity.
- 'Mechanical vibration' means vibration occurring in a piece of machinery or equipment or in a vehicle as a result of its operation.
- 'Health surveillance' means assessment of the state of health of an employee, as related to exposure to vibration.

Displacement, velocity, amplitude, frequency and acceleration for oscillating particles

BASIC CONCEPTS

Vibration is oscillating motion about a fixed point. The characteristics of vibration may be described in terms of the amplitude of the vibration compared to its frequency of oscillation. The amplitude may be measured as the amount of displacement, velocity or acceleration relative to the fixed point. The units used for vibration are similar to those of noise.

These are:

- Frequency measured in Hertz (cycles per second). The term is combined with metric prefixes to denote multiple units such as the kilohertz (1,000 Hz), megahertz (1,000,000 Hz), and gigahertz (1,000,000,000 Hz).
- Displacement measured in metres.
- Velocity measured in meters per second (ms^{-1}).
- Acceleration - the rate of change of velocity in metres per second squared (ms^{-2}).

The amplitude characteristic of most interest in occupational health and safety is acceleration. The exposure limits set in the Control of Vibration at Work Regulations (CVWR) 2005 are expressed as magnitude of acceleration. Vibration can be represented by a waveform reflecting the motion around the fixed datum point and the frequency of that movement.

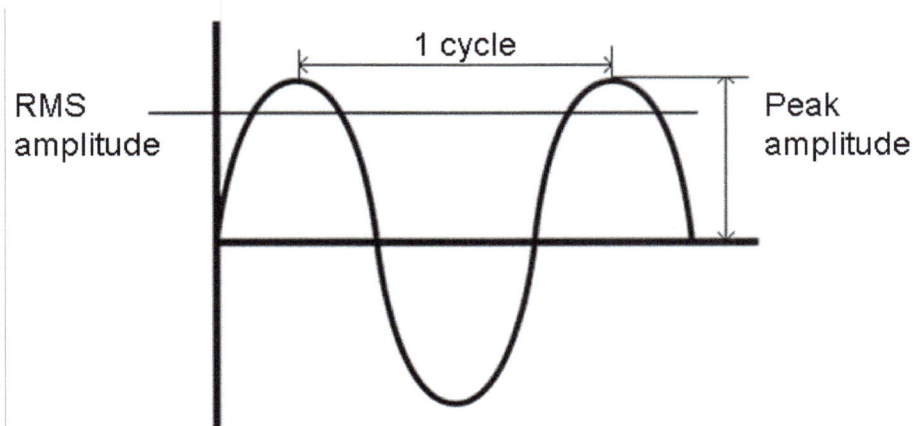

Figure B6-41: Characteristics of a vibration wave form.   Source: RMS.

Peak Amplitude (Pk) is the maximum excursion of the wave from the zero or equilibrium point. Root Mean Square Amplitude (RMS) is the square root of the average of the squared values of the waveform. In the case of the sine wave, the RMS value is 0.707 times the peak value, but this is only true in the case of the sine wave.

The RMS value is proportional to the area under the curve - if the negative peaks are rectified, i.e., made positive, and the area under the resulting curve averaged to a constant level, that level would be proportional to the RMS value.

These areas are equal

RMS Level

Figure B6-42: RMS amplitude of vibration.   *Source: Ambiguous.*

The RMS value of a vibration is an important measure of its amplitude. As mentioned before, it is numerically equal to the square root of the average of the squared value of amplitude. To calculate this value, the instantaneous amplitude values of the waveform must be squared and these squared values averaged over a certain length of time. This time interval must be at least one period of the wave in order to arrive at the correct value. The squared values are all positive, and thus so is their average. Then the square root of this average value is extracted to get the RMS value.

Most mechanical vibration found in the workplace is not simple sinusoidal or pure harmonic motion. For hand-held tools the vibration is complex, being composed of many frequencies at varying amplitudes, thus requiring averaging, such as 'root mean square' levels over a known frequency bandwidth.

Scientists have found difficulty expressing vibration intensity, especially the acceleration unit (ms-2). At low frequencies around 1hz, experienced in large ships and in semi-submersible oil exploration structures, a displacement of several metres is required to generate appreciable acceleration. At frequencies around 30-60hz coinciding with the firing frequencies of pneumatic rock drills or chipping hammers, high acceleration amplitude peak values of around 1,000ms-2 are generated with displacement of the order of fractions of millimetres.

OCCUPATIONAL VIBRATION EXPOSURE

Occupational exposure to vibration may arise in a number of ways, often reaching workers at intensity levels disturbing to comfort efficiency, health and safety.

There are two routes of vibration energy transmission. In the case of whole body vibration it is transmitted to the worker through a contacting or supporting structure which is itself vibrating, for example, a ship's deck, the seat or floor of a vehicle (tractor or tank), or a whole structure shaken by machinery, for example, in the processing of coal, iron ore or concrete where the vibration is intentionally generated for impacting. By far the most common exposure to vibration is through the hands, wrists and arms of the worker, where there is actual contact with the vibrating source - called segmental vibration or hand arm vibration (HAV).

Examples of occupational vibration sources include the use of pneumatic drills, electrically-driven rotary tools, chain saws and the use of grinding and abrading equipment.

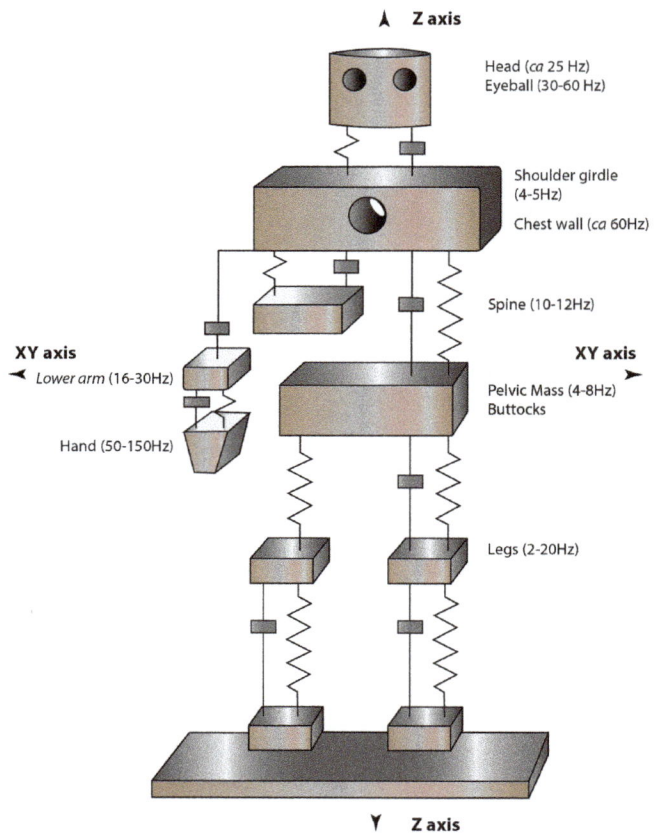

Figure B6-43: Simplified mechanical system representing the human body. *Source: Ambiguous.*

COMFORT LEVELS

The sensory and psycho-physiological response to vibration for humans depends to a marked degree on *resonance*, defined as the tendency of the human body (or parts of it) to act in concert with externally generated vibration. The impinging vibration may be amplified by as much as a factor of four by the resonance of part of a body. For vertical vibration in the region of 4 to 8 Hz, a worker's 'whole body', mainly upper torso, may be in resonance and therefore most susceptible to the harmful effects of the vibration. At this frequency there will be the maximum mechanical vibration energy transfer between the vibration source and the body, with an actual amplification of the incoming vibration signal. The characteristic frequencies and the amplification factors will be determined by the mass, the elasticity and the damping of the organ, for example, the frequency for the head is approximately 25 Hz, eyeball 30-60 Hz and the hand 50-150 Hz, *see figure ref B6-44*.

Vibrations at specific frequencies can therefore have an effect on the comfort levels experienced by the worker:

- 0.3 Hz   Motion sickness.
- 1-2Hz   Relaxing.
- 3-4Hz   Whole body.
- 5-8Hz   Interference.
- 10Hz   Face and cheeks.
- 20Hz   Tingling.

THE CONCEPT OF VIBRATION DOSE

'Dose' is derived from exposure to an amount of something over a period of time. In the concept of vibration dose there is an acceptance that below the nominated action level the risk of harm to the health of the average person is unlikely. Vibration dose is usually referenced over an eight hour period, for example the Control of Vibration at Work Regulations (CVWR) 2005 specifies a daily exposure action value of 2.5m/s^2 A(8) for hand-arm vibration, over an eight hour period. Therefore action is required if this level is exceeded. The general duty to ensure the dose is as far below the action value as is reasonably practicable remains.

The concept of vibration dose uses the principle of equal vibration energy, such that a vibration of a given magnitude for a given time is equally hazardous to a person as one of half the magnitude for twice the time. Thus, where a worker is exposed to a vibration for a shorter period than eight hours the equal energy concept of vibration dose allows an increase in the magnitude of the vibration for that shorter period.

The relationship between time and vibration magnitude is not directly proportional but an exposure can be calculated or the exposure converted into exposure points using the HSE system.

A given exposure value is calculated using the calculation:

$$A(8) = a_{hv}\sqrt{\frac{T}{T_0}}$$

Where a_{hv} is the vibration magnitude in m/s^2, T is the daily duration of exposure to the vibration and T_0 is the reference eight hour period.

Where a worker is exposed to a variety of different exposure values over a work period the values can be combined using the partial exposure formula, where $A_1(8)$, $A_2(8)$, $A3(8)$ etc. are the partial exposure calculations calculated with the formula above:

$$A(8) = \sqrt{A_1(8)^2 + A_2(8)^2 + A_3(8)^2}$$

A worker's daily vibration dose is expressed in acceleration units of m/s^2. The daily exposure is the average vibration exposure balanced over a reference period of a standard working day of eight hours.

To avoid confusion with measured vibration magnitudes A(8) is added after the units, for example, the daily exposure action value is 2.5 m/s^2 A(8).

The chart below gives a range of vibration magnitudes, together with the related exposure times, which would result in exposure at the exposure action value of 2.5 m/s^2 A(8) and the exposure limit value of 5 m/s^2 A(8).

Vibration magnitude (m/s^2)	2.5	3.5	5	7	10	14	20
Time to reach exposure action value (hours)	8	4	2	1	>1	>1	8 min
Time to reach exposure limit value (hours)	>24	16	8	4	2	1	>1

Figure B6-44: Equivalent vibration exposure times.   *Source: HSE, L140.*

The relationship between vibration magnitude and exposure duration is not a direct proportion, *figure ref B6-45* illustrates this and the significance of the exposure action value and exposure limit value.

As part of the package which supports the Control of Vibration at Work Regulations (CVWR) 2005, the HSE have produced a calculator, nomogram and ready-reckoner to assist in calculating exposures for hand-arm vibration and the relative dose of a worker. The calculator and other tools are available online at the www.hse.gov.uk website.

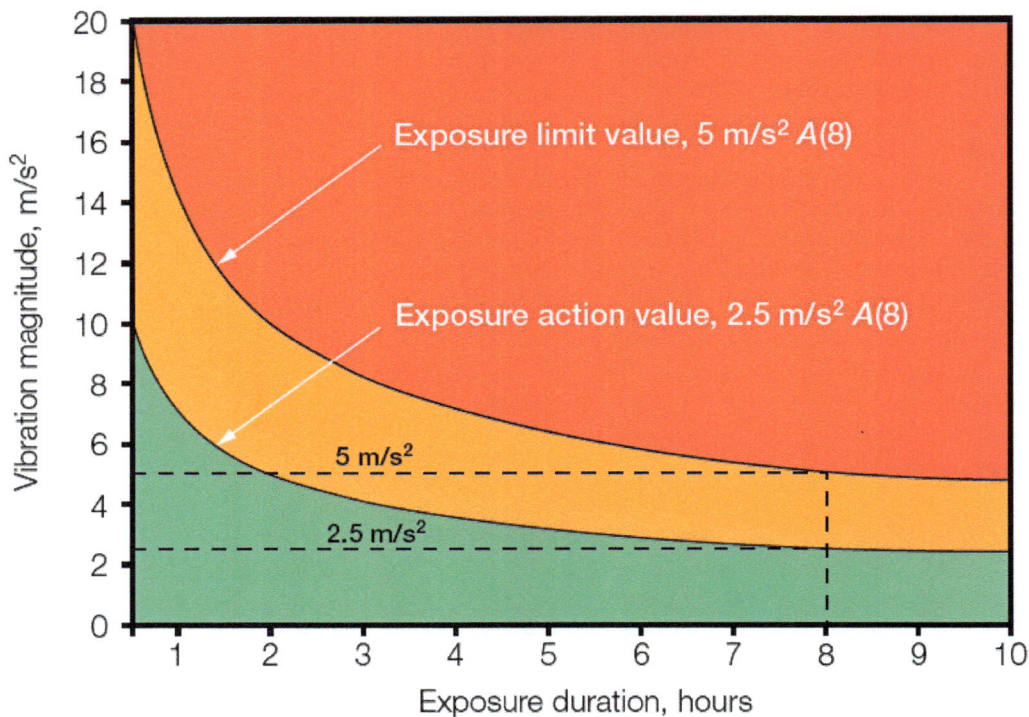

Figure B6-45: Relationship of magnitude and duration of vibration exposure.   *Source: HSE, L140.*

B6.6 - Effects of vibration on the individual

The physiological and ill-health effects of exposure to vibration

WHOLE BODY VIBRATION

The human response to *whole body vibration (WBV)* transmitted through vehicle seats in trucks, farm vehicles, or heavy equipment earth movers, or through ships and aeroplanes depends primarily on resonance of the body within the frequency range 1 to 20 Hz, predominantly in the vertical or Z axis mode.   Unlike segmental vibration, whole body vibration is a general stressor impinging on all the body organs.   The non-specific physiological response is an increase in heart rate, cardiac output and oxygen uptake. If, however, the vibration is intense (up to 50ms^{-2}) gross mechanical interference with the haemodyamics of both central and peripheral blood flow will result.

The principal resonance of a human subject vibrated in the Z axis occurs around 5 Hz. Rough riding vehicles over uneven terrain could therefore disturb cerebral blood flow in drivers resulting in decreased work performance and increased risk of error. The most common problem associated with whole body vibration is, however, back pain and in particular lower back pain.

Such signs as cerebral oedema and generalised debilitating effects, conveyed through the central nervous system, have been described by investigators in the Soviet Union. Displacement of intervertebral discs has been reported, associated with the vibration from off-road trucks, farm tractors and earth moving vehicles. It is, however, difficult to establish clear cause and effect relationships.

This is because many of the surveys reported poor ergonomic design factors, adverse working environments, and fatigue and non-specific stress factors. Even harder to establish are the causes of conditions such as abdominal pain, digestive problems, urinary difficulties, prostatitis, visual disorders, vertigo, headaches and sleeplessness. Discretion, therefore, is required to establish a direct link with vibration especially since many of the behavioural/performance field studies to date have not been controlled under strict epidemiological, statistical conditions.

Studies carried out under laboratory conditions attempting to simulate military environments such as jet aircraft cockpits or navy vessels use young, physically fit military subjects who may not represent a typical factory worker. In these laboratory studies sinusoidal vibration has been used, a characteristic that does not represent the variety of vibrations found in the workplace.

Occupational health physicians are aware that long distance coach drivers and truck drivers are more prone to many of the above ailments, but they have not been able to establish vibration as the cause due to the number of confounding variables present - bad posture, irregular dietary habits and the ever increasing stress of driving on crowded roads.

To rectify this, studies carried out by NIOSH, Cincinnati in 1974-1976 on 3205 long distance truck drivers and 1448 interstate bus drivers concluded that the combined effects of body posture, postural fatigue, dietary habits and whole body vibration contributed to a significant excess of venous, bowel, respiratory, muscular and back disorders compared to three control groups - office workers, the general population and air traffic controllers.

The studies also showed that rest periods and their distribution throughout the work day played a complex role in the overall performance.

HAND-ARM VIBRATION

The specific effects of intense vibration transmitted to the hands and arms by vibrating knobs, controls or similar equipment are long term damage to soft tissues, bones and joints. Thickening of the skin and underlying tissues probably explains **Raynaud's phenomenon** (commonly known as vibration white finger) a form of **hand-arm vibration syndrome (HAVS)** seen in experienced work people who use vibrating hand tools. The compensation award of £125,000 to 7 former miners combined with the later agreement by the government to pay about £500 million to over 30,000 former miners suffering from HAVS indicates the scale of the problem.

The symptoms of hand arm vibration syndrome are progressive and there is no cure or treatment currently available. After a latent period, the fingers 'throb' particularly on cold, wet and windy mornings. This is followed by intermittent tingling and numbness, with the fingers going white or blanching. Continued exposure results in a steady deterioration in symptoms, with sufferers experiencing spasm, 'dead finger' and painful 'hot aches'. Fingers can also become blue (cyanotic, due to lack of circulation).

This seems to occur in response to a change in metabolic demand in the fingers induced, for example, by temperature change. It seems that the blood vessels are unable to dilate rapidly enough or at all, because of thickened tissues, which then become anoxic.

The medical effects of hand arm vibration syndrome may include:

- Vascular changes in the blood vessels of the fingers.
- Neurological changes in the peripheral nerves.
- Muscle and tendon damage in the fingers, hands, wrists and forearms, including carpal tunnel syndrome, leading to reduced grip strength.
- Suspected bone and joint changes.
- Damage to the autonomic centres of the central nervous system in the brain influencing the endocrine, cardiac, vestibular and cochlea functions (not proven).

The classification of HAVS was formerly carried out using the Taylor-Pelmar classification - however this is being replaced by the **Stockholm workshop** scale.

USE OF STOCKHOLM SCALE

Classification of HAVS using the Stockholm Workshop scale:

1 Vascular component*

Stage	(Grade)	Description
0		No attacks.
1V	(mild)	Occasional attacks affecting only tips of one or more fingers.
2V	(moderate)	Occasional attacks affecting distal and middle (rarely also proximal) phalanges or one or more fingers.
3V	(severe)	Frequent blanching attacks affecting all phalanges of most fingers.
4V	(very severe)	As in Stage 3 but with trophic skin changes in fingertips.

Figure B6-46: Vascular component.   *Source: HSE, L140.*

2 Sensori-Neural Component*

Stage	Description
0_{SN}	Vibration exposed. No symptoms.
1_{SN}	Intermittent or persistent numbness with or without tingling.
2_{SN}	As in 1_{SN} with reduced sensory perception.
3_{SN}	As in 2_{SN} with reduced tactile discrimination and manipulative dexterity.

Figure B6-47: Sensori-neural component.   *Source: HSE, L140.*

***The staging is made separately for each hand. Grade of disorder is indicated by the stage (as above) and the number of affected fingers on each hand for example, 'Stage/Hand/No. of digits'.**

The condition appears to be most common with low frequency vibrations; however it has been known to occur at frequencies up to 1,500 Hz. Provided the source of vibration is stopped early in stage 2 then circulation may improve; however it is debatable whether the neurological symptoms can improve. Sufferers are therefore encouraged to change their jobs.

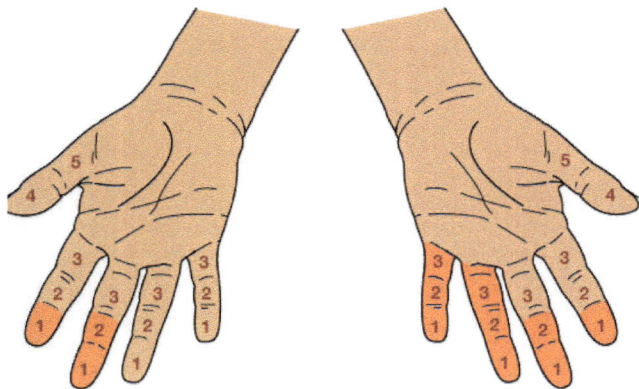

Figure B6-48: Vibration white finger.

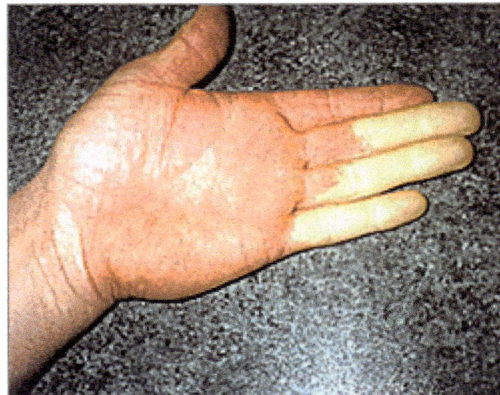

Source: www.whitefinger.co.uk.

AGGRAVATING FACTORS

Factors which will aggravate HAVS include failure to:

- Maintain or replace defective or old tools with low vibration equivalents.
- Replace worn or damaged cutting tools; ensure they are kept sharp so that they remain efficient.
- Use suitable or robust enough tools for the job, including gripping too hard.
- Take sufficient work breaks by doing other jobs in between; fingers and hands should be massaged and exercised at breaks.
- Store tools correctly to avoid very cold handles when next used.
- Encourage good blood circulation by keeping warm and dry.
- Give up or cut down on smoking tobacco; because smoking reduces blood flow to the hands and finger tips.

Groups of workers at risk to WBV and HAV

WBV

Exposure to whole body vibration tends to be due to the feet or seat of a person being in contact with a vibrating item; therefore the groups of workers at risk tend to reflect this. Drivers of some mobile machines, including certain tractors, fork lift trucks and quarrying or earth-moving machinery, may be exposed to WBV, and shocks which are associated with back pain.

Other work factors, such as posture and heavy lifting, are also known to contribute to back problems for drivers and the relative importance of WBV is not clear at present. Other workers at risk would be those that stand or sit on equipment that vibrates; this can be deliberate vibration as part of the process, for example, to sort or grade material.

Equipment that vibrates during its normal operation, perhaps because it has moving parts or it moves, can also present a vibration risk, for example, road laying equipment or food manufacturing equipment. The likelihood of risk may be increased if equipment that people stand or sit on is not maintained adequately and vibration magnitude increases.

Examples of at risk activities include:

- Fast boats travelling across rough seas.
- Movement of the wheels or tracks of a vehicle or mobile machine crossing an uneven surface, or while using mobile machines to excavate holes or trenches in the ground or to load materials such as sand or gravel into vehicles.
- Operating large static compaction, hammering or punching machines, for example, hammer mills and mobile crushers.
- Operating railway vehicles.
- Use of helicopters where the rotation of the rotor blades of a helicopter may transmit WBV through the airframe into the seats.

HAVS

Workers may be exposed to HAVS when operating hand-held power tools such as road breakers or when holding materials being worked by machines such as pedestal grinders. The most well-known health effect is vibration white finger, but other effects include damage to sensory nerves, muscles and joints in the hands and arms.

Vibration that can lead to HAVS is common in a number of industry sectors, including:

- Construction and civil engineering.
- Estate management (for example, maintenance of grounds, parks, watercourses, road and railway verges).
- Forestry.
- Foundries.
- General and heavy engineering, fabrication and metalworking.

- Manufacture of concrete products.
- Mines and quarries.
- Motor vehicle manufacture and repair.
- Road and railway construction and maintenance.
- Shipbuilding and ship repair.
- Utilities (gas, electricity, water, telecommunications etc).

Examples of at risk activities include:

- The use of hand-held chain saws in forestry.
- The use of hand-held rotary tools in grinding or in the sanding or polishing of metal, or the holding of material being ground, or metal being sanded or polished by rotary tools.
- The use of hand-held percussive metal-working tools, or the holding of metal being worked upon by percussive tools in riveting, caulking, chipping, hammering, fettling or swaging.
- The use of hand-held powered percussive drills or hand-held powered percussive hammers in mining, quarrying, demolition, or on roads or footpaths, including road construction.
- The holding of material being worked upon by pounding machines in shoe manufacture.

B6.7 - Measurement and assessment of vibration exposure

Vibration risk assessment

Regulation 5 of the Control of Vibration at Work Regulations (CVWR) 2005 requires the employer to make a suitable and sufficient assessment of the risk created by work that is liable to expose employees to risk from vibration. The assessment must observe work practices, make reference to information regarding the magnitude of vibration from equipment and, if necessary, measurement of the magnitude of the vibration.

When conducting the assessment, consideration must also be given to the type, duration, effects of exposure, exposures limit/action values, effects on employees at particular risk, the effects of vibration on equipment and the ability to use it, manufacturers' information, availability of replacement equipment, the extension of exposure at the workplace (for example, rest facilities), temperature and information on health surveillance.

The assessment should establish which workers are at risk of health effects from vibration, and the level of risk. It will also identify sources of vibration that particularly contribute to the vibration level workers are exposed to, for example, equipment and specific activities. This will enable analysis of the options to control the vibration at source or by other means.

The assessment will also help the employer when marking equipment or zones that present significant risk, giving information, instruction and training to workers and organising health surveillance for those particularly at risk. The risk assessment should be recorded as soon as is practicable after it is made and be reviewed regularly. The recorded, it will help to provide evidence of a structured risk assessment process.

A detailed vibration risk assessment should include consideration of:

- The risk of ill-health.
- The results of health surveillance.
- The identification of sources, tasks.
- The expected vibration emission levels from equipment.
- The expected time of exposure.
- Exposure to cold, nature of the vibration (WBV, HAV etc.).
- Planning (who, how, where, how often).
- Instrumentation for carrying out vibration assessments.
- The importance of calibration.
- The types of measurements to be taken.
- The use of specialist consultants.
- The interpretation and evaluation of results.
- The use of vibration calculators to determine mixed exposures.
- Comparison with legal limits to make control decisions.

The measurement of vibration

INSTRUMENTATION (THE ACCELEROMETER) AND UNITS

The electrical output from a piezo-electric crystal (accelerometer) mounted on the vibration source is fed into a vibration analyser. The signal derived from the accelerometer is 'weighted', a requirement of all the existing *vibration standards*. The acceleration magnitude may be given in decibels (dB), as with noise, or as acceleration in metres per second squared (ms-2).

An acceleration of 1 ms-2 corresponds to 120 dB, and 10 ms-2 to 140 dB. As the exposure action and limit values are expressed as acceleration in ms-2 this is the vibration meter reading most used for workplace measurements. For workplace measurements instruments give direct readings.

This instrumentation is valuable for controlling the vibration levels in the workplace and for testing new tools before issue to employees. The vibration dose is the product of both the magnitude of vibration and the exposure time working with the vibration source.

In order to ensure measurements are reliable the vibration meter should be calibrated with a known source of vibration immediately prior to use.

USE OF MANUFACTURERS' DATA TO ESTIMATE EMPLOYEE'S EXPOSURE TO VIBRATION

Manufacturers work to standard tests, written to address the requirements of the Supply of Machinery (Safety) Regulations 2008, implemented on 29 December 2009.

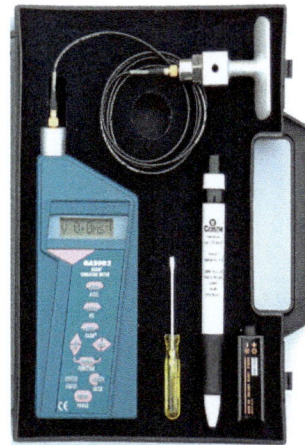

Figure B6-49: Vibration meter. *Source: www.castle.co.uk.*

These standard tests are intended both to facilitate comparison of the vibration emissions of different models of equipment and to be suitable for initial estimates of exposure. Employers still need to check with the manufacturer that the vibration emission declared in the instruction manual is representative of the employer's normal use of the equipment.

Equipment vibration data should be found in the supplier equipment handbook for new equipment. The manufacturer's vibration data should be checked to ensure that it represents the way the equipment is to be used, since some data provided may underestimate workplace vibration levels substantially. Existing equipment, put into use before 2008 may not give accurate data and it may be appropriate to double the figure quoted before using it for estimating daily exposures.

Always give consideration to replacing old equipment, pre 2008, to take advantage of any new improvements in vibration emission.

Tool type	Lowest	Typical	Highest
Road breakers	5 m/s^2	12 m/s^2	20 m/s^2
Demolition hammers	8 m/s^2	15 m/s^2	25 m/s^2
Hammer drills/combi hammers	6 m/s^2	9 m/s^2	25 m/s^2
Needle scalers	5 m/s^2	-	18 m/s^2
Scabblers (hammer type)	-	-	40 m/s^2
Angle grinders	4 m/s^2	-	8 m/s^2
Clay spades/jigger picks	-	16 m/s^2	-
Chipping hammers (metal)	-	18 m/s^2	-
Stone-working hammers	10 m/s^2	-	30 m/s^2
Chainsaws	-	6 m/s^2	-
Brushcutters	2 m/s^2	4 m/s^2	-
Sanders (random orbital)	-	7-10 m/s^2	-

Figure B6-50: Manufacturers' data.   *Source: HSE leaflet INDG175(rev2).*

MAKING WORKPLACE MEASUREMENTS FOR BOTH WBV AND HAV EXPOSURE
Legal requirements

Regulation 5 of CVWR 2005 requires the employer to make a suitable and sufficient assessment of the risk created by work that is liable to expose employees to risk from vibration.

The assessment must observe work practices, make reference to information regarding the magnitude of vibration from equipment and if necessary measurement of the magnitude of the vibration.

Consideration must also be given to the:

■ Type and duration of use.
■ Effects of exposure, exposures limit/action values.
■ Effects on employees at particular risk.
■ Effects of vibration on equipment and the ability to use it, manufacturers' information.

- Availability of replacement equipment.
- Extent of exposure at the workplace (for example, rest facilities), temperature.
- Information on health surveillance.

The risk assessment should be recorded as soon as is practicable after the risk assessment is made and reviewed regularly.

Taking measurements

Where possible the accelerometer should be positioned as near to the centre of where the operator holds or grips the vibration source for HAVS and near the centre of where the body contacts the vibration source for WBV.

During vibration measurements the instrument calculates an RMS value of acceleration over the measurement period. It will be necessary to take readings for a sufficient time to reflect any variability in the vibration and to establish a stable reading, this will be for at least 15 seconds, and readings of between 3 to 10 minutes will provide increased accuracy and repeatability of measurements.

The risk to health from vibration is affected by the frequency content of the vibration. When vibration is measured in accordance with BS EN ISO 5349-1:2001, vibration frequencies between 8 and 16 Hz are given most weight, and frequencies above and below this range make a smaller contribution to the measured vibration magnitude. This process is called frequency weighting. Vibration meters intended for HAV measurement are equipped with a frequency weighting filter, to modify their sensitivity at different frequencies of vibration.

In order to establish a complete measurement of vibration exposure it is necessary to measure the vibration on three axis, x, y and z, as shown in *figure ref B6-51*. X is taken as fore and aft vibration, Y as side-to-side vibration and Z as vertical vibration. The vibration is measured in one axis at a time and once all three axes have been individually measured the combined vibration exposure of the worker can be determined.

This is done by calculating the resultant vector sum of the vibrations and establishing an overall magnitude of acceleration by using the formula:

$$a_{hv} = \sqrt{a_{hwx}^2 + a_{hwy}^2 + a_{hwz}^2}$$

Each overall magnitude of vibration acceleration figure represents part of the worker's exposure to vibration and contributes to their daily exposure; the fraction that this contributes is established by the formula:

$$A(8) = a_{hv}\sqrt{\frac{T}{T_0}}$$

The result of this equation is the contribution to daily exposure and the reason that this is used is to give a measurement to each of a number of processes that may be carried out in one day. To calculate the overall daily exposure from a set of partial exposures, the following equation is used:

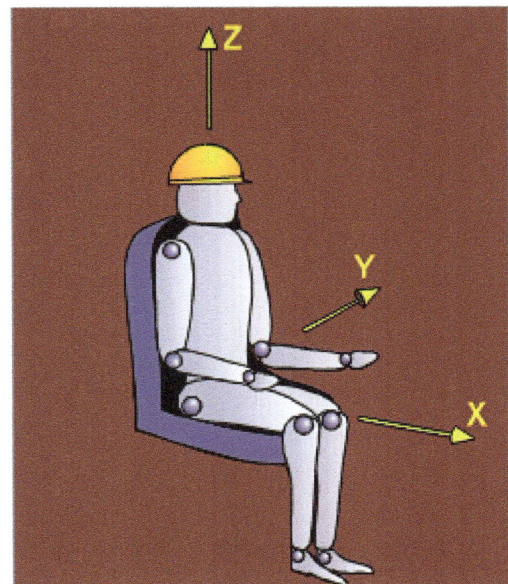

Figure B6-51: Triaxial measurements.   *Source: HSE.*

$$A(8) = \sqrt{A_1(8)^2 + A_2(8)^2 + A_3(8)^2}$$

Where $A_1(8)$, $A_2(8)$, $A_3(8)$ are the partial vibration exposure values for each vibration source the worker is exposed to.

Exposure standards for vibration with reference to legal limit and action values

The Council of Ministers and the European Parliament adopted a joint text for a Physical Agents (Vibration) Directive regarding exposure of workers to the risks arising from vibration on 21 May 2002.

The Directive was published in the Official Journal of the European Communities on 6 July 2002 (L177 Vol 45, p12) as Directive 2002/44/EC on the minimum health and safety requirements regarding the exposure of workers to the risks arising from vibration.

The Directive requirements were enacted in the UK as The Control of Vibration at Work Regulations (CVWR) 2005, made under the Health and Safety at Work etc Act (HASAWA) 1974.

EXPOSURE ACTION VALUES AND LIMIT VALUES

Regulation 4 of The Control of Vibration at Work Regulations (CVWR) 2005 states the personal daily exposure action and daily exposure limits values normalised over an 8-hour reference period.

The exposure action values (EAV) are the daily amount of exposure to vibration above which the employer is required to take certain actions. The greater the exposure level, the greater the risk and the more action employers will need to take to reduce the risk.

When assessing exposure to ensure exposure values are not exceeded, the effect of personal protection provided to the employee can be taken into account.

This will depend on the attenuation provided by the protection, its appropriateness for the type of vibration protection, use and maintenance requirements.

	Daily exposure action value (EAV)	Daily exposure limit value (ELV)
Hand-arm vibration	2.5 m/s^2 A(8)	5 m/s^2 A(8)
Whole body vibration	0.5 m/s^2 A(8)	1.15 m/s^2 A(8)

Figure B6-52: Vibration exposure values.   *Source: CVWR 2005.*

Exposure action value

The daily exposure action value (EAV) is the level of daily exposure to vibration above which the employer must take action as prescribed in regulations 6(2), 7(1)(b) and 8(1)(b) of CVWR 2005. This includes reducing exposure to as low as is reasonably practicable by the implementation of organisational and technical measures. Employees exposed to vibration at levels above the exposure action value must be placed under suitable health surveillance and provided with suitable and sufficient information, instruction and training.

Exposure limit value

The exposure limit values (ELV) are the maximum amount of vibration an employee may be exposed to on any single day, which must not be exceeded. The CVWR 2005 allowed a transitional period for the limit value until July 2010. This only applied to work equipment already in use before July 2007. The exposure limit value could be exceeded during the transitional period as long as all the other requirements of the CVWR 2005 were complied with and all reasonably practicable actions to reduce exposure were taken. As the transition period has now passed all equipment in use in the workplace must conform to the CVWR 2005.

In general, where the exposure action limit is exceeded regulation 6 of CVWR 2005 requires that the employer immediately:

- Reduce exposure to vibration to below the limit value.
- Identify the reason for that limit being exceeded.
- Modify the measures taken in accordance requirements of regulation 6 to control exposure.

Regulation 6 of the CVWR 2005 allows the exposure limit value to be occasionally exceeded in circumstances where the employee's exposure is usually below the limit but it varies significantly from time to time.

This is only allowed where:

"(a) Any exposure to vibration averaged over one week is less than the exposure limit value.

(b) There is evidence to show that the risk from the actual pattern of exposure is less than the corresponding risk from constant exposure at the exposure limit value.

(c) Risk is reduced to as low a level as is reasonably practicable, taking into account the special circumstances.

(d) The employees concerned are subject to increased health surveillance, where such surveillance is appropriate within the meaning of Regulation 7(2)."

HSE vibration calculators and their use to determine simple and mixed exposure

READY-RECKONER

The table below is used for calculating daily vibration exposures. The ready-reckoner covers a range of vibration magnitudes up to 40 m/s^2 and a range of exposure times up to 10 hours.

The exposures for different combinations of vibration magnitude and exposure time are given in exposure points instead of values in m/s^2 A(8).

The exposure points are easier to work with than the A(8) values:

- Exposure points change simply with time: twice the exposure time, twice the number of points.
- Exposure points can be added together, for example where a worker is exposed mixed exposure to two or more different sources of vibration in a day.
- The exposure action value (2.5 m/s^2 A(8)) is equal to 100 points.
- The exposure limit value (5 m/s2 A(8)) is equal to 400 points.

USING THE READY-RECKONER CALCULATOR

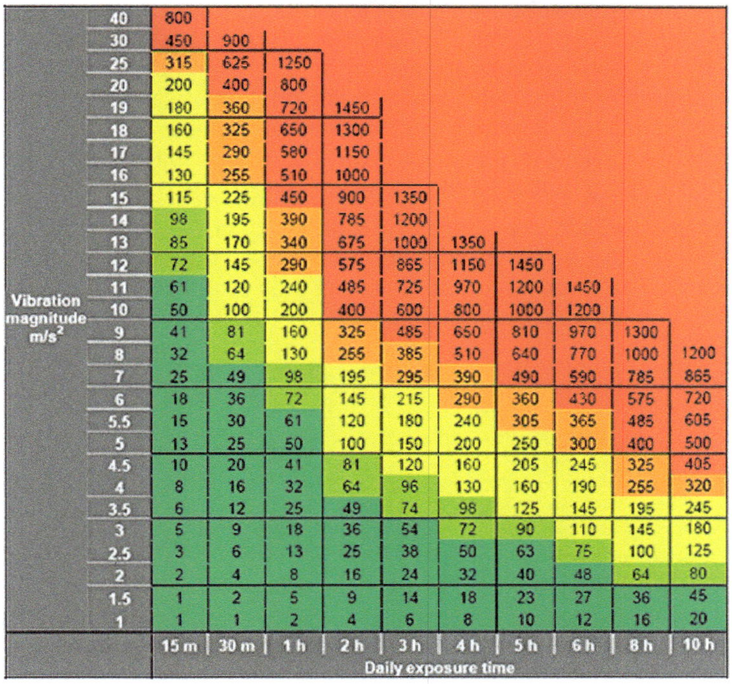

Vibration magnitude m/s²	15 m	30 m	1 h	2 h	3 h	4 h	5 h	6 h	8 h	10 h
40	800									
30	450	900								
25	315	625	1250							
20	200	400	800							
19	180	360	720	1450						
18	160	325	650	1300						
17	145	290	580	1150						
16	130	255	510	1000						
15	115	225	450	900	1350					
14	98	195	390	785	1200					
13	85	170	340	675	1000	1350				
12	72	145	290	575	865	1150	1450			
11	61	120	240	485	725	970	1200	1450		
10	50	100	200	400	600	800	1000	1200		
9	41	81	160	325	485	650	810	970	1300	
8	32	64	130	255	385	510	640	770	1000	1200
7	25	49	98	195	295	390	490	590	785	865
6	18	36	72	145	215	290	360	430	575	720
5.5	15	30	61	120	180	240	305	365	485	605
5	13	25	50	100	150	200	250	300	400	500
4.5	10	20	41	81	120	160	205	245	325	405
4	8	16	32	64	96	130	160	190	255	320
3.5	6	12	25	49	74	98	125	145	195	245
3	5	9	18	36	54	72	90	110	145	180
2.5	3	6	13	25	38	50	63	75	100	125
2	2	4	8	16	24	32	40	48	64	80
1.5	1	2	5	9	14	18	23	27	36	45
1	1	1	2	4	6	8	10	12	16	20

Daily exposure time

Above limit value
Likely to be above limit value
Above action value
Likely to be above action value
Below action value

Figure B6-53: HSE vibration calculator (ready-reckoner). *Source: HSE website.*

- Find the vibration magnitude (level) for the tool or process (or the nearest value) on the grey scale on the left of the table.
- Find the exposure time (or the nearest value) on the grey scale across the bottom of the table.
- Find the value in the table that line up with the magnitude and time. The illustration shows how it works for a magnitude of 5 m/s2 and an exposure time of 3 hours: in this case the exposure corresponds to 150 points.
- Compare the points value with the exposure action and limit values (100 and 400 points respectively). In this example the score of 150 points lies above the exposure action value.
- The colour of the square containing the exposure points value tells you whether the exposure exceeds, or is likely to exceed, the exposure action or limit value.
- If a worker is exposed to more than one tool or process during the day, repeat steps 1 - 3 for each one, add the points, and compare the total with the exposure action value (100) and the exposure limit value (400).

VIBRATION EXPOSURE CALCULATOR

The HSE have produced vibration exposure calculator in the form of an excel document for HAVS and WBV. They have the appropriate formulae for determining single and mixed exposure built in and uses the same point methodology as in the ready-reckoner, without the need to look up the details on the chart.

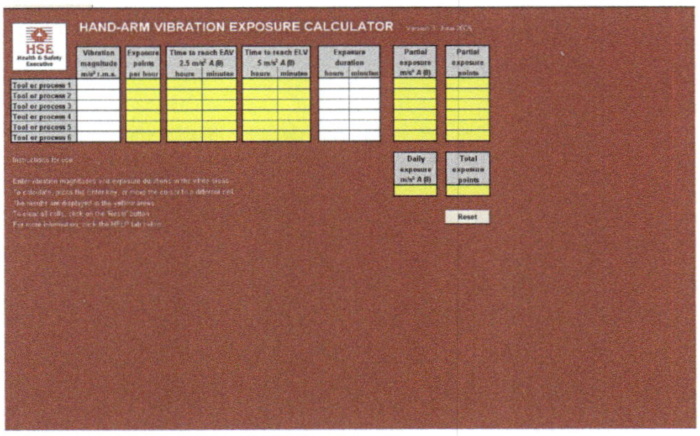

Figure B6-54: Hand arm vibration calculator. *Source: HSE, website.*

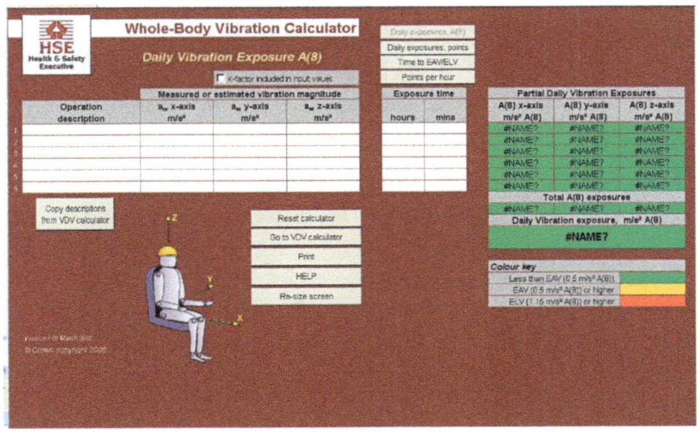

Figure B6-55: Whole body vibration calculator. *Source: HSE, website.*

ROLE OF HEALTH SURVEILLANCE

The role of health surveillance is to provide early detection of work related ill-health, as it will assist with the identification of symptoms of the effects of vibration on health. Health surveillance can therefore prevent or diagnose any health effect linked with exposure to vibration. Health surveillance should be carried out for all workers where there is a risk to their health due to being exposed to vibration.

By conducting health surveillance from the start of a worker working with vibration risks it is possible to detect early signs of the effects of vibration and provide early intervention to limit the continuing effects. Health surveillance can assist with confirming the success of vibration control measures. Surveillance for HAVS usually involves the worker or an occupational health specialist examining the hands to identify early signs of tingling or blanching. For WBV, it can be a simple reporting method or questionnaire relating to experience of lower back discomfort or pain. A record of health shall be kept of any worker who undergoes health surveillance. The employer should, providing reasonable notice is given, provide the worker with access to their health records and provide copies to a relevant competent authority on request.

If health surveillance identifies a disease or adverse health effect, considered by a doctor or other occupational health professional to be a result of exposure to vibration, the employer should ensure that a qualified person informs the worker and provides information and advice. The employer should ensure they are kept informed of any significant findings from health surveillance, taking into account any medical confidentiality.

In addition, the employer should also:
- Review risk assessments.
- Review the measures taken to comply.
- Consider assigning the worker to other work.
- Review the health of any other worker who has been similarly exposed and consider alternative work.

B6.8 - Controlling vibration and vibration exposure

Legal requirements and duties to manage exposure to vibration

Regulation 6 of CVWR 2005 states that the employer must seek to eliminate the risk of vibration at source or, if not reasonably practicable, reduce it to as low a level as is reasonably practicable. Where it is not reasonably practicable to take preventive measures which eliminate the risk at source and the personal daily exposure action value is likely to be reached or exceeded the employer must reduce exposure by implementing a programme of organisational and technical measures.

These precautionary measures include automation, the use of other methods of work, improved ergonomics/equipment, maintenance of equipment, design and layout, purchasing policy, job rotation, rest facilities, information, instruction and training, limitation by schedules and breaks and the provision of personal protective equipment to protect from cold and damp. Measures must be adapted to take account of any group or individual employee whose health may be at particular risk from exposure to vibration.

Practical control measures to prevent or minimise exposure to both WBV and HAV

GENERAL CONTROL MEASURES FOR VIBRATION

As with many controls to prevent or minimise harm, the preferred option would be to prevent all exposure; this may be achieved in a small number of situations where processes can be **automated** so that the use of vibration creating tools or processes do not involve an operator directly.

Some work **methods** may be changed to prevent exposure of a worker to high levels of vibration, for example, instead of using a hand-held vibration concrete breaker (which produces whole body and hand arm vibration) a machine with a breaker attachment may be used. This would put the worker at a distance to the vibration and remove their hands from close contact with the vibrating parts. Though vibration may be an aspect of this work, levels should be greatly reduced. In order to identify opportunities to change work methods, a detailed ergonomic assessment of high risk tasks should be carried out, and this should include work methods creating the risk, duration and frequency of the task.

It is important to use the most **suitable equipment** for the task, for example, using the lowest vibration tool as much as possible and choosing equipment that will do the task effectively and quickly so that exposure duration is reduced. Where possible, older equipment, that has have been manufactured to different standards, should be replaced with newer ones. The need for such equipment should be reviewed and a replacement plan put in place to enable the elimination of unnecessary high risk operations through the use of improved equipment.

The vibration characteristics of specific equipment in use should be assessed and reference made to the BSI and ISO guidelines. Even though the equipment has been purchased more recently its vibration characteristics may present a significant risk after use for a time. Equipment that presents a higher risk should be identified, controlled and its use avoided where possible by issuing alternative equipment that presents a lower risk first.

In order to reduce whole body vibration, vehicles should be fitted with suspension seats that have appropriate vibration-damping characteristics.

In order to reduce the exposure as far as is reasonably practicable it is important that the **purchasing policy** sets standards for the vibration characteristics of new equipment at least as low as those set to meet the Supply of Machinery (Safety) Regulations 2008. Where possible, purchase of equipment that provides better than the minimum protection should be made.

It is the combined responsibility of the employer and the equipment manufacturer to reduce vibration levels on hand tools to safe limits. Foremost in this drive have been the chain saw manufacturers. Sweden and Finland have succeeded in reducing the relevance of HAVS from 50-60% to 5-10% of the workforce in their forestry operations. Some progress has been made with pneumatic, percussive drills and chipping hammers.

The EC Machinery Directive (89/392/EEC) requires that instructions for hand-held portable machinery should indicate whether or not the operator will be subjected to vibrations where the accelerations exceed 2.5 ms^{-2}.

"Discomfort, fatigue and psychological stress faced by the operator must be reduced to a minimum taking ergonomic principles into account."

Figure B6-56: Quote - ergonomic principles.   *Source: CVWR 2005.*

In order to control exposure levels equipment should be subject to planned preventative **maintenance** which ensures its efficient operation and limits the vibration generated by parts wearing out. Equipment and tools should be maintained to ensure their optimum performance, thereby reducing the creation of vibration to a minimum, for example, for example, the bearings of grinders and blades of cutting equipment.

A simple way to reduce vibration exposure is to ensure **job rotation**, so that equipment that is known to create a high magnitude of vibration is not operated by the same person all the time. Tasks should be subject to rotation in order that each worker limits their exposure and is afforded some recovery period.

The work schedule should be examined to reduce duration of vibration exposure and magnitude, either by alternating with non-vibration work or avoiding continuous vibration by, for example, scheduling ten minute breaks every hour.

Care should be taken to organise work schedules so that rest periods from this high risk work happen naturally in the process. Where they do not, it may be necessary to use reminders in the form of timed alarms or supervision.

It is essential that workers are provided with **information, instruction and training** that enables the correct use of equipment, so that the manufacturers expected exposure levels can be achieved. Information may be provided in a manufacturer's handbook, and this should be provided to those that are to use the equipment in a form that they can understand and act on.

Instruction and training should be given to all employees exposed to significant levels of vibration, and this should include the proper use of equipment and the minimisation of exposure.

It should be remembered that when considering HAVS, the greater the coupling (hand and tool interface), the more vibration energy enters the hand. Increasing grip force increases the coupling. There are working techniques for all equipment and the expertise developed over time justifies an initial training period for new starters.

Operators of vibrating equipment should be trained to recognise the early symptoms of HAVS and WBV and how to report them.

Health surveillance is an integral and supporting part of the controls for vibration, and this can include personal and professional health surveillance. It will provide early warning of possible harm and symptoms like tingling in the hands may be identified at an early stage of over exposure, enabling intervention action.

The role of health surveillance is to provide early detection of symptoms of the effects of vibration on health. By conducting health surveillance from the start of an employee's employment it is possible to detect early signs of the effects of vibration and provide early intervention to limit the continuing effects. Health surveillance can assist with confirming the success of vibration control measures.

Surveillance for HAVS usually involves the worker, or an occupational health specialist, examining the hands to identify early signs of tingling or blanching. For whole body vibration, it can be a simple reporting method or questionnaire relating to experience of lower back discomfort or pain.

Workers with established HAVS should avoid exposure to cold and thus minimise the number of blanching attacks. Workers with **advanced** HAVS whose condition is **deteriorating** (as measured by annual medical checks) should be removed from further exposure. The medical priority is to prevent fingertip ulceration (tissue necrosis).

Regulation 7 of CVWR 2005 states that health surveillance must be carried out if there is a risk to the health of employees liable to be exposed to vibration. This is in order to prevent or diagnose any health effect linked with exposure to vibration.

A record of health shall be kept of any employee who undergoes health surveillance. The employer shall, providing reasonable notice is given, provide the employee with access to their health records and provide copies to an enforcing officer on request.

If health surveillance identifies a disease or adverse health effect, considered by a doctor or other occupational health professional to be a result of exposure to vibration, the employer shall ensure that a qualified person informs the employee and provides information and advice. The employer must ensure they are kept informed of any significant findings from health surveillance, taking into account any medical confidentiality.

In addition the employer must also:

- Review risk assessments.
- Review the measures taken to comply.
- Consider assigning the employee to other work.
- Review the health of any other employee who has been similarly exposed and consider alternative work.

Personal protective equipment may provide some absorption of vibration, provided it has been selected for this purpose. It may have a secondary helpful effect of keeping the hands warm and improving circulation, which helps limit the effects of hand arm vibration.

Wearing gloves is recommended for safety and for retaining heat. They will not absorb a significant fraction of the vibration energy which lies within the 30-300 Hz range. There is also the danger that the absorbent material used in glove manufacture may introduce a resonance frequency and therefore the total energy input to the hands may be increased (Bednall, Health and Safety Executive 1988).

There have been recent reports of patented composite materials that, when moulded into hand-grips and fitted onto vibrating power tools, reduce vibration by 45%.

The HSE state in their publication L140 - Hand-arm Vibration; the Control of Vibration at Work Regulations 2005 Guidance on Regulations that *"It is not usually possible to assess the vibration reduction provided in use by anti-vibration gloves, so you should not generally rely on them to provide protection from vibration."*

Warm clothing can help workers exposed to hand-arm vibration maintain a good body core temperature, which will assist circulation to the hands. It may be necessary for workers to be provided with a warm location for rest breaks or periods when cold is affecting their circulation. Workers with established HAVS should avoid exposure to cold to minimise the number of blanching attacks.

Finally, regarding the redesigning tools, rescheduling work methods, or automating the process - a continuous review should be maintained until such time as the risks associated with vibration are under control. As with any management system, the controls in place for vibration should include audit and review.

CONTROL MEASURES IN RESPECT OF WHOLE BODY VIBRATION

Although no single technique is best for all situations, the following guidelines will be helpful in most instances for occupations such as farmers, construction workers and drivers:

- Provide suspension seats that have appropriate vibration-damping characteristics. Obtain appropriate advice when replacing a vehicle seat to ensure the seat is matched to the vehicle's vibration characteristics, to avoid making vibration exposure worse.
- Maintain vehicle suspension systems correctly, including the cab, seat suspension and tyre pressures.
- Adjust the driver weight setting on suspension seats, where it is available, to minimise vibration and to avoid the seat suspension 'bottoming out' when travelling over rough ground.
- Maintain paved surfaces and site roadways, for example, fill in potholes, level ridges, fill in joints between concrete slabs, and remove debris.
- Replace solid tyres on equipment such as fork-lift trucks, sweepers and floor scrubbers before they reach their wear limits.
- Ensure the vehicle speed is appropriate for the ground conditions, to avoid excessive bumping and jolting.

CONTROL MEASURES IN RESPECT OF HAND ARM VIBRATION

In addition to the general controls the following specific guidelines will be helpful in reducing the risk of a hand arm vibration syndrome.

- Change fabrication methods, for example, use adhesives or welding to avoid using pneumatic riveting hammers.
- Design metal castings production processes to eliminate or reduce hand finishing (fettling).
- Use machine-mounted breakers, mobile road-cutting machines and/or trenching machines instead of hand-operated road breakers for cable laying, water and mains repairs and similar work.
- Establish replacement programmes for tools and their consumables to encourage their replacement when worn, for example, chisels, abrasive discs, drill bits and cutters.
- Ensure targeted HAV reduction maintenance takes place, including keeping cutting tools sharp, dressing grinding wheels, balance rotating equipment, tune and adjust engines, and replace defective vibration dampers.
- Use jigs and similar aids that have anti-vibration mounts, to avoid the need to grip vibrating surfaces.
- Identify daily limits on actual equipment use, sometimes called 'trigger time', to help employees manage their own HAV exposure.

Radiation

On completion of this element, candidates should be able to demonstrate understanding of the content through the application of knowledge to familiar and unfamiliar situations. In particular, they should be able to:

B7.1   Outline the nature of the different types of ionising and non-ionising radiation.

B7.2   Explain the effects of exposure to non-ionising radiation, its measurement and control.

B7.3   Outline the effects of exposure to ionising radiation, its measurement and control.

B7.4   Outline the different sources of lasers found in the workplace, the classification of lasers and the control measures.

Content

Relevant statutory provisions

Control of Artificial Optical Radiation at Work Regulations (CAOR) 2010

Health and Safety at Work etc, Act (HASAWA) 1974

Ionising Radiations (Medical Exposure) Regulations (IRMER) 2000

Ionising Radiations Regulations (IRR) 1999

Management of Health and Safety at Work Regulations (MHSWR) 1999 (as amended)

Management of Health and Safety at Work Regulations (Northern Ireland) 2000

Workplace (Health, Safety and Welfare) Regulations (Northern Ireland) 2000

Workplace (Health, Safety and Welfare) Regulations (WHSWR) 1992

Directive 2013/35/EU on the minimum health and safety requirements regarding the exposure of workers to the risks arising from physical agents (electromagnetic fields) (or equivalent regulations)

Sources of reference

Reference information provided, in particular web links, was correct at time of publication, but may have changed.

Guidance for Employers on the Control of Artificial Optical Radiation at Work Regulations (AOR) 2010, HSE
http://www.hse.gov.uk/radiation/nonionising/employers-aor.pdf

Laser radiation: safety advice, https://www.gov.uk/government/publications/laser-radiation-safety-advice/laser-radiation-safety-advice

Public Health England, The UK reference site on radon from Public Health England, http://www.ukradon.org/

Radon in the workplace, http://www.hse.gov.uk/radiation/ionising/radon.htm#testingradon

Safety of laser products; equipment classification and requirements BS 60825-1:2014, ISBN: 978-0-580779-69-5

Workplace health, safety and welfare, Workplace (Health, Safety and Welfare) Regulations 1992, ACOP, L24, HSE, ISBN: 978-0-717604-13-5, http://www.hse.gov.uk/pubns/priced/l24.pdf

Working with ionising radiation, Approved Code of Practice and guidance, L121, HSE, ISBN: 978-0-717617-46-3 http://www.hse.gov.uk/pubns/priced/l121.pdf

The above web links along with additional sources of reference, which are additional to the NEBOSH syllabus, are provided on the RMS Publishing website for ease of use - www.rmspublishing.co.uk.

B7.1 - The nature and different types of ionising and non-ionising radiation

The distinction between ionising and non-ionising radiation

The part of the electromagnetic spectrum with greatest energy is where the gamma rays and the X-rays are. They are so powerful that they are capable of removing the electron from an atom of material they encounter, a process that is called ionisation.

These powerful rays are known as ionising radiation. The radiation from the lower energy part of the electromagnetic spectrum, ultra-violet (UV), visible light, infra-red (IR), microwaves and radio waves, do not contain enough energy to ionise and are therefore known as non-ionising radiation. Radiation can therefore be classified in terms of its energy into two types: **ionising** and **non-ionising**.

There is also a type of ionising radiation that is not electromagnetic waves, but is particulate. The most common particles of ionising radiation are known as alpha (α), beta (β) and neutron radiation. There are other particles of ionising radiation, but consideration of them is beyond the scope of this Element.

Ionising radiation can be defined as that radiation, typically alpha and beta particles and gamma and x-rays, which has sufficient energy to produce ions by interacting with matter. Whereas non-ionising radiation can be defined as that radiation that does not possess sufficient energy to cause the ionisation of matter.

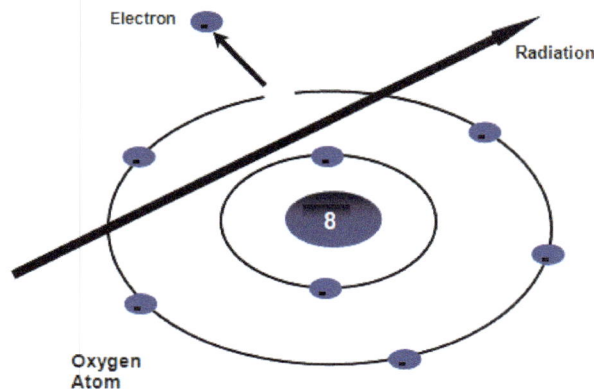

Figure B7-1: Ionisation.   Source: RMS.

Ionising radiation has sufficient energy to dislodge electrons from atoms thus changing the chemical properties of the substance it interacts with. Ionising radiation includes alpha particles, beta particles, gamma rays, x-rays and cosmic rays.

Figure B7-2: Definition of ionising radiation.   Source: London Health Observatory.

Non-ionising radiation (NIR) is the term given to the part of the electromagnetic spectrum where there is insufficient quantum energy to cause ionisations in living matter. It includes static and power frequency fields, radiofrequencies, microwaves, infra-red, visible and ultraviolet radiation.

Figure B7-3: Definition of non-ionising radiation.   Source: Health Protection Agency.

The electromagnetic spectrum

BASIC CONCEPTS - WAVELENGTH, ENERGY, FREQUENCY

Radiation is a form of energy. The types of radiation are grouped and labelled according to the nature of the radiation and the amount of energy they have. One of these types of energy is transmitted by waves called electromagnetic waves.

An electromagnetic wave travels, or propagates, in a direction that is oriented at right angles to the oscillations of both the electric (E) and magnetic (B) fields, transporting energy away from the radiation source.

The two oscillating energy fields are mutually perpendicular and act in phase, as a sine wave.

Electric and magnetic field vectors are not only perpendicular to each other, but are also perpendicular to the direction of wave propagation.

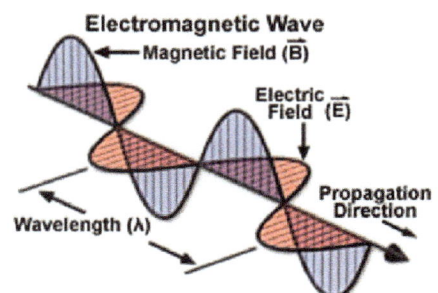

Figure B7-4: Electromagnetic wave. Source: Olympus.

By convention, and to simplify illustrations, the vector representing the electric oscillating fields of electromagnetic waves are often omitted. The resulting diagrams show a single sinusoidal electromagnetic wave.

The distance between electromagnetic wave peaks is the 'wavelength'. The number of wave peaks passing a given point in one second is the 'frequency'. The greater the frequency, the greater the associated energy, which results in increasing potential to harm the exposed human body.

Electromagnetic radiation can be arranged according to its frequency or wavelength into a series called the **electromagnetic spectrum**. This is represented by the following diagram (not to scale):

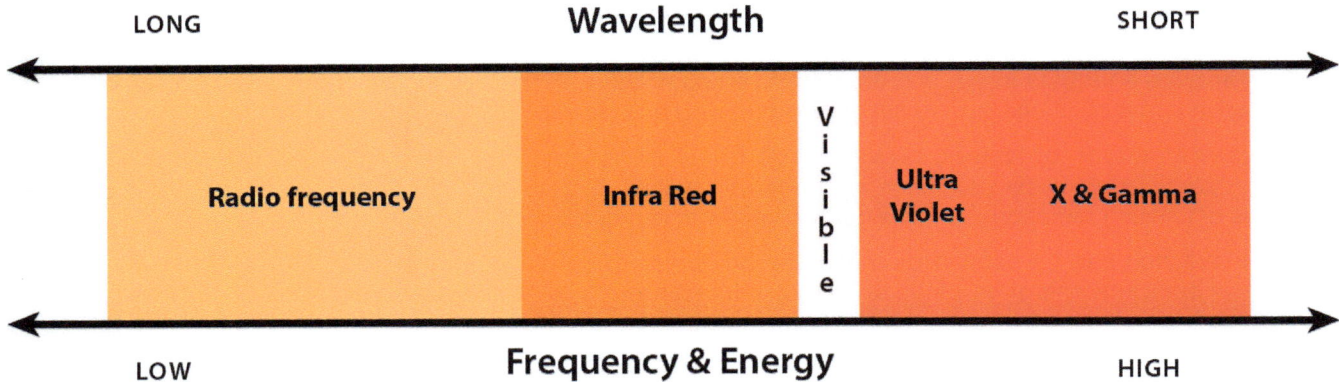

Figure B7-5: Electromagnetic spectrum - wavelength, frequency and energy.   Source: RMS.

The longer wavelength electromagnetic waves are radio waves, which have a wavelength in the order of 10^3 metres, a lower frequency (approximately 10^4 Hz) and relatively low energy. In the middle of the spectrum is visible light, which has a wavelength of approximately 0.5×10^{-6} metres and frequency in the order of 10^{15} Hz.

The shorter wavelength electromagnetic waves are X and gamma waves, which have a wavelength in the order of 10^{-10} to 10^{-12} metres, they have a higher frequency (approximately 10^{20} Hz) and very high energy.

The nature and types of ionising radiation

All matter is composed of atoms. Different atomic structures give rise to unique elements. Examples of common elements, which form the basic structure of life, are hydrogen, oxygen and carbon.

Atoms form the building blocks of nature and cannot be further sub-divided by chemical means. The centre of the atom is called the nucleus, which consists of protons and neutrons. Electrons take up orbit around the nucleus.

Protons: Have a unit of mass and carry a positive electrical charge.

Neutrons: These also have mass but no charge.

Electrons: Have a mass about 2,000 times less than that of protons and carry a negative charge.

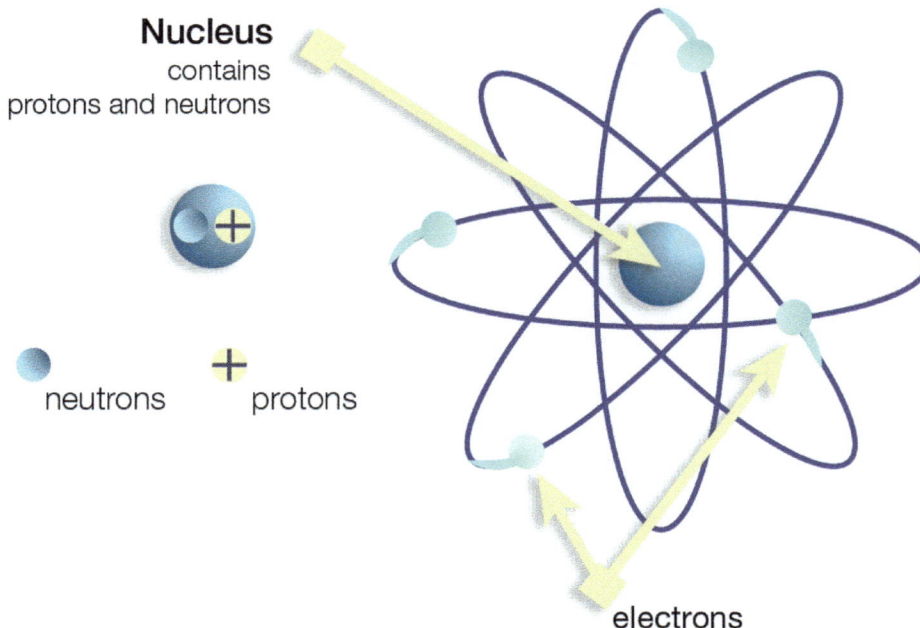

Figure B7-6: Structure of an atom.   Source: RMS.

Ionising radiation is produced by unstable atoms, which differ from stable atoms because they have an excess of energy or mass or both. Unstable atoms are radioactive. In order to reach stability, these atoms emit, the excess energy or mass.

The types of ionising radiation are:

- Particulate - mass emitted by the atom as a particle with the energy of motion - alpha, beta and neutron.
- Electromagnetic - energy emitted by the atom in a combined electrical and magnetic wave - X-ray and gamma rays.

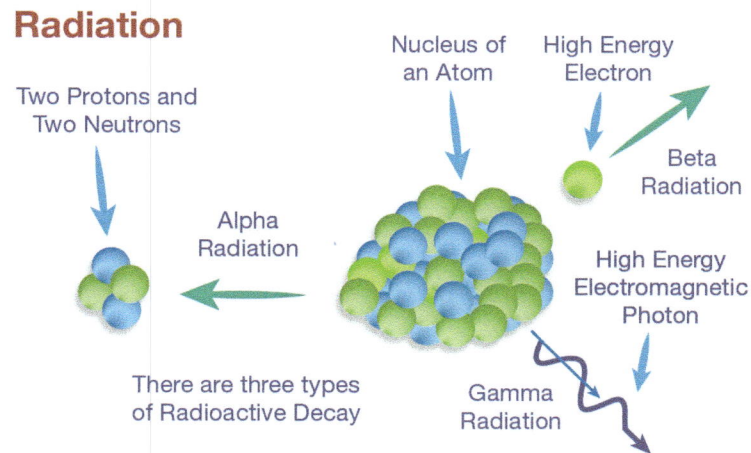

Radiation

Two Protons and Two Neutrons

Alpha Radiation

Nucleus of an Atom

High Energy Electron

Beta Radiation

High Energy Electromagnetic Photon

Gamma Radiation

There are three types of Radioactive Decay

Figure B7-7: Ionising atomic decay.

Source: RMS.

Note: other types of radiation exist, such as positrons (β^+), thermal neutrons, fast neutrons and protons, but are beyond the scope of this Element.

ALPHA PARTICLES

Alpha particles (α) are particularly stable groups of two protons and two neutrons. Alpha particles are comparatively large, travel short distances in dense materials, and are unlikely to penetrate living skin tissue, as the dead layer of skin at the surface of the epidermis will prevent entry. The principal risk is through ingestion or inhalation of a source of alpha particles, which might place the material close to vulnerable tissue. When this happens, the high-localised energy effect will destroy tissue of the organ/s affected. A source of alpha particles is smoke detectors.

BETA PARTICLES

Beta particles (β^-) are electrons, which are negatively charged particles. Beta particles are much faster moving than alpha particles. They are smaller in mass than alpha particles, but have longer range, so they can pass through several metres of air and penetrate the skin. Whilst they have greater penetrating power than alpha, beta particles are less ionising and take longer to affect the same degree of damage. An occupational source of beta particles can be found in thickness gauges. Materials such as paper and plastics can be measured by the amount of Beta radiation that is transmitted through the material.

GAMMA RAYS

Gamma rays (λ) are photons of electromagnetic radiation that are emitted by the nucleus of an unstable atom. Gamma radiation is emitted when an energetic or 'excited' nucleus loses excess energy. This may result from the ejection of α and $\beta-$ particles, which means that both alpha and beta radiation may be accompanied by the release of gamma rays.

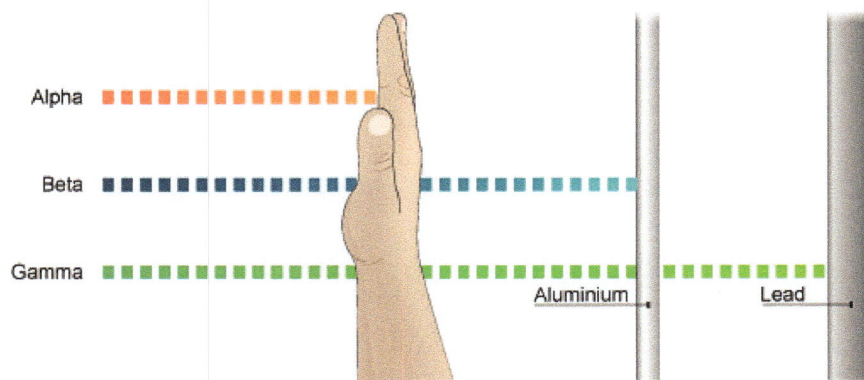

Alpha

Beta

Gamma

Aluminium

Lead

Figure B7-8: Relative penetration power of several types of ionising radiation.

Source: BBC.

Gamma rays are 'packets' of electromagnetic energy that possess a characteristic wavelength similar to, but shorter than, X-rays. They have great penetrating power. Gamma radiation passing through a normal atom will sometimes force the loss of an electron, leaving the atom positively charged, and the resultant atom is called an ion. Occupational sources of gamma rays would be the use of industrial radiography in non-destructive testing, i.e. welding seams on an oil pipeline. It can also be used to kill cancer cells. Frequency Range: 10^{20} - 10^{24} Hz. Wavelength Range: <1pm.

X-RAYS

X-rays are photons of electromagnetic radiation that are either produced artificially or occur as part of radioactive decay, when atoms re-arrange themselves to release energy and become more stable. X-rays are produced by the sudden acceleration or deceleration of a charged particle, usually when high-speed electrons strike a suitable target, usually made of dense material, under controlled conditions. A charged particle, such as an electron, will experience a force when placed in an electromagnetic field and can be accelerated to very high energy.

In an X-ray generation machine a hot cathode emits electrons that are attracted to a tungsten anode. When the electrons hit the anode target most of the energy is transferred to heat, but some is converted to X-rays. The electrical potential required to accelerate electrons to speeds where X-ray production will occur is a minimum of 15,000 volts.

High voltage (greater than 15 kV) equipment, such as that used with radar, has the potential to produce parasitic (unintentional) X-rays. X-rays have high energy, and high penetration power through fairly dense material. In low-density substances, including air, they may travel long distances.

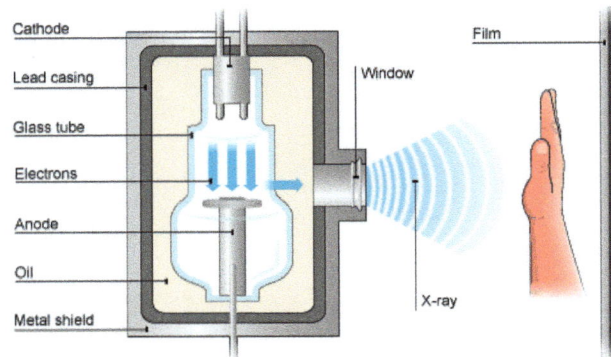

Figure B7-9: X-ray generation machine.   *Source: BBC.*

NEUTRONS

Neutrons (n or n°) are single particles emitted during certain nuclear processes, for example, nuclear fission. Neutrons have great penetrating power and are able to pass through relatively dense materials such as 50 mm of lead. Their particular penetration characteristics are used in meters for measuring the moisture content of soil.

The nature and types of non-ionising radiation

Non-ionising radiation relates to the part of the electromagnetic spectrum covering two main regions, optical radiation (ultraviolet, visible and infrared) and electromagnetic fields (power frequencies, microwaves, and radio frequencies). They are defined according to their wavelength:

Radiation	Wavelength
Ultraviolet (UV)	100 - 400 nm
Optical / Visible light	400 - 780 nm
Infrared (IR)	780 nm - 1,000,000 nm (1 mm)
Microwave and radio waves	> 1 mm

Figure B7-10: Non-ionising radiation wavelengths.   *Source: RMS.*

Optical radiation is another term for radiation derived from light, covering ultraviolet (UV) radiation, visible light radiation, and infrared radiation (IR).

ULTRAVIOLET RADIATION

Ultraviolet (UV) radiation is emitted from very hot bodies, for example, electric arcs and the sun. It can also be created by an electric discharge through gases. It has a wavelength of between 100 to 400 nm. This part of the electromagnetic spectrum is further subdivided into three regions according to wavelength.

Region	Wavelength nanometre (nm)
UV-A	315 - 400
UV-B	280 - 315
UV-C	100 - 280

Figure B7-11: UV radiation wavelengths.   *Source: NRPB (HPA), Advice on protection against ultraviolet radiation.*

Possible occupational sources

There are many possible sources of ultraviolet (UV) radiation to which people may be exposed at work, for example:

- The sun.
- Electric arc welding and cutting.
- Sunbeds and sunlamps.
- Crack detection equipment.
- Water treatment.
- Adhesive curing equipment.
- Forgery detectors.
- Some lasers.
- Mercury vapour lamps.
- Tungsten halogen lamps.

VISIBLE LIGHT

Visible light is that part of the electromagnetic spectrum visible to the human eye. The human eye is sensitive to a range of wavelengths from ~ 400 nanometres (the blue end) to ~ 750 nm (the red end), which sets the limits of the range of the electromagnetic spectrum that relates to visible light. It can therefore be described by its wavelength (colour) and its amplitude (intensity). In common with the rest of the electromagnetic spectrum, it is a form of energy that has the characteristics of wave motion.

Possible occupational sources

Any high intensity source of visible light can cause problems. Possible sources of visible light at work include:

- Lasers operating in the visible wavelength, for example, in surveying or level alignment equipment.
- Furnaces or fires.
- Molten metal.
- Welding and cutting.
- High intensity light beams and light bulbs.
- Other high intensity lights, such as in photocopiers and printers.

The danger may be due to direct or reflected radiation.

INFRARED (IR)

Infrared radiation (IR) relates to the range of invisible radiation wavelengths from about 780 nanometres, just longer than red in the visible spectrum, to wavelengths of 1 millimetre, on the border of the microwave region of the electromagnetic spectrum. This part of the electromagnetic spectrum is further subdivided into three regions according to wavelength.

Region	Wavelength nanometre (nm)
IR-A	780 - 1,400
IR-B	1,400 - 3,000
IR-C	3,000 - 1,000,000 (1 mm)

Figure B7-12: IR radiation wavelengths.   *Source: HPA.*

Infrared radiation is emitted from high temperature sources, the higher the temperature the shorter the wavelength of the infrared radiation emitted.

Possible occupational sources

Anything that glows is likely to be a source of infrared radiation, for example:

- Furnaces or fires.
- Molten metal or glass.
- Burning or welding.
- Heat lamps.
- Some lasers.
- Intruder alert systems.

MICROWAVE

Microwaves are the name given to the electromagnetic radiation between the infrared and radio wave region of the electromagnetic spectrum, with wavelengths typically in the 1 mm to 10 cm range. Microwaves are actually just radio waves of shorter wavelength and therefore higher frequencies. As they constitute the highest frequency radio waves they have significant energy related to them.

Possible occupational sources

Radio frequency radiation is produced by radio/television transmitters and high voltage electricity cables. It is used industrially for induction heating of metals and is often found in intruder detectors. Microwave radiation is used in communication systems (mobile phones) and in cooking equipment.

RADIO WAVE

This part of the electromagnetic spectrum ranges from a wavelength of approximately 1 mm to thousands of kilometres, there is no defined upper limit to the radio wave wavelength category. If microwaves are considered to be in a separate category the remaining part of the radio waves range from approximately 10 cm upwards.

The radiation relating to these electromagnetic waves contains little energy. At the longer wavelengths of this range the photon character of the energy is not apparent and the waves appear to transfer energy in a smooth manner.

Particulate and non-particulate types of ionising radiation

PARTICULATE IONISING RADIATION

Particulate ionising radiation refers to the radiation energy carried by moving particles. Ionising radiation in the form of a particle includes very small (sub-atomic) particles such as electrons, protons and neutrons. When radioactive material transforms by decay, it can emit 'beta particles' (electrons), 'alpha particles' (two protons and two neutrons) and/or single neutrons. When this happens, particulate radiation is emitted by the atom and a new element is produced. The radionuclide uranium-238, found in all rocks, soil, and the food we eat, emits (predominantly) alpha particles.

The amount of radiation produced by radioactive material is governed by the number of disintegrations occurring per second and is called *activity*. In general, this depends on two things: the speed with which atoms decay (i.e. how unstable they are) and the number of radioactive atoms present (i.e. the amount of material). The SI unit of activity is the *Becquerel (Bq).* 1 Bq equals 1 disintegration per second *(see units of radioactivity in the following section).*

OCCUPATIONAL SOURCES OF IONISING RADIATION

Ionising radiation occurs as either electromagnetic rays, for example, gamma rays or X-rays, or in particles, for example, alpha and beta particles. A wide range of sources and appliances used throughout industry, medicine and research emits radiation. It is also a naturally occurring part of the environment.

Half-life

The speed with which atoms radioactively decay varies from one substance to another and is characterised by radioactive decay at a fixed rate called the *half-life*.

The *half-life* of the radioactive material is defined as the time taken for activity in the form of emission of particles and energy to be reduced by one half. Therefore, half the nuclei originally present having changed into a different type.

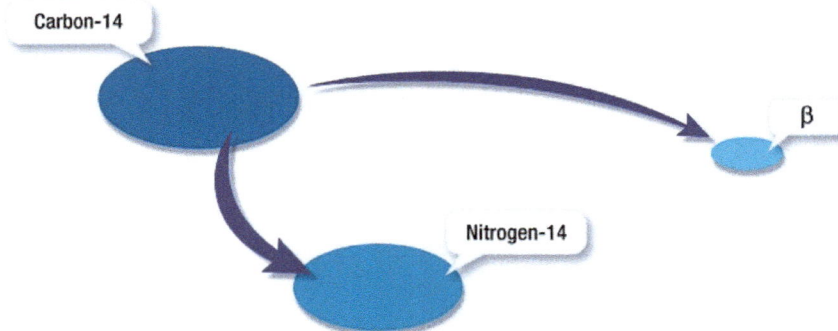

Figure B7-13: Carbon 14 radioactive decay.   Source: RMS.

After two half-lives the radioactivity will be one-quarter of the original value. Half-lives for different radioactive elements can range from fractions of a second to millions of years. The more unstable the element, the shorter the half-life.

Examples of half-lives are:

Element	Half-life
Uranium 235	718 million years
Carbon 14	5,730 years
Caesium 137	30 years
Strontium 90	50 days
Radon 216	4.5×10^{-7} seconds

Figure B7-14: Examples of half-lives.   Source: RMS.

NON-PARTICULATE IONISING RADIATION

Non-particulate ionising radiation refers to the radiation energy carried by electromagnetic waves. Electromagnetic waves can vary in energy, frequency and wavelength.

Quantum mechanics predicts that very short wavelength electromagnetic waves behave as uncharged particles, called photons. Therefore, the distinction between waves and particles at short wavelengths, as with X-ray and gamma rays, is blurred.

Figure B7-15: Behaviour of short wavelength radiation.   *Source: National Physics Laboratory.*

These high-energy electromagnetic waves have the capability of acting like a particle and causing ionisation of material they met and are therefore classed as ionising. X-rays and gamma rays are non-particulate ionising radiation.

Name	Symbol	Type
Alpha	α	Particulate
Beta	β^-	Particulate
Neutron	n or n^0	Particulate
X-ray	X-ray	Electromagnetic
Gamma	λ	Electromagnetic

Figure B7-16: Common types of ionising radiation.   *Source: RMS.*

The role of the radiological protection organisations

ROLE OF THE INTERNATIONAL COMMISSION FOR RADIOLOGICAL PROTECTION

The International Commission on Radiological Protection (ICRP) was founded in 1928 to advance for the public benefit the science of radiological protection.

The ICRP provides recommendations and guidance on protection against the risks associated with ionising radiation from artificial sources, as widely used in medicine, general industry and nuclear enterprises, and from naturally occurring sources. Reports and recommendations are published four times each year on behalf of the ICRP as the journal **Annals of the ICRP.**

ROLE OF PUBLIC HEALTH ENGLAND (PHE)

As the UK's primary authority on radiation protection, Public Health England (PHE) carries out research to advance knowledge about protection from the risks of radiation.

PHE provides a range of radiation protection services, including technical and monitoring, training and compliance; also provides expert information and fulfils a significant advisory role to regulators, government, the public and others.

ROLE OF HEALTH PROTECTION SCOTLAND

Health Protection Scotland (HPS) was established by the Scottish Government in 2005 to strengthen and co-ordinate health protection in Scotland. HPS is organised into specialist groups with expertise provided by a multi-disciplinary workforce and includes doctors, nurses, scientists and information staff.

The work of the HPS included the monitoring of the effects of ionising, electromagnetic, solar radiations, including mapping and advice on radon. The specialist groups take an active monitoring role and monitor such things as the possible effects of radioactivity on medicines sourced from Japan following their tsunami which caused nuclear disaster.

ROLE OF THE ENVIRONMENT AGENCY/SCOTTISH ENVIRONMENT PROTECTION AGENCY

The Environment Agency and the Scottish Environment Protection Agency (SEPA) are the environmental regulators for England and Wales and Scotland respectively. Their main role is to protect and improve the environment. This is done by assisting organisations to understand their environmental responsibilities, enabling them to comply with legislation and good practice.

The agencies protect communities by regulating activities that can cause harmful pollution and by monitoring the quality of air, land and water. The regulations they implement also cover the keeping and use, and the accumulation and disposal, of radioactive substances.

ROLE OF THE OFFICE FOR NUCLEAR REGULATION

The Office for Nuclear Regulation (ONR) independently regulates safety and security at 37 licensed nuclear sites in the UK. These include the existing fleet of operating reactors, fuel cycle facilities, waste management and decommissioning sites and the defence nuclear sector.

Their role is to provide efficient and effective regulation of the nuclear industry, holding it to account on behalf of the public. In addition, they regulate the design and construction of new nuclear facilities and the transport and safeguarding of nuclear and radioactive materials.

B7.2 - Non-ionising radiation

Sources of non-ionising radiation

ULTRAVIOLET

Workplaces

Major workplace sources of ultraviolet radiation relate to lighting provided in the workplace. This includes incandescent lamps (particularly tungsten halogen lamps) and gas discharge lamps, for example, fluorescent lights (particularly compact fluorescent lights, known as low-energy light bulbs), xenon arc lamps and solid state light sources (light emitting diodes).

Other workplace sources are:

- Curing with UV (for example, inks, coatings on floor and wall coverings, timber panels, fibre optics, etc.).
- UV sources in photocopiers and laser printers.
- UV Lasers.
- Welding.
- Germicidal lamps used in water treatment, research and food processing.
- Diagnostic lighting such as foetal/neonatal transilluminators.

Another occupational group at risk from UV exposure are dentists and dental technicians. This is due to certain dental procedures such as the hardening of fillings used for cavities (photo-polymerisation of methacrylates). In this process each tooth typically requires 250 seconds of exposure to UVA 350-380 nm at a high power density.

UVA sources are also used in workplaces that provide tanning facilities to customers.

Naturally

Sunlight is a natural source of ultraviolet radiation, which can affect outdoor workers and those involved in leisure activities. This natural exposure to UV radiation has increased in recent years due to environmental factors such as ozone depletion and the reduction of particulates in the air, due to the reduced burning of fossil fuels. This cleaner, more transparent, air has increased natural radiation levels. UV radiation reflects off water, sand, snow, concrete or any light-coloured surface and this reflected radiation may increase the exposure a person receives.

The amount of UV radiation received from the sun increases in intensity with altitude as the distance from the source will be less and less radiation will be absorbed as it will have passed through fewer molecules in the air before reaching the person exposed to it.

VISIBLE LIGHT

Visible light is not normally a problem, but where it is intense (for example, in lasers) it can potentially cause damage to the cornea and retina of the eye, the 'blue light' hazard. Workplace and leisure sources higher intensity visible light which includes:

- Lasers.
- Spotlights, used in theatres, exhibition halls and to illuminate sports grounds.
- Image display projectors.
- Flash lights used in photography.
- Welding and high voltage switching equipment that create arcs.
- Photolithography.

As with UV, the sun is a natural source of high intensity visible light, which presents a hazard when looked into directly or reflectively.

This can affect outdoor workers and those involved in leisure activities that look at the sun directly when sighting objects or receive intense reflected light from polished surfaces (for example, a photovoltaic cell array), water or light-coloured surfaces.

INFRARED

Outdoor workers are exposed to the sun, a natural source of infrared radiation. Infrared is a component of radiant heat, and so is a significant factor in foundries and smelters. Other workplace and leisure sources are:

- Lasers, including those used in manufacturing and light shows at public festivals.
- Material at a high temperature, for example, molten glass, kiln heated bricks and ceramics, flames in a furnace or boiler.
- Infrared lamps for treatment of medical conditions and for space heating or drying processes.

MICROWAVES AND RADIO FREQUENCY RADIATION

Micro waves and radio frequency radiation are mainly used as a source of heat or in the communications industry:

■ TV, FM radio and radar transmitters.
■ Mobile communications (mobile telephones, 'CB' radios, walkie-talkies).
■ Satellite communications.
■ Proximity and security detection.
■ Dielectric heaters for plastic sealing, glue curing, particle and panel board production.
■ Induction heaters for hardening, tempering, forging, etc.
■ Plastic welders.
■ Microwave ovens.
■ Diathermy in medical clinics.
■ Electronic article surveillance.
■ Wire free connection of equipment such as computers, printers and cordless telephones.

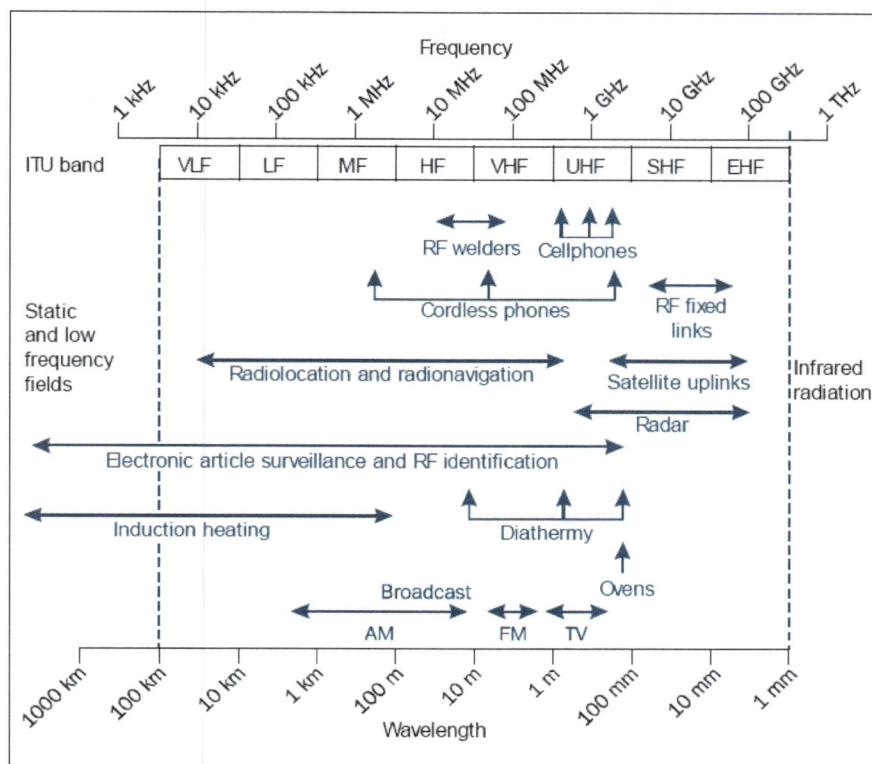

Figure B7-17: RF Spectrum and sources.   *Source: NRPB (HPA).*

Radio frequency radiation is also produced as a result of the transmission of electricity, particularly at high voltages, and as fields produced by the operation of equipment, such as display screen equipment.

Exposure routes for non-ionising radiation

Normal routes of exposure to non-ionising radiation include direct exposure to unprotected skin and eyes or reflections from shiny surfaces and mirrors. Exposure to high levels of microwaves may lead to the radiation passing through the skin and causing harm to the internal organs of the body. Risks may be increased if the radiation is concentrated through a lens, which increases the power density and heating effect significantly.

Effects of exposure to non-ionising radiation (acute and chronic)

ULTRAVIOLET

Ultraviolet (UV) radiation has a relatively low penetrating power and its biological effects are restricted to the surface organs such as the skin, the eyes and the lining of the mouth - arising from certain dental procedures. Certain drugs, such as diuretics, oral contraceptives, anti-diabetic drugs and some antibiotics can magnify the effects of UV due to their photosensitising properties. Principal parts of the body particularly at risk are the eyes, hands and buccal lining of the mouth.

The direct potential radiation hazards to health arise from UVR (UV radiation) with wavelengths greater than 180nm, since UVR of lower wavelengths is strongly absorbed in air and common materials.

Figure B7-18: UV wavelengths hazardous to health.   *Source: NRPB (HPA), Advice on protection against ultraviolet radiation.*

A summary of UV radiation effects on biological systems include:

- Bactericidal (destructive to bacteria).
- Keratitis (eye's cornea becomes inflamed).
- Erythema (redness of the skin caused by engorgement of skin tissue vessels with oxygenated blood 'sun burn').
- Carcinogenesis (formation of a cancer).
- Skin pigmentation.

The skin

A significant amount of UV radiation in the range 280-315 nm is absorbed into the epidermis. While a small amount of UV radiation may be beneficial, because it allows for the formation of vitamin D_3 by the skin, excessive amounts can affect cellular metabolism.

An acute effect of exposure to UV radiation, for example, sunlight in the 280-315 nm (UV-B) region, is reddening of the skin (erythema), which is a form of sunburn. Mild sunburn shows erythema a few hours after exposure, while longer exposure causes thickening of the skin (oedema), blistering, burning, tenderness and irritation of skin. The skin has a delayed protection mechanism called melanogenesis (i.e. suntan).

One chronic effect on the skin of continued exposure to UV radiation is premature ageing, due to a decrease in the skin's elasticity. Although UV can only penetrate the superficial layers of the skin, it causes damage to both RNA and DNA molecules, which can lead to cutaneous tumours that may be malignant (skin cancer). Chronic exposure to UV radiation can cause cheilitis, which is the inflammation, cracking, and dryness of the lips. Groups prone to this include seamen, agricultural workers and arc welders.

The eye

Most damage to the eye is thought to occur at the 210-315 nm wavelength range, which includes UV radiation mainly in the UV-B and UV-C region.

Excessive exposure leads to:

- Photokeratitis (inflammation of the cornea).
- Photoconjuctivitis (inflammation of the conjunctiva).
- The induction or promotion of cataracts (the lens of the eye becomes opaque).
- Damage due to an acceleration of the normal ageing of cells in the retina.

Cornea

Ultraviolet radiation of wavelengths shorter than 300 nm (actinic rays) can damage the corneal epithelium (a layer of tissue that covers and protects the front of the cornea from tears and bacteria). This is most commonly the result of specific workplace exposures as well as exposure to the sun at high altitude and in areas where shorter wavelengths are readily reflected from bright surfaces such as snow, water, and sand.

Acute effects include photokeratitis which is a condition that often occurs in welding operations where it is given the name 'arc eye' or 'welder's flash'. There is a latent period of a few hours (about 6-12 hours) between exposure and the onset of symptoms. Symptoms begin with pain which feels like 'grit in the eyes'. There is an aversion to bright light. The conjunctiva and cornea become inflamed. The severity of the condition depends on duration, intensity and wavelength of UV radiation exposure. The symptoms normally abate after approximately 36 hours and permanent corneal damage is unusual, provided that the corneal tissue has time to recover. The eye tends to become more sensitive to UV radiation after repeated exposure.

Chronic effects on the eye of exposure to UV radiation arise from UVA penetrating the cornea, where it is absorbed by the aqueous humour and in the lens, which may produce a harmless, transient fluorescence - chronic exposure may lead to yellowing of the lens.

Lens

Wavelengths of 300-400 nm are transmitted through the cornea, and 80% are absorbed by the lens, where they can cause cataracts. Epidemiologic studies suggest that exposure to solar radiation in these wavelengths, near the equator, is correlated with a higher incidence of cataracts.

They also indicate that workers exposed to bright sunlight in occupations such as farming, truck driving and construction work appear to have a higher incidence of cataract than those who work primarily indoors. Experimental studies have shown that these wavelengths cause changes in the lens protein, which lead to cataract formation in animals.

VISIBLE LIGHT

Visible light has a wavelength spectrum of 400-750 nm. If the wavelengths of this spectrum penetrate fully to the retina, they can cause thermal, mechanical, or photopic (affecting the cell cones in the retina which enables the eye to perceive colour in daylight) cell injuries.

Thermal injuries/retinal thermal injuries

They are produced by light intense enough to increase the temperature in the retina by 10-20°C. Lasers used in therapy can cause this type of injury. The light is absorbed by the retinal pigment epithelium, where its energy is converted to heat, and the heat causes photocoagulation of retinal tissue.

Mechanical injuries

They can be produced by exposure to laser energy from a Q-switched or mode-locked laser, which produces sonic shock waves that disrupt retinal tissue.

Photopic injuries/retinal photo-chemical injury

Photopic injuries are caused by prolonged exposure to intense light. This effect is most pronounced at wavelengths around 435-440 nm, which relates to that part of the visible wavelength spectrum related to blue light, and so it is sometimes called the 'blue-light hazard'.

Chronic exposure to high ambient levels of visible light can produce varying degrees of photochemical damage to the cells of the retinal macula (photoretinitis), without a significant increase in the temperature of the tissue.

Sun gazing is the most common cause of this type of injury, but prolonged unprotected exposure to a welding arc can also damage the retinal macula ('yellow spot'), which contains the largest concentration of cone cells in the eye and is responsible for central, high resolution vision.

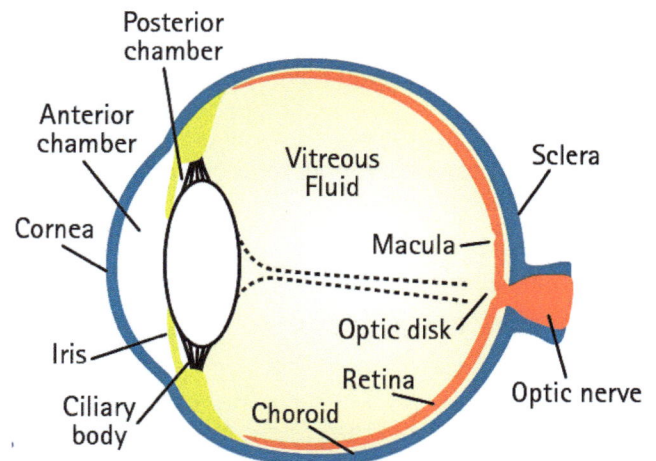

Figure B7-19: The macula.   *Source: Guild of Television Cameramen.*

Damage will result in a permanent decrease in visual acuity, poor colour and night vision. The intensity and wavelength of the light, length of exposure, and age are all-important factors. Older people are more sensitive, as are those who have had cataract surgery, because filtration of light by the lens is impaired. Another effect is that pulsing or stroboscopic visible light can cause fits in susceptible people.

INFRARED

The most damaging infrared wavelengths for organs, such as the eyes, are below 1400 nm. Infrared is emitted from material which is at a high temperature and the proportion of radiation with wavelengths below 1,400 nm increases as the temperature rises. For example, at 1,000°C ~ 5% of energy is at wavelengths below 1,400 nm, whereas at 2,000°C this rises to 40%.

Two different effects can be identified from infrared (IR) radiation. These are:

- Acute effects on the thermo-regulatory system by the radiant energy absorbed (for example, heat stress).
- Effects on specific organs, particularly the skin and eyes.

The skin responds to IR radiation by vasodilation (inflammation of the skin); skin temperatures above 45°C cause burns.

Chronic effects of exposure to high doses of IR radiation can cause acute pain, through damage to the cornea. Blood vessels grow into the normally clear cornea and it becomes opaque due to the ulcers caused by burning. A delayed effect can occur due to the cumulative effects of IR exposure.

'La cataracte des verriers' (glassblower's cataract) is an example of an IR radiation injury that damages the anterior lens capsule among unprotected artists. Denser cataractous changes can occur in unprotected workers who observe glowing masses of glass or iron for many hours a day.

Steel workers have been recorded as developing cataracts 15-20 years after first exposure. A link has been established between the effects of industrial exposure and ageing with many workers developing cataracts over the age of 60. Chronic exposure of the skin to IR radiation produces pigmentation.

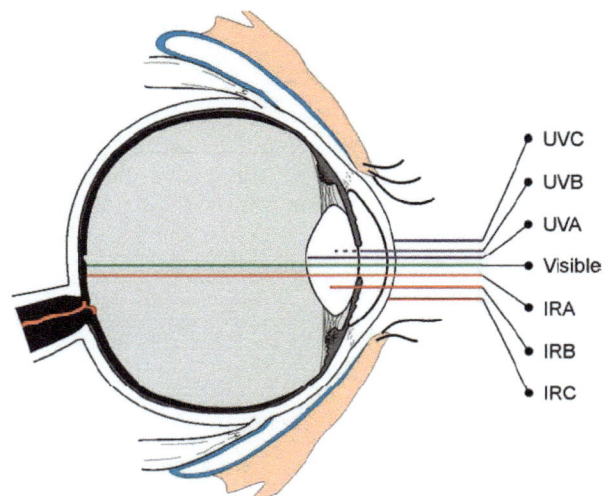

Figure B7-20: Penetration of optical radiation.   *Source: HPA*

An important factor in the effects of IR radiation is the distance between the worker and the source of radiation. In the case of arc welding, infrared radiation decreases rapidly as a function of distance, so that further than 1 metre away from where welding takes place, it does not pose an ocular hazard anymore, but ultraviolet radiation still does. That is why welders wear tinted glasses and surrounding workers only have to wear clear ones.

The effects of IR-A (780 - 1,400 nm)

Effects on skin

IR-A is the shortest wavelength in the range of IR radiation and penetrates several millimetres into tissue, which is, well into the dermis. Because of this and the energy it has, it can produce the same thermal effects as visible radiation.

Effects on the eyes

Like visible radiation, IR-A is also focused by the cornea and lens and transmitted to the retina. There, it can cause the same sort of thermal damage as visible radiation can. However, the retina does not detect IR-A, and so there is no protection from natural aversion responses. Chronic exposure to IR-A may also induce cataracts. IR-A does not have sufficiently energetic photons for there to be a risk of photochemically induced damage, which UV radiation causes.

The effects of IR-B (1,400 - 3,000 nm)

Effects on the skin

IR-B penetrates less than 1 mm into tissue, but it can cause similar thermal effects as IR-A.

Effects on the eyes

At wavelengths around 1400 nm, the aqueous humour is a very strong absorber of electromagnetic radiation and longer wavelengths are attenuated by the vitreous humour; therefore the retina is protected from IR-B sufficiently that harm should not result.

Heating of the aqueous humour and iris can raise the temperature of the adjacent tissues, including the lens, which does not have any veins and so cannot control its temperature. This, in conjunction with direct absorption of IR-B by the lens, induces cataracts.

The effects of IR-C (3,000 nm - 1 mm)

Effects on the skin

IR-C penetrates only to the uppermost layer of dead skin cells. Though penetration is limited, heat strain and discomfort from thermal stress may take place.

Effects on the eyes

IR-C is absorbed by the cornea, and so the main hazard is corneal burns, though the likelihood of this occurring and its effects are usually moderated by the dissipation of the heat to the surrounding tissue and environment.

A summary of Infrared radiation effects include:

- The skin is transparent to this radiation.
- Causes permanent damage to the lens. A significant proportion of the energy is transmitted by the cornea and penetrates the eye.
- Causes cataracts (clouding of the transparent lens of the eye).
- Radiation is absorbed by water in surface layers of skin where energy is dissipated as heat - thermoregulatory system.

MICROWAVES AND RADIO FREQUENCY RADIATION

Electromagnetic radiation can interact with objects (or people) in three different ways. The energy waves can pass through an object without being changed, like light through a window. It can be reflected, like light off a mirror, or it can be absorbed and cause the object to heat up, like a pavement in the sun. The health hazards of electromagnetic radiation are related only to the absorption of energy. The effects of absorbed energy depend on many different factors such as its wavelength and frequency, its intensity and duration. Different materials also absorb energy differently. The greatest danger lies with the 10-20 cm wavelength microwaves, as deep tissue heating may occur without the warning sensation of heat.

Wavelength	Effect
< 3 cm	Microwaves are absorbed in the outer surface of the skin.
3-10 cm	Microwaves penetrate more deeply into the skin.
10-200 cm	Microwaves penetrate deeply with the potential to damage internal organs.
> 200 cm	The body is thought to be transparent to these microwaves.

Figure B7-21: Summary of microwave radiation effects. Source: RMS.

The acute effects of exposure to high levels of microwave or radio frequency radiation occur when the energy is absorbed by body tissues, leading to localised or spot heating. The increased temperature can damage tissue, especially those tissues with poor temperature control, such as the lens of the eye. At low frequencies (including ultra-low) the effects will tend to be on the central nervous system.

Chronic effects, such as cataracts, which are the clouding of the lens of the eye, may occur at the very high energy levels encountered close to radiating radar antennas. Heat damage to tissue is caused, after a short period of time, by high levels of exposure.

The health effects of low levels of exposure to radio-waves or microwaves for long periods of time are much harder to find and to prove. Some scientific studies show health effects from long-term, low level exposure, other studies do not. The following list includes health effects which some researchers suspect may be related to long-term, low level radio frequency/microwave exposure:

- Psychological changes, for example, insomnia, irritability, mood swings.
- Depression.
- Headaches.
- Nervous system abnormalities.
- Hormonal changes.
- Miscarriages and birth defects.
- Male Infertility.
- Altered immunity.
- Leukaemia.

Many of these health effects are relatively common, and most people having these problems have not necessarily been affected by excessive exposure to radio frequency/microwave radiation. Exposure to low-frequency electric and magnetic fields normally results in negligible energy absorption and no measurable temperature rise in the body. However, exposure to electromagnetic fields at frequencies above about 100 kHz can lead to significant absorption of energy and temperature increases. As regards absorption of energy by the human body, electromagnetic fields can be divided into four ranges:

Frequencies		Effect on the body
20 MHz to less than 100 kHz		Absorption in the trunk decreases rapidly with decreasing frequency, and significant absorption may occur in the neck and legs.
20 MHz to 300 MHz		Relatively high absorption can occur in the whole body and to even higher values if partial body (for example, head) resonances are considered.
300 MHz to several GHz		Significant local, non-uniform absorption occurs.
Frequencies above about 10 GHz		Energy absorption occurs primarily at the body surface.

Figure B7-22: Effects on the body of exposure to electromagnetic fields. Source: Durney et al. 1985.

Units and methods of measurement

POWER DENSITY

Power density (or irradiance) is a measure of the power per unity area, commonly used when expressing exposure to radio frequency and optical radiation. Typically the unit of measure is given in watts per square metre or centimetre (W/m^2 or W/cm^2).

SPECIFIC ABSORPTION RATE (SAR)

The Specific Absorption Rate (SAR) is defined as the radio frequency (RF) power absorbed per unit of mass of an object, and is measured in watts per kilogram (W/kg). The SAR describes the potential for heating of the patient's tissue due to the application of the RF electromagnetic field. Hot spots occur in exposed tissue as the absorbed energy is applied to one location. SAR is usually averaged either over the whole body, or over a small sample volume (typically 1g or 10g of tissue). The value cited is then the maximum level measured in the body part studied over the stated volume or mass.

CENELEC is the European Committee for Electrotechnical Standardization and is responsible for standardization in the electrotechnical engineering field. CENELEC prepares voluntary standards, which help facilitate trade between countries and support the development of a single European Market.

RADIANT EXPOSURE

Radiant exposure is the energy (measured in Joules) that reaches a surface area (measured in cm^2) due to an irradiance (measured in watts per cm^2) maintained for a time period. Radiant exposure is measured in Joules per centimetre squared or J/cm^2. Often when discussing a light source, such as a laser or light from a distant source like the sun, the term irradiance is used to characterise the source. When the light source falls on an area, the reference would be the radiant exposure.

EXPOSURE LIMIT VALUES

The Health Protection Agency, Radiation Protection Division, funded by the European Commission Employment, Social Affairs and Equal Opportunities DG, has produced a non-binding Guide to the Artificial Optical Radiation Directive 2006/25/EC. The requirements of the Directive are implemented into UK legislation through the Control of Artificial Optical Radiation at Work Regulations (CAOR) 2010. The guide gives detailed advice on sources and types of optical non-ionising radiation in the workplace and provides interpretation of the Exposure Limit values (ELV's) set out in Annex 1 (for non-coherent optical radiation) and 2 (for coherent radiation - lasers) of the Directive. The CAOR 2010 refers to the Directive regarding ELVs. Exposures limit values are based on power density measurements over a given time.

The use of the ELVs for non-coherent optical radiation is generally more complex than for laser radiation. This is due to worker exposure potentially being to a range of wavelengths instead of a single wavelength. Three weighting functions are provided in Annex I to the Directive. The weighting function $S(\lambda)$ (biological spectral effectiveness) applies from 180 to 400 nm and is used to take account of the adverse health effects dependant on wavelength to the eye and the skin.

When a weighting function has been applied, the data is referred to as effective irradiance (the amount of light or other radiant energy striking a given area of a surface; illumination) or effective radiant exposure.

The peak value for $S(\lambda)$ in this range is 1.0 at 270 nm. A simple approach is to assume that all of the emission between 180 nm and 400 nm is at 270 nm (since the $S(\lambda)$ function has a maximum value of 1, this is equivalent to simply ignoring the function altogether).

If the irradiance of the source is known it is possible to use the following table in *figure ref B7-24* to see the maximum time a worker can be exposed if they are not to exceed the ELV, i.e. 30 J m^{-2}.

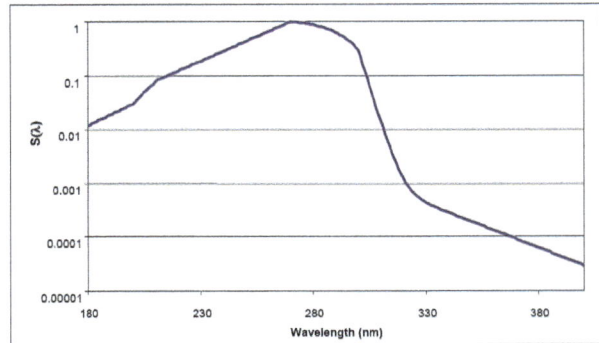

Figure 5.1 – Weighting function $S(\lambda)$

Figure B7-23: Weighting function for non-coherent optical radiation. *Source: HPA.*

Duration of exposure per 8 hour day	Irradiance (effective) - W m^{-2}
8 hours	0.001
4 hours	0.002
1 hour	008
30 seconds	1.0
10 seconds	3,0
1 seconds	30
0.1 seconds	300

Figure B7-24: Exposure time related to exposure limit value for wavelength range 180 nm to 400 nm.   *Source: HPA Guide to Directive 2006/25/EC.*

Other weighting function and worst case exposure levels are established for other wavelength ranges. Table 1.1 of Annex I of the Directive provides the ELVs for different wavelengths. In some wavelength regions, more than one exposure limit will apply.

The Control of Artificial Optical Radiation at Work Regulations (CAOR) 2010, Regulation 6, 'health surveillance and medical examinations' requires the employer to ensure that a medical examination is made available to an employee if an employee:

- Has been exposed to levels of artificial optical radiation which exceed the *exposure limit values*.
- Is found to have an identifiable disease or adverse health effects to the skin which is considered by a doctor or occupational health professional to be the result of exposure to artificial optical radiation.

Exposure limit values are given under each type of optical radiation as set out in the following text.

EXPOSURE TO ULTRAVIOLET RADIATION

Exposure limit values

The exposure limit values for occupational exposure to ultraviolet radiation, incident upon the skin or eye are measured over an 8-hour period.

Exposure of the eyes

Ultraviolet radiant exposure in the spectral region:

- 180 to 400 nm incident upon the unprotected eye(s) should not exceed 30 J m^{-2}.
- 315 to 400 nm should not exceed 10^4 J m^{-2}.

Exposure of the skin

For the most sensitive skin phototypes (known as 'melanoma-compromised'), ultraviolet radiant exposure in the spectral region of 180 to 400 nm should not exceed 30 J m^{-2}.

Methods of measurement

Ultraviolet radiation can be measured using a variety of instruments such as photoelectric cells, photo-conductive cells, photo-voltaic cells or photochemical detectors. Most instruments also use a filter system.

Radiometers

Radiometers are mainly used to measure the intensity of ultraviolet radiation (i.e. UVC, UVB and UVA). In addition to wavelength range, radiometers carry specifications such as sensor diameter, power range, accuracy, resolution, operating temperature range, and humidity range.

The wavelength range specifies the detection range for radiometers. The sensor diameter measures the active part of the sensor. Power range is the power per unit area. Radiometer accuracy is usually given as a percentage. The resolution of the detector is a measure of radiometer sensitivity.

Operating temperature and humidity range are also important specifications to consider when selecting radiometers.

EXPOSURE TO VISIBLE LIGHT RADIATION

Methods of measurement

Photometric sensor

Photometry refers to the measurement of visible light radiation with a sensor having a spectral response curve equal to the average human eye. This curve is known as the CIE Standard Observer Curve (photopic curve).

Photometric sensors are used to measure lighting conditions where the eye is the primary receiver, such as illumination of work areas, interior lighting, television screens, etc. Photometric measurements have been used in the past in plant science and irradiance, as the preferred measurement.

EXPOSURE TO INFRARED RADIATION

The three infrared spectral bands, A, B and C, roughly distinguish between different penetration depths of infrared radiation into tissue. Penetration is strongly dependent upon water absorption. IR-A radiation penetrates several millimetres into tissue. IR-B penetrates less than 1 mm. IR-C does not penetrate beyond the uppermost layer of the dead skin cells, the stratum corneum.

Exposure limit values

Calculation of potential thermal hazards from intense incoherent optical sources of radiation normally includes consideration of the contributions of IR-A and IR-B. IR-C is seldom considered for most light sources, including the sun or molten metals, since the contribution of IR-C is marginal.

Exposure of the eyes

IR-B has a greater penetrative effect on the ocular structures of the eye than IR-A. However, this has only a minor effect on the final temperature rise resulting from exposure to a continuous source, once thermal equilibrium is achieved.

The current ICNIRP (1997) guidelines for infrared radiation and recommendations for their application to wavelength ranges are as follows.

To avoid thermal injury of the cornea and possible delayed effects on the lens of the eye (cataractogenesis), infrared radiation 770 nm should be limited (E_{IR}):

$E_{IR} \leq 10$ mW cm^{-2} for t >1,000 s   (lengthy exposures)

$E_{IR} \leq 1.8t^{-3/4}$ W cm^{-2} for t <1,000 s   (short exposures)

Exposure of the skin

To protect the skin from thermal injury from optical radiation in the wavelength range of 400 nm to 3 μm, the radiant exposure H, for durations less than 10 seconds should be limited to:

$H = 2\ t^{1/4}$ J cm^{-2}

There is no limit for longer exposure, depending upon initial skin temperature, ambient temperature, humidity, and air movement.

Normal pain response and avoidance behaviour will impose exposure tolerance limits and tend to avoid overexposure.

Radiant heating tends to play a greater role for indoor work environments, since sunlight provides the principal radiant heat load on the body for outdoor workers and skin reflects significant solar radiation, which is predominantly visible IR-A radiation.

By contrast, the IR-B radiation is not significantly reflected by the skin, *see figure ref B7-26*.

The two curves show the range of reflectance from dark (lower curve) to light skins (upper curve).

IR-B is typically emitted by industrial work environments involving high temperature, such as working with molten metals or glass.

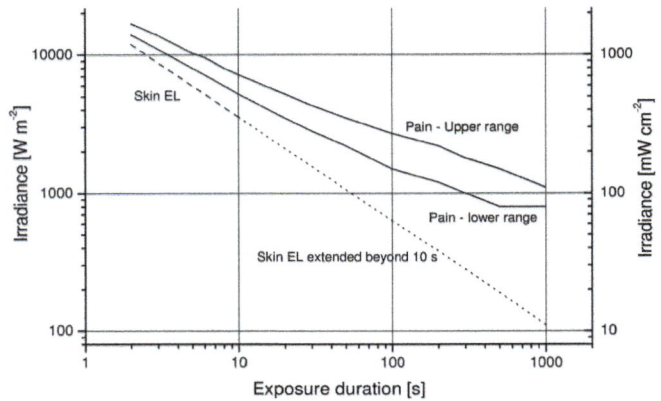

Figure B7-25: Skin exposure duration to radiation.   *Source: ICNIRP.*

Methods of measurement

Radiometers (discussed earlier) may also be used to measure infrared radiation.

Combined measuring equipment

A portable photo-radiometer data logger can be used for the measurement of non-coherent optical radiation, including ultraviolet, visible light and infrared.

The instrument consists of a series of sensors to cover different portions of the spectrum and a small laser that is used to indicate the analysed source.

The various sensors operate in the following spectral fields:

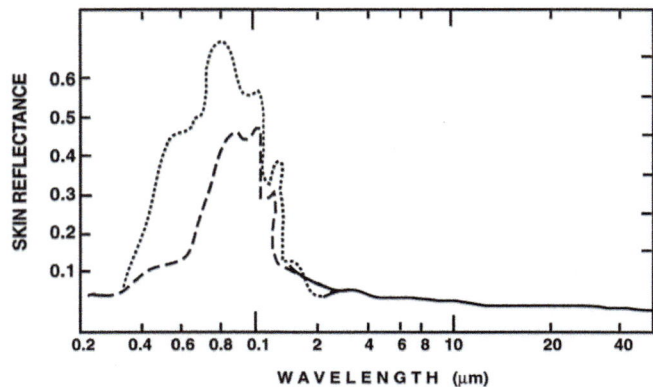

Figure B7-26: Radiation reflected by the skin.   *Source: ICNIRP.*

- Photometric sensor for measuring the illuminance (lux meter) in the spectral range 380 - 780 nm.
- Radiometric sensor for the UV band (220 - 400 nm) with spectral weighing factor $S(\lambda)$.
- Radiometric sensor for UVA band (315 - 400 nm).
- Radiometric sensor for the band 400- 700 nm (blue) with spectral weighing factor $B(\lambda)$.

EXPOSURE TO ELECTRIC AND MAGNETIC FIELDS

Restrictions on exposure to electric and magnetic fields are covered by Health Protection Agency (formerly NRPB) guidelines. The International Commission on Non-ionising Radiation Protection (ICNIRP) has also published guidelines.

In the electromagnetic fields, the limits are there mainly to protect against the body heating. This is in terms of 'specific absorption rate' (SAR).

Exposure limit values

The HPA (NRPB) guidelines give a whole body SAR restriction of 0.4 W kg-1, localised exposure in the head and trunk restricted to 10 W kg-1 and in the limbs 20 W kg-1.

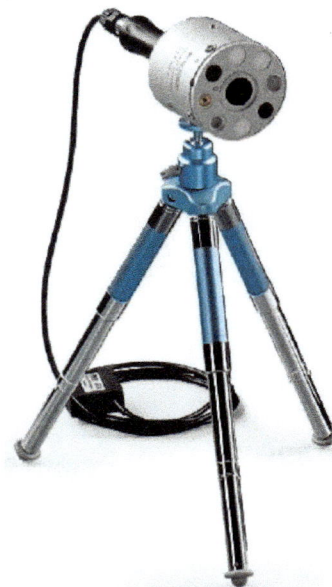

Figure B7-27: Portable photo-radiometer. *Source: DeltaOhm.*

Although there is no specific legislation on this, the HSE would consider compliance with these restrictions as fulfilling the general duty of care on the operators of equipment that generate electromagnetic fields.

Radiation risk assessment

The Control of Artificial Optical Radiation at Work Regulations (CAOR) 2010 requires an employer to conduct a specific risk assessment and to eliminate or reduce risks. Optical radiation includes ultraviolet, visible light and infrared.

Exposure limit levels are set for this form of radiation. Information and training is to be provided to those that may be affected, which includes employees and others carrying out work on behalf of the employer. Medical examination and health surveillance is to be provided for those employees that receive over exposure.

The purpose of undertaking a radiation risk assessment is to identify the measures needed to restrict the exposure to ionising radiation of anyone who might be affected by it; including:

- Workers.
- People working in the area.
- Cleaning/maintenance staff.
- Members of the general public.

These may be required for licensed nuclear sites, hospitals, universities, ports, airports, etc.

Control measures to prevent or minimise exposure to non-ionising radiation

EXPLANATION OF ASSESSMENT OF RISKS

The risk of being damaged by non-ionising radiation is dependent on the power of the radiation, i.e. the frequency of the electromagnetic waves. It also depends on the likelihood of coming into contact with the radiation, the number of people exposed and the presence of any vulnerable people. Ultra violet radiation is the most powerful of the non-ionising radiations. As with other forms of energy, the controls should be based on time, distance and shielding.

PRACTICAL CONTROL MEASURES TO MINIMISE EXPOSURE

Protection against over-exposure to non-ionising radiation may be achieved by a combination of:

- Engineering control measures, i.e. the design should consider where sources are sited, the distance from people and the need for engineering restrictions to prevent exposure, for example, interlocks.
- Administrative control measures, i.e. instruction, training, understanding the hazards and controls and the use of safe systems of work, such as permits, to restrict access to high risk areas.
- Personal protection, i.e. for the eyes and the skin.

Engineering control measures

The use of engineering controls can provide physical shielding against the effects of radiation. In addition to **well-designed** equipment, appropriately **sited** to minimise exposure, with **direction control** limited to the area required for use, the following can be used to further reduce exposure.

- **Enclosures** that provide sealed housings to contain the radiation source.
- **Screening** of sources using partitions or curtains, for example, to protect from welding UV.
- Use of non-reflective surfaces to ensure the **reduction of stray fields**.
- Use of **distance** - to ensure a safety factor (inverse square law), for example, microwave transmitters located on roofs.

Administrative control measures

The use of administrative controls can reduce an individual's exposure to radiation, by controlling the proximity and time of any interaction between an individual and radiation.

- **Safe systems of work** to limit access, use of interlocks, signage, decontamination facilities, and safe maintenance procedures.
- **Instruction and training** to provide hazard awareness, details on the use of hazard controls, restricted access procedures and the importance of limitation of exposure time.

Personal protection equipment

Personal protective equipment (PPE) is used to limit exposure where other methods have not reduced it to an acceptable level.

Protection of the skin - cover exposed areas, i.e. backs of hands, face, neck and forearms. Protective gloves are suitable for hands (for example, PVC). Forearms protected by long sleeves made from poplin or flannelette. Some barrier creams contain a UV filter, but their effectiveness depends on an adequate film thickness.

Protection of eyes - is normally achieved through the use of eye and face protectors (glasses, goggles, shields and helmets. Most standard ophthalmic lenses offer a degree of UV attenuation (for example, sunglasses and spectacles).

This is achieved through the incorporation of UV absorbing additives such as iron oxides into the glass used for lenses or by the addition of reflective coatings. The degree of protection achieved is however variable depending on factors such as the material the lens is made from, its thickness, the tints used and the radiation wavelength (for example, UV-A, UV-B).

Standards for goggles, spectacles or face shields offering protection in welding or similar operations are specified in a series of standards including:

- BS EN 169:2002 - personal eye-protection. Filters for welding and related techniques. Transmittance requirements and recommended use.
- BS EN 170:2002 - personal eye protection. Ultraviolet filters. Transmittance requirements and recommended use.
- BS EN 171:2002 - personal eye protection. Infrared filters. Transmittance requirements and recommended use.

Summary of control measures

- Eliminate as far as possible by exploring other technologies.
- Choose equipment emitting less radiation.
- Design, location and layout of workplaces.
- Time, distance and shielding will protect against the range of non-ionising radiations.
- Complete enclosure will shield, as will screens.
- Other shielding could be the use of PPE: goggles with a filter, sunglasses, and skin covering.
- Suntan creams and lotions can be used to filter out the harmful effects of UV.
- Clothing and screens will protect against IR, for example, for foundry workers.
- Total enclosure by design, for example, a microwave oven, which incorporates an interlock device that will only allow the oven to work with the door shut, will protect against tissue damage from microwaves.
- Reducing the time of exposure will help to prevent skin damage.
- The further the distance radiation has to travel, the more likely it is to lose its harmful energy. The further away the person is from the source, the less likely the energy will be enough to do any individual harm.
- PPE correctly specified, correctly fitted and worn.
- Signage to warn of danger.
- Provision and use of specialist advice.

Legal requirements to manage exposure to non-ionising radiation

CONTROL OF ARTIFICIAL OPTICAL RADIATION AT WORK REGULATIONS (CAOR) 2010

The Control of Artificial Optical Radiation at Work Regulations (CAOR) 2010, implemented in the UK the European Union Directive 2006/25/EC on the minimum health and safety requirements regarding the exposure of workers to risks arising from artificial optical radiation, one of the physical agents. The regulations impose duties on employers to protect both employees who may be exposed to risk from exposure to artificial optical radiation at work and other persons at work who might be affected by that work.

Regulation 3 imposes a duty to carry out a specific form of risk assessment where an employer carries out work which could expose its employees to levels of artificial optical radiation that could create a reasonably foreseeable risk of adverse health effects to the eyes or skin and where those risks have not already been eliminated or controlled. Where a risk assessment is necessary the CAOR 2010 also impose duties on the employer to:

- Eliminate, or where this is not reasonably practicable, to reduce to as low a level as is reasonably practicable the risk of adverse health effects to the eyes or skin of the employee as a result of exposure to CAOR 2010.
- Devise an action plan comprising technical and organisational measures to prevent exposure to artificial optical radiation exceeding the exposure limit values.
- Take action in the event that the exposure limit values are exceeded despite the implementation of the action plan and measures to eliminate or reduce so far as is reasonably practicable the risk of exposure.
- Limit access to, and provide for appropriate signs in those areas where levels of artificial optical radiation are indicated in the risk assessment as exceeding the exposure limit values.
- Provide information and training if the risk assessment indicates that employees could be exposed to artificial optical radiation which could cause adverse health effects to the eyes or skin of the employee.
- Provide health surveillance and medical examinations in certain cases.

PHYSICAL AGENTS (ELECTROMAGNETIC FIELDS) DIRECTIVE

On 29th June 2013, the European Commission published Directive 2013/35/EU on the minimum health and safety requirements for the protection of workers from risks to their health and safety arising, or likely to arise, from exposure to electromagnetic fields during their work. The Directive deals only with health and safety at work, and applies to work activities where workers are exposed to risks from electromagnetic fields.

In summary, the Directive places a number of duties on employers, main ones being that it:

- Places a duty on the employer to conduct a risk assessment and calculate EMF strengths.
- Places a duty on the employer to eliminate or reduce as low as possible the risk of exposure; and where risk can't be eliminated that measures are devised by the employer to reduce the risk of exposure below ELV.

- Requires the employer to provide: the risk assessment to the nominated person responsible for health surveillance.
- Requires an investigation and medical examination where an employee is 'detected' as having been exposed.
- Records of Health surveillance activities are kept.

See also - Relevant statutory provisions section.

B7.3 - Ionising radiation

Sources of ionising radiation

WORKPLACE SOURCES OF IONISING RADIATION

Man-made sources

Man-made sources of radiation account for approximately 14 per cent of the average annual personal radiation dose in the UK and are dominated by the use of radiation in healthcare. Radioactivity is also present in our environment due to nuclear weapons testing, accidents at nuclear facilities and the authorised discharge of radioactive wastes from nuclear and other facilities. An increasing variety of uses for ionising radiation are being developed for the workplace.

Nuclear industry

The nuclear industry includes nuclear fuel fabrication, nuclear ***power generation***, nuclear fuel reprocessing, radioactive waste (treatment, decommissioning, storage and disposal) and research.

Other uses and applications

Ionising radiation is widely used throughout the working environment. The most obvious place is in the nuclear industry; however it is also used for non-destructive testing (NDT), thickness/level measurement, security equipment, in smoke detectors, archaeological dating, baggage inspection, research and teaching and for a variety of medical purposes, including diagnosis, treatment and sterilisation of medical appliances.

The following table ***figure ref B7-28*** gives some common uses and applications:

Uses	Application - notes.
Radiography (X-rays)	***Healthcare*** (hospital, dentistry and veterinary), security and NDT.
Tracing	The use of isotopes to trace movement of a substance in a system, in medical, manufacturing and engineering situations.
Ionisation effect smoke detectors	Mainly alpha radiation. Widespread use at home and at work. Optical and photo-electric detectors are non-radioactive.
Thickness gauges, flow gauges and level gauges	Penetrating radiations (for example, beta and gamma rays) needed to highlight differences in flow, level, thickness in the ***manufacturing*** industry, for example, measuring thickness in the paper industry and in the food processing to measure the content of cans and bottles.
Non-destructive testing	Gamma and X-ray radiation is also used for NDT of metal equipment, for example, to measure the quality of welds in boilers and pipelines.
Sterilisation	In the food, medical and research to sterilise food containers, medical instruments and laboratory equipment.
Research	Biomedical diagnostic or therapeutic research, age dating of materials.
Moisture/density gauges	Used in portable devices. Contain two radioactive sources.
Radioactive materials	Radioactive tracer isotopes used for medical purposes; very dense depleted uranium used for weight in aircraft and yachts as well as for military purposes; thoriated (element thorium) alloys used in high performance magnesium alloys (for example, for aircraft engines).
Luminous dials, badges, gauges and markers	Radium was used until the 1960's in luminous paints used by watch and instrument manufactures. Radium was replaced by the much safer tritium (trim 'phone dials) a beta emitter, used to provide permanent illumination (when mixed with phosphor) in complete darkness. Radium may still be found in old equipment and watches; radium based luminous paints etc. are very active and long lived.

Figure B7-28: Common sources of radiation.

Source: RMS.

NATURAL SOURCES OF IONISING RADIATION

Natural sources of ionising radiation account for approximately 86 per cent of the annual average radiation dose.

Radon is a naturally occurring radioactive gas produced from the uranium that is present in varying amounts in all rocks and soils. Radon accounts for the largest proportion, approximately 56 per cent, of the average annual personal radiation dose in the UK and Ireland.

Radon

Radon occurs naturally from decaying uranium, and it is particularly abundant in regions with granite bedrock. Exposure is high in England (Cornwall, Devon and Somerset), because of these counties' underlying geology, and there are also 'hotspots' in Wales, the Cotswolds and the Pennines. However, the gas disperses outdoors so levels are generally very low.

Radon enters buildings from the ground and can sometimes build up to unacceptable levels. Once inhaled into the lungs, the gas decays into other radioactive isotopes of lead, bismuth and polonium. Some decay products emit alpha particles that, when breathed in, can cause harm to the sensitive cells of the lungs. Radon is the term used as shorthand to describe Radon gas's short-lived decay products (isotopes).

a)

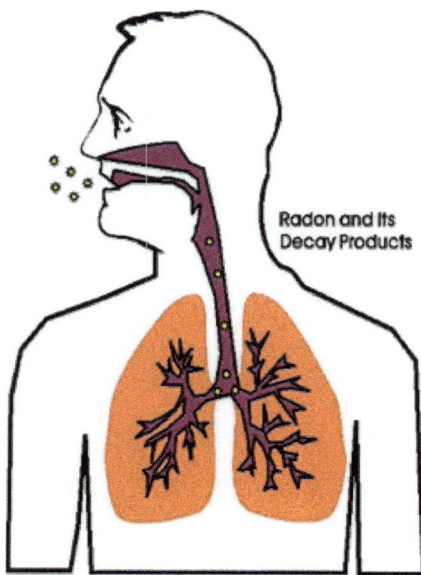

b)

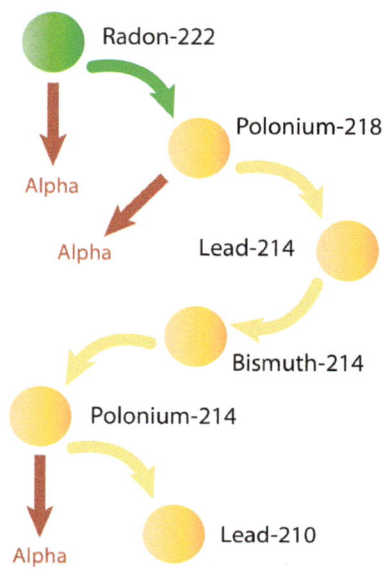

Figure IB7-29: Radon decay a) and b).

Source: Administration and Business Portal.

Study	Source	Type of mine	Number of exposed miners	Mean total WLM	Mean exposure duration (years)	Number of lung cancer deaths	% increase in age-specific risk of lung cancer per WLM[i] (with 95% CI)
Yunnan, China	Xuan et al, 1993	Tin	13,649	286.0	12.9	936	0.16 (0.1–0.2)
West Bohemia, Czech Republic	Tomasek and Placek, 1999	Uranium	4,320	196.8	6.7	701	0.34 (0.2–0.6)
Colorado, USA [ii]	Roscoe, 1997	Uranium	3,347	578.6	3.9	334	0.42 (0.3–0.7)
Ontario, Canada [iii]	Kusiak et al, 1993	Uranium	21,346	31.0	3.0	285	0.89 (0.5–1.5)
Newfoundland, Canada	Morrison et al, 1988	Fluorspar	1,751	388.4	4.8	112	0.76 (0.4–1.3)
Malmberget, Sweden	Radford and St Clair Renard, 1984	Iron	1,294	80.6	18.2	79	0.95 (0.1–4.1)
New Mexico, USA	Samet et al, 1994	Uranium	3,457	110.9	5.6	68	1.72 (0.6–6.7)
Beaverlodge, Canada	Howe and Stager, 1996	Uranium	6,895	21.2	1.7	56	2.21 (0.9–5.6)
France	Tirmarche et al, 1993	Uranium	1,769	59.4	7.2	45	0.36 (0.0–1.2)
Port Radium, Canada	Howe et al, 1987	Uranium	1,420	243.0	1.2	39	0.19 (0.1–0.6)
Radium Hill, Australia	Woodward et al, 1991	Uranium	1,457	7.6	1.1	31	5.06 (1.0–12.2)
Total [iv]			60,606	164.4	5.7	2,674	

Figure B7-30: The Beir V1 report.

Source: Beir Committee, 1999.

Epidemiological studies of miners established many years ago that exposures to radon increased the risk of lung cancer. The US National Academy of Sciences Committee on health risks from exposure to radon published its report in 1999, known as the Bier V1 report, **see figure ref B7-30**. The report considered eleven major studies, covering a total of 60,000 miners in Europe, North America, Asia and Australia, among whom 2500 deaths from lung cancer had occurred. Eight of the studies were of uranium miners, the remainder were of miners of tin, fluorspar, or iron. Radon is the second most common cause of lung cancer in the UK, tobacco smoking being the most common cause.

Cosmic radiation

The earth is continuously bombarded by high-energy radiation from either the sun (solar radiation) or from outside the solar system (galactic radiation). Collectively this is termed cosmic radiation. Radiation doses from cosmic radiation are greater at higher altitudes and those who fly regularly, such as air crew, receive an additional dose. Radionuclides of uranium, thorium and potassium are relatively abundant in rocks and soils. Radiation emitted from these radionuclides gives a gamma radiation dose.

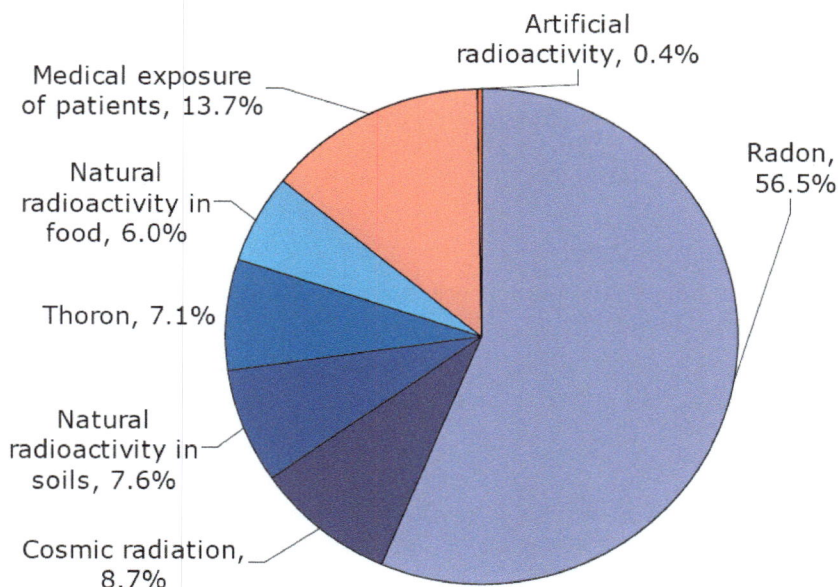

Figure B7-31: Distribution natural and man made ionising sources.   Source: RPII.

Units of radioactivity, radiation dose and dose equivalent

UNITS OF RADIOACTIVITY

Most of the international community measure ionising radiation using the SI ('System Internationale') system of units, a uniform system of weights and measures that evolved from the metric system. In the USA the older, conventional units of measurement are still widely used.

	Radioactivity	Absorbed dose	Dose equivalent	Exposure
SI Units	Becquerel (Bq).	Gray (Gy).	Sievert (Sv).	Coulomb/kilogram (C/kg).
Conventional Units (USA)	Curie (Ci).	Rad.	Rem.	Roentgen (R).

Figure B7-32: International systems of units for radiation measurement.   Adapted from: Guidance for Radiation Accident Measurement.

The following is a list of prefixes and their meanings that are often used in conjunction with SI units:

Multiple	Prefix	Symbol
10^{12}	Tera	T
10^{9}	Giga	G
10^{6}	Mega	M
10^{3}	Kilo	k
10^{-2}	Centi	c
10^{-3}	Milli	m
10^{-6}	Micro	µ
10^{-9}	Nano	n

Figure B7-33: SI prefixes.   Source: Guidance for Radiation Accident Measurement.

Different units of measurement are used depending on what aspect of radiation is being measured. The amount of radioactivity of a source is measured by the SI unit **Becquerel** (Bq). The radiation dose absorbed by a person, the amount of energy deposited in human tissue by radiation, is measured using the SI unit **gray**

(Gy). The biological risk of exposure to radiation is measured using the SI unit *sievert* (Sv). The amount of radiation exhibited by an x-ray or gamma ray source, that which someone is exposed to, is measured in coulomb/kilogram (C/kg).

Units for measuring radioactivity

A radioactive atom gives off or emits radioactivity because the nucleus has too many particles, too much energy, or too much mass to be stable. The nucleus breaks down, or disintegrates, in an attempt to reach a non-radioactive (stable) state. As the nucleus disintegrates, energy is released in the form of ionising radiation.

When the amount of ionising radiation being emitted or given off the unit of measure used is the SI unit Becquerel (Bq), a much smaller more practical unit that has now replaced the Curie (Ci). The Bq is used to express the number of disintegrations of radioactive atoms in a radioactive material over a period of time. One Bq is equal to one disintegration per second.

The Ci or Bq may be used to refer to the amount of radioactive materials released into the environment. For example, during the Chernobyl power plant accident that took place in the former Soviet Union, an estimated total of 81 million Ci of radioactive cesium (Cs), a radioactive isotope Cs-137 was released (one Ci is equal to 37 billion disintegrations per second).

Units for measuring biological risk

A person's biological risk, the risk that a person will suffer health effects from an exposure to radiation, is measured using the SI unit the sievert (Sv).

The equivalent dose is used to quantify the *effective dose* (this is the sum of the equivalent doses) to the exposed organs and tissues weighted by the appropriate tissue-weighting factor. The equivalent and effective dose are expressed in sieverts (Sv).

RADIATION DOSE

When a person is exposed to radiation, energy is deposited in the tissues of the body. The amount of energy deposited per unit of weight of human tissue is called the *absorbed dose.* Absorbed dose is measured using the SI unit Gray, Gy. Biological damage does not just depend on the absorbed dose. It also depends on the type of radiation. For example, one Gy of alpha (α) radiation in tissue can be much more harmful than one Gy of beta (β) radiation.

DOSE EQUIVALENT

This refers to radiation dose to the whole body, or single organ that has been adjusted to make it equivalent in risk of cancer to the amount of dose from gamma radiation that would cause the same risk of cancer. The weighting factor for gamma (γ) radiation, X-rays and beta (β) particles is set at 1. For alpha (α) particles it is set at 20. The equivalent dose is expressed in a unit called the sievert (Sv). The measurement is usually in millisieverts (mSv), the mSv being a thousandth of a Sv.

DOSE RESPONSE

A *dose-response curve* is a simple graph relating the intensity of radiation to the response of the part of the body under study. The response may be a physiological or biochemical response, or even death. The first point along the graph where a response above zero is reached is referred to as a threshold-dose. At higher doses, undesired side effects appear and grow stronger as the dose increases. The greater the level of exposure, the steeper this curve will be.

High doses of radiation tend to kill cells, while low doses tend to damage or change them. High doses can kill so many cells that tissues and organs are damaged. The dose-response relationships will depend on the exposure time.

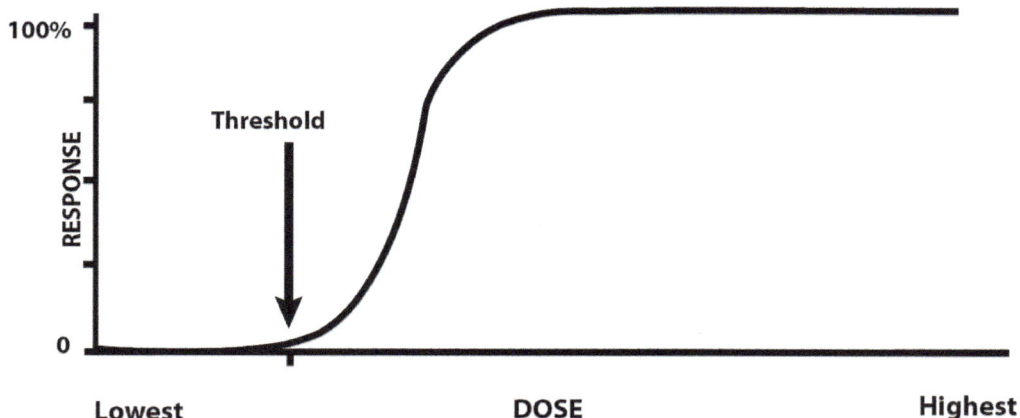

Figure B7-34: Dose-response curve.

Source: RMS.

DOSE EFFECT

Dose effect is used in radiological protection, to compare the stochastic (random) risk of a non-uniform exposure to ionising radiation, with the risks caused by a uniform exposure of the whole body. The stochastic risks are carcinogenesis and hereditary effects. It is not intended as a measure for acute or threshold effects of radiation exposure, such as redness of the skin (ethema), radiation sickness or death.

IRRADIATION

Irradiation is the process of exposure to ionising radiation, for example, in medical terms to a level of radiation that will serve a specific purpose, such as an x-ray, rather than radiation exposure to normal levels of background radiation or abnormal levels of radiation due to accidental exposure.

Routes of exposure for ionising radiation

EXTERNAL EXPOSURE

External exposure includes *absorption* of ionising radiation arising from cosmic rays, from the sun, background radiation from historic weapon testing, nuclear accidents, power generation, burning of fossil fuels, industrial and medical use, such as industrial and clinical x-ray, mining and conduction activities.

INTERNAL EXPOSURE INCLUDING INHALATION, INGESTION, INJECTION, ABSORPTION

Internal exposure includes that related to *inhalation* of dust particles from mining operations contaminated with short-lived radon isotopes (alpha emitters).

Many medical applications use short-lived tracers isotopes for clinical examination of patients, administered by either through *ingestion, injection* or absorption.

Common isotopes include:

■ Ingestion of barium isotopic salts as tracers to show up soft tissue on x-ray.
■ Ingestion of isotopic iodine for partial removal of the overactive thyroid gland.
■ Injection of isotopes prior to bone marrow transplants to destroy malignant blood cells.
■ Absorption of soluble salts of isotopes' such as tritium oxide (super heavy water ^3H_2O) can occur through the skin.

> An isotope is each of two or more forms of the same element that contain equal numbers of protons but different numbers of neutrons in their nuclei, and hence differ in relative atomic mass but not in chemical properties; in particular, a radioactive form of an element.

Figure B7-35: Definition of isotope.   Source: Oxford dictionary.

Effects of exposure

THE EFFECTS OF THE VARIOUS TYPES OF IONISING RADIATION

All ionising radiation can, by definition, remove orbital electrons from atoms, causing them to become electrically charged.

When this occurs in a living cell, various types of damage may occur. With quite low doses of radioactivity no effect may occur, but higher doses can cause molecular changes relating to the cells structure and function.

The possible damage includes direct interference with the cell's operation so that it no longer functions as it should, for example, the ability of the cell to reproduce itself, or the cell may be destroyed. Ionising radiation may affect the DNA without destroying the cell, when molecular chains are broken within the DNA structure; this will cell reproduction and may result in tumours.

In some cases, instead of directly interfering with the cell the ionising radiation can act indirectly, by producing highly reactive chemicals (free radicals) which then go on to damage the DNA.

Ionising radiation can therefore affect the cell in a number of ways:

■ Complete recovery.
■ Loss of reproductive capability (cell division is prevented).
■ Genetic, DNA, changes (mutations).
■ Cell death.

The different forms of ionising radiation have different effects, influenced by their nature and the energy they have. Alpha particles are comparatively large, consisting of two protons and two neutrons, and travel only short distances in dense materials.

Therefore they may be stopped by, or penetrate only a short way into, the skin of a persons outer body.

However, as the lungs and gastrointestinal tract are not protected by the thick dead skin layers that cover a person's outer body alpha particles have sufficient penetrative ability to cause harm when breathed in or ingested.

The remaining forms of ionising radiation, beta, gamma and X-ray, all have sufficient penetrative ability to pass into and through the human body, with the ability to cause harm. Whilst beta particles have greater penetrating power than alpha particles they are less ionising and take longer to cause the same degree of damage as alpha particles.

Gamma rays create ionisation by indirect means, creating harmful free radicals. Because neutrons do not have an electric charge, they do not interact directly with the electrons of the atom. Instead, they scatter from and collide with the atomic nuclei of the atoms, causing ionisation of the atom.

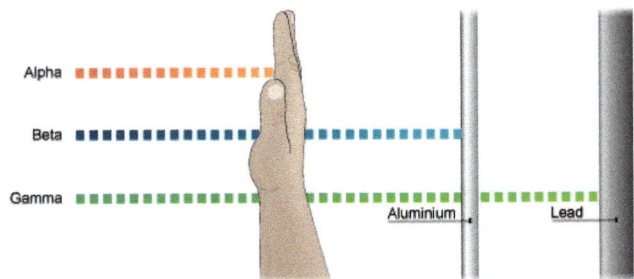

Figure B7-36: Relative penetration power of several types of ionising radiation. *Source: BBC.*

Effects of exposure to each type of radiation

ACUTE EXPOSURE

Early effects of a single acute dose (apparent within 60 days of exposure) vary from slight blood changes, for exposures between 0.25 and 1 Sv, to death, which is the probable outcome of exposures exceeding 6 Sv.

Delayed effects may occur some time after recovery from early effects. They include loss of hair and temporary or permanent infertility. Late effects such as leukaemia may occur many years after the exposure, their probability increasing as the dose increases.

CHRONIC EXPOSURE

Practically all occupational radiation involves chronic exposures, i.e. small weekly doses (for example, < 1 mSv) occurring over many months or years. Somatic effects of an accumulated chronic exposure are less serious than those of an equivalent acute exposure.

In a lifetime of occupational exposure, without any observable effect, an individual's total dose can be large enough so that an equal dose given in a few hours would be seriously disabling or fatal.

From chronic exposures there are no delayed effects and the probability of late effects is minimal. If chronic occupational doses are within the International Commission on Radiological Protection (ICRP) dose limiting recommendations, the probability of late effects is so small that it has not been possible to establish whether any such effect exists.

The clinical symptoms of chronic exposure can take the form of radiation sickness, cataracts or the development of cancer in the longer term. The effects of exposure to ionising radiation can be classified as somatic and genetic.

SOMATIC EFFECTS

Somatic cells are all cells in the human body other than those responsible for reproduction. The somatic effect is the effect on somatic cells of exposure to ionising radiation. The effects become evident in the individual during the lifetime of the exposed individual.

GENETIC EFFECTS

This is the effect where the cells responsible for reproduction, male sperm and female ova, are affected by exposure to ionising radiation. The effects become apparent in the descendants of the exposed individuals.

Potential health effects

The effects on the body of exposure to ionising radiation will depend on the type of radiation, the frequency and duration of exposure. Acute effects will include nausea, vomiting, diarrhoea and burns (either superficial skin burns or deep, penetrating burns causing cell damage). Long term (chronic) effects such as dermatitis, skin ulcers, cataracts and cancers can also be expected.

STOCHASTIC EFFECTS AND NON-STOCHASTIC

The biological effects of radiation can also be defined as either ***stochastic (random)*** or ***non-stochastic***.

Stochastic

Stochastic effects are usually associated with exposures to low levels of ionising radiation over a long period of time, for example, years. The term stochastic means 'random', the implication being that low levels of radiation exposure are not certain to produce an effect.

The stochastic effects are therefore those for which the probability of the effect occurring, rather than its severity, is a function of dose, for example, cancer and genetic effects. There is no threshold dose below which a stochastic effect will definitely not occur.

The probability of experiencing a stochastic effect is directly proportional to the dose, for example, doubling the dose doubles the probability of experiencing cancer or a genetic effect.

Non-stochastic

Are those for which the severity of the effect varies with dose and for which a threshold may therefore exist, for example, damage to blood forming tissues, damage to the lining of the gastrointestinal tract, cataracts of the lens of the eye, damage to blood vessels, impairment of fertility.

Non-stochastic effects are associated with much higher levels of radiation exposure, usually incurred over a much shorter period of time (fractions of a second to tens of days) than is the case for stochastic effects.

SUMMARY OF EFFECTS

The following flow chart summarises the biological effects of exposure to ionising radiation:

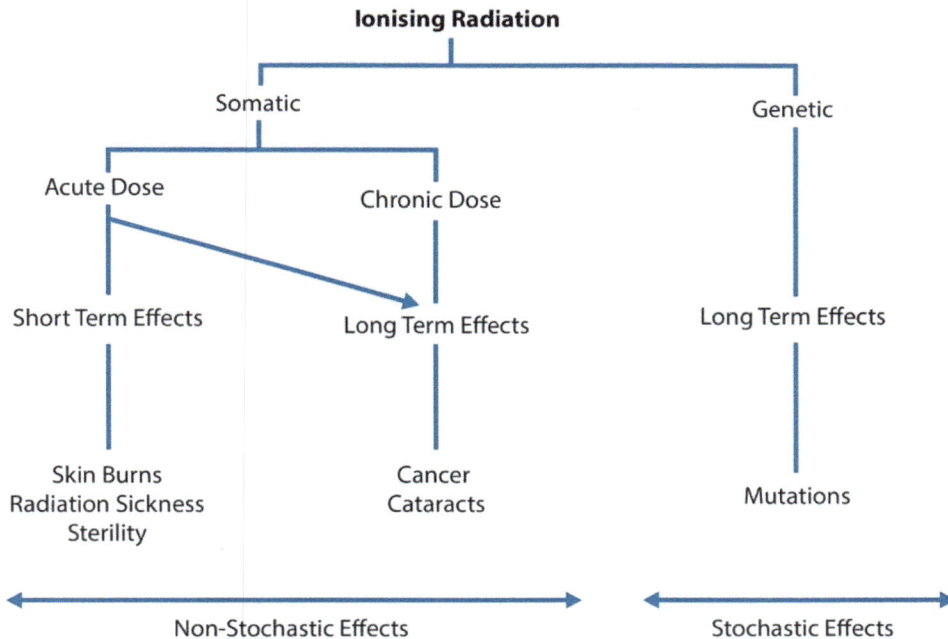

Figure B7-37: Biological effects.   Source: RMS.

Methods of measuring ionising radiation

Radiation cannot be detected by human senses. A variety of instruments are available for detecting and measuring radiation; they may be used to:

- Monitor personal radiation exposure.
- Monitor an area or perimeter.
- Detect radiation leaks and contamination.
- Ensure regulatory compliance.

- Monitor changes in background radiation.
- Demonstrate principles of nuclear physics.
- Check for radioactive minerals in the earth.

One of the most common types of radiation detector is a Geiger-Mueller (GM) tube, also called a Geiger counter.

IONISING CHAMBERS

Geiger-Muller tube

The Geiger-Muller tube is an instrument for measuring radioactivity that uses an ionisation tube. The instrument has a sealed tube containing a low pressure gas (usually argon with a little bromine). In the tube are two high voltage electrodes. An ionisation between the two electrodes produces a current (an 'avalanche of charge') between the electrodes. This current change is displayed on some form of scale or digital counter.

An ionising particle will produce a pulse of charge of almost constant size. The size of the pulse does not vary with the energy or amount of ionisation produced by the ionising particle. The number of pulses represents the number of ionising particles coming into the tube.

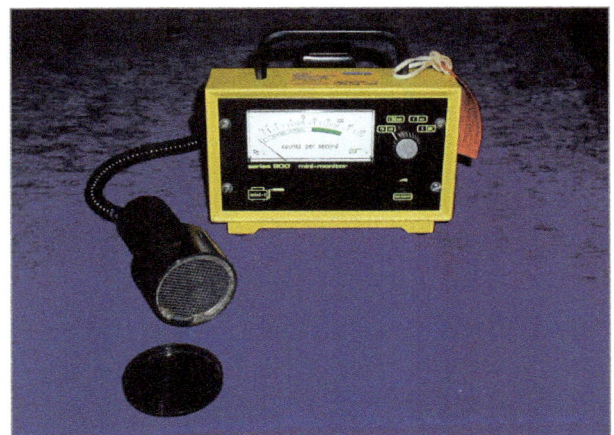

Figure B7-38: Geiger counter.   Source: RMS.

Geiger-Müller tubes do not distinguish between one kind of particle and another, or between a more energetic particle and a less energetic one, provided the particle enters the tube (which is difficult for alpha radiation) and does not pass right through (as most gamma rays do).

SCINTILLATION DETECTORS

A popular method for the detection of gamma rays involves the use of crystal scintillators. The general description of a scintillator is a material that emits low-energy (usually in the visible range) photons when struck by a high-energy charged particle. When used as a gamma-ray detector, the scintillator does not directly detect the gamma rays. Instead, the gamma rays produce charged particles in the scintillator crystals, which interact with the crystal and emit photons. These lower energy photons are subsequently collected by photomultiplier tubes (PMTs).

Figure B7-39: A scintillation detector.   *Source: AMETEK.*

When gamma and x-ray radiation passes through matter it can cause three processes to take place:

1) Photoelectric absorption - low energy - one of the inner electrons of the atom absorbs the energy of the gamma ray and is ejected from the atom, leaving a positively charged ion and a free electron.
2) Compton scattering - medium energy - gamma and x-ray exposure results in an increase in wavelength and loss of energy (from the rays), part of which is absorbed by the matter releasing electrons and the matter then becomes ionised.
3) Pair production - high energy - results when a gamma or x-ray strikes an atom causing the simultaneous release of an electron and proton of equal and opposite charge.

Each of these processes can create high-energy electrons or anti-electrons (positrons) which interact in the scintillator as charged particles. By adding up the energy collected in the surrounding photomultiplier tubes the energy of the gamma ray can be determine.

Scintillators can be made of a variety of materials, depending on the intended applications. The most common scintillators used in gamma ray detectors that are made of inorganic materials are usually an alkali halide salt, such as sodium iodide (NaI) or cesium iodide (CsI). To improve the efficiency of response of the salts, a small quantity of impurity is often added. This material is called an 'activator'. Thallium and sodium are often used for this purpose. Detectors of this type are often described as NaI(Tl), which means it is a sodium iodide crystal with a thallium activator, or as CsI(Na), which is a cesium iodide crystal with a sodium activator.

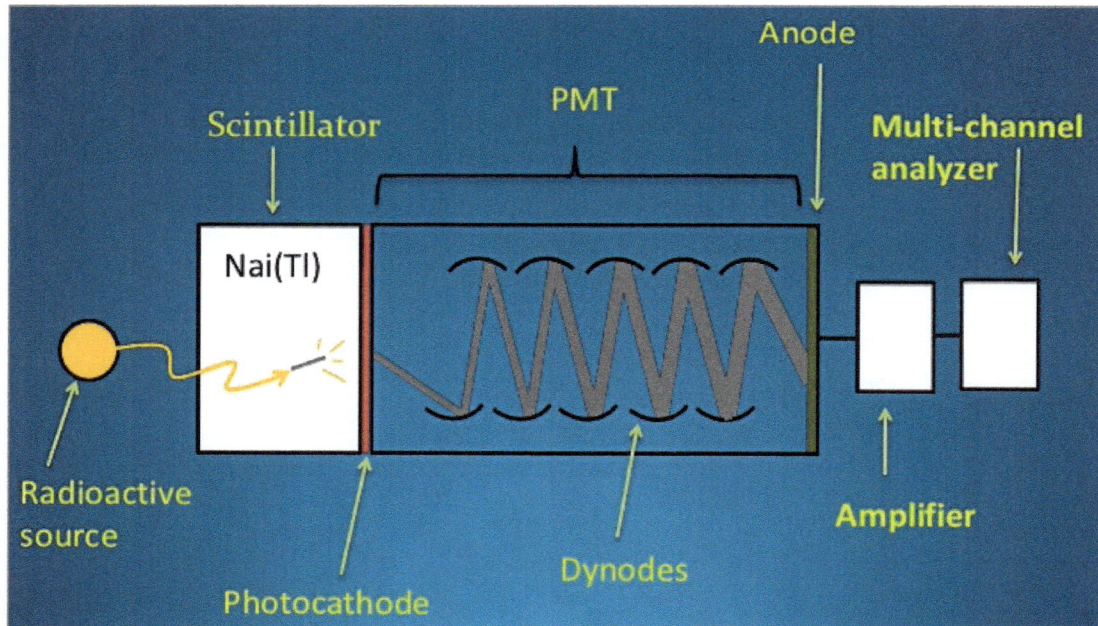

Figure B7-40: Parts of scintillation detector.   *Source: Hemn Rahman.*

MEASUREMENT OF WORKPLACE EXPOSURE

Where work comes under the Ionising Radiation Regulations (IRR) 1999 the employer must make an assessment of the work being carried out, or to be carried out, in any particular area to determine likely exposure levels, to determine the status of the area and the people that work in it.

The areas should be designated as controlled if, after consulting the Radiation Protection Advisor, it appears to be necessary to set up special procedures for persons in the area, whether for routine procedures or for accidents, in order to prevent significant exposures and in any event if persons are likely to receive more than 6mSv per year.

Any area where the employer finds it necessary to keep conditions under review to check if the area should be designated as controlled, or if the area is such that persons may receive more than 1mSv per year, should be designated as supervised.

An area should also be designated as controlled if there is significant risk of spreading radioactive contamination outside the working area.

Controlled areas should be physically demarcated or suitably delineated. Signs should be displayed in controlled and supervised areas to indicate that the area is controlled and they should also provide information about the nature and risks of the radiation sources. Persons entering a controlled area should either be classified or be entering in accordance with 'written arrangements' designed to ensure that doses will not exceed, in cases of persons over 18, levels which would otherwise require the persons to be classified. In the case of other persons the relevant dose limit is applicable.

Thermo-luminescent dosimeters and approved dosimetry services

THERMOLUMINESCENT DOSIMETERS (TLD)

The thermo-luminescent dosimeter badges contain small pieces of material such as lithium fluoride or manganese sulphate, which absorbs the radiation in such a fashion that some of the electrons in the material remain in excited or high energy states for a long time. When radiation hits the thermo-luminescent material electrons are freed from some atoms and move to other parts of the material, leaving behind 'holes' of positive charge. Subsequently, when the thermo-luminescent material is heated the electrons and the 'holes' re-combine and release the extra energy in the form of light.

They release light in quantities related to their radiation exposure; therefore light intensity can be measured and related to the amount of energy initially absorbed through exposure to the radiation.

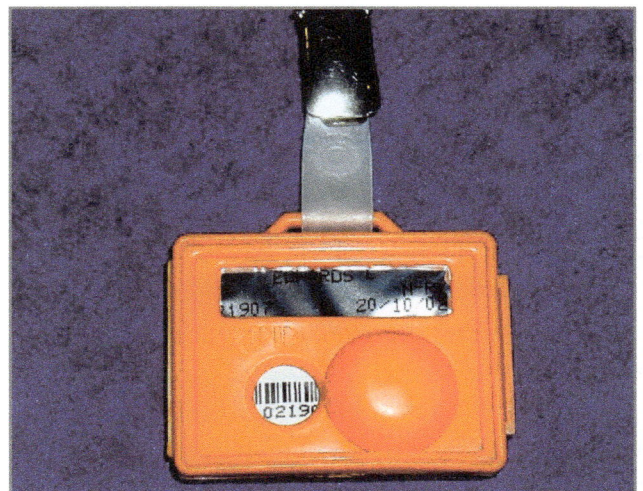
Figure B7-41: Personal dose monitor.   Source: RMS.

The thermo-luminescent dosimeters can be reused, are relatively rugged and give reliable information over long periods of time. Also, the badges are generally smaller than film badges.

APPROVED DOSIMETRY SERVICES

Those who provide dosimetry services must be approved in accordance with the Ionising Radiations Regulations (IRR) 1999 Regulation 35 or the Radiation (Emergency Preparedness and Public Information) Regulations (REPPIR) 2001 Regulation 14. The HSE is given power under Regulation 35 of IRR 1999 to approve suitable dosimetry services, or to specify another Approval Body for this purpose.

Practical measures to prevent or minimise exposure

EXTERNAL RADIATION

External radiation hazards are those from sources outside the body, such that there is no possibility of these entering into the body as particulate contamination. In general, these sources can be shut off, removed or a person can leave an area in which they arise.

There are three general guidelines for controlling exposure to ionising radiation: minimising exposure time, maximising distance from the radiation source, and shielding the person from the radiation source.

Time

Time is an important factor in limiting exposure to the public and to radiological emergency responders. The shorter the period of time one stays in a radiation field, the smaller the dose received. The maximum time to be spent in the radiation environment is given as the exposure time. The exposure time can be calculated using the following equation:

Exposure Time = Dose Limit (Sv)/Dose Rate.

Because of this time factor, it is very important to carefully plan the work to be done prior to entering the radiation environment. Working as quickly as practicable once there and rotating the workers who are in the radiation area will help minimise exposure of individuals. These principles work equally well for workers that may work in a radiation exposure area.

Distance

Distance can be used to reduce exposures. A dramatic reduction in effective dose can be obtained by increasing the distance between the worker and the radiation source. The decrease in exposure rate, as the worker moves away from the source, is greater than one might expect. Doubling the distance from a point source of radiation decreases the exposure rate to 1/4 of its original value.

This relationship is called the inverse square law. The word inverse implies that the exposure rate decreases and the distance from the source increases. Square suggests that this decrease is more rapid than just a one-to-one proportion.

Radiation exposure levels decrease as distance from a non-point source increase, but not in the same mathematical proportions as the inverse square law suggests. In radiological emergencies or as an additional workplace precaution, where the radiation exposure rates are very high, some shielding may be necessary.

Shielding

Shielding is the placement of a radiation 'absorber' between the worker and the radiation source. A radiation absorber is a material that reduces the number of particles or photons travelling from the radiation source to the worker.

Alpha, beta and neutron radiation can all be stopped by different thickness of absorbers.

There is no absorber shield that can stop all gamma rays. Instead, introduction of a shield of a specified thickness will reduce the radiation intensity by a certain fraction.

Addition of more shielding will reduce the intensity further.

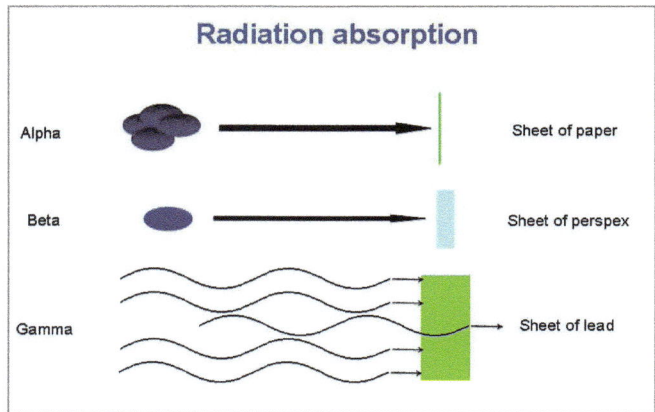

Figure B7-42: Shielding.   Source: RMS.

INTERNAL RADIATION (INHALATION PREVENTION, INGESTION, INJECTION, ABSORPTION)

Internal radiation hazards can occur where it is possible for radioactive materials to be taken into the body from contamination on articles or from airborne contamination. This may be by inhalation, ingestion, through the skin or through a break in the skin. Once inside the body such contamination will continue to irradiate the person until it has been excreted, decayed away or otherwise removed.

The internal hazard is minimised by limiting the intake of contaminated air and drinking water and the consumption of contaminated foods. Breathing apparatus will be required if the atmosphere is contaminated and whole body protection to prevent entry through the skin. Lead lined clothing can form a barrier against ionising radiation. In the planned introduction of radiation sources into the body, it is important to keep radiation doses to the minimum necessary for the process and to match the dose delivered to the person's size and health.

Where exposure to radon is likely this can lead to exposure to internal radiation when the radon is breathed in. Control of radon exposure in new buildings can be done by installing a 'radon proof membrane' within the floor structure. In areas more seriously affected by radon it may be necessary to install a 'radon sump' to vent the gas into the atmosphere. A radon sump has a pipe connecting a space under a solid floor to the outside.

A small electric fan in the pipe continually sucks the radon from under the house and expels it harmlessly to the atmosphere. Modern sumps are often constructed from outside the building so there is no disruption inside. In existing buildings it is not usually possible to provide a radon proof barrier, so alternative measures are used to control the build-up of radon in the building and subsequent exposure to it.

Such measures include a mixture of active and passive systems such as improved under floor and indoor ventilation in the area, positive pressure ventilation of occupied areas, installation of radon sumps and extraction pipework and sealing large gaps in floors and walls in contact with the ground.

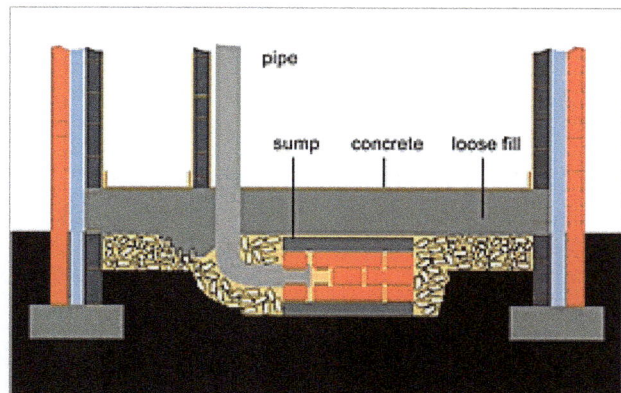

Figure B7-43: Radon sump.   Source: HSE.

Legal requirements to minimise occupational exposure

The Ionising Radiation Regulations (IRR) 1999 impose duties on employers to protect employees and other persons against ionising radiation arising from work with radioactive substances and other sources of ionising radiation, and also impose certain duties on employees.

The IRR 1999 are divided into 7 Parts. Part 2 sets out the general principles and procedures to manage non-medical exposure to ionising radiation. These are covered in Regulations 5 to 12, and:

- Prohibit the carrying out of specified practices without the authorisation of the Health and Safety Executive (HSE).
- Require specified work with ionising radiation to be notified to the HSE.
- Require radiation employers to make a prior assessment of the risks arising from their work with ionising radiation, to make an assessment of the hazards likely to arise from that work and to prevent and limit the consequences of identifiable radiation accidents.
- Require radiation employers to take all necessary steps to restrict so far as is reasonably practicable the extent to which employees and other persons are exposed to ionising radiation.
- Require respiratory protective equipment used in work with ionising radiation to conform with agreed standards and require all personal protective equipment and other controls to be regularly examined and properly maintained.
- Impose limits (specified in Schedule 4 of IRR 1999) on the doses of ionising radiation which employees and other persons may receive, including employees of 18 years and over, trainees under 18, women of reproductive capacity and 'other persons'.
- Require in certain circumstances the preparation of contingency plans for radiation accidents which are reasonably foreseeable.

The IRR 1999 sets out requirements to establish arrangements for the management of radiation protection, encompassing:

- Consulting a radiation protection advisor with regard to arrangements set out in Schedule 5 of IRR 1999, for example, the implementation of requirements for controlled/supervised areas and testing of engineering controls.
- The provision of radiation protection training and sufficient information and instruction for them to know the risks, precautions and compliance with aspects of the IRR 1999, including medical requirements.
- Establishing co-operation between employers where the ionising radiation work of one may affect the other.

An important part of the IRR 1999 and management of exposure is the requirement to establish designated areas as 'controlled' or 'supervised'.

A ***controlled*** area is one where an assessment has identified that special procedures are required to restrict significant exposure to ionising radiation or where any worker, aged 18 or more, working in the area is likely to be exposed to an effective dose greater than 6 mSv a year (or 30% of any relevant dose limit in Schedule 4 of IRR 1999).

A ***supervised*** area is one that is not a controlled area but where it is necessary to keep conditions under review in case it may need to be designated a controlled area or where any worker, aged 18 or more, working in the area is likely to be exposed to an effective dose greater than 1 mSv a year (or 10% of any relevant dose limit in Schedule 4 of IRR 1999).

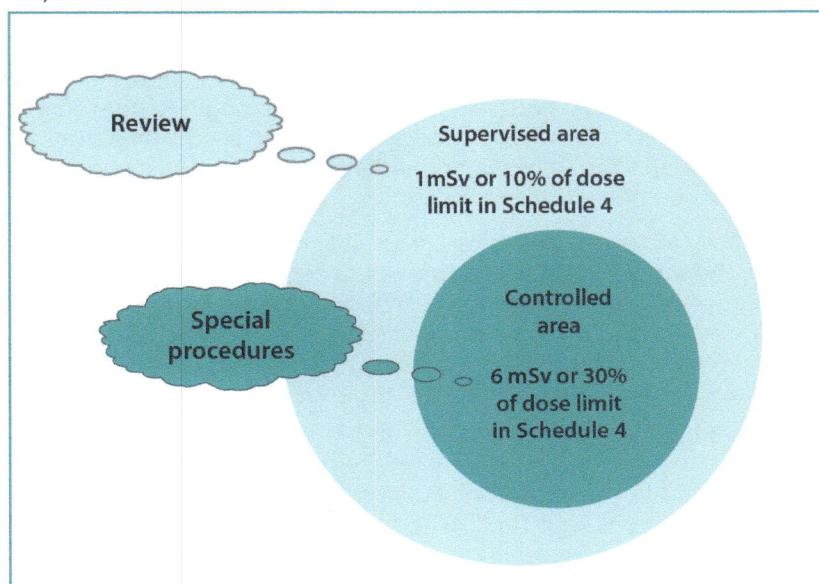

Figure B7-44: Controlled and supervised areas. *Source: RMS.*

The management of control of exposure, set out in IRR 1999, includes the classification and monitoring of those exposed to ionising radiation. This requires the employer to establish which workers are classified persons, ensure they are under medical surveillance and, where they are overexposed, investigate the reasons, notify the HSE and limit further doses they receive.

B7.4 - Lasers

Definition of laser

Laser stands for the 'Light Amplification by the Stimulated Emission of Radiation'.

The *laser* process is where electrons are stimulated to produce photons that all have the same wavelength and are precisely in step so that the peaks and troughs of their electromagnetic wave exactly match each other in a coherent way.

Stimulation is encouraged by trapping this radiation, so that further emissions occur within a resonant cavity. Here the radiation bounces back and forth, stimulating further emissions in the desired way, before the light is allowed to escape. This 'coherent' light possesses the ability to carry information and can have a high energy density.

This 'coherent' light possesses the ability to carry information and can have a high energy density. An example of this coherent light is the use of a laser to hit a mirror target on the moon (the original mirror was placed there by Apollo 11 astronauts) that is then reflected back to earth to measure the distance between the moon and the earth. The distance from the earth to the mirror is 384,467 kilometers.

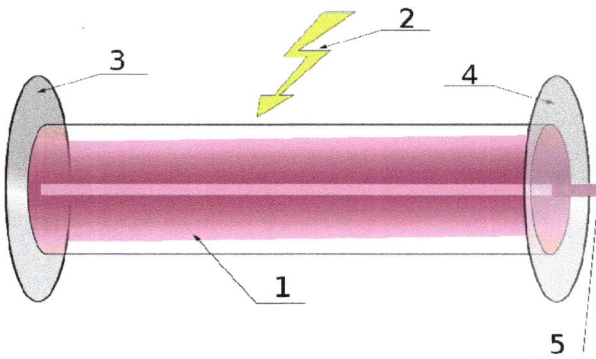

Figure B7-45: Schematic laser.   *Source: Tatoute.*

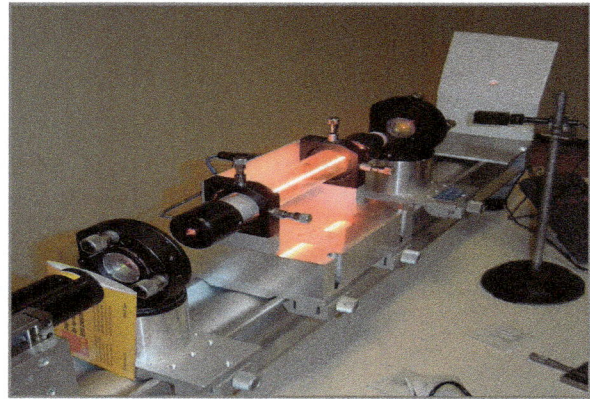

Figure B7-46: Helium-neon laser.
Source: Kastler-Brossel Laboratory Paris.

Principal components of a schematic laser are shown and labelled in *figure ref B7-45*:
1) Gain medium - a material with properties that allow it to amplify light by stimulated emission for example, helium.
2) Laser pumping - energy required for the amplification.
3) High reflector - consists of two or more mirrors arranged to bounce the light back and forth.
4) Output coupler - a partially transparent mirror, through which the laser beam is emitted.
5) Laser beam.

The wavelength of the radiation emitted by a laser may be in the invisible ultraviolet/infrared light regions or in the visible light region of the electromagnetic spectrum. The wavelength of the emitted radiation is related directly to the chemical composition of the substance that is used as the medium to generate the beam. Therefore, lasers are generally referred to in terms of their lasing medium, which may be a solid, liquid or gas, for example, helium/neon, argon, carbon dioxide, ruby etc.

Unlike other sources of optical radiation, the laser produces radiation at a specific, coherent wavelength (monochromatic). The output beam of a laser source may be a continuous emission, which is called continuous wave (CW), or a pulsed emission. They are classified according to their emitted radiation power. Because the laser generally produces an almost parallel beam of radiation, the intensity will not fall off with the square of the distance as in the case of conventional radiation sources. Reflections, even from dark surfaces, are also hazardous, because the beam does not lose energy by divergence.

Typical laser sources in workplaces

The beam produced by most lasers is very small and maintains its size and direction over very large distances. This sharply focused beam of coherent light (optical radiation) is suitable for a wide variety of applications, including laser displays connected with indoor and outdoor *entertainment* venues and public places.

Lasers are used in *manufacturing* in materials processing (cutting, drilling and welding), laser marking, and photolithography and for inspecting optical equipment.

Optical measurement uses include distance measurement, surveying, laser velocimeters, electronic speckle pattern interferometers, optical fibre hydrophones, high-speed imaging and particle sizing. In *healthcare* they are used in surgical operations, cosmetic applications (tattoo removal), dermatology, vascular surgery, dentistry, medical diagnostics and vision correction. They are widely used in the *communication* industry, related to fibre optics and satellite communications.

Work is being carried out to develop lasers in a manner similar to radio transmission; the transmitted light beam is modulated with a signal and is received and demodulated some distance away. The *retail and distribution* industries make wide use of lasers in the form of barcode scanners. The field of holography (holograms) is based on the fact that wave-front patterns, captured in a photographic image of an object illuminated with laser light, can be reconstructed to produce a three-dimensional image of the object. This is used for both entertainment and *information storage*. Lasers are widely used in equipment used in *commercial* operations, such as scanners, printers and CD/DVD equipment.

Lasers have been used in several kinds of scientific *research and analysis*, including spectroscopy for substance identification. Lasers have opened a new field of scientific research, non-linear optics, which is concerned with the study of such phenomena as the frequency doubling of coherent light by certain crystals. One important result of laser research is the development of lasers that can be tuned to emit light over a range of frequencies, instead of producing light of only a single frequency. Lasers have also been used in plasma physics and chemistry.

Exposure routes for lasers

Normal routes of exposure to lasers include direct exposure to unprotected skin and eyes or reflections from shiny surfaces and mirrors. Risks from exposure to lasers may be increased if the radiation is concentrated through a lens, increasing the power density and heating effect significantly.

Hazard classification of lasers (BS EN 60825-1:2014)

Laser products are classified to take account of the amount of laser beam to which a person can get access when the product is in normal use or during routine user maintenance.

A laser product may contain a laser of a higher class and this may be accessible during more complex servicing and maintenance by a specialist.

Labels on the product should provide guidance on the laser beam hazard, their class and appropriate warnings.

Full details about the classification scheme can be found in BS EN 60825-1:2007 'Safety of laser products. Equipment classification, requirements'. A brief description of each laser class is shown in *figure ref B7-47.*

Figure B7-47: Warning sign - laser. *Source: HSE, L64.*

Class	Characteristics
Class 1	Inherently safe laser products. Safety is achieved either due to their low power or by their total enclosure, for example, laser printers, CD players.
Class 1M	Either a highly divergent beam or a large diameter beam, therefore only a small part can enter the eye. Harmful if viewed through magnifiers, for example, fibre optic communication systems.
Class 2	Maximum output of 1 milliwatt (mW), wavelength between 400 and 700 nm. Damage to the eye is prevented by blinking and averting the head, an instinctive response. Repeated deliberate exposure is not safe, for example, laser pointers and barcode scanners.
Class 2M	Either a highly divergent beam or a large diameter beam, therefore only a small part can enter the eye, limited to 1mW. Harmful if viewed through magnifiers or for long periods, for example, civil engineering applications - level and orientation instruments.
Class 3R	Higher powered than classes 1 and 2. Max output of 5mW, exceeds max permissible exposure for accidental viewing and can cause eye injuries, but the actual risk of injury following a short exposure is small. This class replaces the former 3A and lower part of 3B, for example, some laser pointers and some alignment products for home improvement work.
Class 3B	Output power of 500 mW (half a watt). Causes eye damage from direct beam exposure and

Class	Characteristics
	reflections. Extent of injury depends on radiant power entering eye and duration of exposure, for example, lasers for physiotherapy treatment and research lasers.
Class 4	Output greater than 500 mW, with no upper restriction. Causes injury to eye and skin and a fire hazard for example, lasers in displays, laser surgery and cutting metals.

Figure B7-48: Laser classes and appropriate warnings.   *Source: HPA/RMS.*

'M' in Class 1M and Class 2M is derived from magnifying optical viewing instruments.

'R' in Class 3R is derived from reduced, or relaxed, requirements: reduced requirements both for the manufacturer (for example, no key switch, beam stop or attenuator and interlock connector required) and the user.

The 'B' for Class 3B has historical origins.

Figure B7-49: Laser classification letters.   *Source: HPA, Non-binding Guide to Artificial Optical Radiation Directive.*

EXPOSURE LIMITS

The Control of Artificial Optical Radiation at Work Regulations (CAOR) 2010 specifies that the employer must put in place actions to limit exposure to artificial optical radiation so that the exposure limit values (ELVs) are not exceeded.

They establish that for the purposes of CAOR 2010 the exposure limit values for laser (coherent) radiation are those set out in Annex II of Directive 2006/25/EC of the European Parliament and of the Council on the minimum health and safety requirements regarding the exposure of workers to risks arising from physical agents (artificial optical radiation).

Wavelength range (nm)	Limiting aperture (mm)	Exposure duration (s)	ELV ($W\,m^{-2}$)	Maximum power through aperture (W)	Maximum power through aperture (mW)
180 to 302,5	1	10	3,0	0,000 002 4	0,002 4
≥ 302,5 to 315	1	10	3,16 to 1 000	0,000 002 5 to 0,000 79	0,002 5 to 0,79
305	1	10	10	0,000 007 9	0,007 9
308	1	10	39,8	0,000 031	0,031
310	1	10	100	0,000 079	0,079
312	1	10	251	0,000 20	0,20
≥ 315 to 400	1	10	1 000	0,000 79	0,79
≥ 400 to 450	7	0.25	25,4	0,000 98	0,98
≥ 450 to 500	7	0.25	25,4	0,000 98	0,98
≥ 500 to 700	7	0.25	25,4	0,000 98	0,98
≥ 700 to 1050	7	10	10 to 50	0,000 39 to 0,001 9	0,39 to 1,9
750	7	10	12,5	0,000 49	0,49
800	7	10	15,8	0,000 61	0,61
850	7	10	19,9	0,000 77	0,77
900	7	10	25,1	0,000 97	0,97
950	7	10	31,6	0,001 2	1,2
1000	7	10	39,8	0,001 5	1,5

Figure B7-50: Laser ELVs.   *Source: HPA, Non-binding Guide to Artificial Optical Radiation Directive.*

The ELVs set out in the Directive take account of the biological effectiveness of the optical radiation at causing harm at different wavelengths, the duration of exposure to the optical radiation and the target tissue. The ELVs are based on the Guidelines published by the International Commission on Non-Ionizing Radiation Protection (ICNIRP). If the ICNIRP modify their Guidelines this could lead to the ELVs in the Directive being subsequently modified.

The ELVs for laser radiation are generally simpler to determine than that for non-coherent radiation, because the emission is usually at a single wavelength. However, for laser products that emit laser radiation at more than one wavelength, or for exposure scenarios involving multiple sources, it may be necessary to take account of additive effects.

The ELVs are expressed in terms of irradiance (W m^{-2}) or radiant exposure (J m^{-2}), Annex II provides ELVs for short duration eye exposure (<10 seconds), longer eye exposure and skin exposure. The time of exposure will depend on whether it is accidental or not. For accidental exposures of the eye 0.25 s is taken as the time of exposure for lasers in the wavelength 400 to 700 nm (the visible light part of the electromagnetic spectrum) because of the natural aversion reaction.

For all other wavelengths and exposure of the skin it is taken to be 10 or 100 s; the duration before pain will cause a response. Using the information in the Annex II it is possible to calculate the maximum power of the laser through the stated aperture before the ELV is exceeded. The HPA's 'Non-Binding Guide to the Artificial Optical Radiation Directive' provides worked examples of this.

The effect of exposure to each class of laser

EFFECTS ON THE EYE

The fact that eye damage may result from exposure to high levels of ultraviolet, visible and infrared radiation (optical radiation) has been discussed in the previous sections.

As with all nerves when damaged, the retina cannot be repaired and the cells will not regenerate. This damage can be caused naturally by staring into the sun, which would cause a level of retinal illumination about a million times greater than normal. A 1mW continuous wave visible laser can give an order of magnitude higher than the direct viewing of the sun.

The type of eye damage depends on the particular tissue affected, which in turn depends on the ocular transmission characteristics of the type of laser used. Both the far infrared (1400 nm - 1 mm) and far ultraviolet (200 - 295 nm) radiation will affect only the cornea at the front of the eye, where they will be absorbed.

Near ultraviolet radiation (295 - 400 nm) will penetrate some depth into the eye as far as the lens, whereas the near infrared (700 - 1400 nm) and all visible light radiation (400 - 700 nm) will penetrate to the retina. Near infrared lasers pose a particular hazard, as they cannot be perceived, although they enter the eye and are focused onto the retina.

Both visible light and near infrared radiation is focused onto a very small spot (~ 10 μm diameter) on the retinal surface. Exposure can produce varying degrees of photochemical damage to the cells of the retinal macula (photoretinitis), without a significant increase in the temperature of the tissue. The retinal macula ('yellow spot') contains the largest concentration of cone cells in the eye and is responsible for central, high resolution vision.

EFFECTS ON THE SKIN

Because there is no focusing effect, the skin is more resistant to damage when irradiated than the eye. Acute exposure to laser radiation from Class 4 lasers may give rise to skin damage, which may range from mild erythema (skin reddening) through to deep burns (tissue charring).

Control measures to prevent or minimise exposure to lasers used in workplaces

LEGAL REQUIREMENTS

The CAOR 2010 requires the employer to eliminate any risk of adverse health effects to the eyes and skin. Where this is not reasonably practicable they should be reduced to as low a level as is reasonably practicable.

Measures to eliminate or reduce risk must be based on the general principles of prevention set out in the Management of Health and Safety at Work Regulations (MHSWR) 1999.

If the risk assessment indicates that employees will be exposed to levels above the exposure limit values, the employer must devise and implement an action plan that takes account of:

"a) Other working methods.

b) Choice of appropriate work equipment emitting less artificial optical radiation.

c) Technical measures to reduce the emission of artificial optical radiation including, where necessary, the use of interlocks, shielding or similar health protection mechanisms.

d) Appropriate maintenance programmes for work equipment, workplaces and workstation systems.

e) The design and layout of workplaces and workstations.

f) Limitation of the duration and level of the exposure.

g) The availability of personal protective equipment.

h)   The instructions of the manufacturer of the equipment where it is covered by relevant European Union Directives.

i)   The requirements of employees belonging to particularly sensitive risk groups."

The employer must also ensure that areas where this exposure is likely to be above the ELVs are:

- Demarcated and access restricted.
- Identified by appropriate signs.

CONTROL MEASURES TO PREVENT OR MINIMISE EXPOSURE

The use of engineering controls can provide physical shielding against the effects of lasers.

In addition to **well-designed** equipment, appropriately **sited** to minimise exposure with **direction control** limited to the area required for use, the following can be used to further reduce exposure:

- Engineering control measures, i.e. the design should consider minimum power requirements, siting, the distance from people, interlocks, emergency stops, remote operation and time delay switches.
- Administrative control measures, i.e. instruction, training, understanding the hazards and controls and safe Systems of work such as permits, i.e. to access restricted areas.
- **Reduction of stray beams** - elimination by use of non-reflective surfaces.
- **Screening and enclosures** - sealed housings, screened areas using partitions or curtains, protected viewing windows or remote systems.
- Use of **distance** - to ensure a safety factor.
- **Safe systems of work** - limitation of access, use of interlocks, signage, and safe maintenance procedures.
- **Instruction and training** - hazard awareness training and use of hazard controls and restricted access procedures; the importance of limitation of exposure time.
- **Personal protective equipment**, i.e. for the eyes and the skin.

Personal protection

Protection of the skin - cover exposed areas, i.e. backs of hands, face, neck and forearms. Protective gloves are suitable for protection of the hands. Forearms should be protected by long sleeves made from poplin or flannelette.

Protection of eyes - is normally achieved through the use of laser eye and face protectors (glasses, goggles, shields and helmets).

Electrical equipment and systems

- Always be aware of the high risk of injury and fire in laser operations, because of the presence of electrical power sources. The use of lasers should be considered as part of the fire risk assessment process.

Lighting

- Adequate lighting is necessary in controlled areas. If lights are extinguished during laser operation, provide control switches in convenient locations or install a radio controlled switch. When natural light is not sufficient for safe egress from a laser area during an electrical power failure, emergency lighting should be installed.

	Class 1	Class 1M	Class 2	Class 2M	Class 3R	Class 3B	Class 4
Description of hazard class	Safe under reasonably foreseeable conditions	Safe for naked eye;, may be hazardous if the user employs optics	Safe for short exposures; eye protection is afforded by aversion response	Safe for naked eye for short exposures,; may be hazardous if the user employs optics	Risk of injury is relatively low, but may be dangerous for improper use by untrained persons	Direct viewing is hazardous	Hazardous for eye and skin; fire hazard
Controlled area	Not required	Localised or enclosed	Not required	Localised or enclosed	enclosed	Enclosed and interlock protected	Enclosed and interlock protected
Key control	Not required	Not required	Not required	Not required	Not required	required	required
Training	Follow manufacturer instruction for safe use	Recommended	Follow manufacturer instruction for safe use	Recommended	Required	Required	Required
PPE	Not required	Not required	Not required	Not required	May be required – subject to the findings of the risk assessment	required	required
Protective measures	Not necessary under normal use	Prevent use of magnifying, focusing or collimating optics	Do not stare into the beam	Do not stare into the beam. Prevent use of magnifying, focusing or collimating optics	Prevent direct eye exposure	Prevent eye and skin exposure to the beam. Guard against unintentional reflections	Prevent eye and skin exposure from direct and diffuse reflection of the beam.

Figure B7-51: Summary of laser controls.   *Source: HPA, Non-binding Guide to Artificial Optical Radiation Directive.*

■ A laser operation may involve ionising radiation that originates from the presence of radioactive materials or the use of electrical power in excess of 15 kV. Microwave and radio frequency (RF) fields may be generated by laser systems or support equipment.

The legal requirements to manage exposure to lasers

The Control of Artificial Optical Radiation at Work Regulations (CAOR) 2010 place duties on the employer who carries out any work that could expose its employees to levels of artificial optical radiation, which could create a reasonably foreseeable risk of adverse health effects to the eyes or skin. The employer must make a suitable and sufficient assessment of the risk from exposure to lasers, for the purpose of identifying the measures the employer needs to take to meet the requirements of CAOR 2010.

The employer must, as part of the risk assessment, assess, and if necessary, measure or calculate, the levels of artificial optical radiation to which employees are likely to be exposed.

The employer must take steps to eliminate or reduce risks, including the development and implementation of an action plan for controls. This must be supported by information and training for employees, relating to the outcome of the risk assessment and include:

"a) *The technical and organisational measures taken in order to comply with the requirements of Regulation 4.*

b) *The exposure limit values.*

c) *The significant findings of the risk assessment, including any measurements taken, with an explanation of those findings.*

d) *Why and how to detect and report adverse health effects to the eyes or skin.*

e) *The circumstances in which employees are entitled to appropriate health surveillance.*

f) *Safe working practices to minimise the risk of adverse health effects to the eyes or skin from exposure to artificial optical radiation.*

g) *The proper use of personal protective equipment."*

The management of laser risks will also include setting out local rules, controlled areas where risks are significant and the rules apply, an 'authorised user' register, formal arrangements for maintaining and testing control measures and detailed contingency plans. The appointment of a laser protection advisor should be considered. The employer must also provide suitable health surveillance and medical examinations for those at risk.

Role, competency and training for a 'Laser Protection Adviser'

LASER PROTECTION ADVISOR

The term Laser Protection Adviser (LPA) is a term used for a person who can be consulted on complex laser safety management issues. This person may be employed by the organisation with the laser(s) or may be an external consultant. The term may also be applied to an organisation.

Figure B7-52: Laser Protection Advisor.   *Source: HPA.*

It is recommended that organisations using Class 3B and Class 4 lasers, and intense pulsed light sources, should have access to advice from a certificated LPA. This is a non-mandatory (voluntary) scheme.

The Health Protection Agency (HPA) operates one of the Laser Protection Adviser (LPA) certification schemes. The LPA is required to complete a **rigorous training programme**, which includes peer-reviewed assessment of their competency to give advice. They produce a portfolio of evidence, which is assessed by an external assessor, before undertaking an oral examination by the external assessor. If satisfactory, a recommendation is made to the Director of the Radiation Protection Division of the HPA. The Director makes the final decision on **certification**. After certification, the LPA will continue to be subject to peer review and re-certification is every three years.

The LPA will provide the specialised expertise needed when undertaking the initial risk assessment, determining the laser safety controls and procedures that are necessary, and in providing training to staff, additional and specialised expertise will be needed. The role of the LAP is also to ensure correct procedures are in place and to provide advice on the use of lasers and eye protection.

The HPA acts as LPA to a range of users of lasers. A number of the HPA staff are certified LPAs under the comprehensive certification scheme operated by the HPA.

LASER SAFETY OFFICER

BS EN 60825-1:2007 'Safety of laser products: equipment classification, requirements' recommends that where lasers in Class 3B or Class 4 are in use, a Laser Safety Officer (LSO) (often called a Laser Protection Supervisor in healthcare environments) should be appointed to take day-to-day responsibility, on behalf of the employer, for maintaining safe laser use. Employers are also advised to consider the appointment of a LSO where Class1M and 2M lasers are used.

The degree of training and competence of the LSO may depend on the circumstances of use of the laser equipment and their decided role. Where circumstances are static or simple, the work of the LSA may mean a LSO is not required. Where the work is complex and more dynamic, for example, in some aspects of research, the appointment of an LSO may be advised by the LSA. The British Standards document PD CLC/TR 50448:2005, 'Guide to levels of competence required in laser safety' provides information on competence of the LSO.

Mental ill-health and dealing with violence and aggression at work

Learning outcomes

On completion of this element, candidates should be able to demonstrate understanding of the content through the application of knowledge to familiar and unfamiliar situations. In particular, they should be able to:

B8.1   Explain the effects and causes of common types of mental ill-health within the workplace.

B8.2   Explain the identification and control of workplace mental ill-health with reference to legal duties and other standards.

B8.3   Explain the scope, effects and causes or work-related violence/aggression.

B8.4   Explain the identification and control of work-related violence/aggression with reference to legal duties.

Content

Relevant statutory provisions

Criminal Law Act (CLA) 1967, Section 3, (reasonable force)

Employment Rights Act (ERA) 1996

Public Order Act (POA) 1986

Reporting of Injuries, Diseases and Dangerous Occurrence Regulations (RIDDOR) 2013

Working Time Regulations (WTR) 1998

Sources of reference

Reference information provided, in particular web links, was correct at time of publication, but may have changed.

Drug misuse at work – a guide for employers, INDG91, HSE, http://www.hse.gov.uk/pubns/indg91.pdf

HSE Stress Management Standards, HSE, http://www.hse.gov.uk/stress/standards/

Managing the causes of work-related stress; A step by step approach to using the management standards, HSG218, http://www.hse.gov.uk/pubns/priced/hsg218.pdf

Violence at work – a guide for employers, HSE, http://www.hse.gov.uk/pubns/indg69.pdf

The above web links along with additional sources of reference, which are additional to the NEBOSH syllabus, are provided on the RMS Publishing website for ease of use - www.rmspublishing.co.uk.

DECIDED CASES

Sutherland v Hatton and others [2002] EWCA Civ76;

Walker v Northumberland County Council [1995] IRLR 35

Barber v Somerset County council [2004] UKHL 13

Intel Corporation (UK) Limited v Draw [2007] EWCA Civ 70

O'Toole v First Quench [2005]

Mitchell and Others v United Co-operatives Ltd. [2012]

B8.1 - Extent, scope, effects and causes of mental ill-health at work

The prevalence of mental ill-health within the workplace

Mental disorders are health conditions that are characterised by alterations in thinking, mood, or behaviour (or a combination of all three) associated with distress and/or impaired functioning. The main issues are associated with lack of concentration, feelings of isolation and in some situations inability to interact with others.

The majority of those suffering from mental ill-health conditions have what are classed as common mental health disorders, such as anxiety and depression, which tend to range in intensity from 'mild to moderate'.

Findings from the Royal College of Psychiatrists (2008) describe how people with mental health problems can be divided into three broad groups; those with:

1) Sleep problems, fatigue, irritability and worry.

2) Depression and anxiety.

3) Psychotic and/or severe mental illness (for example, bipolar disorder, schizophrenia, or severe depression).

The prevalence of work-related mental ill-health is particularly difficult to determine with some degree of accuracy, as the condition does not reveal itself as readily as other work-related harm. In addition, people suffering may not be aware of it or not readily want to admit it.

Work related stress depression and anxiety continue to represent a significant ill-health condition in the workforce of Great Britain. Work related stress accounts for 35% of work related ill health and 43% of days lost, in 2014/15. The occupations and industries reporting the highest rates of work related stress remain consistently in the health and public sectors of the economy. The reasons cited as causes of work related stress are also consistent over time with workload, lack of managerial support and organisational change as the primary causative factors.

The characteristics and causes of common types of mental ill-health observed within the workplace

The causes of work-related mental ill-health relating to organisation, job and individual include:

- Organisation of work: working hours, long hours, shift work, unpredictable hours, changes in working hours.
- Workplace culture: communication, organisational structure, resources, support.
- Working environment: space, noise, temperature, lighting, etc.
- Job content: work load, time pressures, boredom, etc.
- Job role: clarity, conflict of interests, lack of control, etc.
- Relationships: bullying and harassment, verbal/physical abuse.
- Home-work interface: travel to/from work, childcare issues, relocation, etc.

The four common types of mental ill-health are depression, anxiety, bipolar and schizophrenia.

DEPRESSION

People with mental health problems are more likely to develop physical health problems and vice versa. Furthermore, people with mental health problems can present to their doctor or employer complaining of physical symptoms that have no physical cause (Royal College of Psychiatrists, 2008). Depression is a broad and varied diagnosis and can be mild, moderate or severe. The main features are depressed mood and/or loss of pleasure in most activities, a feeling of helplessness: a cognitive state of mind.

Certain physical illnesses can trigger depression in people of any age, but conversely people with depression may have symptoms that they think are caused by a physical illness, but are actually caused by depression. The severity of depression is determined by the number of symptoms, as well as the degree of functional impairment to undertake everyday tasks such as shopping, cleaning and going to work.

Psychological symptoms include:	Physical symptoms include:	Social symptoms include:
■ Continuous low mood or sadness. ■ Feelings of hopelessness and helplessness. ■ Low self-esteem. ■ Tearfulness. ■ Feelings of guilt. ■ Feeling irritable and intolerant of others. ■ Lack of motivation and little interest in things. ■ Difficulty making decisions. ■ Lack of enjoyment.	■ Slowed movement or speech. ■ Change in appetite or weight (usually decreased, but sometimes increased). ■ Constipation. ■ Unexplained aches and pains. ■ Lack of energy or lack of interest in sex. ■ Changes to the menstrual cycle. ■ Disturbed sleep patterns (for example, problems going to	■ Not doing well at work. ■ Taking part in fewer social activities and avoiding contact with friends. ■ Reduced hobbies and interests. ■ Difficulties in home and family life.

Psychological symptoms include:	Physical symptoms include:	Social symptoms include:
■ Suicidal thoughts or thoughts of harming yourself. ■ Feeling anxious or worried.	sleep or waking in the early hours of the morning).	

Figure B8-1: Symptoms of depression.   *Source: NHS.*

ANXIETY

Anxiety and panic disorders are common, but can cause extreme distress to individuals if left untreated. Symptoms that may manifest themselves at work may include loss of interest, poor concentration, low mood and irritability. The employee may turn down a promotion or other opportunity because it involves travel or public speaking; make excuses to get out of office parties, staff lunches, and other events or meetings with co-employees; or be unable to meet deadlines.

BIPOLAR

Bipolar is a psychotic condition (mental illness) that is characterised by fluctuations in mood, which can sometimes affect concentration and behaviour. These moods may range from feelings of severe depression to abnormally elevated mood swings (mania). According to the World Health Organisation it is the sixth leading cause of disability in the world. During the 'manic' phases, sufferers of bi-polar disorder can lose their sense of judgement and perspective of what is 'normal', they may develop fixed ideas and illusions of grandeur, for example, believing that they are related to Royalty. The depressive phase is characterised by severe sadness and sometimes suicidal tendencies. This condition can affect people from all walks of life and it may go undiagnosed and treated in sufferers where the symptoms are not so obvious.

SCHIZOPHRENIA

Schizophrenia results in a dramatic disturbance in an individual's thoughts and feelings. The person with schizophrenia may begin to experience the world differently to others and their behaviour may change and seem bizarre to others. Other symptoms may include an inability to concentrate, apathy, depression, delusions, hallucinations and ultimately social withdrawal. The individual may be affected by episodes of 'positive' symptoms, for example, delusions, and 'negative' symptoms, for example, social withdrawal. The delusions that people suffering from schizophrenia experience can be very powerful beliefs and even if evidence is produced to the contrary it will not alter the sufferer's belief. The person may also experience thought disturbances where they believe their thoughts are not their own and are being put into their mind by someone else. Another major symptom of schizophrenia is experiencing hallucinations, commonly in the form of hearing voices telling them to do things, though hallucinations can be in visual (see), tactile (feel) or olfactory (smell) form. Those suffering from schizophrenia may also present an apparent lack of emotional sensitivity, for example, not responding in an appropriate manner when hearing tragic news, but laughing instead.

The meaning of work-related stress

There are a number of definitions of stress in existence, but for the purposes of this Element the Health and Safety Executive's (HSE) definition from the publication 'Managing the Causes of Work-Related Stress' (HSG218) will be used:

> *"The adverse reaction people have to excessive pressure or other types of demand placed on them".*

Figure B8-2: Definition of stress.   *Source: HSE HSG218.*

However, other terms which broaden understanding of the nature of stress, include:

- Any force which interferes with a person's ability to control their emotional and physical state within a comfortable range and which prevents them from producing a control strategy for that force.
- A broad range of problems which unduly test a person's psychological, physiological or social system and the response of that system to the problems.
- A physiological state in which the mental and physical energy expended to cope with pressure exceeds the body's ability to replace that energy.

Individuals have different reactions to pressure and varying methods of controlling those reactions. For example, a mundane and routine job, which may be the preferred choice of an employee who is comfortable with a familiar and undemanding role, may cause stress in another employee who would prefer an unpredictable, challenging, fast-paced role. The reverse of this situation would be true for other individuals.

Problems may arise when the pressure faced by the individual appears overwhelming and uncontrollable. The individual may consider that they are not in possession of the skills necessary to control the stress and therefore feel unable to cope. Stress, whatever the cause, is brought about and made worse by an individual's inability to cope and it is stress which results in physical and mental harm.

The subject of stress is complex, varied and complicated by the fact that many people suffer from stress caused at work, outside of work or a combination of both. Many people come to work against a background of problems at home which may involve family/relationship, financial, health and legal issues, to name but a few. These problems, which are outside the employer's responsibility, may cause an employee to be more vulnerable to work-related stress. In addition, the problems associated with stress are often made worse by sceptical employers and colleagues.

The prevalence of work-related stress

Work-related stress is recognised worldwide as a major challenge to workers' health and the healthiness of their organisations. The prevalence of work-related stress is particularly difficult to determine with some degree of accuracy, as the condition does not reveal itself as readily as other work-related harm. In addition, people suffering from stress may not be aware of it or not readily want to admit it.

Stress can cause numerous minor disorders that create discomfort, but which may also lead to serious ill-health. There are a number of physical, emotional and behavioural signs of stress which may become apparent to colleagues, supervisors and managers.

Recognising these signs or symptoms is an important part of any strategy to manage work-related stress illnesses. Stress will manifest itself in a variety of ways with different people but the following are common symptoms:

PHYSICAL SIGNS OF STRESS

Physical signs of stress can include:

- Heart and circulatory problems, for example, palpitations, pain/tightness in the chest, heart attack, stroke.
- Repeated colds, flu or other infections.
- Menstrual pattern changes.
- Rapid weight gain/loss.
- Headaches, shaking.
- Tiredness, fainting.
- Skin complaints, for example, eczema, psoriasis, sweating and baldness.
- Digestive system problems, for example, indigestion, nausea/vomiting, stomach cramps, irritable bowel syndrome (IBS).

PSYCHOLOGICAL EFFECTS OF STRESS

Psychological effects of stress may be manifested in both emotional and behavioural signs.

Emotional signs of stress

- Mood swings/irritability.
- Cynicism.
- Anxiety, nervousness, apprehension.
- Loss of confidence.
- Lack of self-esteem.
- Lack of concentration/lack of enthusiasm.
- Panic attacks.

Behavioural signs of stress

- Poor quality work.
- Increased smoking, alcohol/drug use.
- Insomnia.
- Loss of appetite or overeating.
- Poor time management.
- Accident proneness.
- Impaired speech.
- Too busy to relax.

Certain symptoms are important warning signs that action should be taken to identify and tackle the cause of the stress. Medical advice should be sought if symptoms such as some of these listed below are experienced.

- Frequent heartburn, diarrhoea, inability to swallow.
- Memory/concentration impairment.
- Inability to make decisions.
- Difficulty in problem solving.
- Recurring headaches/migraines.
- Feeling of faintness.
- Prone to illness.
- Palpitations/chest pain.
- Frequent use of self-prescribed drugs.
- Anger, irritation, tearfulness and frustration are common emotions.

It should be noted that many of the symptoms listed are experienced by individuals who are not suffering from stress. However, if the symptoms are uncharacteristic of that individual, begin to occur in combination or occur at a time when the individual is known to be experiencing pressures at home or at work, they could be indicative of stress related illness.

POST-TRAUMATIC STRESS DISORDER

Post-traumatic stress disorder (PTSD), paraphrased from Merriam-Webster's Medical Dictionary, is a psychological reaction that occurs after experiencing a highly stressing event such as physical violence, air, road or rail disaster, natural disaster or wartime combat, which is outside the range of normal experience. The term is usually reserved for those more extreme events that can cause stress. Though many would see bereavement of a close relative or being diagnosed with a life-threatening illness as traumatic, they are not usually associated with PTSD, but may lead to stress in the usual way. However, being witness to the death of a worker in the workplace would usually be seen as sudden and severe enough to be classed as a traumatic event. PTSD is characterised by depression, anxiety, hyper-vigilance, emotional numbing, unexplained physical symptoms (sweating, shaking and stomach pain), 'flashbacks', recurrent nightmares and avoidance of reminders of the event. When a person perceives themselves or others to be in danger they produce hormones, including adrenaline. Adrenaline helps to deaden the senses and dull pain. Studies have shown that people with PTSD continue to produce high levels of adrenaline even when the danger has passed, which may account for their numbed emotions and feeling of detachment. When high levels of adrenaline are produced the hippocampus, the part of the brain responsible for memory and emotions, can stop working properly. This can cause flashbacks and nightmares to be continually repeated.

Causes of work-related mental ill-health

Stress in workers can be caused by a range of issues that relate to the organisation they work in and issues that are external to the organisation but influence the worker while they are at work, for example, personal relationship problems at home. What may cause one worker to become stressed may not have the same effect on others. Work pressure may be a normal part of the work conducted. Pressure perceived as acceptable to one worker may help to retain focus and motivation; however, when that pressure becomes excessive or unmanageable it can lead to stress. The causes of stress in the workplace may be grouped under two categories: causes related to the work content and causes related to the work context. The causes relating to work context include job content, workload and pace, participation and control.

Some causes are very specific and predictable, such as during times of change, for example, when mergers or redundancies are imminent. Others are more general and may exist over a longer period of time, such as poor working relationships, long working hours, boredom, poor workstation design, harassment and bullying.

ORGANISATION OF WORK

Working hours

Working excessive hours, typically 60 or more hours per week, can be a significant stressor. The results of a joint study conducted by the International Stress Management Association UK and Royal and Sun Alliance concluded that there are four main causes of stress at work:

- *Long working hours* - typical for many people at work at present.
- *Poor work/life balance* - family/private life suffers due to long working hours, work tasks taken home.
- *Excessive workload/deadline pressures* - which may be caused by 'down-sizing' or unrealistic targets being set.

In addition:

- The working hours may not be long but may be considered to be anti-social, in that they occur outside the general part of the day - this can be a factor in retail work.
- The hours worked may be *unpredictable*, with employees having to stay on at work for rush orders or to cover for a co-worker, which can lead to disruption of plans and family arrangements.
- *Changes* in working hours may be a stressor, particularly when hours are cut along with pay, leading to money worries. The change in working hours may involve changes in the days worked, possibly involving Saturday and Sunday, which involves disruption to weekend activities.
- *Shift work*, while acceptable to some, is a stressor to many and they suffer from problems with body clock adjustment. This may be caused by constant night shift work or regular shift changes. In some cases, shifts are split over the day causing the worker to have to work on two separate occasions in one day. This can be a factor in the hotel and catering industry.

Workplace culture

A workplace that is characterised by poor organisational factors is likely to generate unacceptable levels of stress in employees. Important issues include:

- *Communication* - poor communication between the different functions of an organisation, no information about the plans of the organisation inappropriate communication allowing rumours being allowed to circulate and too much communication, such as being copied in to every e-mail and spending time trying to sort out the relevant parts.
- *Organisational structure* - stress can be created by insufficient staff, vacant posts, excess staff resulting in spare time and boredom, lack of variety of tasks, lack of responsibility, inconsistency in approach by employer/manager, shift work/piece work/bonus systems, emphasis on competitiveness.

- **Resources** - money not made available for training, new equipment and/or suitable equipment, not enough time may be given to complete tasks safely.
- **Support** - lack of support from subordinates/peers/employer, line managers failing to support workers' genuine grievances, failing to support workers with health issues or family issues.

JOB

Working environment

Poor physical working conditions and inadequate work equipment are common causes of stress.

In the case of working conditions, the following are common stressors:

- Lack of **space**/privacy.
- Excessive **noise/vibration**.
- Unreasonable **temperature**.
- Inadequate **lighting/ventilation**.
- Dirty and untidy workplaces.
- Hazardous working conditions.
- Poor welfare facilities.

In the case of work equipment, the following are common stressors:

- Poor working order.
- Design of equipment hinders the work.
- Uncomfortable or difficult to use.
- Work equipment unsuitable for location/task/operator.
- Prone to breakdowns.
- Associated with production of noise/fumes/vibration.

Job content

- **Work load** - the work load may increase beyond the worker's capabilities, perhaps because of a co-worker being absent, unfilled job roles or the management failing to manage a non-productive member of the team. In addition, work orders may increase, but new staff may not be employed because of the perceived economic climate.
- **Time pressures** - targets may be unreasonable and the worker may have to stay longer at work to achieve them. The shortage of time can lead to workers compromising their values, cutting corners and not completing the task satisfactorily or safely.
- **Boredom** - the job may require basic skills that are far below the capabilities of the worker. It may be repetitive and give no job satisfaction at all. There may be too much time available, because there are few orders and the person starts to feel unnecessary to the company.

Job role

Stress can occur when job roles are not thought through by management.

- **Clarity** - the workers may be uncertain about what is expected of them. They may feel that they have to wait to see what sort of job they will be given at any one time, because they don't have a job description. There may be no feedback from managers, so it is unclear if the person is achieving what is required.
- **Conflict of interests** - this can occur, for example if a production manager is given responsibility for health and safety in their department and they are very much unsupported in the role. This could lead to them trying to enforce safety rules while trying to increase production.
- **Lack of control** - the work may not be supervised and each worker tends to work in their own way. They do not know what the rules are until they inadvertently break one.

INDIVIDUAL

Relationships

Stress is frequently associated with work and social relationships. At work, the following stressors may cause problems:

- **Bullying and harassment** - A worker may feel that they are being 'picked on' by their line manager and that no matter what they do, it is wrong.
- Isolation from or rejection by colleagues - also seen as bullying, even though it may involve being ignored rather than anything being said.
- Personality conflicts - This can occur when a person does not get on with a co-worker or a line manager. They may feel it makes the work situation too unpleasant to cope with.
- Racial/sexual harassment - this may be direct racial or sexual harassment, with constant nasty comments, or it may be considered by the perpetrator as 'banter'. It can make working life a misery for the recipient.
- **Verbal/physical abuse** - This can involve swearing, threats and actual physical contact such as slaps and punches or the use of a weapon. It may happen between co-workers, but is more likely during contact with members of the public.

Stressors that cause common mental health problems

There are many factors that contribute to stress within the workplace such as job insecurity, poor relationship with managers and peers, bullying and harassment, work overload or under load.

However, there may be other external factors that have a direct influence on whether a worker experiences stress, such as caring for children, elderly relatives and financial worries.

The UK's Health and Safety Executive (HSE) identify *six* primary sources of stress at work. These are:

1) *Demands* - this includes issues such as workload, work patterns and the work environment.

2) *Control* - how much say the person has in the way they do their work.

3) *Support* - this includes the encouragement, sponsorship and resources provided by the organisation, line management and colleagues.

4) *Relationships* - this includes promoting positive working to avoid conflict and dealing with unacceptable behaviour.

5) *Role* - whether people understand their role within the organisation and whether the organisation ensures they do not have conflicting roles.

6) *Change* - how organisational change (large or small) is managed and communicated in the organisation.

However, often there is no single cause of work-related stress. Although work-related stress can be triggered by sudden, unexpected pressures, it is often the result of a combination of stressful factors that build up over time.

Sometimes people may suffer from stress that is caused by external issues rather than work-related issues or the work-related issues add to already well established external stressors, which leads to stress.

Common external causes of stress include:

- *Bereavement of a relative or friend* - said to be one of the greatest stressors.
- *Relationship/marital problems* - divorce is also a high stressor, even for the individual who wanted the divorce. Marital and relationship problems may leave the individual emotional and vulnerable at work.
- *Medical conditions* - again, these are worries that the individual carries around with them. The worries are exacerbated if the individual is worried that they may not be able to carry on working, which would leave them isolated and with money problems.
- *Commuting* - it is a modern problem that people travel long distances to get to work, traffic hold-ups may mean leaving the house at very early in the morning to avoid them. Being stuck in traffic is a stressor, but twice a day every day might be more than the individual can cope with. Those who travel into cities by train have to contend with packed trains and train cancellations.
- *Childcare issues* - arranging for childcare that a parent feels confident in is difficult, but care becomes a greater problem if the child is ill. This can also lead to conflict between the parents as to who should stay at home with the sick child.
- *Caring for elderly relatives* - including placing them into a care or residential home.
- *Relocation* - as companies are trying to save money by closing down certain sites, workers are left with a dilemma, whether to accept they are out of work or relocate and take another job. This is a very stressful situation as the other site may be some distance away and require the individual to either stay away from home all week or move the family to a new location. Moving could mean that children have to change schools and the family will move away from friends and relatives, some of whom usually help with the childcare.

Trying to leave personal worries at home rather than taking them to work is not possible for most people.

B8.2 - Identification and control of work-related mental ill-health

How people with mental health problems can be supported at work by employers

Mental health problems may cause fatigue, impaired cognitive ability and lead to poor concentration. Despite this most people can continue to work effectively and there is strong evidence that work is beneficial for their health and well-being. It may be necessary to provide the worker with enough time and rest periods in order to complete work, but the work should be done to a good standard. In the same way, in taking account of cognitive ability and concentration the worker may be provided with aids that prompt the right action or a co-worker that can assist when needed.

When considering mental health, account must be taken of any factors that may have contributed to a worker's mental ill-health and/or emotional distress, either from the worker's personal circumstances and/or the workplace, for example, work overload and poor co-worker relations. Workers may perceive that it is their workplace that has contributed to their poor mental status. Therefore a full assessment of the nature of their work and workplace may be required, leading to some form of temporary adjustment to support and enable the worker to remain in work.

> *"Work may be stressful and potentially psychologically detrimental to people with mental health problems. However, the evidence broadly shows that work is therapeutic for people with mental health problems (as for any other form of disability) in terms of symptom management, self-esteem, and self-identity, 'normalisation' of activities and participation, improved social functioning and quality of life".*

Figure B8-3: Work is therapeutic. *Source: Gordon Waddell, Kim Burton "Concepts of rehabilitation for the management of common health problems".*

PRINCIPLES OF THE GOOD MANAGEMENT OF AN INDIVIDUAL WITH MENTAL ILL-HEALTH IN THE WORKPLACE

The role of managers in supporting individuals with mental ill-health at work should not be underestimated. It is important that they are trained and equipped with the necessary skills and knowledge to enable them to effectively provide support. Employers have a duty under health and safety legislation to protect the health of their employees, including mental health. Mental ill-health is widespread within the workforce and the way they are managed is an important responsibility for managers. An organisation with a good reputation for supporting its workers when they are experiencing personal difficulties may offer an advantage over competitors; skilled staff will not be lost and staff morale may be enhanced when employees know that they will be sympathetically supported should they themselves experience any mental health or other problems. Managers need to recognise that physical and mental health can be interrelated and the worker may be absent from work with a physical problem, whereas the underlying cause may be psychological and vice versa. The worker's general practitioner (GP) doctor has a key role, but the management of workers with a mental health problem, is not a matter for healthcare professionals alone. There is strong evidence that medical treatment by itself has little impact on work outcomes, which emphasises the importance of the role of managers. Proactive organisational approaches to sickness, together with the temporary provision of modified work are effective and cost effective, particularly in large organisations. Early intervention is central to supporting workers to stay in work, preventing long term incapacity and early rehabilitation. The longer a person is absent from work the more difficult it becomes to return them to full employment. Mental ill-health is not a barrier to work and rehabilitation is about helping workers with health problems stay at, return to and remain in work. This should start with effective interventions, within the first six weeks, as most people with common mental health problems can be helped to return to work by following a few basic principles. These can be in the form of:

- Temporary modifications to work.
- Supportive environment.
- Directing them to appropriate help and considering the need for the provision of external resources, such as counselling, that may not be easy to access in a timely fashion via the NHS.

The HSE/Shift guidance document 'Line Managers' Resource. 'A practical guide to managing and supporting people with mental health problems in the workplace' suggests that the approach to managing mental health problems should be to:

- Focus on mental well-being. A holistic approach to promoting the mental and physical well-being is most effective and beneficial. By presenting the issue of mental health in terms of well-being it is more likely to overcome barriers around stigma and to achieve involvement and commitment. *Figure ref B8-5* suggests ideas of a holistic approach.
- Engage with people. Over-emphasis on definitions and diagnoses may prove unhelpful, as a diagnostic 'label' can lead to preconceptions of what a person can or cannot do. The better approach, as with other health issues, would be to talk to the person and get a clear understanding of what they can do, rather than what they cannot do, and work on the basis of the person's capabilities.

The main point behind *figure ref B8-5* is that mental health needs to be approached at different levels, corporate for policies and local interactions for the employee with the line manager.

> *The way forward is to bring mental well-being within the boundaries of normal working life , rather than focusing on it as out of the ordinary and thereby something 'different' or stigmatised.*

Figure B8-4: Approach to managing ill-health.   *Source: HSE/Shift Line Managers' Resource.*

SOURCES OF EXTERNAL SUPPORT AVAILABLE

The extent to which external agencies will be involved in the management and support of individuals with mental ill-health will depend on the reason they are required. Occupational health specialists will be able to provide independent advice on the impact of the worker's mental ill-health on their employment and can advise management on the intervention and support that is required to enable a return to the workplace.

In addition there are a number of organisations such as *Mind* and *Mindful Employer*® who can provide information and support to managers on how to support an employee with mental health problems.

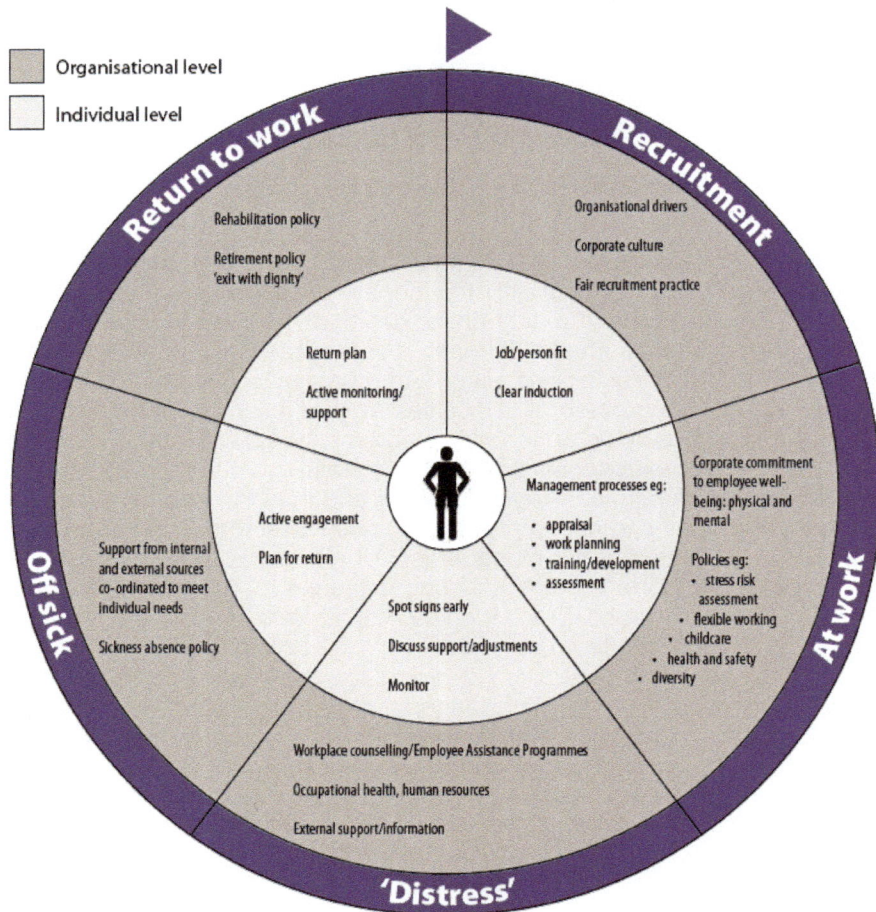

Figure B8-5: A holistic approach to managing an individual with mental ill-health. *Source: HSE/Shift "Line Managers' Resource".*

Identification and assessment of work-related mental ill-health at individual and organisational level

Identifying the main individual and organisational sources of work-related stress and post-traumatic stress disorders is the most effective first step in stress management. This provides the basis for targeting intervention strategies at an individual and organisational level.

USING QUALITATIVE AND QUANTITATIVE DATA TO IDENTIFY STRESS

The main technique used to help identify an actual or potential stress related problem is risk assessment, during which a combination of qualitative and quantitative data is used. Risk assessments can be effectively used to determine the extent and sources of workplace stress. However, although some sources of stress may be easy to identify such as increased working hours and traumatic working conditions during prolonged emergency incidents, others may be more difficult to identify, such as bullying.

Most risk assessments utilise *questionnaires*, which cover work activities, health/well-being and perceived sources of stress. The process must have guaranteed confidentiality or inaccurate/dishonest responses may result. The length of questionnaires varies greatly, but importantly must not be too long that there is an unwillingness to complete it, nor too short that it fails to generate useful data. Online questionnaires have the advantage of ease of data collection and analysis, but must be secure to prevent unauthorised access.

Other qualitative and quantitative data that is frequently used includes that listed below.

Quantitative methods include the use of:

- *Sickness/absence data* such as increased absenteeism-especially short-term.
- Accident data including increasing frequency or significant patterns in accidents.
- Productivity data such as reduced performance, quality and efficiency.
- Staff turnover records analysed for reasons for departure.
- Staff *surveys* to ascertain attitudes, concerns, etc.

Qualitative methods include the use of:

- Informal talks with staff sometimes using independent third-parties to '*interview*'.
- Staff performance appraisals giving staff opportunities to raise issues.
- Focus groups to discuss and raise concerns in a group environment.
- Return to work interviews to ascertain possible links between stress and absence from work.

RETURN TO WORK POLICY

It is also important to enable an employee to return to work following a period of absence through a stress related condition. The worker may be particularly vulnerable on returning to the work environment and so will need support from the employer. This support and the return to work process should be laid down in a return to work policy. The policy may cover a phased return to work so the employee can get used to coming into work and learn coping strategies for work demands etc. The policy should also outline the support that can be provided by occupational health services.

ASSESSMENT OF RISK OF STRESS

It is important that the data used to identify the sources, patterns of incidents and perceived hazard are evaluated and the likelihood of stress arising considered. The assessment should consider what controls are in place and try to determine the general level of risk, at risk activities, at risk groups of people and at risk individuals.

Practical control measures to reduce and manage work-related stress

Reducing the level of occupational stress is by any standards a major undertaking if attempted in isolation. An organisation that is forward thinking and progressive will have in place systems to apply and monitor health and safety along with policies and procedures to cover all aspects of staff development and the correct implementation of appropriate regulations.

DEMAND

Control measures to meet this include:

- Balancing the demands of the work to the agreed hours of work, consider shift working, the amount of additional hours worked and unsocial hours.
- Provision of regular and suitable breaks from work and rest periods.
- Matching worker skills and abilities to the job demands.
- Designing jobs so they are within the capability of workers.
- Minimising the work environment risks, such as noise and temperature.

CONTROL

Control measures to meet this include:

- Providing, where possible, workers with control over their pace and manner of work, consider reducing the effects of repetitive and monotonous work by job rotation.
- Encouraging workers to use their skills and initiative to do the work.
- Encouraging workers to develop to enable them to do more challenging or new work.
- Providing workers with opportunity to influence when breaks are taken.
- Consulting workers regarding work patterns.

SUPPORT

Control measures to meet this include:

- Establishing policies and procedures that provide support, particularly where workers may feel other factors are putting them under pressure.
- Provide systems that enable and encourage managers to identify where workers need support, consider where workers deal with the public in demanding environments, where new workers are introduced and times of high demand.
- Provide systems that enable and encourage managers to provide support to workers, consider particularly those working remotely by virtue of their location or time of working and new workers.
- Encourage co-workers to support each other.
- Ensure workers understand what resources and support is available and how they access it.
- Provide regular constructive feedback to workers.

WORK RELATIONSHIPS

Control measures to meet this include:

- Promote positive behaviour that avoids conflict and leads to fairness; consider co-workers, customers and suppliers.
- Establish policies and procedures that resolve unacceptable behaviour that leads to conflict.
- Encourage managers to deal with unacceptable behaviour, such as harassment, discrimination or bullying.
- Encourage workers to report unacceptable behaviour.
- Establish systems that ensure communication with managers and workers, consider timeliness of communication to those that work isolated by location or time.
- Establish systems that ensure involvement and consultation, such as regarding the process of conducting risk assessments.

ROLE

Control measures to meet this include:

- Ensure workers have the knowledge, skill and experience to conduct their role or are being supported appropriately; consider workers undertaking new or difficult work.
- Ensure role requirements are compatible, for example, that the need to manage costs does not conflict with health and safety.
- Ensure role requirements are clear.
- Ensure workers and their managers understand the roles and responsibilities.
- Provide systems to enable workers to raise concerns about role uncertainty or conflict, particularly consider the work life balance for those workers that provide care for others outside their work.

CHANGE

Control measures to meet this include:

- Provide workers with timely information to help them understand the change, reasons for it and timing of effects.
- Ensure worker consultation on proposed changes.
- Provide workers with information on likely impacts of change on their jobs.
- Provide training and support through the period of change.

In addition to the six specific factors that relate to the HSE management standards for stress at work there are a number of general control measures that should form part of a stress reduction and management strategy. These include:

- Introducing a stress policy to demonstrate to employees, trades unions and the HSE that the organisation recognises stress as a serious issue worthy of a commitment to manage the problem. This should include or link to a *return to work policy*.
- Promoting general health awareness initiatives within the organisation such as diet, exercise and fitness programmes.
- Addressing the issue of work-life balance which may include the consideration of job-share, part-time work, voluntary reduced hours, home-working, flexi-time etc.
- Providing training and support for employees, supervisors and all levels of management in the form of stress awareness and stress management training, as appropriate.
- Providing access to occupational health practitioners, *counselling* support or employee assistance programmes.

The HSE stress management standards

The Health and Safety Executive (HSE) Management Standards for stress define the characteristics or culture of an organisation where stress is being managed effectively. As far as complying with the Standards is concerned, the HSE is trying the voluntary approach first, but will enforce against failures to assess the risks of stress if necessary.

The HSE Management Standards for stress are based upon the six main stress factors of demands, control, change, relationships, role and support. Each standard defines a desired state (best practice) to be achieved in several areas.

DEMAND

Demand includes issues like workload, work patterns and the work environment. The standard is:

'Employees indicate they can cope with the demands of the job and there are systems in place locally to respond to any individual concerns'.

CONTROL

Control includes how much say the person has in the way they do their work. The standard is:

'Employees indicate that they are able to have a say about the way they do their work and there are systems in place locally to respond to any individual concerns'.

SUPPORT

Support includes the encouragement, sponsorship and resources provided by the organisation, line management and colleagues. The standard is:

'Employees indicate that they receive adequate information and support from their colleagues and superiors and there are systems in place locally to respond to any individual concerns'.

WORK RELATIONSHIPS

Relationships includes promoting positive working to avoid conflict and dealing with unacceptable behaviour.

Improving work relationships and attempting to modify people's attitudes and behaviour is a difficult and time-consuming process. Effective strategies include regular communication with staff, provision of accurate and honest information on the effect of organisational changes on them, adopting partnership approaches to problems, provision of support. The onus is on employers to promote a culture that respects the dignity of others - if it is left to employees to do this it will not happen. The standard is:

'Employees feel able to indicate that they are not subjected to unacceptable behaviours, for example, bullying at work and there are systems in place locally to respond to any individual concerns'.

ROLE

Role includes whether people understand their role within the organisation and whether the organisation ensures that the person does not have conflicting roles. The standard is:

'Employees indicate that they understand their role and responsibilities and there are systems in place locally to respond to any individual concerns'.

CHANGE

Change includes how organisational change (large and small) is managed and communicated in the organisation. The standard is:

'Employees indicate that the organisation engages them frequently when undergoing organisational change and there are systems in place locally to respond to any individual concerns'.

Legal requirements for employers to manage work-related stress

Employer's legal obligations are outlined in the following:

- Health and Safety at Work etc. Act (HASAWA) 1974.
- The Management of Health and Safety at Work Regulations (MHSWR) 1999.
- The Working Time Regulations (WTR) 1998.
- Employers' liability under Common Law.
- Equality Act (EA) 2010.

HEALTH AND SAFETY AT WORK ETC ACT (HASAWA) 1974

Under HASAWA 1974 Section 2, the statement: *"It is the duty of every employer to ensure, so far as is reasonably practicable, the health, safety and welfare of all employees"*, applies to an employee's mental as well as physical health. The employer has a duty to prevent or control the risk of physical or mental harm caused by exposure to excessive pressure at work.

Under HASAWA 1974 Section 7, the statement: *"All employees shall take reasonable care of themselves and others who might be affected by their acts or omissions"* suggests that employees should perhaps take steps to inform the employer of stress affecting their health (taking reasonable care of themselves). This section also implies that employees should not act in ways that may cause stress for others, for example, through bullying (that is, **not** taking reasonable care of others).

THE MANAGEMENT OF HEALTH AND SAFETY AT WORK REGULATIONS (MHSWR) 1999

MHSWR 1999 Regulations 3 and 4 require employers to make suitable and sufficient assessments of risks to the health and safety of employees who may be affected by their work activities. Control measures can then be based upon such assessments. Therefore, stress should be managed using the methods employed for other hazards at work:

- Identify the hazards.
- Assess who may be harmed and how.
- Evaluate the level of risk and decide if current precautions are adequate or whether more should be done.
- Record the findings.
- Review and revise the assessment as necessary.

MHSWR 1999 Regulation 13 requires every employer to take employees' capabilities into account when allocating tasks. Therefore, an assessment of the capabilities of an employee should be made prior to allocating tasks to the individual. The physical and mental capabilities of the individual should be considered along with their experience, training and knowledge.

MHSWR 1999 Regulation 14 requires employees to inform their employer of shortcomings in the employer's health and safety arrangements. This requirement implies that employees should inform their employer of the stress that they are suffering or otherwise make it obvious that there is impending harm to their health due to work-related stress.

THE WORKING TIME REGULATIONS (WTR) 1998

The Working Time Regulations (WTR) 1998 directs the management of working hours and breaks from work that workers receive. As such, this can influence one of the main contributing factors to the demands put on workers that can lead to stress. The main principles of the Working Time Regulations (WTR) 1998 are:

- Workers cannot be forced to work more than 48 hours a week on average.
- Employees can choose to 'opt out' of the 48 hour week, and employers are required to keep a record of all employees who have opted out.
- There must be regular **rest breaks.**

- Where the adult worker works more than six hours the rest break must be not less than 20 minutes.
- Where a young worker works more than four and a half hours the rest break must be at least 30 minutes.
- **Daily rest** must be provided.
  - An adult worker is entitled to a daily rest period of not less than eleven hours in each 24 hour period.
  - A young worker is entitled to a daily rest period of not less than twelve hours in each 24 hour period.
  - The minimum rest period may be interrupted where the pattern of work is broken up into short periods over the 24 hours.
- **Weekly rest** periods must be provided (conditions not explained here apply to this provision).
  - An adult worker is entitled to an uninterrupted rest period of not less than 24 hours in each seven day period.
  - A young worker is entitled to an uninterrupted rest period of not less than 48 hours in each seven day period.
- Workers are entitled to a minimum of 4 weeks paid leave per year.
- Night workers have special rules. Night time is defined as between 11pm and 6am. The rules state that **night workers** should not work more than 8 hours in 24 hours on average.
- Employers must offer a health assessment to night workers before they commence night working. This can be in the form of a questionnaire, with a follow up medical assessment if required. The employee does not have to take up this offer.
- There are special rules for the employment of **young people**; they may not work more than 8 hours a day or 40 hours a week, and are entitled to regular work rest breaks.

The working time regulations provide information on what is classed as 'working time', this includes:

- Working lunches, such as business lunches.
- When an employee has to travel as part of their work for example, sales people, repair/maintenance workers.
- Specific job related training (this does not include non-job related training such as night school or day release).
- Time spent working abroad if the employers' main base is in Great Britain.

EMPLOYERS' LIABILITY UNDER COMMON LAW

Under English law, a claimant would usually bring a case for work-related stress under the common law tort of negligence and would be required to prove:

- The employer owed a duty of care.
- That duty of care was breached.
- As a direct result of the breach of duty, harm was caused.

In addition, there is the question of 'foreseeability'. Foreseeability will depend upon what an employer knows or ought reasonably to know about the individual employee. An employer can assume that an employee can cope with the normal pressures of a job unless the employer knows of some particular problem or vulnerability.

In the landmark case, **Lancaster v Birmingham City Council (1999),** liability was admitted when an employee had taken three periods of sick leave and had then been medically retired. Birmingham City Council conceded that it failed to act upon complaints from Mrs Lancaster and had failed to give her adequate training and guidance to do her job, which had changed from a technical role in a backroom to a pressurised job on a front desk dealing directly with housing issues.

Previous periods of ill-health leave meant that it was foreseeable that failure of Birmingham City Council to act would result in injury.

Sutherland v Hatton and others [2002] EWCA Civ 76 [common law negligence and reasonableness in relation to harm from stress at work]

Facts - The Court of Appeal heard four appeals by employers against compensation awards to employees who had suffered stress-induced psychiatric illness.

Decision - Three of the appeals were allowed. The Court ruled that the general principle was that employers should not have to pay compensation for stress-induced illness unless such illness was reasonably foreseeable. Employers are normally entitled to assume that employees can withstand the normal pressures of a job.

The Court set out a number of practical propositions for future claims concerning workplace stress. These are as follows:

- Employers do not have a duty to make searching inquiries about employees' mental health. They are entitled to take what they are told by employees at face value unless they have good reason to disbelieve the employees' statements.
- Where an employee wishes to remain in a stressful job and the only alternative is demotion or dismissal, the employer is not in breach of duty in allowing the employee to continue.

- Indications of impending harm to health at work must be clear enough to show an employer that action should be taken, in order for a duty on an employer to take action to arise.
- An employer is in breach of duty where he fails to take reasonable steps bearing in mind the following: the size of the risk; the gravity of the harm; the cost of preventing the harm; any justification for taking the risk.
- No type of work may be regarded as intrinsically dangerous to mental health.
- Employers, who offer confidential counselling advice services, with access to treatment, are unlikely to be found in breach of their duty of care in relation to workplace stress.
- Employees must show that their illness has been caused by a breach of duty and not merely by occupational stress.
- The amount of compensation will be reduced to take account of pre-existing conditions or the chance that the employee would have become ill in any event.

The Court of Appeal dealt with the following cases:

1) Penelope Hatton, a schoolteacher who had been awarded £90,000 compensation for depression and debility. Her employer's appeal **was allowed** on the grounds that her workload was no greater than her colleagues' and her absences could be put down to reasons other than workplace stress.

2) Olwen Jones, a local authority employee who had suffered from depression and anxiety as a result of overwork. It was foreseeable that her workplace conditions would cause harm, therefore the employer's appeal against an award of £150,000 damages **was dismissed**.

3) Leon Barber, a teacher, developed symptoms of depression. He was awarded £100,000 compensation. The employer's appeal **was allowed** on the grounds that the claimant had not told the employers about his illness until he suffered a breakdown.

4) Melvyn Bishop, a factory worker awarded £7,000 compensation following a nervous breakdown and attempted suicide. The employer's appeal **was allowed** because the Court ruled that the demands of his work had not been excessive.

Source: Croner's Health and Safety Case Law 2003.

The Court of Appeal stated that an employer who offers a confidential advice service, with referral to appropriate counselling and treatment services, is unlikely to be found in breach of duty.

This may be acceptable under the tort of negligence in that a court would consider the employer to have taken reasonable steps to discharge the duty of care. *See also - Court of Appeal's decision in Intel Corporation (UK) Ltd v. Daw [2007] EWCA Civ. 90; [2007] IRLR 355 later in this element.*

However, in order to discharge statutory duties, an employer would be expected to proactively identify causes of work-related stress, undertake risk assessments and implement measures, so far as reasonably practicable, to prevent or control the risk of physical or mental harm. In other words, the duty of care under civil law can be discharged more easily than the criminal law duty of care contained in Section 2 HASAWA 1974 and Regulation 3 MHSWR 1999.

alker v Northumberland County Council [1995] IRLR 35 [employers' duty of care in relation to mental ill-health arising from excessive workload]

Facts - Mr Walker was employed as a social worker dealing with cases of child abuse. His workload increased steadily over the years and in 1986 he had a nervous breakdown. When he recovered and returned to work, he was promised additional resources to help him with his workload, but they failed to materialise. He had a second breakdown six months later and had to retire. Mr Walker sued the Northumberland County Council claiming they were in breach of their duty of care to provide a safe working environment.

Decision - The council were not held liable for the first breakdown as they could not reasonably have foreseen Mr Walker was exposed to a significant risk of mental illness through his job. They were, however, liable for the second breakdown, given that the same circumstances were there that caused the first. (After the first breakdown, the council had notice of the particular risk facing the claimant and could have taken steps to reduce the stress, by reducing his workload and providing greater assistance). The second breakdown was a reasonably foreseeable risk. The court accepted that this could have caused some disruption to other services provided by the council, but this did not outweigh the obligation to protect the claimant against a serious risk to his health. The council were found to have failed in their duty of care by not providing effective support to alleviate Mr Walker's suffering. The decision is indistinguishable, but what matters is the view that an employer can be under a duty of care to provide an employee with assistance, of uncertain scope and duration, to enable him to perform his contractual duties.

House of Lords decision in Barber v Somerset County Council [2004] UKHL 13 [employers' liability in damages for the mental breakdown; duty owed when problem known or should have been known]

Facts - One of the co-joined cases heard by the Court of Appeal in February 2002 *(see Sutherland v Hatton and others [2002] EWCA Civ 76 earlier)* was taken to the House of Lords in April 2004. Maths teacher Leon Barber was the only one to appeal the decision to the House of Lords. The council employed Mr Barber, a 52 year old schoolteacher, as head of mathematics in a comprehensive school; he worked long hours about which he complained of 'work overload'. Following a period of sickness, because he was 'overstressed and suffered from depression', he suffered a mental breakdown at school.

Decision - On a majority of four to one, the appeal was allowed and he was awarded damages of almost £37,000. The decision was based upon the following important facts:

- The employer should have taken action after Barber separately informed each member of the management team of the pressure he was experiencing at work.
- Barber was treated unsympathetically after a three week absence.
- No attempt was made to reduce his workload upon return to work and he subsequently suffered a nervous breakdown.
- Lack of action by the employer breached the duty of care requirement.
- Barber's ill-health was foreseeable.

The school owed Mr Barber a duty of care, and their breach of that caused the claimant's nervous breakdown. The employer's duty to take some action arose when Mr Barber informed separately each member of the school's senior management team of his problems. However, nothing was done to help him. The senior management team should have made inquiries about Mr Barber's problems and seen what they could have done to ease them, instead of brushing him off unsympathetically or sympathising, but simply telling him to prioritise his work. (Stokes v Guest, Keen and Nettlefold (Bolts and Nuts) Ltd [1968] applied).

Intel Corporation (UK) Ltd v. Daw [2007] EWCA Civ. 90; [2007] IRLR 355 [an employee who does not resign when stresses at work become excessive, does not necessarily lose the right of action against their employer and providing a counselling service does not necessarily discharge the employer's duty of care]

Facts - Mrs Daw suffered a breakdown in 2001 and had not worked since. She suffered from chronic depression. She claimed negligence against her employer because, she asserted, the demands made on her in her job were totally unreasonable and the possible damage to her health was clear. She had already suffered two episodes of post-natal depression and had been absent from work because of depression. It was in her medical records and she was quite open about it. It was claimed that there were many opportunities to deal with the issues, but the employer did not, other than provide a counselling service.

The employer appealed against the High Court decision that they were liable for personal injury because of work-related stress, on the grounds that they provided a counselling service and to have acted any sooner than they did would have imposed too high a burden on them.

Decision - The appeal by the employer was rejected. The judge said that reference in:

'Hatton' to providing a counselling service did not make it a panacea by which employers could discharge their duty of care in all circumstances.

The management of Intel knew what steps should have been taken, without the counselling service backing it up. Also, they had many opportunities to deal with the situation, but they left it too late. Mrs Daw was awarded in excess of £134,000.

EQUALITY ACT (EA) 2010

An employee suffering from a stress related illness may be protected by the EA 2010.

The Equality Act (EA) 2010 aims to protect disabled people and prevent disability discrimination. It provides legal rights for disabled people in the areas of:

- Employment.
- Education.
- Access to goods, services and facilities including larger private clubs and land based transport services.
- Buying and renting land or property.
- Functions of public bodies, for example, the issuing of licences.

The Equality Act (EA) 2010 defines that a person has a disability if:

- They have a physical or mental impairment.
- The impairment has a substantial and long-term adverse effect on their ability to perform normal day-to-day activities.

For the purposes of the Equality Act (EA) 2010, these words have the following meanings:

- 'Substantial' means more than minor or trivial. A severe disfigurement is to be treated as having a substantial adverse effect on the ability of the person concerned to carry out normal day-to-day activities. This includes impairments controlled by measures that mean, but for the measure, the impairment would prevent them conducting day to day activities; this includes medical treatment.
- 'Long-term' means that the effect of the impairment has lasted or is likely to last for at least twelve months (there are special rules covering recurring or fluctuating conditions - treated as continuing if it is likely to recur), or for the rest of the life of the person affected.
- 'Normal day-to-day activities' include everyday things like eating, washing, walking and going shopping.

People who have had a disability in the past that meets this definition are also protected by the Equality Act (EA) 2010.

Most stress cases are associated with mental impairments of one form or another; it is reasonable to deduce that a clinically diagnosed stress condition would be recognised as mental impairment. The impairment also has to be 'long-term', that is, lasting or likely to last 12 months or more. Temporary or short-term stress-related conditions do not attract the protection of the EA 2010. It is also important to note that the impairment must be such that it affects normal day-to-day activities.

Many cases where people experience stress will not come under this classification. Those that do come under the classification of disabled and are afforded protection from direct or indirect discrimination, harassment and victimisation. In addition, the employer must make 'reasonable adjustments' to enable the employee to continue with their existing job. Reasonable adjustments may include:

- Altering the content of the job.
- Shorter or different working hours.
- Rehabilitation following absence from work.
- Additional holidays/unpaid leave.
- Provision of extra support.
- Permitting absence for medical treatment.
- Higher than usual tolerance of absence.

If continuation in the existing job is impossible, redeployment of the employee may need consideration. If an employer is considering dismissing the employee, the dismissal must be approached in the correct way to avoid claims of both unfair dismissal and disability discrimination. Factors which may be taken into account in deciding whether dismissal was justified include the length of absences, any health and safety risks associated with the job, whether the employee is an important worker or not and the scale of any reasonable adjustments already made by the employer.

B8.3 - Scope, effects and causes of work-related violence/aggression

Meaning of work-related violence/aggression

DEFINITION OF 'VIOLENCE'

> *'Any incident, in which a person is abused, threatened or assaulted in circumstances relating to their work'.*

Figure B8-6: Definition of violence.   Source: HSE in their publication "Violence at work: a guide for employers" (INDG 69).

> *The definition includes violence to employees at work by members of the public, whether inside a workplace or elsewhere, when the violence arises out of the employees' work activity. For example, this might include violence to teachers from pupils, to doctors/nurses from patients, to peripatetic employees whose work involves visiting the sick, or collecting payments, to security staff or to officials enforcing legislation. It would not include violence to persons when not at work, for example, when travelling between home and work or violence outside their normal working hours, even though where such risks were significant, employers might wish to take action to safeguard their employees.*

Figure B8-7: Definition of work-related violence (WRV).   Source: HSE INDG 69 (Rev).

Physical violence can involve incidents which:

- Require first aid treatment.
- Require medical assistance.
- Cause injury.

The HSE definition extends this to include incidents which:

- Involve a threat, even if no physical injury results.
- Involve *verbal* abuse.
- Involve non-verbal abuse (for example, stalking).
- Involve other *threatening behaviour*.

> *"Any incident where persons are abused, threatened or assaulted in circumstances related to their work, involving an explicit or implicit challenge to their safety, well being or health". (Wynne et.al.(1996).*

Figure B8-8: Extended definition of work-related violence (WRV).   Source: HSE review of workplace-related violence (CRR143).

This extends the dictionary definition to include verbal abuse and threats as well as physical attack. Physical attacks are comparatively rare; abuse and threats are most common. In addition, definitions are widening their scope from 'workplace' to 'work-related', to include incidents connected with work, but which may not take place in the workplace. This gradual widening of definitions can lead to an increase in subjectivity in deciding what is to be included.

Employees may have a different perspective of what constitutes abuse or threat. The inclusion of incidents involving verbal abuse and behaviour perceived as threatening moves the scope and numbers of potential incidents from the relatively small number of objective incidents involving physical injury to a larger number of subjective incidents involving perceived threat.

There are many views on definitions for work-related violence, but these too are a matter of perspective. The HSE review of workplace related violence reports their view that 'any working definition should be produced locally'.

This perspective being influenced by Lamplugh and Pagan (1996) who noted:

> *"A definition of a problem is not an end in itself but an initial collective effort to raise the awareness of the problem in order to promote effective strategies. A working definition should hence take into account the context and culture of the organisation, and should be developed as a flexible tool for understanding, building commitment, and developing polices, procedures and working practices". Lamplugh and Pagan (1996).*

Figure B8-9: Note on a working definition of work-related violence (WRV).   *Source: HSE review of workplace-related violence (CRR143).*

Organisations and employees working for them will have a wide variety of opinions of what is an acceptable definition and as expressed by Lamplugh and Pagan the creation of an acceptable working definition for any organisation is an excellent opportunity for employee consultation. Through the involvement of employees the policies, procedures and practices, including appropriate reporting, are more likely to be effective.

Physical and psychological effects of violence and aggression

CONSEQUENCES FOR ORGANISATIONS AND INDIVIDUALS

Organisations have become increasingly concerned about violence at work. This can lead to both poor morale and a poor image for the organisation, with the consequent difficulties in recruiting or retaining staff and the effect on confidence in the business and its profitability. It can also be expensive with extra costs relating to:

- Absenteeism.
- Higher insurance premiums.
- Civil compensation payments.

For the individual employee, physical violence can cause pain, distress and even disability or death. Verbal abuse or threats, particularly if they are serious or persistent, can affect the individuals' health through the mental anguish caused by stress or anxiety.

The effect that an incidence of violence has on an individual will depend on many factors, such as the degree of threat and the resilience of the individual.

Physical injuries will often heal before psychological ones do. Victims of violence often withdraw from social interactions, become hyper vigilant or demonstrate a lack of confidence.

Anxiety and depression may follow (which are closely related to stress) as the employee tries to cope with the trauma caused. Often, the feeling of being at risk is enhanced and the employee feels vulnerable.

> *The percentage of people at work who suffer violence varies with profession:*
>
> *Health professionals - 3.8*
>
> *Health and Social Welfare professionals - 2.6*
>
> *Protective Service professionals - 9.0*
>
> *All - 1.4*

Figure B8-10: Violence percentages.   *Source: Report 'Violence at work: Findings from the 2010/2011 BCS'.*

Prevalence/extent of work-related violence/aggression and consequences

Crime Survey for England and Wales

The latest statistics on violence at work were published for 2014/2015 in the report of the Crime Survey conducted by the Home Office and the UK Health and Safety Executive (HSE).

The number of violent incidents at work has declined over the last decade, with the incident rate remaining stable over the last five years.

Findings from the 2014/15 Crime Survey for England and Wales (CSEW) show that:

- The risk of being a victim of actual or threatened violence at work is similar in 2014/15 to the last few years, with an estimated 1.2 per cent of working adults the victims of one or more violent incidents at work.
- In 2014/15 285,000 adults of working age in employment experienced work related violence including threats and physical assault.
- There were an estimated 569,000 incidents of violence at work according to the 2014/15 CSEW, comprising of 308,000 assaults and 261,000 threats. This compares to an estimated 583,000 incidents in 2013/14. This change is not statistically significant.
- The 2014/15 CSEW found that 1.1% of women and 1.3% of men were victims of violence at work once or more during the year prior to their interview.
- Strangers were the offenders in 54% of cases of workplace violence. Among the 46% of incidents where the offender was known, the offenders were most likely to be clients or a member of the public known

through work. The survey found 58% per cent of violence at work resulted in no physical injury, of the remaining 42% of cases, minor bruising or a black eye accounted for the majority of the injuries recorded.

■ RIDDOR reported 4,810 injuries to employees, where the 'kind of accident' was 'physical assault/act of violence' in Great Britain (England, Wales and Scotland). This represents 6.3% of all reported workplace injuries. Of this figure, there were two deaths (RIDDOR, 2014/15).

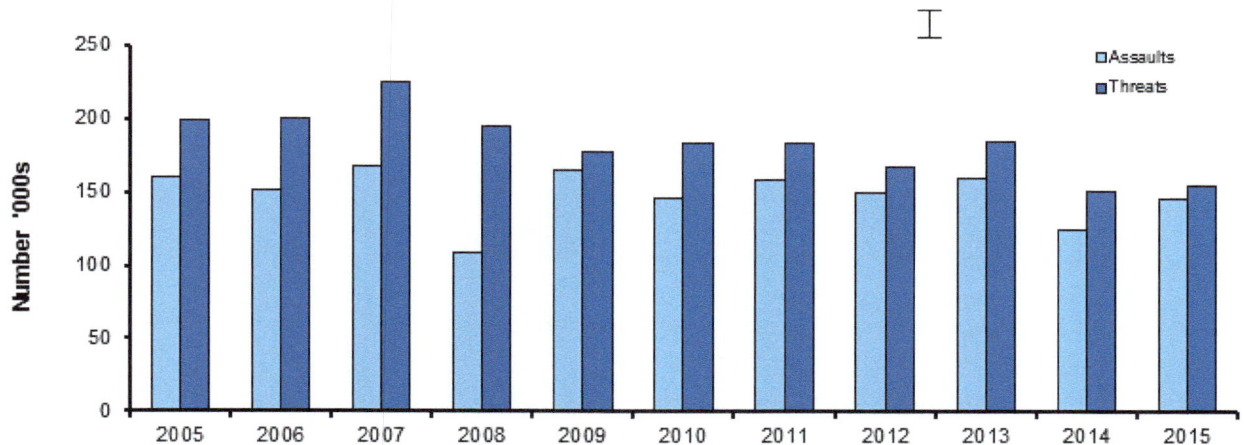

Figure B8-11: Figure 1 Number of incidents of violence at work for adults of working age in employment 2004/05 to 2014/15. *Source: CSEW.*

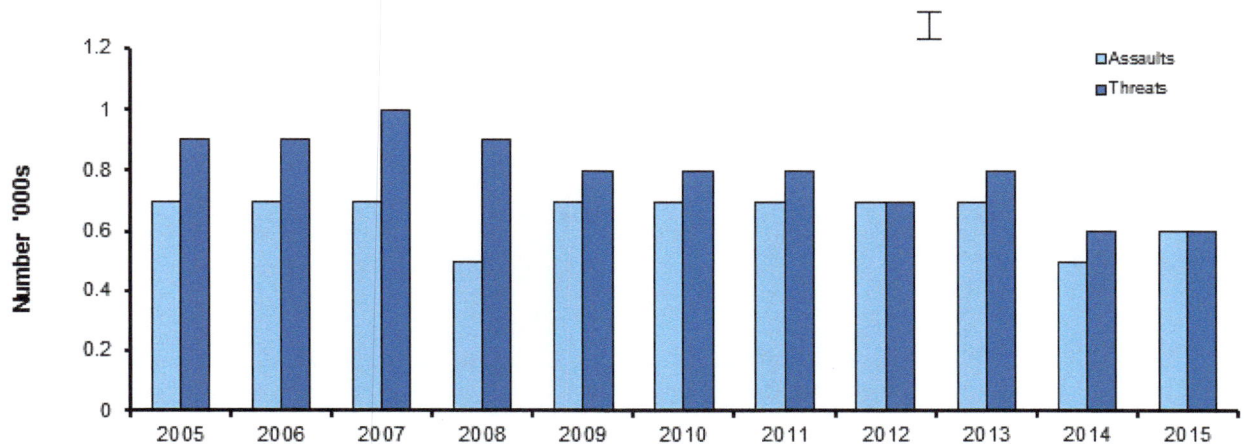

Figure B8-12: Figure 2 Number of victims of violence at work for adults of working age in employment 2004/05 to 2014/15. *Source: CSEW.*

OCCUPATIONAL VARIATIONS

National level estimates of violence at work mask variation in risk among workers with different occupational characteristics. Previous research has shown that not all workers share the same risk of violence at work (Mayhew et al., 1989, Jones et al., 1997, Budd, 1999 and 2001, Upson 2004, Webster et al., 2008, Buckley et al., 2010, Packham, 2011, Buckley 2013).

The CSEW shows that there is large variation in the risks at work across occupational groups.

Table 1 lists the occupational groups most at risk of assaults or threats at work. Overall, respondents in protective service occupations (such as police officers) faced by far the highest risk of assaults and threats while working, at 9.6% - 8 times the average risk of 1.2%.

Additionally, health care professionals and health and social care specialists had higher than average risk at 3.1% and 3.4% respectively. These professions have consistently had higher than average risk rates over the last number of years.

Other professions with higher risk include transport and mobile machine drivers at 3.0%.

Examples of workers least at risk include workers in elementary trades, agriculture plant and storage-related occupations, science and technology professionals and associate professionals and workers in administrative occupations.

REPEAT VICTIMISATION

In 2014/15, of those respondents who were the subject of assault or threat in the workplace, 58% were assaulted or threatened once whilst a further 20% reported being threatened or assaulted twice and a further 21% reported being assaulted or threatened three or more times.

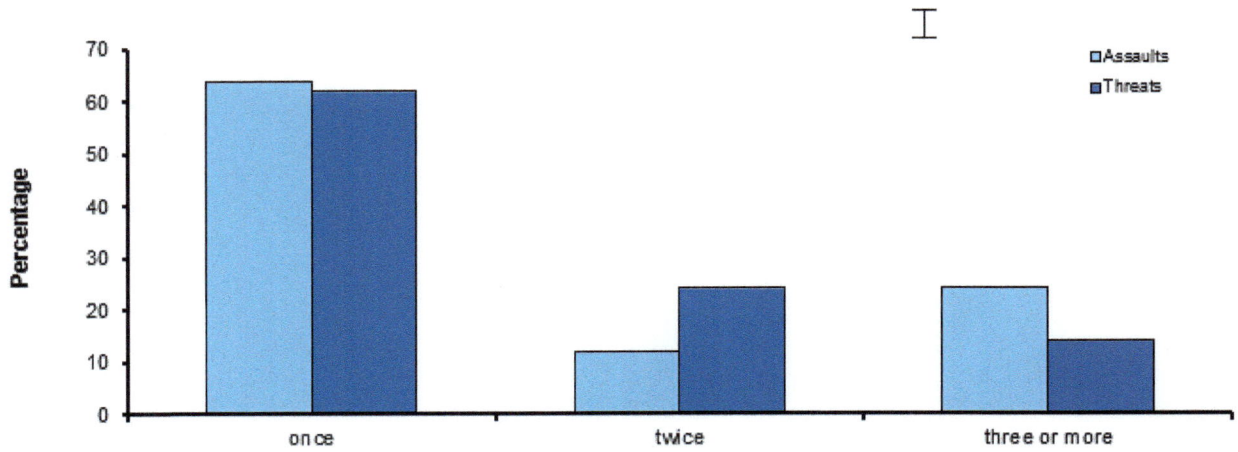

Figure B8-13: Figure 3 Repeat victimisation at work for adults of working age in employment 2013/14.   *Source: CSEW.*

OFFENDER - VICTIM RELATIONSHIP

According to the 2014/15 CSEW, the offender was unknown to the victim in 54% of all work related violence incidents, whilst in 46% of incidents the offender was known to the victim. In cases where the offender was known, they were most likely to be either a client, customer or work colleague.

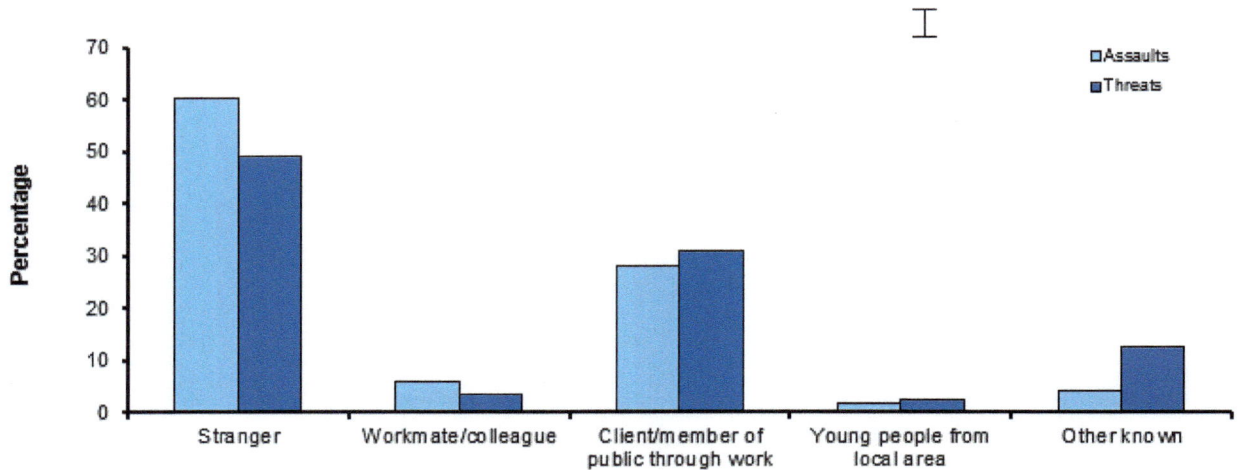

Figure B8-14: Figure 4 Offender - victim relationship for adults of working age in employment 2014/15.   *Source: CSEW.*

CONSEQUENCES OF VIOLENCE BY INJURY TYPE

Experiencing violence at work can have both physical and emotional consequences for victims and worry about workplace violence may impact upon people's health (Chappell and Di Martino (2006). Looking at the physical consequences of assault in the workplace, 58% resulted in no physical injury, whilst 42% resulted in a physical injury. Minor bruising or a black eye was the most common physical injury, at 30%. Severe bruising from heavier trauma was suggested in 13% of physical assaults. Cuts and scratches were estimated at 14% and 9% respectively. Other injuries accounted for 3%. This category includes broken bones, broken nose, broken, lost or chipped teeth, concussion or loss of consciousness, facial or head injuries or other injuries.

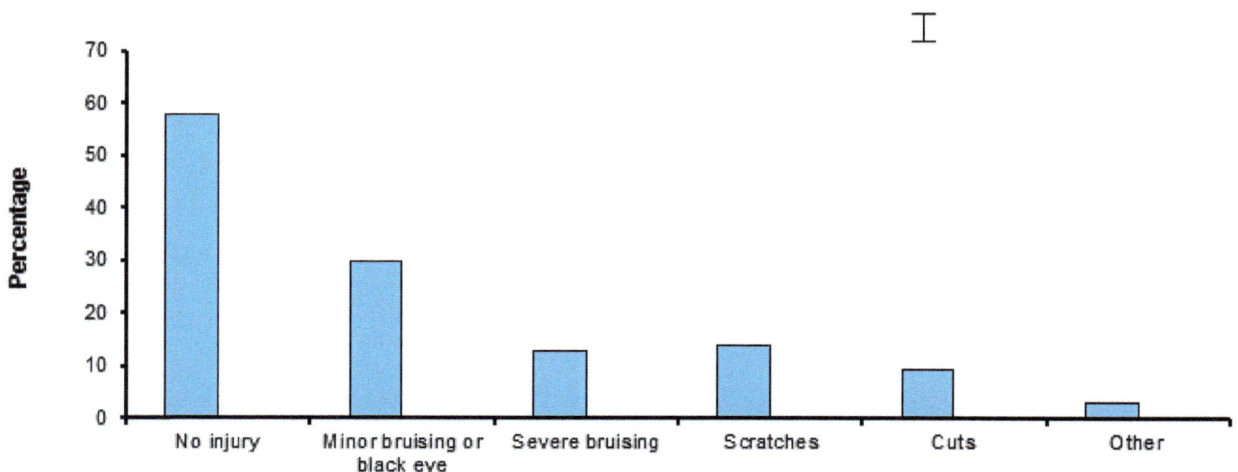

Figure B8-15: Figure 5 Percentage of violence at work incidents by injury type 2014/15.   *Source: CSEW.*

Injuries to workers as notified under The Reporting of Injuries, Diseases and Dangerous Occurrences Regulations 2013

Employers have a legal duty to report certain workplace accidents, required under RIDDOR (The Reporting of Injuries, Diseases and Dangerous Occurrences Regulations 2013). RIDDOR applies where an accident to a worker results in death; in a non-fatal specified injury (typically most bone fractures or amputations); or in over-7-days' off work.

In relation to RIDDOR, an accident is a separate, identifiable, unintended incident, which causes physical injury. This specifically includes acts of non-consensual violence to people at work. Hence 'verbal' assault is excluded, even if it results in time off work. Physical assaults in the workplace that are not work-related are also excluded (for example, an assault over a domestic matter that takes place at work, but is not over a work matter). Suicides and self-harm are excluded.

Non-fatal injuries reported under RIDDOR are known to be substantially under-reported. Currently it is estimated that just under half of all types of non-fatal reports that should be made, are actually reported. There is no separate estimate of whether violence-related RIDDOR incidents are differently reported than all types of injury.

Industry (SIC)	Number of injuries
Human health activities	1,529
Residential care activities	989
Public administration and defence; compulsory social security	909
Education	427
Social work activities without accommodation	249
Security and investigation activities	149
Land transport and transport via pipelines	124
Retail trade, except of motor vehicles and motorcycles	109
Food and beverage service activities	60
Accommodation	43
Other personal service activities	37
Warehousing and support activities for transportation	24
Wholesale trade, except of motor vehicles and motorcycles	20
Office administrative, office support and other business support activities	20
Services to buildings and landscape activities	16
Gambling and betting activities	15
Waste collection, treatment and disposal activities; materials recovery	11
Postal and courier activities	10
Other industries	69
All Industries	**4,810**

Figure B8-16: Table 2 Reported injuries to employees in Great Britain, due to 'act of violence/physical assault', by main industry, 2014/15. *Source: RIDDOR.*

Occupation (SOC)	Number of injuries
Nursing auxiliaries and assistants	828
Nurses	640
Care workers and home carers	535
Welfare and housing associate professionals, not elsewhere classified	423
Prison service officers (below principal officer)	386
Police officers (sergeant and below)	372
Teaching assistants	204
Security guards and related occupations	202
Other occupations	1,220
All occupations	**4,810**

Figure B8-17: Reported injuries to employees in Great Britain, due to 'act of violence/physical assault', by occupation, 2014/15.   *Source: RIDDOR.*

Identification and assessment of risks of work-related violence/aggression

MHSWR 1999 Regulations 3 and 4 require employers to make suitable and sufficient assessments of risks to the health and safety of employees who may be affected by their work activities. Control measures can then be based upon such assessments. Therefore, violence/aggression should be managed using the methods employed for other hazards at work:

- Identify the hazards.
- Assess who may be harmed and how.
- Evaluate the level of risk and decide if current precautions are adequate or whether more should be done.
- Record the findings.
- Review and revise the assessment as necessary.

RISK IDENTIFICATION

Use of staff surveys

To find out if there is a problem then the hazard of violence/aggression in work activities should be identified. This is the first step of the risk assessment process. The management perception may be that violence is not a problem in the workplace or that incidents are rare. A staff survey carried out by a major oil company of forecourt employees showed that they believed that increased customer violence was the most serious threat to their personal health and safety.

A staff survey can be carried out by informal discussion between workers and managers, supervisors or health and safety representatives. A more formal approach would be the use of a questionnaire to find out whether employees felt threatened in their work activities. The results of the survey should be publicised in order to reassure staff that the problem is recognised.

Incident reporting

Incident reporting should be encouraged and detailed records kept of all events - including verbal abuse and threats. It may be useful to record the following information:

- An account of what happened.
- Details of the victim(s), assailant(s) and any witnesses.
- The outcome, including working time lost to both the individual(s) affected and to the organisation as a whole.
- Details of the location of the incident.

There are many reasons why employees may be reluctant to report incidents of aggressive behaviour that make them feel threatened or worried. For example, they may feel that accepting abuse is part of the job. Records should be kept in order to allow the full extent of the problem to be built up. Employees should be both encouraged and expected to report incidents promptly and fully.

Incidents should be classified in order to aid analysis. Classifications could include:

- Place.
- Potential severity.
- Time.
- Who was involved.
- Type of incident.
- Possible causes.

It is important that each incident is examined in order to establish whether there could have been a more serious outcome. Again, a simple classification system could be used to decide on the possible or actual severity of incidents:

- Fatal injury.
- First aid injury (including for emotional shock).
- Counselling.
- Emotional trauma (feeling of being at risk or distressed).
- Major injury.
- Out-patient treatment.
- Absence from work (record number of days).

While it may be easy to classify major injuries, it may be more difficult to define serious or persistent verbal abuse. The classification guidelines should be detailed enough to ensure consistency in classifying all incidents that worry staff.

Useful information can also be gathered from outside the organisation. Trade and professional organisations and trade unions can provide information about patterns of violence linked to certain work situations. Articles in the local, national and technical press might contain details of relevant incidents and potential problem areas.

ASSESSMENT OF RISK OF VIOLENCE/AGGRESSION

Having defined the extent of the problem, it is important to assess the level of risk of violence/aggression and decide what action to take. It is important that the data used to identify the sources, patterns of incidents and perceived hazard are evaluated and the likelihood of violence/aggression arising considered. The assessment should consider what controls are in place and try to determine the general level of risk, at risk activities, at risk groups of people and at risk individuals. This involves deciding who might be harmed and how.

The first task is to identify those employees at risk. For example, people whose job entails personal contact with the public are normally vulnerable. If necessary, potentially violent people should be identified in advance so that the risks from them can be minimised.

Once data has been gathered then analysis can be carried out in order to identify patterns. Common causes such as areas or times can be established. Appropriate control measures can then be targeted where they are needed most. For example, a Trade Union survey found that, after 12 separate shop robberies, each incident occurred between 5 and 7pm. This could be used to improve the security measures for late night opening shops. Risk assessment entails the consideration of existing control measures. If the existing arrangements and precautions already in place are not adequate then more controls will be required. Factors which can be influential include:

■ The level of training and information provided.
■ The work environment.
■ The design of the job.

Factors likely to increase the risk of work related violence

GENERAL ISSUES

Reports on work-related violence consider age, gender and occupation, the most significant of these being occupation. Though there is no statistically significant difference in incidence of violence to men or women there is a significant decrease with increased age, for example, 1.8% for women 16-24 compared with 1.6% for women 25-34. Though other factors than age may influence the incidence it tends to illustrate a higher risk for younger women.

Violence at work can be inflicted by:

■ Clients or customers.
■ Members of the general public.
■ Co-workers and fellow workers (including bullying).

While bullying at work is increasingly recognised as a significant problem, the main emphasis of this unit is on other forms of violence. The great majority of violent incidents occur in situations where the victim is providing (or not as the case may be) a service and the aggressor is the client or customer for that service.

Additional risk factors to consider include:

■ When incidents occur.
■ Where incidents occur.
■ Influence of factors such as alcohol, drugs and mental health.
■ The relationship with the person likely to be violent.

Percentage very/fairly worried	Assaults
Health and social welfare associate professionals	36
Protective service occupations	33
Health professionals	26
Sales occupations	23
Transport and mobile machine drivers and operatives	21
Managers and proprietors in agriculture and services	21
Caring personal service occupations	18
Leisure and other personal service occupations	16
Teaching and research professionals	15
Elementary administration and service occupations	15
All	13
1. Source 2001/02 and 2002/03 BCS	
2. Based on adults of working age, in employment	
3. Full details of the SOC occupations within each of the groups are given in Appendix B	

Figure B8-18: Occupations with a high level of worry about assaults at work.   *Source: Report 'Violence at work: findings from the 2002/2003 BCS'.*

PEOPLE WORKING WITH THE PUBLIC

Health care

Acts of violence at work have long been of real concern in the health services due to the necessarily close interaction with the public, some of whom may be mentally disturbed or suffering from alcohol and other drug abuse problems.

The 1996 British Crime Survey reported the following incidents in relation to health workers:

Occupation	Incidents per 10,000 workers
Medical practitioners	762
Nurses and midwives	580
Other health-related occupations	830
All survey subjects	251

Figure B8-19: Risk of work-related violence to health workers.   Source: Home Office Research and Statistics Directorate, British Crime Survey 1996.

As previously stated the definition of violence includes: wounding, common assault, robbery and snatch theft, excluding incidents committed by partners, ex-partners, relatives or other household members occurring while the victim said that they were working. Key risk factors for healthcare staff include:

- Working alone.
- Working after normal working hours.
- Working and travelling in the community.
- Handling valuables or medication.
- Providing or withholding a service.
- Exercising authority.
- Working with people who are emotionally or mentally unstable.
- Working with people who are under the influence of drink or drugs.
- Working with people under stress.

Education sector/teaching professions

Teachers, support staff and volunteers in the education sector may be at risk of violence from pupils and their parents, and from other visitors. Many of those working in the education sector consider violence at work to be one of the most serious problems that they face. This has been exacerbated by incidents such as the shootings at Dunblane Primary School, the machete attack in Wolverhampton and an attack with a home-made flame thrower on pupils taking examinations in Northern Ireland. These incidents have involved mainly intruders and Lord Cullen's Public Inquiry into the Dunblane shootings emphasised that employers need to consider the protection of the school population as a whole from intruders.

The HSE publication 'Violence in the Education Sector' deals with many of these problems. While it retains an emphasis on schools, the general principles outlined can be applied throughout the education sector. The issue of pupil on pupil violence, other than where this creates a risk to staff, is not addressed nor does it consider the particular problems experienced by high-security research departments within universities, although some of the general advice given may be relevant.

Others working with the public

This publication identifies the following activities, and those who undertake them, as situations where the risk of violence may be increased:

Activities	People
Looking after premises (caretaking).	Site supervisors (for example, caretakers), porters, security staff.
Lone working.	Cleaning staff, library staff, head-teachers, principals, teachers, lecturers, site supervisors, maintenance and administrative staff.
Home visiting, off-site working.	Education welfare officers, education social workers, teachers, researchers, psychologists etc.
Evening working.	Teaching staff, library staff, cleaning staff, education welfare officers, youth and community workers, site supervisors, bursars, students.
Running licensed premises in education establishments.	Bar staff.
Looking after animals/research with animals.	Animal house technicians, teaching staff, research workers.
Working with pupils with	Teachers, educational psychologists, day-care helpers, nursery nurses.

Activities	People
behavioural difficulties.	
Looking after money.	Bursars, clerical and finance staff, headteachers, teachers, school secretaries.
Disciplining/supervising/students.	Headteachers, teachers, lunchtime supervisors, lecturers.
Dealing with angry parents or relatives of pupils.	Head teachers, teachers, principals, lecturers, school secretaries, bursars, receptionists.

Figure B8-20: Risk of violence.   *Source: Violence in the Education Sector, HSE.*

WORKING WITH PSYCHIATRIC CLIENTS OR ALCOHOL/DRUG IMPAIRED PEOPLE

Workers who come into contact with psychiatric clients or alcohol/drug impaired people are at increased risk of experiencing violence when working. Both of these characteristics can have an effect on a client's perception, sense of values or restraint, and their co-ordination. These factors, coupled with a potential for restricted ability to communicate, can cause the client to feel frustrated, threatened or to misunderstand a situation, particularly if this is an offer of help that brings the worker in close proximity with the client. In other situations, the client might be one whose psychiatric tendencies make them prone to commit violence. In the case of alcohol/drug impaired people, the simple fact that their control over their own body is diminished may mean that they could cause injury to workers without intention.

WORKING ALONE

Lone workers are those who work by themselves without close or direct supervision. Examples of lone workers include:

Persons who work in a static location:

- Only one person works on the premises (for example, small workshops, petrol stations, kiosks, shops and those who work at home).
- People work separately from others (for example, in factories, warehouses, some research and training establishments, leisure centres or fairgrounds).
- People work outside normal hours (for example, cleaners, security, special production, maintenance or repair staff).

Persons who work away from their base (mobile workers):

- On construction, plant installation, maintenance and cleaning work, electrical repairs, lift repairs, painting and decorating, vehicle recovery.
- Agricultural and forestry workers.
- Service workers, for example, rent collectors, postal staff, social workers, home helps, district nurses, pest control workers, drivers, engineers, architects, estate agents, sales representatives and similar professionals visiting domestic and commercial premises.

HOME VISITING

As identified above, home visiting is frequently an activity that is done alone and carries particular risks of violence based on the fact that the work takes place alone at a person's home. The fact that it is a person's home means that there is usually little control over it as a workplace. It is a possibility that the home visit may be attended by a number of people at a time, for example, relatives of a sick person.

Though many home visits are routine there are times when there may be an increased tension related to the visit, due to the anxiety or frustration of the people being visited. A home contains many ordinary, everyday items that may be used against a worker in a violent act. This increases the degree of harm that may be caused and in some cases may make violence more likely to move from the verbal to the physical. The layout of homes being visited, such as a single entrance/exit flat, makes it more difficult for the worker to leave the home and avoid a potentially violent situation before it escalates.

HANDLING MONEY/VALUABLES

Workers involved in handling money/valuables are at risk of being involved in planned or opportunist theft. Theft accompanied by actual violence or a threat of violence is commonplace. Organisations involved in industries that know they are targets have generally evaluated the risks and taken steps to minimise them, for example, in petrol stations, banks or security transfer activities. Those organisations where money or valuables are seen as a secondary or incidental activity of their operation are at particular risk, as they will not have well developed strategies to deal with the risks. Examples might be charities handling fund-raised money, a person emptying coin operated launderette machines, shop keepers banking takings or organisations that pay wages in cash.

In addition to money, employers have to consider workers that handle valuables, though the term 'valuable' has a wide definition and what is not valuable to an organisation or worker may be valuable to a person who wants it. Obvious valuables that may place someone at risk include drugs for medical treatment, a mobile phone, laptop or even the company car they drive. All of these may represent an increased risk of violence to obtain them from a worker.

INSPECTION AND ENFORCEMENT DUTIES

Workers involved in inspection and enforcement duties are at increased risk because, despite any good intentions, they will be seen by some people in an adversarial perspective. Distrust, anxiety and frustrations may accompany the execution of these duties, leading to a heightened tension and an increased risk of violence. These duties may be carried out on a forced, rather than planned, basis; which in itself could lead a person to show offensive behaviour and possibly become violent. If the duties are carried out by an individual alone similar issues as those referred to with regard to lone working apply. Increased risks exist when carrying out these duties with regard to work that makes use of equipment that could be used as a weapon of violence. For example, a shotgun for farmers, dogs for scrap yards, knives for catering or butchery trades.

RETAIL AND LICENSED TRADE

Workers in the retail trade are at risk, because of money being exchanged on the premises and from the possible attempted theft of goods. They may also work in 24 hour outlets which makes them especially vulnerable during the night hours.

Workers in the licensed trade are at risk of violence because they regularly have to deal with clients who are drunk. They may have to refuse to serve a drunken client, which can lead to violent verbal and physical abuse.

Legal requirements to manage work-related violence/aggression

RELEVANT STATUTORY PROVISIONS

Health and Safety at Work etc. Act (HASAWA) 1974

An employer's responsibilities under HASAWA 1974 Section 2 extends to protecting staff from violence. In the decided case West Bromwich Building Society v Townsend [1983] IRLR 147, a principal environmental health officer served an improvement notice on the building society requiring the installation of screens to protect staff against attack. Although in the particular circumstances of that case, the High Court found for the Society, in his summing up Mr Justice McNeill stated that protecting staff from violent attack was not included within an employer's responsibilities under HASAWA 1974 Section 2.

HASAWA 1974 is a wide ranging Act and could be applied to almost any incident of violence at work. However, it is HSE policy not to enforce the Act where there is other more specific legislation. In practice, it is intended that attention should focus on protecting employees where there is a significant risk of violence from members of the public. An exception to this general rule would be where skylarking or other incidents between employees warranted action under HASAWA 1974 Section 7.

Violence between employees should normally be dealt with as a personnel matter by the employer, but reasonably practicable measures need to be taken by employers to satisfy their legal duty under HASAWA 1974, particularly so where the employer is aware of the likelihood of violence occurring.

The Management of Health and Safety at Work Regulations (MHSWR) 1999

The Regulations require employers to assess risks to employees and make arrangements for their health and safety by effective planning, organisation, control, monitoring and review. Where appropriate, employers must assess the risks of violence to employees and, if necessary, put in place control measures to protect them.

Although there is no general legal prohibition on higher risk factors such as working alone, the broad duties of the Health and Safety at Work etc Act (HASAWA) 1974 and the Management of Health and Safety at Work Regulations (MHSWR) 1999 still apply.

Therefore, employers need to ensure, so far as is reasonably practicable, the health, safety and welfare at work of employees. Risk assessments and the resultant appropriate planning, organisation, control and monitoring and review arrangements are also required.

The Reporting of Injuries, Diseases and Dangerous Occurrences Regulations (RIDDOR) 2013

Employers must notify their enforcing authority in the event of an accident at work to any employee resulting in death, major injury or incapacity for more than seven days. This includes any act of non-consensual physical violence done to a person at work.

The Safety Representatives and Safety Committees Regulations (SRSCR) 1977 and the Health and Safety (Consultation with Employees) Regulations (HSCER) 1996

Employers must inform, and consult with, employees in good time on matters relating to their health and safety. Employee representatives, either appointed by recognised trade unions under SRSCR 1977 or elected under HSCER 1996 may make representations to their employer on matters affecting the health and safety of those they represent.

Equality Act (EA) 2010

Under the EA individuals with a 'relevant protected characteristic' are protected from harassment. The EA defines harassment as where:

"A person (A) harasses another (B) if:

a) *A engages in unwanted conduct related to a relevant protected characteristic.*

b) *The conduct has the purpose or effect of:*

    (i) Violating B's dignity.

    (ii) Creating an intimidating, hostile, degrading, humiliating or offensive environment for B".

Protected characteristics are specified as - age; disability; gender reassignment; marriage and civil partnership; pregnancy and maternity; race; religion or belief; sex; sexual orientation.

The EA 2010 particularly emphasises intimidation and hostility as aspects of harassment, this may typify the sort of behaviour that may be classed as violence or aggression.

In deciding whether the conduct had the effect referred to in the EA 2010 account is taken of:

- The perception of the person affected.
- The other circumstances of the case.
- Whether it is reasonable for the conduct to have that effect.

Section 40 of the EA specifically relates to employees harassed in the course of their employment. The employer has specific responsibility to take reasonably practicable steps to prevent a third party from harassing an employee.

The EA 2010 clarifies that the employer must take action where the employer knows the employee has been harassed on at least two other occasions by a third party, whether it was the same or different third party on the previous occasions.

Claims made under EA 2010 are heard by an Employment Tribunal. This could result in compensation being paid to the individual and or a remedy of the situation.

Employers' common law duties

An employer's duties under common law and employment law have been interpreted as including a duty to protect staff from violent attack from the public.

If an employee considers the employer has failed in his/her duty of care, redress is available on 2 fronts. If injury has occurred the employer can be sued for negligence.

If, however, no injury has occurred, but the employee has left the job because they considered the risk of injury unacceptable, then it might be possible in some cases for the employer to be sued.

O'Toole v First Quench [2005]

In the Scottish case of Collins v First Quench Retailing Limited [2003] the Claimant won over £100,000 in damages when the trial Judge came to the dubious conclusion that the risk of the robbery would have been significantly reduced had a second member of staff been present. Fortunately this decision was not followed in the case of O'Toole v First Quench in August 2005 when an English Court found that it was unreasonable to 'double staff' purely as a deterrent against robberies.

Mitchell and Others v United Co-operative Ltd [2012] Court of Appeal Stress - psychiatric injury - staff in retail store at risk of violence

The facts

Mrs Mitchell, Mrs Benton and Mrs Goodwin worked in the Co-op in Shaw Road, Heaton Moor between Stockport and Manchester. One day robbers came in and stole cigarettes and cash. The ladies were frightened and developed post-traumatic stress and anxiety states as a result. They brought a claim against the Co-op for damages in respect of psychiatric injury. A risk assessment had been undertaken in 2003 had concluded that the overall risk rating in respect of robbery or shop lifting was high.

In the 11 years before the Co-op bought the shop there had been two robberies. Between February 2000 and December 2005 there had been 10. The previous owner of the shop had security screens around the tills and the areas containing high value goods. The Co-op removed those screens when they bought the shop. In the course of those robberies a shot gun had been used on one occasion, batons on another and a knife on a third. Mrs Mitchell, Mrs Benton and Mrs Goodwin had been in the shop on some of those occasions.

The Co-op had taken steps as a result of these robberies including CCTV monitoring both outside and inside the shop; panic alarms connected to a control centre; video surveillance; FOB operated door locks; minimising the amount of cash in the tills; the provision of smoke notes (notes which admitted a dye when passing the transmitter and the door way); staff training to avoid confrontation; provision of a part time security guard for a short period after a robbery had occurred and provision of a mobile security response team.

The judge accepted the evidence of Brian Edwards who had served for 30 years for the Metropolitan Police and who had given expert evidence for the Co-op to the effect that the Co-op's policy compared favourable with that of other retailers.

The decision

The starting point was the duty of care owed by an employer to an employee to take reasonable care to keep the employees safe as expressed by Swanwick J in Stokes v Guest and recently proved by Lord Mance in Baker v Quantum Clothing Group. The Co-op had taken reasonable steps to deter robberies; no employer could be expected to go so far as to prevent any robbery taking place.

The direct link that the claimants had alleged between the removal of the screens and the incident of robberies had not been made out. The judge had been entitled to conclude that although the screen might have had a deterrent affect it carried risks for the staff which outweigh that benefit. The issue was not only what deterrent effect the screen would have had upon a robbery taking place but also what deterrent effect the screens would have to guard the employees against psychiatry injury. There was some evidence that the shop was running at a loss of about £60,000 per annum. The cost of provision of a full time guard was likely to be in the sum of £30,000 per annum. A proper approach required a balance to be struck against the probable effectiveness of a precaution that could be taken and the expense that it involved. By implication the judge had held such a balance and come down against requiring full time security. Having regard to the expert evidence his conclusion that the failure to provide full time guarding did not amount to a failure to take reasonable care was a conclusion he was entitled to reach.

Appeal dismissed.

Public Order Act (POA) 1986

The Public Order Act (POA) 1986 is enforced by the police and covers threats and abuse as well as physical assault. There could be some interplay between the POA 1986 and HASAWA 1974 where cases of violence occur. However, police action under the POA 1986 will be directed against the perpetrator of the violence, whereas HSE's and Local Authorities' activities will be directed towards assessing whether the employer complied with his/her general duties under HASAWA 1974. The POA 1986 also only applies after an offence has been committed and cannot be used to require preventive measures beforehand.

LEGAL CONSTRAINTS REGARDING 'REASONABLE FORCE'

A person who is subjected to violence is entitled to protect himself/herself if they are put in fear of their life or the safety of their person. They must not, however, use more force than is necessary or reasonable in the circumstances. This is the basis for the defence of Self-Defence (also known as Private Defence) against a charge related to assault and battery. Legally, *assault* is a threat to apply force immediately to the person of the victim and battery is the actual application of force. If accepted this is a complete defence because it negates the unlawful nature of the assault carried out in self-defence.

B8.4 - Identification and control of workplace violence/aggression

Identification of practical control measures

PHYSICAL CONTROL MEASURES

Cash free systems

The handling of money is a high risk factor associated with violence. Where possible, effort should be made to use cash free systems. Technology has enabled this for many quite ordinary activities. For example, drink and food machine vending organisations can work in association with companies to provide swipe card systems that automatically debit a person's card when purchases are made. Even the company does not need to handle money as they can make adjustment to the person's salary to account for the cost of credit on the swipe card. Many organisations do not pay staff using cash, which avoids the drawing and distribution of cash to employees, a high risk for the employer and for the employee that needs to make their way from their workplace carrying a significant sum of money.

Layout of public areas and design of fixtures and fittings

The provision of better seating, decor, lighting in public waiting rooms and more regular information about stress factors such as delays can be beneficial in reducing and managing violent situations. Potential weapons/missiles should be identified and then modified or removed. It must not be forgotten that weapons can include ones which pose a biological threat (for example, used syringe needles).

Care should be taken when deciding on controls. In one housing department it was found that protective screens made it difficult for staff and the public to speak to each other. This caused tension on both sides. Management and health and safety representatives agreed a package of measures including taking screens down, providing more comfortable waiting areas and better information on waiting lists and delays. This package of measures reduced tension and violent incidents.

Employees are likely to be more committed to the measures if they help to design them and put them into practice. A mix of measures often works best. Concentrating on just one aspect of the problem may make things worse in another. An overall view must be taken and the risks to employees balanced against any possible reaction of the public. An atmosphere that suggests employees are worried about violence can sometimes increase its likelihood.

Use of cameras, protective screens and security-coded doors

The following physical measures can be taken to reduce the risk of violence such as:

- Video cameras or alarm systems.
- Coded security locks on doors to keep the public out of staff areas.
- Wider counters and raised floors on the staff side of the counter to give staff more protection.
- Protective screens.
- Escape routes.
- Barriers.
- Security patrols.
- Emergency call systems such as panic buttons and personal alarms.

These measures act both as visual deterrents to people that might commit violent acts and to provide comfort and assurance to workers.

Use of panic buttons and personal alarms

Where workers are in close proximity to people that may be violent, such as in interviews for benefit claims, it is important to anticipate this likelihood. Workers should have a clear perspective of what constitutes potentially violent behaviour and have ready access to assistance. This may be best provided by equipping them with an alarm, on their person or at the workstation, for them to call for help when they feel threatened.

ORGANISATIONAL CONTROL MEASURES

The threat of violence does not stop when the work period has ended. It is good practice to make sure that employees can get home safely. For example, where employees are required to work late, employers might help by arranging transport home or by ensuring a safe parking area is available.

Communication systems

Passing on information on risks from individual clients

Risk assessments should include the identification of increased risk from individual clients, situations and areas. In some situations, it may not be acceptable for an individual to visit high risk clients by themselves. Alternatively it may not be appropriate for certain employees (for example, a woman or a young person) to carry out certain visits. Information gathered from visits that indicates an increased risk of violence/aggression should be compiled in a systematic way and communicated on a timely basis to those affected, particularly on a cross-department or service basis. This can involve a number of means but methods that promptly and clearly identify the changed status of the client should be adopted, systems such as 'marker flags' that indicate the increased risk of violence may be used.

Recording of staff whereabouts and recognition when staff are overdue

Where people work away from a fixed point in higher risk activities, such as community nurses that visit in the home to administer drugs to patients, it is usual to establish simple systems where their visits are recorded and balanced against a proposed time sheet for the day. This helps to establish where they should be at any point in time. Systems for calling in to confirm progress and the worker's well-being enable progress sheets to be updated and to observe when a worker is overdue for an appointment at a given location. An agreed protocol needs to be in place to act promptly to deal with an observation that a worker is overdue. Usually this will include a call to the person, if they are equipped to receive one, and the escalation to involve the police if necessary.

Use of mobile communication equipment

Clearly, individuals who work alone visiting clients should not be at more risk than other employees. This may require extra control measures, which should include precautions to take account of both normal work and foreseeable emergencies. Employers should consider the use of mobile communication equipment for workers that may need to gain assistance rapidly or for those that need to confirm their safe condition intermittently. Equipment could include:

- Regular telephone or radio contact between the lone worker and supervision.
- The use of automatic warning devices that operate if specific signals are not received periodically from the lone worker.
- Other alarm devices for use in the event of an emergency and which are operated manually or automatically by the absence of activity.
- GPS systems that lets a distant supervisor track the exact whereabouts of the person.

BEHAVIOURAL CONTROLS

Staff training

Training should be provided to all employees at risk from violence. It may also be appropriate to train managers so that they can recognise the problems associated with violent and aggressive incidents, and how to manage them.

The objective of training is to bring about a reduction in both the number and seriousness of incidents. Further benefits include:

- A reduction in the psychological effects of incidents.
- An improved response to incidents.
- An improvement in staff morale.

A training programme might include:

- **Theory:** recognising and understanding aggression and violence in the workplace.
- **Prevention:** assessing danger and taking precautions including causes of violence.
- **Interaction:** with aggressive people. This includes the recognition of warning signs, relevant interpersonal skills and details of working practices and control measures.
- **Post-incident action:** incident reporting procedures, investigation, counselling and other follow-up.

There are various levels of training ranging from basic through to the more advanced skills required to defuse, de-escalate and avoid incidents. Some workers, for example, those working in mental health, may need training in breakaway, control and restraint techniques.

Detailed training records should be kept.

Recognition of situations where violence could result

It is important that workers are trained to identify the situations where violence could result. This means the work activities, locations and specific features of the work that could exacerbate the likelihood of violence. When dealing with clients this may mean non-routine work which the client will not like or agree with. It may be a change in a client's circumstances, such as the return of a violent relative to the home. For some it may simply be the fact that they are working late during winter hours and need to make their way to their car or train station to get home.

Interpersonal skills to defuse aggression

Workers that may be exposed to the risk of violence should be trained in the use of interpersonal skills to defuse an aggressive situation. The training does not need to be overly complicated but will often develop a heightened awareness of how the workers' behaviour can add to or reduce the risk of violence, frequently providing a better perspective of how the other person sees things. If this consideration is taken into account at recruitment or selection it may help to prevent workers that do not have these skills being placed in work with a risk of violence.

Use of language and body language

Two of the critical factors to consider when analysing and preventing violence are the language and body language of the worker and violent person. In the first instance it is possible for the worker to precipitate violence not because of what they say but because of how they say it. In addition, many situations that are moving towards a violent situation can be identified by the person's language and body language. This provides an opportunity to re-assess if the worker is contributing to the move towards violence and use the interpersonal skills to defuse the situation.

Guidance to staff on dealing with an incident

Aside from possible physical injuries, someone involved in a violent incident, can suffer severe distress. In some cases, the psychological effects can be long term and debilitating. This can include post-traumatic stress disorder (PTSD). Planning is needed to ensure a quick response in order to avoid long-term distress to victims. The following factors should be considered:

- Debriefing: victims will need to talk through their experience as soon as possible after the event. It should be remembered that verbal abuse can be just as upsetting as a physical attack.
- Time off work: individuals will react differently and may need differing amounts of time to recover. In some circumstances they might need specialist counselling.
- Legal help: in serious cases legal help may be appropriate.
- Other employees: may need guidance and/or training to help them to react appropriately.

Support for staff post-incident including training for managers in counselling

The need for counselling varies between individuals. Reaction to a violent incident ranges from anxiety attack, phobias, guilt and self-blame through to post traumatic stress disorder (PTSD). The severity of the violence is not necessarily related to the victim's response. Personality has much to do with the psychological impact. Some people who are said to have 'an external locus of control', are classified as 'Type A' personalities. Type A's are thought to suffer a more exaggerated psychological impact than that which non-type A's would suffer.

Victims are thought to suffer three distinct phases:

1)	The impact phase:	The victim experiences emotions ranging from shock, fear, vulnerability; sleep loss, fatigue and anger are experienced.
2)	The recoil phase:	The victim tries to make sense of the incident (why me?).
3)	The reorganisation phase:	The victim eventually regains control of their emotions.

Support on return to work should be considered, especially if the aggressor is still within the working environment. If an employee suffers an injury, loss or damage from a crime then useful information can be obtained from the Home Office leaflet 'Victims of Crime' which gives then useful advice - including how to apply for compensation. It should be available from libraries, police stations, Citizens Advice Bureaux and victim support schemes.

This page is intentionally blank

Musculoskeletal risks and controls

On completion of this element, candidates should be able to demonstrate understanding of the content through the application of knowledge to familiar and unfamiliar situations. In particular, they should be able to:

B9.1 Outline types, causes and relevant workplace examples of injuries and ill-health conditions associated with repetitive physical activities, manual handling and poor posture.

B9.2 Explain the assessment and control of risk from repetitive activities, manual handling and poor posture.

Content

Relevant statutory provisions

Health and Safety (Display Screen Equipment) Regulations (DSE) 1992

Management of Health and Safety at Work Regulations (Northern Ireland) 2000 (MHSWR)

Management of Health and Safety at Work Regulations 1999 (MHSWR) (as amended)

Manual Handling Operations Regulations (MHOR) 1992

Workplace (Health, Safety, Welfare) Regulation (WHSWR) 1992

Sources of reference

Reference information provided, in particular web links, was correct at time of publication, but may have changed.

Assessment of Repetitive Tasks (ART) tool, HSE, http://www.hse.gov.uk/msd/uld/art/

Ergonomic Guidelines for Manual Material Handling, NIOSH, http://www.hse.gov.uk/msd/uld/art/

Manual handling assessment charts (MAC Tool), HSE, http://www.hse.gov.uk/msd/mac/

Manual Handling, Manual Handling Operations Regulations 1992 (as amended), Guidance on Regulations, L23, HSE, http://www.hse.gov.uk/pubns/priced/l23.pdf

Rapid Upper Limb Assessment Tool (RULA), http://www.rula.co.uk/

Seating at work, HSG57, third edition 2002, HSE, http://www.hse.gov.uk/pubns/priced/hsg57.pdf

The law on VDUs – An Easy Guide, HSG90, HSE, ISBN: 978-0-717626-02-1 http://www.hse.gov.uk/pubns/priced/hsg90.pdf

The health and safety toolbox, How to control risks at work, HSG268, HSE, ISBN: 978-0-717665-87-7 http://www.hse.gov.uk/pUbns/priced/hsg268.pdf

Understanding ergonomics at work, HSE, INDG90(rev2), http://www.hse.gov.uk/pubns/indg90.pdf

Upper Limb Disorders in the Workplace - A Guide, second edition 2002, HSG60, HSE, ISBN: 978-0-717619-78-8, http://www.hse.gov.uk/pubns/priced/hsg60.pdf

Variable manual handling assessment chart (V-MAC) tool, http://www.hse.gov.uk/msd/mac/vmac/

Workplace health, safety and welfare, Workplace (Health, Safety and Welfare) Regulations 1992, ACOP, L24, HSE, ISBN: 978-0-717604-13-5, http://www.hse.gov.uk/pubns/priced/l24.pdf

Work with display screen equipment: Health and Safety (Display Screen Equipment) Regulations 1992 as amended by the Health and Safety (Miscellaneous Amendments) Regulations 2002: Guidance on Regulations, L26, http://www.hse.gov.uk/pubns/priced/l26.pdf

The above web links along with additional sources of reference, which are additional to the NEBOSH syllabus, are provided on the RMS Publishing website for ease of use - www.rmspublishing.co.uk.

B9.1 - Types, causes and examples of musculoskeletal injuries and ill-health

Human musculoskeletal system including bones, tendons, ligaments, nerves and muscles

Anatomy is the study of the structure of the body and physiology is the study of its function.

One of the systems within the body is the musculoskeletal system; this system provides support, stability, movement and form. It is created from tendons, ligaments, nerves and muscles, etc. which form the muscular system and from bones of the skeleton system.

THE SKELETON

The skeleton is the framework of the body consisting of bones held together by cartilage and ligaments. A new-born baby has over 300 bones, whereas on average an adult human has 206 bones, though these numbers can vary slightly from individual to individual. The difference comes from a number of small bones that fuse together during growth, such as the sacrum and coccyx of the vertebral column. The framework of bones supports the soft body tissues and protects vital organs. The long bones, arms, legs, sternum, contain the bone marrow where red blood cells are formed, which are vital for carrying oxygen around the body.

Bones and muscles together give stability and mobility to the body and allow for a certain amount of movement in all planes.

The skull

The skull consists of the bones of the cranium, the rounded part which protects the brain, and the bones of the face and jaws. Passing through the base of the skull are blood vessels and the nerve fibres of the spinal cord. The bones of the face are fixed together except for the lower jaw which is hinged to the sides of the base of the skull. The palate is the roof of the mouth.

The trunk

The spinal column consists of thirty-three bones called vertebrae.

These bones are given a letter and a number depending on where they are located in the spinal column:

- C (cervical) followed by a number from 1 to 7, refers to the vertebrae in the neck.
- T (thoracic) followed by 1 to 12, refers to the thoracic spine (where the 12 ribs are attached). Sometimes the vertebrae in the thoracic spine are referred to as dorsal with the letter D.
- L (lumbar) followed by 1 to 5, refers to the lumbar section of the spine.
- S (sacral) followed by 1 to 5, refers to the lowest vertebrae, although these vertebrae are fused together, forming the sacrum.
- Coccyx (or tail bone), formed out of 4 fused vertebrae at the very bottom of the spinal column.

Vertebrae are separated by inter-vertebral discs that act as shock absorbers. They are made up of a soft jelly like substance (the nucleus), which is held inside a tough, elastic and fibrous outer casing (the annulus). The spine is not straight, but is actually an 'S' shape. Not all backs are the same 'S' shape, but they are usually curved with a hollow in the base of the neck and another in the lower part of the back.

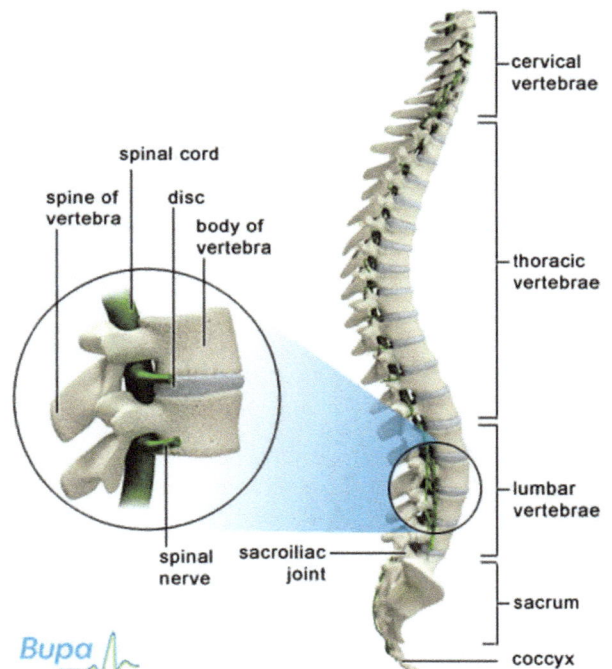

Figure B9-1: The bones of the spine.   *Source: BUPA.*

The spinal cord runs through the spinal canal, which is an arch formed by the bones. The vertebrae have a small gap (called the 'foramen') through which the spinal nerves run. The spinal nerves, which are part of the central nervous system, run all the way from the base of the brain to the bottom of the spinal column. The nerves exit the spinal column at the level where they need to be, for example, the nerves that go to your arms, exit the spinal column in the neck area (cervical), and the nerves going to your legs exit much lower and run along the whole length of the spinal column.

The spine is bound together and supported by ligaments, tendons and muscles. These provide strength and stability to the 'chain' of vertebrae and discs. The muscles are connected to your bones with tendons; when a muscle contracts the forces are passed on to the skeletal system via the tendons. This ensures that a muscle contraction results in a movement of a certain body part. The ligaments provide stability to joints, but are also somewhat flexible so they can stretch or contract when the joint moves.

The ribs extend from the thoracic or dorsal vertebrae round to the front and the upper seven pairs (of twelve pairs) are joined to the breastbone (sternum).

The next three pairs of ribs are joined to ribs above them by cartilage. The lower two pairs of ribs are not attached at the front (floating ribs). The rib cage protects the lungs, heart, the liver, stomach and spleen.

Upper limbs

The collar bone (clavicle) lies above and in front of the first rib and maintains the position of the upper limb away from the chest. The shoulder blades (scapula) lie at the back of the chest and form joints with the collar bone and with the upper arm. The upper arm bone is the humerus. The forearm consists of two bones, the radius on the thumb side and the ulna on the other. The wrist, carpus, is composed of eight bones. The hand has five bones called metacarpals. The fingers have three bones each called phalanges and the thumb two.

Lower limbs

The pelvis consists of the two hip bones (innominate bones) and the sacrum, which together form a 'basis'. The pelvis provides the deep sockets for the hip joints. It supports the abdomen and protects the contents. The thigh bone (femur) runs from the hip to the knee; it fits into the hip socket at one end and forms part of the knee joint at the other.

The knee cap (patella) is a small flat bone protecting the knee joint. The lower leg consists of two bones, the shin bone (tibia) and the fibula. The fibula does not form part of the knee joint. The ankle and foot:

- The ankle bones are called tarsus or tarsal bones of which there are seven.
- In front of the instep are five metatarsal bones.
- The bones of the toes are the phalanges.

Joints

Joints are either mobile or immobile. Mobile joints are ball and socket joints (shoulder and hips), hinge joints where the bones are shaped to give movement in one direction only (elbow and knee) or joints that give restricted movement only (wrist, foot, spine, etc.). Immobile joints are to be found in the bones of the skull.

The ends of bones of moveable joints are covered with smooth cartilage to minimise friction and are held together by ligaments. The joint is encapsulated and lubricated by secreted fluid (synovial fluid).

LIGAMENTS AND TENDONS

Tendons and ligaments are soft collagenous connective tissues. Ligaments connect bone to bone and tendons connect muscle to bone. Ligaments and tendons play a significant role in musculoskeletal biomechanics.

Tendons are tough, flexible, fibrous bands of tissue that attach muscles to bones. Without these connections movement would not be possible. The biggest is the Achilles tendon, named after the Greek hero of the Trojan War.

Ligaments are composed of long, stringy collagen fibres; they create short bands of tough fibrous tissue that connect bone to bone. Whereas tendons help initiate movement, ligaments work to restrict movement - by limiting how far you can move and in which direction - they therefore help to stabilise joints.

NERVES

Nerves provide a vast communication network. They carry the electrical impulses that initiate movement and pass messages from outside the body to the spinal cord and brain, such as heat, cold and damage to the body. The network is denser in certain parts of the body than in others, for example, it is very dense at the ends of the fingers.

The brain receives the message relating to damage to the body as pain. Some reactions to pain are reflex actions, i.e., they happen automatically, such as the reaction to touching something hot. Other reactions to pain depend on the person's pain threshold and may vary from living with the discomfort to not being able to move because of the pain intensity, such as with a trapped nerve in the spinal region.

MUSCLES

A muscle is a bundle of many cells called fibres. Muscle fibres may be thought of as long cylinders and, compared to other cells in your body, muscle fibres are quite big. A muscle fibre contains many myofibrils, which are cylinders of muscle proteins. These proteins allow a muscle cell to contract. The muscular system is under the control of the nervous system and can be subdivided into three sections:

- Voluntary.
- Involuntary.
- Cardiac.

Voluntary muscles

About forty percent of the mass of the body consists of voluntary muscles. The main function of voluntary muscles in the body is that of motion. Muscles are paired so that when a prime mover contracts there is an equal and opposite action of the antagonist, i.e. one group of muscles contracts and the group of muscles with which it is paired relaxes.

Involuntary muscles

Involuntary muscles are not under the power of will, but are controlled by the involuntary or autonomic nervous system. They continue to work even when a person is asleep, i.e. the heart, the respiratory system, stomach, bowels.

Cardiac

Cardiac muscles are the muscles of the heart and are a special type of involuntary muscle.

Types of injury and ill-health conditions

There are a number of types of injury and ill-health conditions that result from poor ergonomic design, repetitive physical activities, manual handling or poor posture. These include almost the entire range of both physical and mental harm that can occur to persons at work.

- Physical harm in terms of general discomfort, fatigue and musculoskeletal disorders, (tenderness, aches and pain, stiffness, weakness, tingling, numbness, cramp, or swelling), eye and eyesight effects.
- Mental harm as evidenced by increased stress, lack of motivation and an increase in error leading to accidents or incidents.

WORK-RELATED UPPER WRULDS (WORK RELATED UPPER LIMB DISORDERS)

Work-related upper limb disorders (WRULDs) were first defined in medical literature as long ago as the 19th century as conditions caused by **forceful, frequent, twisting and repetitive movements**. The body will be affected to a varying degree by tasks that involve bending, reaching, twisting, repetitive movements and poor posture.

WRULDs is a collective term for a group of occupational diseases that comprise musculoskeletal disorders caused by exposure in the workplace affecting the muscles, tendons, ligaments, nerves, blood vessels, joints and bursae (a small fluid-filled sac which provides a cushion from injury between bones and tendons and/or muscles around a joint) in the upper limbs such as the neck, shoulders, arms, wrists, hands and fingers. They are often called repetitive strain injuries (RSI), cumulative trauma disorder or occupational overuse syndrome.

WRULDs include well-known conditions such as tennis elbow, flexor tenosynovitis, tendinitis and carpal tunnel syndrome. The disorder is usually caused by repetitive tasks or movements and aggravated by excessive workloads, inadequate rest periods and sustained or constrained postures.

This can result in pain, soreness or inflammatory conditions of soft tissues, such as muscles and the synovial lining of the tendon sheath. Present approaches to treatment are largely effective, provided the condition is treated in its early stages. Clinical signs and symptoms are local aching, pain, tenderness, swelling or crepitus (a grating sensation in the joint).

Some common WRULDs are described below:

Carpal tunnel syndrome (CTS)

CTS occurs when tendons or ligaments in the wrist become enlarged, often from inflammation, after being aggravated. The narrowed tunnel of bones and ligaments in the wrist pinches the nerves that reach the fingers and the muscles at the base of the thumb.

The first symptoms usually appear at night due to the wrists being flexed during sleep. Symptoms range from a burning, tingling numbness in the fingers, especially the thumb and the index and middle fingers, to difficulty gripping or making a fist, to dropping things.

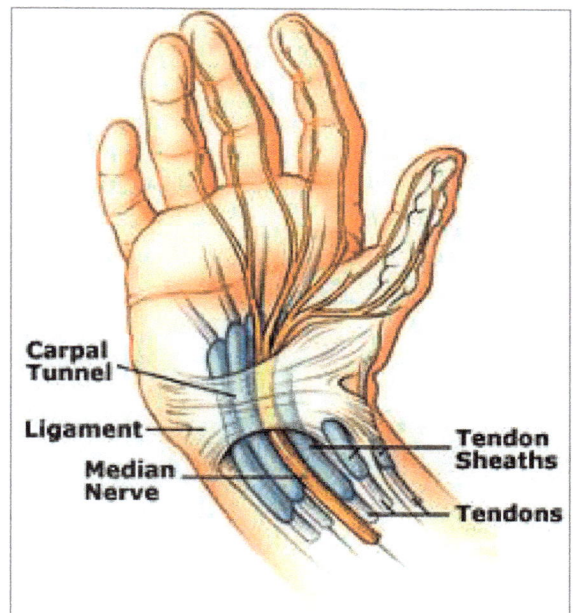

Figure B9-2: Carpal tunnel syndrome.   Source: www.sodahead.com.

Tenosynovitis

An irritation of the tendon sheath. It occurs when the repetitive activity becomes excessive and the tendon sheath can no longer lubricate the tendon. As a result, the tendon sheath thickens and becomes aggravated. Symptoms include pain and swelling, usually near where the tendon attaches to the bone.

Tendinitis

Tendinitis involves inflammation of a tendon, the fibrous cord that attaches muscle to bone. It usually affects only one part of the body at a time and lasts a short time.

However, in some cases it involves tissues that are continuously irritated. It may result from an injury, activity or exercise that repeats the same movement. Rotator cuff tendinitis is one of the most common causes of shoulder pain.

Peritendinitis

Peritendinitis is an inflammation of the sheath that surrounds a tendon. It can be associated with tendinitis, an inflammation of the tendon itself. This condition is most commonly seen in overuse injuries, such as those involving continually repeated tasks (for example, data input into a computer) or for those people who do not get enough rest between tasks or following injury. The primary treatment is resting in order to allow the inflamed tissue to heal.

Epicondylitis

Commonly known as 'tennis elbow' or lateral epicondylitis, this is a condition that occurs when the outer part of the elbow becomes painful and tender, usually because of a specific strain, overuse, or a direct impact. Sometimes no specific cause is found. Tennis elbow is similar to golfer's elbow (medial epicondylitis), which affects the other side of the elbow.

Trigger finger (Stenosing tenosynovitis)

A painful condition that affects the tendons in the hand, which may involve one or more fingers, usually in the dominant hand. It may occur through repetitive movements at work. Trigger finger occurs if there is a problem with the tendon or sheath, such as swelling, which means the tendon can no longer slide easily through the sheath and it can become bunched up to form the nodule. This makes it harder to bend the affected finger or thumb. If the tendon gets caught in the opening of the sheath, the finger can click painfully as it is straightened.

The aches, pains and fatigue suffered doing certain tasks will eventually impair the worker's ability and lead to degradation in performance. It is therefore essential to consider the task, in order to match it to the individual, so the level of general comfort is maximised.

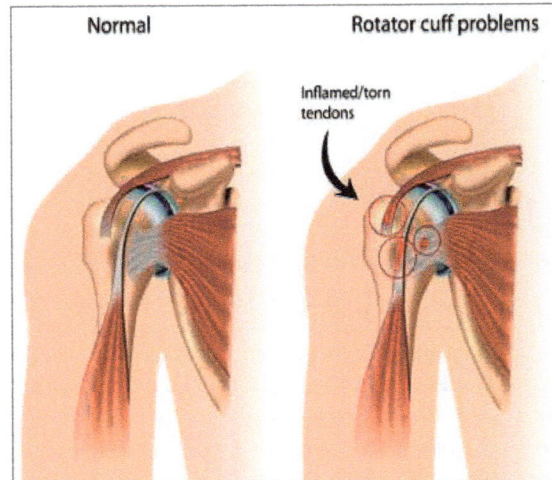

Figure B9-3: Tendinitis. *Source: Beth Israel Deaconess Medical Centre.*

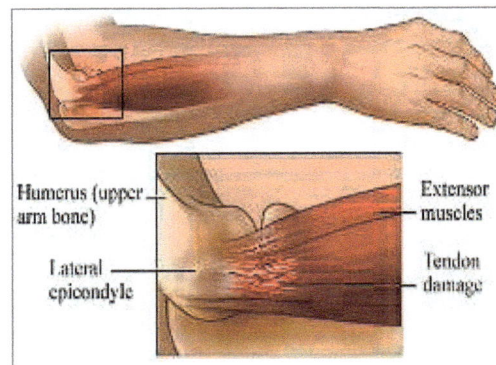

Figure B9-4: Epicondylitis. *Source: Myhealth.alberta.ca.*

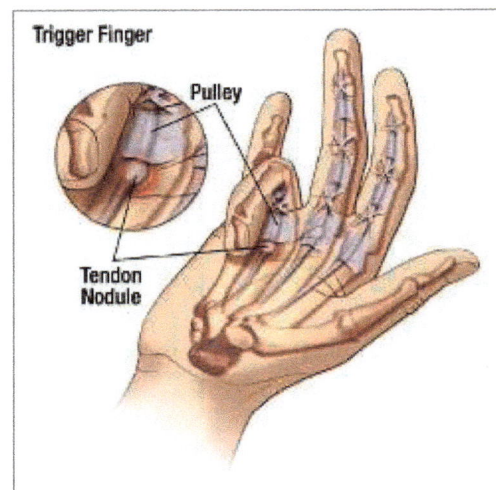

Figure B9-5: Trigger finger. *Source: 3pointproducts.*

For example, when carrying out manual handling assessments it is important to look at the relationship between the individual, the task, the load and the environment.

MUSCULOSKELETAL INJURY AND DISCOMFORT

Many cases of musculoskeletal injury and discomfort can be related to excessive periods of continuous intensive work, for example at a keyboard, combined with a badly designed workstation. This could result in a range of musculoskeletal problems, such as pain in parts of the spine, joint injuries, muscle strain and pain. Musculoskeletal injuries and discomfort are therefore often caused by poor workplace or job design. The risk factors for musculoskeletal injuries and discomfort may act alone or in combination (when the risk is generally increased), and include:

- Awkward and uncomfortable working postures (for example, bending, stretching and static postures).
- Manual handling.
- Repeating an action too frequently.
- Exerting too much effort.

- Poor posture.
- Working too long without adequate breaks.
- Adverse working conditions (for example, hot, cold).
- Psycho-social factors (for example, high job demands, time pressures and lack of control).
- Employers not receiving or acting on reports of worker symptoms quickly enough.

Musculoskeletal injury and discomfort can include backache, sore shoulders or elbows and numb or tingling wrists and hands. Such health problems manifest themselves in different ways, such as:

- Cases of injury to backs and limbs.
- Aches and pains.
- Frequent worker complaints and rest stops.
- Sickness absence.
- Individuals modifying workstations and tools, for example, seat padding.
- Workers wearing bandages, splints, copper bracelets or using rub-on ointments.
- Poor product quality.
- High material waste.
- Low output.

Work activities may cause a worker to develop poor posture, such as repeatedly reaching to pick up items from a conveyor system, holding a tool to fix things, bending underneath something to work on it or operating display screen equipment that has a poor layout. Some general musculoskeletal risks can also cause muscle tension and fatigue that, ultimately, leads to poor posture. The effects of working with a poor posture include:

- Rounded shoulders.
- Potbelly.
- Bent knees when standing or walking.
- Head that either leans forward or backward.
- Back pain.
- Body aches and pains.
- Muscle fatigue.
- Headache.

Incorrect (or poor) posture places stress on your joints, muscles, bones, nerves and tendons. It forces your body's muscular system to adapt to improper alignment. The body's postural mechanisms include:

- Slow-twitch and fast-twitch muscle fibres.
- Muscle strength and length.
- Nervous system feedback on the body's position in space.

Slow-twitch and fast-twitch muscle fibres

Skeletal muscle is made up of two types of muscle fibre, i.e. slow-twitch and fast-twitch. Generally, slow-twitch muscle fibres are found in the deeper muscle layers. They help us to maintain posture without too much effort, and contribute to balance by 'sensing' our position and relaying this information to the brain. Fast-twitch muscle fibres are used for movement and activity.

Slow-twitch fibres burn energy slowly and can keep working for a long time without tiring. However, fast-twitch fibres quickly use up their energy and tire. Poor posture causes muscle fatigue because it calls on the fast-twitch fibres instead of slow-twitch fibres to maintain the body's position.

Muscle strength and length

Over time, poor posture that demands support from fast-twitch fibres causes the deeper supporting muscles to waste away from lack of use. Weak, unused muscles tend to tighten, and this shortening of muscle length can compact the bones of the spine (vertebrae) and worsen posture.

Nervous system feedback on the body's position in space

The deeper layers of muscle are concerned with 'sensing' our position in space and relaying this information to the brain. If this function is taken over by muscles that mainly contain fast-twitch fibres, the brain gets an incomplete picture.

The brain assumes that the body needs to be propped up to counteract the effects of gravity, so it triggers further muscle contraction. This adds to the general fatigue and pain felt by the person with poor posture.

BACK PAIN

Back pain usually affects the lower back. It can be a short-term problem, lasting a few days or weeks, or continue for many months or even years. Eight out of ten adults will have some form of back pain at some stage in their life.

In many cases there is not a specific, underlying problem or condition that can be identified as the cause of the pain.

However, there are a number of factors that can increase the risk of developing back pain, or aggravate it once it starts. These include:

■ Standing, sitting or bending down for long periods.
■ Lifting, carrying, pushing or pulling loads that are too heavy, or going about these tasks in the wrong way.
■ Operating equipment that subjects the person to whole body vibration.
■ Having a trip or a fall.
■ Being stressed or anxious.
■ Being overweight.
■ Having poor posture.

One of the causes of lower back pain is the strain of back muscles or other soft tissue (ligaments or tendons) connected to the vertebrae.

Sometimes it is the intervertebral disc that is strained, causing it to bulge out (called slipped, prolapsed or herniated) and presses on nearby nerves.

The discs are made from a tough, fibrous case, which contains a softer, gel-like core. A slipped disc, as it is generally known by non-medical people, occurs when the outer part of the disc ruptures allowing the gel inside to bulge and protrude out between the vertebrae.

The damaged disc can put pressure on the spinal cord or part of it, causing pain and inflammation in the area of the spinal cord contacted. Pain may also be felt in the area of the body that the part of the spinal cord affected relates to, as in sciatica. If the sciatic nerve is contacted by the damaged disc pain will be felt in the leg, which is the part of the body that the sciatic nerve controls, this is called a referred pain.

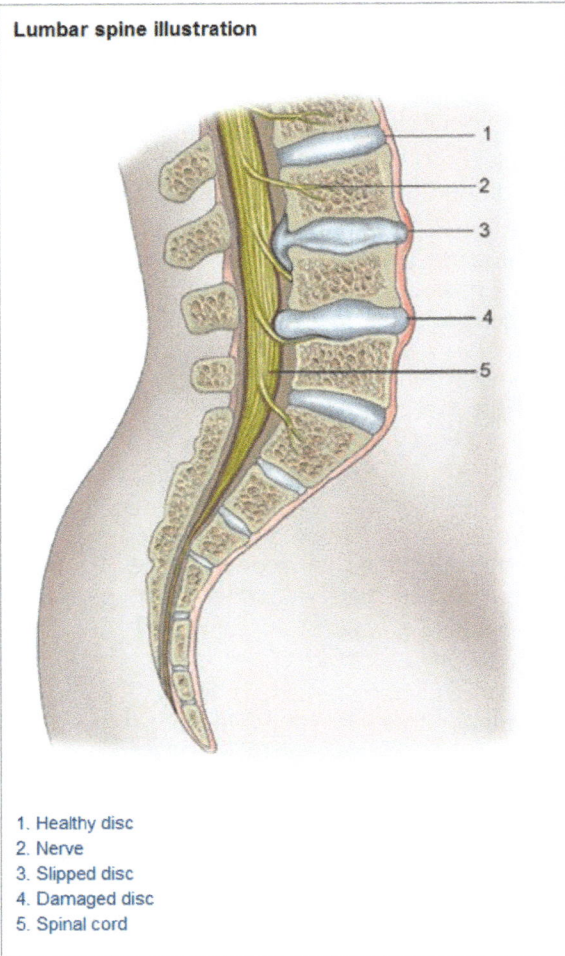

Lumbar spine illustration

1. Healthy disc
2. Nerve
3. Slipped disc
4. Damaged disc
5. Spinal cord

Figure B9-6: Lumber and slipped disc.   *Source: NHS.*

The damage to the disc is usually caused by too much pressure being applied to it during lifting and handling operations that involve 'top-heavy' bending. This is where the knees are not bent sufficiently and the head and upper body are bent over, causing the spine to bend in a curved manner (sometimes called stooping).

This bending creates a high degree of leverage force on the base of the spine, which leads to the extreme pressure exerted on the disc. The force and therefore pressure on the disc is accentuated by the carrying of heavy loads. Though top-heavy bending is a particular cause of this damage, poor posture due to leaning over for a sustained period could lead to similar damage.

EYE AND EYESIGHT EFFECTS

Where a poor posture is adopted by a worker while doing tasks that require the worker to read a display screen, machine controls or other information requiring visual acuity, this can lead to eye strain, the result of which could cause fatigue to the muscles of the eye. This may accentuate any normal acuity difficulties the worker has and make it difficult to focus on the item in question for more than short periods. Work requiring a fixed posture of attentiveness can particularly accentuate this, for example, monitoring a display screen relating to security surveillance or air traffic control.

It should also be remembered that a worker's eyesight acuity limitations may cause a posture to be adopted that can lead to musculoskeletal harm.

FATIGUE

Workers may experience fatigue of muscles of the body because they have to hold a postural position for a long time, manually handle heavy loads or repeatedly conduct a task. The muscles used to carry out work lose their ability to contract when there is insufficient adenosine triphosphate (ATP) available. This is a substance that serves as an energy source for many metabolic processes, like use of muscles.

In periods of heavy exertion, the body produces ATP by the aerobic breakdown of glucose and glycogen, using oxygen supplied in the blood. The cardiovascular system is limited in its ability to supply blood and oxygen to keep muscles working without becoming fatigued. Build-up of lactic acid from exertion is also said to contribute to muscle fatigue, by blocking the signals to the muscles. This slows the muscle action, in order to balance the demands on the body with the amount of oxygen that it can supply. This allows sufficient oxygen to be supplied to the muscles for the reduced amount of work they are doing.

STRESS

The repetitive actions, difficult posture and heavy exertion of some work can act as a constant pressure on the physical and mental condition of the worker. Being in a constant state of fatigue or discomfort can lead to the mental condition of stress.

SPRAINS/STRAINS, FRACTURES AND LACERATIONS

Around 25% of all injuries reported to the appropriate enforcing authority have been attributed to the manual lifting and handling of loads. The most common injuries are sprains, strains, fractures and lacerations.

Sprains and strains

Sprains and strains often occur in the back or in the arm and wrists. Though injuries in the legs can also occur where the leg has been hyperflexed when kneeling or handling loads beyond the worker's limit.

- A *sprain* is an injury to a ligament, the tough fibrous tissue that connects a bone to another bone. Ligament injuries involve a stretching or a tearing of this tissue.
- A *strain* is an injury to either a muscle or a tendon, the tissue that connects muscles to bones. Depending on the severity of the injury, a strain may be a simple overstretch of the muscle or tendon or it can result in a partial or complete tear. Rupture of the muscles in a section of the abdominal wall can cause a hernia.

These types of injuries tend to occur where a person's body has been over loaded due to a large steady load being applied, a sudden smaller load being applied without the opportunity of the ligament or tendon to stretch or a part of the body being forced to move in an unusual way. This can be because a person is reaching over to an extreme level (sometimes without picking up a load as their own body weight may be sufficient to cause damage), two people picking up a load together in an uncoordinated way or a load may slip and someone move into an awkward position to prevent it falling completely.

Cumulative muscle strain

Prolonged use of muscles can lead to strain, which involves tightening of the muscles, that persists after exertion has ceased. This can be caused by the work being done over-stretching the muscle fibres a small amount beyond their normal limit. With insufficient recovery time the strain on the muscles can become cumulative. If sustained over-stretching occurs it can lead to progressive damage and scarring of muscle fibres, leading to progressively restricted movement.

Hernia

A common injury resulting from lifting is a hernia, which develops in the abdomen when a weakness in the abdominal wall evolves into a localised hole, or distortion, through which tissue or abdominal organs may protrude.

A hernia is the protrusion of an organ or the fascia (a layer of fibrous tissue) of an organ through the wall of the cavity that normally contains it.

There are different kinds of hernia, each requiring a specific management or treatment.

Fractures

A fracture is a break in a bone, which is usually the result of trauma where the physical force exerted on the bone is stronger than the bone itself.

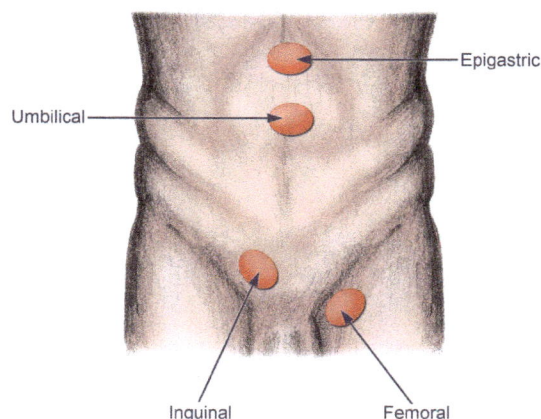

Figure B9-7: Variations of hernias type and location.
Source: img.medscape.com.

Older people, whose bones are more brittle, are more likely to get fractures than younger people, a feature we have to have regard to as the age of the working population increases. A fracture can also be made more likely by an acquired disease of bone such as osteoporosis (low bone mineral density, which leads to an increased risk of fracture) or by abnormal formation of bone in a disease such as osteogenesis imperfecta ('brittle bone disease').

Fractures are classified according to their character and location. Fractures can be broadly described as:

- Closed (simple) fractures are ones in which the skin remains intact.
- Open (compound) fractures are ones where the bone breaks through and exposes the bone.
- Displacement (fracture gap in alignment) and angulations (out of alignment); if displacement or angulations is large, reduction (manipulation) of the bone may be required.
- Complete fracture is where bone fragments separate completely.
- Compression fractures usually occur in the vertebrae, for example, when the front portion of a vertebra in the spine collapses.

Fractures of the hands and feet are the most likely type of fractures to arise from manual handling work and may be due to the load that is being lifted inadvertently being dropped.

Lacerations

Lacerations are often caused by the unprotected handling of loads with sharp corners or edges, or slippage of grip when trying to prevent a dropped load.

Examples of jobs and workplace situations that give rise to risks of these injuries and ill-health conditions

There are a wide range of jobs and workplace situations that may give rise to risks relating to WRULDs.
Various factors have the potential to cause WRULDs, such as:

- Repetitive work.
- Uncomfortable working postures.
- Sustained or excessive force.
- Carrying out tasks for long periods without suitable rest breaks.
- Poor working environment and organisation.

These are found within a wide range of workplace situations, for example, production/assembly lines, working in restricted work spaces, use of computers/laptops, manual handling of objects and people, etc.

PRODUCTION/ASSEMBLY LINES

When working on production or assembly lines it will introduce ergonomic risk factors including:

- Potential repetitive nature of the task involving frequent movements of the upper body.
- Poor posture may be adopted by the operators, for example, standing and reaching from one conveyor to a workbench or vice versa.
- There may be an imposed work rate and the speed of the production or assembly line(s).
- Height or width of the production or assembly line(s), for example, conveyor to work-bench.
- Patterns of work in relation to the number and length of the breaks allowed.
- Manual handing of loads on the production or assembly line(s).

Figure B9-8: WRULDs caused by conveyor work (repetitive twisting).
Source: Bremerhaven.

WORKING IN RESTRICTED WORK SPACES

Working in restricted spaces often forces employees to adopt poor posture for long periods of time. Workers in excavations doing repair or maintenance work on water pipes might find themselves bending down to do work at ankle height. Restricted space greatly limits the number and type of mechanical devices such as cranes, hoists, forklifts, etc. that could be used to reduce the muscular demands on the worker.

USE OF DISPLAY SCREEN EQUIPMENT (DSE)

Key board operations may have many thousands of movements an hour, which will create repeated movement of the fingers. Where the operator is not fully proficient in using all fingers to operate the keyboard, many of the movements are centred on a small number of fingers. Where the use of the display screen equipment requires the use of a mouse this can also accentuate the use of a small number of fingers centred on the users dominant hand. Because of the amount of repeated movements, sometimes without taking breaks, the risks of WRULD are high and the effects are cumulative.

HANDLING OBJECTS

Bricklaying

Many building site trades are associated with WRULDs. Plasterers, joiners, electricians and bricklayers are commonly affected. Bricklayers are often paid by the number of bricks they lay and therefore cannot afford to lose time off from work or take regular rest breaks.

There is a risk that they may carry on working and ignore the warning signs related to musculoskeletal disorders. The size of the bricks being laid, the number of bricks they are expected to lay (often several hundred per day), weather conditions and the position of the wall or structure (which may mean reaching overhead) are factors to consider when assessing the risk.

Checkout operators

Checkout operations involve a large number of lifting and tracking movements that sweep from left to right or right to left. The item held in the operator's hands is frequently held at an extended distance from the body and will involve a large variety of sizes, shapes and weights of object.

The pace of work can be particularly high when working at peak times. This repeated movement involves the hands, arms, shoulders and lower back. Some of the movements will be particularly awkward and strenuous as it can involve large heavy items.

HANDLING PEOPLE AND ANIMALS

People handling tasks are recognised as the primary cause of musculoskeletal disorders (MSDs) among ambulance crews, carers for the disabled/elderly, nursing staff and those involved with animals, such as vets and animal rescue.

A variety of handling tasks are performed within the context of such care, such as lifting, transferring, and repositioning people, and are often performed manually. Lifting and manoeuvring people (and animals) is unlike other manual handling tasks which only involve inanimate objects.

Staff working in a busy 24 hour Accident and Emergency Department will be exposed to health risks in relating to manual handling:

- Patients may have to be moved when they are often unconscious or incapacitated and cannot help themselves or when their movements are unpredictable or unhelpful particularly if they have consumed alcohol or substances.
- Equipment may have to be moved to the patient before assistance is available and often mechanical aids are not immediately to hand.
- In an Accident and Emergency Department, staff need to work quickly.

Therefore suitable control measures will include:

- Provision of training in handling techniques appropriate for accident and emergency situation; this will include following the guidance from The Guide to the Handling of People.
- Undertaking suitable patient handling risk assessments - using Rapid Entire Body Assessment (REBA).
- Working in teams of at least two people.
- Ensuring that equipment was always at hand to assist in any handling or lifting operation that has to be carried out.

B9.2 - Assessing risks from repetitive physical activities, manual handling and poor posture

Legal requirements

All employers have a common law duty of care to ensure their employees' health and safety at work. This is reinforced by the Health and Safety at Work etc. Act (HASAWA) 1974. In particular, Section 2(2)c of HASAWA 1974 sets out the requirement that the employer, for the benefit of employees, must provide:

> *"Arrangements for ensuring, so far as is reasonably practicable, safety and absence of risks to health in connection with the use, handling, storage and transport of articles and substances".*

Figure B9-9: Section 2(2)c of HASAWA 1974.   *Source: Health and Safety at Work etc. Act (HASAWA) 1974.*

The other, wider duties of Section 2 of HASAWA 1974 establish obligations that mean arrangements have to be in place that would manage repetitive activities and minimise poor posture. These duties establish the need of the employer to assess risks and take action with regard to repetitive physical activities, manual handling and poor posture.

In addition, the Management of Health and Safety at Work Regulations (MHSWR) 1999, place specific duties on the employer to assess the risks associated with all work activities. Other regulations, the Health and Safety (Display Screen Equipment) Regulations (DSE) 1992 and Manual Handling Operations Regulations (MHOR) 1992, also contain requirements for risk assessment specific to the hazards and risks they cover. Where an employer is assessing a work situation or activity for the first time, a general assessment is particularly useful to identify where a more detailed risk assessment is needed to fulfil the requirements of specific regulations. Where employers have already carried out assessments under other regulations, they may not need to repeat those assessments as long as they remain valid, but they do need to ensure that they cover all significant risks.

MANAGEMENT OF HEALTH AND SAFETY AT WORK REGULATIONS (MHSWR) 1999

The Management of Health and Safety at Work Regulations (MHSWR) 1999 are designed to encourage employers to take a holistic approach to risk assessment. The regulations contain general duty to risk assess and manage risks that will relate to all work activities that could lead to musculoskeletal conditions. This will accommodate repetitive physical activities that are not covered by the more specific requirements of MHOR 1992, for example, use of tools for sweeping/mopping or tasks requiring manipulation, such as grinding or polishing.

The MHSWR 1999 may also apply to situations where posture is an issue or the combination of risks may need to be considered because they may create additional concerns for the individuals involved. Risk assessments are conducted to satisfy more specific regulations and will normally be sufficient to fulfil the general requirements of the MHSWR 1999 for the specific issue considered.

MANUAL HANDLING OPERATIONS REGULATIONS (MHOR) 1992

The Manual Handling Operations Regulations (MHOR) 1992, as amended in 2002, requires employers to assess the risks to employees from manual handling operations.

The regulations establish a clear hierarchy of measures:

■ Avoid hazardous manual handling operations so far as is reasonably practicable.

■ Make a suitable and sufficient assessment of any hazardous manual handling operations that cannot be avoided.

■ Reduce the risk of injury so far as is reasonably practicable.

The assessment must have regard to the questions set out in Schedule 1 of MHOR 1992 relating to the task, load, working environment, individual and other factors specified. The MHOR 1992, Regulation 4, was amended in 2002 to include an additional requirement:

"(3) In determining for the purposes of this regulation whether manual handling operations at work involve a risk of injury and in determining the appropriate steps to reduce that risk regard shall be had in particular to:

(a) The physical suitability of the employee to carry out the operations.

(b) The clothing, footwear or other personal effects he is wearing.

(c) His knowledge and training.

(d) The results of any relevant risk assessment carried out pursuant to regulation 3 of the Management of Health and Safety at Work Regulations 1999.

(e) Whether the employee is within a group of employees identified by that assessment as being especially at risk.

(f) The results of any health surveillance provided pursuant to regulation 6 of the Management of Health and Safety Regulations 1999".

Employees also have duties, in addition to the general duties imposed on them by the sections 7 and 8 of the HASAWA 1974 or MHSWR 1999, and they must:

■ Make full and proper use of any equipment or system of work provided by his employer.

■ Inform his employer about any physical condition suffered by him which might reasonably be considered to affect his ability to undertake manual handling operations safely.

See also - Relevant statutory provisions - Manual Handling Operations Regulations (MHOR) 1992.

HEALTH AND SAFETY (DISPLAY SCREEN EQUIPMENT) REGULATIONS (DSE) 1992

The Health and Safety (Display Screen Equipment) Regulations (DSE) 1992, as amended in 2002, aim to reduce risks from display screen equipment work, include WRULDs, such as carpal tunnel syndrome. They apply where there are employees who 'habitually use display screen equipment as a significant part of their normal work'. Such people are called users.

Employers are required to:

■ Analyse workstations to assess and reduce risks.

■ Ensure that workstations used for the purposes of the employer's undertaking meet specified minimum requirements set out in the schedule to the regulations, which relate to requirements of the EU Directive on display screen equipment. This includes the equipment, environment and interface between computer and user.

■ Plan work activities so that they include breaks or changes of activity.

■ Provide eye and eyesight tests on request, and special spectacles if needed.

■ Provide information and training, training must take place before the person becomes a user.

■ The employer must also assess the need for and provide appropriate rest breaks.

WORKPLACE (HEALTH, SAFETY AND WELFARE) REGULATIONS 1992

The Workplace (Health, Safety and Welfare) Regulations (WHSWR) 1992 aim to reduce risks in the workplace, including those relating to musculoskeletal risks. They consider the workplace environment and the management of musculoskeletal risks through consideration of workstations and seating.

This will encompass those tasks that are not covered in DSE 1992, for example, sitting at a work bench soldering circuit boards. The requirements of WHSWR 1992 most relevant to managing risks of musculoskeletal disorders are:

■ Lighting is to be suitable and sufficient for the type of work, for example, no glare or reflections, no dark areas where people are handling loads.

■ Room dimensions and space must be adequate, so that people have sufficient room to manoeuvre and workstations are not too close together for comfort; a minimum recommended space of 11 m^{-3} per person.

■ Workstations and seating have to be suitable for the person and the work being done.

■ Cleanliness and tidiness of the workplace (to prevent slips, trips and falls).

Ergonomic principles as applied to the control of musculoskeletal risks

The study of ergonomics is essential to good job design. It is the applied science of equipment design intended to maximise productivity effectiveness by reducing worker fatigue and discomfort.

The meaning of ergonomics can be defined as the study of the interaction between workers and their work, in terms of the design of the workplace, work equipment and work methods.

Ergonomic principles are:

1) Designing the workplace, work methods and work equipment to suit the worker.

2) Ensuring a good fit between the person and their work as far as tools, equipment and workstation are concerned.

Figure B9-10: Workstation design. *Source: RMS.*

Anthropometry, is defined as the scientific measurement of the human body and its movement. It is a broad area of study that includes the disciplines of psychology, physiology, anatomy and design engineering.

Ergonomics has the human being at the centre of the study, where individual capabilities and fallibilities are considered in order to, ultimately, eliminate the potential for human error and harm to effectiveness and efficiency. It is also the study of ways to prevent 'ergonomic illnesses'.

This includes the minimisation of such things as work-related musculoskeletal disorders that are caused by poorly designed machines, tools, tasks or the workplace.

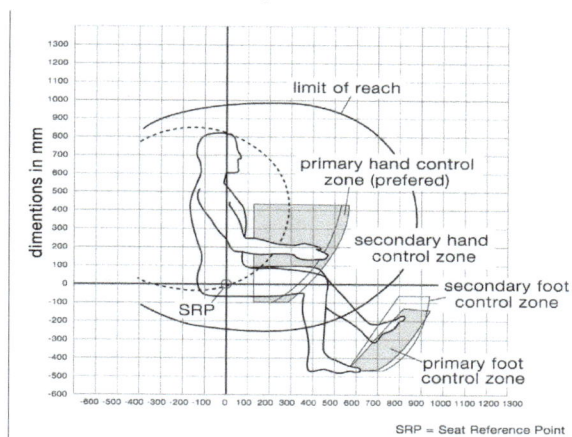

Figure B9-11: Reach distance to controls. *Source: McPhee, 2005.*

The aims of ergonomics are to design the equipment and the working environment to fit the needs and capabilities of the individual, i.e. fitting the task to the individual, and to ensure that the physical and mental well-being of the individual is being met. This involves the consideration of psychological and physical factors, including the work system, body dimensions (for example, reach distance to controls), capability (consideration of fatigue reduction), competence (individual knowledge, understanding and skill) and the work environment (layout, noise, temperature and lighting).

Individuals have different physical capabilities due to height, weight, age and levels of fitness. They also have different mental capabilities, memory retention and personalities. All these factors can influence ergonomic choices and the successful matching of the workplace to the individual.

Practical measures that can be undertaken to prevent WRULDs from poor ergonomic design, include:

■ Designing working space and equipment to suit individuals and not 'average' persons.
■ Providing equipment and furniture that are easily adjustable.
■ Fitting controls that are clearly marked, standardised and easily accessible.
■ Introduction of automation of part or all of a system.
■ Eliminating or reducing the need for employees to twist, stretch and bend.
■ Reducing the weight of loads that need to be moved manually.
■ Designing tools to minimise the application of force.
■ Changing work patterns.
■ Improving the working environment particularly as far as temperature, lighting and noise are concerned.
■ Improving the presentation of information using signs, displays and colour coding.

Consideration of task, load, force, working environment, equipment and individual capability

When assessing the risks from repetitive physical activities, manual handling and poor posture the employer should have regard to the following factors:

■ The task.
■ The load.
■ Force.
■ The working environment.

- Equipment.
- Individual capability.

The detailed consideration of each factor is necessary to achieve a suitable and sufficient risk assessment.

The process of risk assessing includes observing the task as it is actually done; recording the factors that contribute to risk; assessing the level of risk that each factor represents (taking account of the circumstances and controls in place); and considering if the risks are different at different times and for different people.

Each factor, in turn, should be considered for assessment to determine whether there is a risk of injury. Not all factors will be relevant to all situations, for example, the main factors to consider in manual handling are the task, load, environment and individual.

When this has been completed, the information can then be processed to give a *suitable and sufficient* risk assessment.

Factors	Questions	Level of Risk:		
		High	Med	Low

Task	Does it involve:
	
	■ Holding load at distance from trunk?
	■ Unsatisfactory bodily movement or posture?
	  • Twisting the trunk.
	  • Stooping.
	  • Reaching upwards.
	■ Excessive movement of load?
	  • Excessive lifting or lowering distances.
	  • Considerable carrying distances.
	  • Positioning the load precisely.
	  • Excessive pushing or pulling distances.
	  • Frequent or prolonged physical effort.
	■ Insufficient rest or recovery periods?			

Load	Is it: ■ Heavy? ■ Bulky or Unwieldy? ■ Difficult to grasp? ■ Unstable, or with contents likely to shift? ■ Sharp, hot or otherwise potentially damaging?			
Working Environment	Are there: ■ Space constraints preventing good posture? ■ Uneven, slippery or unstable floors? ■ Variations in level of floors or work surfaces? ■ Extremes of temperature, humidity or air movement? ■ Poor lighting conditions?			
Individual Capability	Does the job: ■ Require unusual capability? ■ Call for special information/training? Particular consideration should be given to employees who: ■ Have recently been pregnant. ■ Are known to have a history of back, knee or hip trouble, hernia or other health problems which could affect their manual handling capability or physical or learning difficulty. ■ Young workers.			

Figure B9-12: Manual handling risk assessment.
Source: (HSE Manual handling) Manual Handling Operations Regulations (MHOR) 1992 Guidance L23.

TASK

Task related factors include:

- Repetition - the more the task is repeated the greater the risk.
- Working posture - if the posture is awkward and/or held for prolonged periods in a fixed position the risk will increase.
- Duration - the greater the duration without rest the greater the risk. This includes the duration of a given task, the cumulative duration during a shift and the cumulative duration over time.

LOAD

The main factors related to load, identified in *figure ref B9-12*, are:

- Weight.
- Bulky or unwieldy.
- Difficulty in grasping the load.
- Unstable loads, or with contents that are likely to shift when it is moved.
- Sharp, hot or otherwise potentially damaging surfaces.

FORCE

When assessing the influence of force consideration should be given to the:

- Force necessary to handle or move an object.
- Force necessary to hold a tool or material being worked on.
- Speed of application of the force.
- Amount of force necessary to conduct the task: is it necessary to generate excessive force by use of muscles?
- Force necessary to grip or operate something that is done repetitively.
- Reactive forces effect, from the equipment being gripped, such as pliers or a screwdriver digging into the hand.
- Duration that the force is applied.

WORKING ENVIRONMENT

Consideration of the environment will include factors like:

- Space requirements, for example, a workstation should have sufficient space for the worker to change position and vary movements.
- Layout of workplace, for example, workstation design may not provide easy access to equipment, such as display screen equipment key boards or components used to assemble goods, resulting in poor posture and positioning of the hands giving rise to back, carpal tunnel and tenosynovitis injuries.
- Lighting, for example, good general and local lighting that provides appropriate contrast between what is being viewed and the background environment. This might include when viewing a DSE screen or hazards that could affect moving loads. Glare and reflections should be considered, for example, artificial light sources or daylight from a window.
- Noise and vibration should not exacerbate the risk.
- Temperature should not be unreasonable, such that the worker can maintain a good core body temperature while working.
- The humidity should not be so low that it affects the eyes of DSE operators nor so high that it prevents effective sweating to maintain core body temperature.

EQUIPMENT

The equipment available to conduct the activity can greatly influence the level of risk of musculoskeletal disorders. The provision of load handling equipment can minimise the amount of physical effort necessary to lift and move loads; equipment of this type can include the use of a sack truck, conveyors and pneumatic load lifting equipment. Tools that provide leverage or additional grip can reduce the forces necessary.

Tools that require a very firm grip can also create upper limb problems if the handles are narrow and dig into the soft tissues of the hands. The problem is compounded where the pressure has to be held for several seconds at a time, or where the application of pressure is very frequent.

Display screen equipment used in the workplace can significantly affect musculoskeletal risks and consideration should be made to the following factors:

- Display screen.
- Keyboard.
- Work desk or work surface.
- Work chair.

INDIVIDUAL CAPABILITY

The factors to consider include:

- The physical capabilities required by the work.

- The variety of body sizes of workers, such as height.
- Whether the work requires unusual capability.
- Whether special information or training is required.
- Information gained from medical surveillance conducted.
- The individuals eye sight.

Particular consideration should be given to:

- Those women who are or have recently been pregnant.
- Those known to have a history of back, knee or hip trouble, hernia or other health problems that can affect musculoskeletal risks.
- Those with physical or mental disability.
- Young workers.

Guidelines for assessment - handling operations

LIFTING

The Manual Handling Operations Regulations (MHOR) 1992 set no specific requirements such as weight limits. An assessment based on a range of relevant factors should be used to determine whether there is a risk and whether there is a need for any remedial action. The guidelines to MHOR 1992 set out an *approximate* boundary within which manual handling operations are unlikely to create a risk of injury sufficient to warrant more detailed assessment.

This should enable assessment work to be concentrated where it is most needed. Even operations lying within the boundary should be avoided or made less demanding wherever it is reasonably practical to do so.

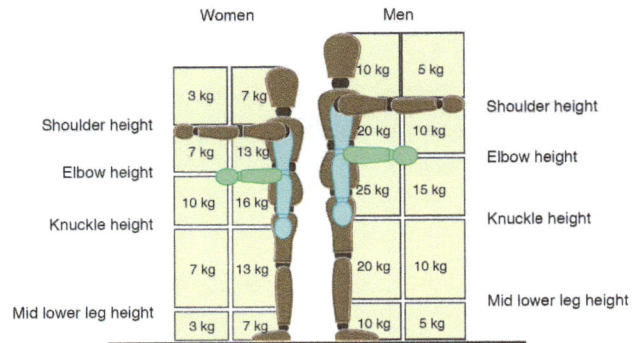

Figure B9-13: Guideline figures. Source: HSE: Guidance L23.

The guideline figures are not weight or force limits. They may be exceeded where a more detailed assessment shows it is safe to do so. However, even for a minority of fit, well-trained individuals working under favourable conditions the guideline figures should not normally be exceeded by more than a factor of about 2.

The guideline figures for weight and force will give reasonable protection to nearly all men and between one half and two thirds of women. To provide the same degree of protection to nearly all working women the guideline figures should be reduced by about one third.

CARRYING

The guideline figures for manual handling operations involving carrying are similar to those given for lifting and lowering. It is assumed that the load is held against the body and is carried no further than about 10 metres without resting. If the load is carried over a longer distance without resting the guideline figures may need to be reduced.

Where the load can be carried securely on the shoulder without attendant lifting, for example, unloading sacks from a vehicle, a more detailed assessment may show that it is safe to exceed the guideline figure.

HANDLING WHILE SEATED

The guideline figures for handling operations carried out while seated are shown in *figure ref B9-14* and apply only when the hands are within the box zone indicated. If handling beyond the box zone is unavoidable a more detailed assessment should be made.

TWISTING

The basic guideline figures for lifting and lowering should be reduced if the handler twists to the side during the operation. As a guide the figures should be reduced by about 10% where the handler twists through 45° and by about 20% where the handler twists through 90°.

These guideline figures should not be regarded as precise recommendations and should be applied with caution, noting particularly the assumptions on which they are based.

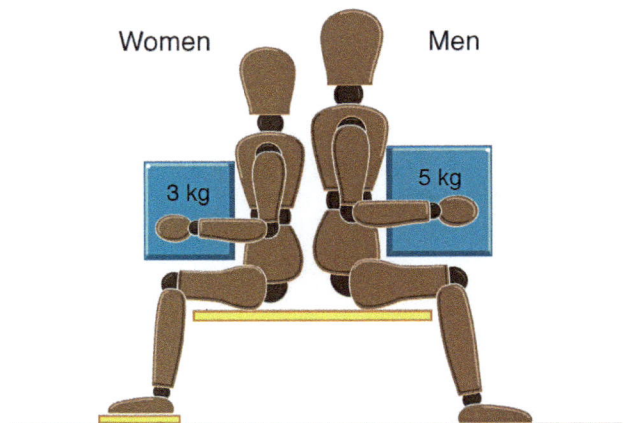

Figure B9-14: Handling while seated. Source: HSE Guidance L23.

ASSUMPTIONS

The UK guideline standard figures for lifting and lowering should not be regarded as absolute limits or allowances and should be applied with caution, noting particularly that they are based on the following assumptions:

- The handler is standing or crouching in a stable body position with the back substantially upright.
- The trunk is not twisted during the operation.
- Both hands are used to grasp the load.
- The hands are not more than shoulder width apart.
- The load is positioned centrally in front of the body and is itself reasonably symmetrical.
- The load is stable and readily grasped.
- The work area does not restrict the handler's posture.
- The working environment (heat, cold, wet, condition of floor) and any personal protective equipment used do not interfere with performance of the task.

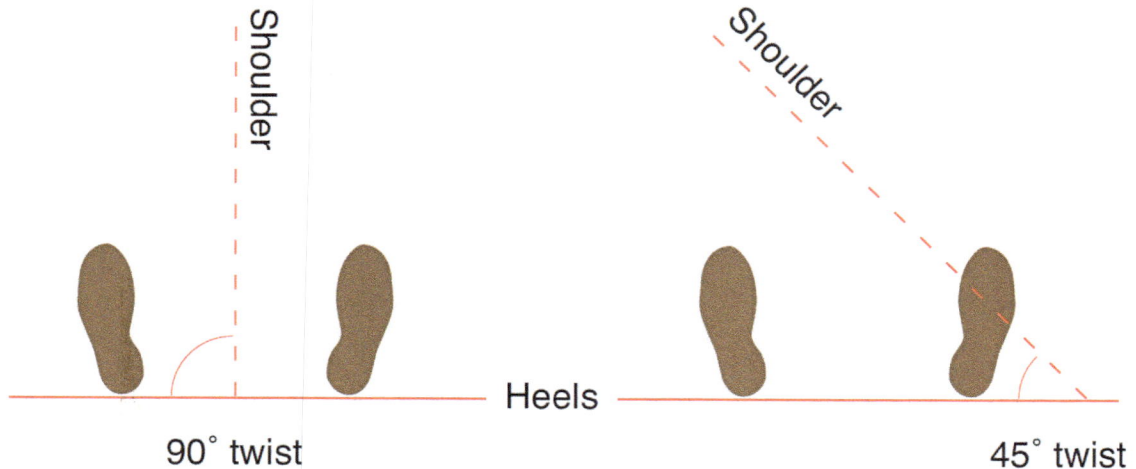

Figure B9-15: Twisting.

Source: HSE, Guidance L23.

ASSESSING THE TASK

Headroom

Stooping to move an item is a bad movement, but stooping to lift a load is even more likely to cause injury. The highest risk move is to stoop, lift and twist. This is a common combination, such as in removing items from under racking, boxes of paper from low shelves and in maintenance of vehicles, conveyors, machinery etc.

Where stooping and twisting cannot be designed out of the task, then help may be required with the lifting element of the task.

Working on different levels

Carrying loads up steps, stairs, and ladders can be problematic:

- Climbing ladders requires at least one hand to grip the ladder.
- Stairs, even in good condition, can cause trips, often dependent on other factors such as type of shoes, height or weight of package, wearing bifocal spectacles or passing others going in the opposite direction. Carrying a load up or down stairs makes it difficult for the person to see their feet, making a stumble and sudden movement more likely.
- Doors at the top and bottom of stairs may well open in the opposite direction to travel and may be spring-loaded.

Height of storage

The height of storage of loads is important in that the objective will be to eliminate unnecessary lifting where possible. The best height for storage or benches is around waist-height, particularly for heavy items.

Where pallets of heavy items, for example, up to 50 kg each, are used it may be sensible to put the pallet on top of two other empty pallets, with the objective of ensuring that the operator does not have to stoop too low to lift or place items.

Holding loads away from the body

Holding loads away from the body causes additional and mostly unnecessary stress on the back. At arms' length, the load that can be handled may be reduced by as much as 80%. Loads should be held close to the body where possible, as this also allows the body and its clothing to give frictional help in stabilising the load. In addition, it allows the arms to be brought into the sides and thus reduce stress on the neck, shoulders and arms.

The provision of protective clothing may encourage people to hold loads close to their bodies. This can be a problem in an office environment where people handle heavy, often dusty, materials infrequently (without the benefit of protective clothing) increasing their risk of injury by holding the load away from their body in order to keep their clothing clean.

Posture

Poor posture injuries that are received as a result of carrying out activities that include manual handling operations need not necessarily arise solely from lifting large, awkward or heavy items. Poor posture can greatly increase the likelihood of suffering manual handling injuries. Examples of poor posture can include over-stretching, twisting, lifting with the spine (in bending position) or lifting whilst seated. Many construction and maintenance tasks can encourage the worker to take up a poor posture so they are bent over for a period of time, for example, laying a floor.

Due consideration should also be given to clothing, as tight fitting clothing will restrict the person from adopting the correct lifting posture.

Static loads

Holding static loads can rapidly cause fatigue in muscles, particularly when the arms are extended above the head. In fact, all work above the head has a similar effect. The pains and any immobility arising from this work are called work related upper limb disorders.

Lifting loads whilst seated

There is usually an element of twisting involved when lifting whilst seated. This problem may be compounded by reaching and bending with the added possibility of the seat moving or tipping.

Personal Protective Equipment (PPE)

The wearing of PPE can add additional stress to workers carrying out a manual handling task and care should be taken to ensure that the PPE does not become a hazard. Comfort and mobility are the two areas that need assessing.

Often the combination of work rate and PPE results in excessive temperature, stress, early fatigue, and could, with an unfit person, endanger their health.

Rest

When considering the work task, regard should be given to the provision of suitable rest breaks for individuals engaged in highly repetitive manual handling such as that involving frequent twisting or turning.

Similarly, those carrying loads over a distance further than 10 metres will need to plan a rest break. If the need for rest breaks is frequent, then consideration may need to be given to the use of more than one person to conduct the work using job rotation, alternating between high fatigue work and less demanding work.

ASSESSING THE LOAD

Assessing the load means not only considering its weight but:

- Physical size, whether it can be lifted between the knees from the ground by one person. If not, then assistance may be required.
- The centre of gravity of a load, for example, whether it will twist when lifted.
- Flexible loads, such as bags of cement, may shift and strain the back of the worker.
- Hot loads, be they tar for a roof repair or food in a container, require protective equipment and great care.
- Loads that can cut or pierce the hands or body require protective equipment and a good system of work.
- Extremely cold loads offer hazards that range from the lack of grip to 'frost burn' of the skin if protective clothing is not used.
- Loads may contain toxic or corrosive substances and may require assessment under other regulations such as the Control of Substances Hazardous to Health Regulations (COSHH) 2002.
- Loads with fragile or damaged packaging have the potential to damage feet and legs, or even the back of the worker, if a package breaks suddenly.

ASSESSING THE FORCE

Pushing and pulling (force)

Guideline figures for manual handling operations involving pushing and pulling, whether the load is slid, rolled or supported on wheels are as follows. The guideline figure for starting or stopping the load is a force of about 250 Newtons (i.e. a force of about 25 kg as measured on a spring balance).

The guideline figure for keeping the load in motion is a force of about 100 Newtons. It is assumed that the force is applied with the hands between knuckle and shoulder height. If this is not possible the guideline figures may need to be reduced. No specific limit is intended as to the distances over which the load is pushed or pulled provided there are adequate opportunities for rest or recovery.

ASSESSING THE WORKING ENVIRONMENT

Assessing the working environment, in the main, is looking at an area and really seeing what is there.

The difficulty is that this takes considerable experience and training, because over familiarity with a poor environment makes it difficult to perceive the hazards when assessing the risks.

Floors

It may be appropriate to assess the floor first, because the results of trips, slips and falls can be compounded when someone is lifting or carrying a load. Uneven or slippery floors create unpredictability and make the person carrying the load tense, which in itself causes fatigue. A partial slip or trip under these circumstances will cause a person's balance to be lost and a sudden reflex of the muscles may cause soft tissue injuries.

Housekeeping is an important factor when it comes to slips and trips. Pallets, items of plant or equipment, rubbish bins, trucks and trolleys may be left on the route of those handling loads.

Studs often protrude from the floor long after items of plant have been removed, duck-boards get moved into passages and packaging material is left to be moved later.

Distance between storage areas

The carrying of loads should be restricted to a minimum and due consideration should be given to using mechanical means where distances are more than a few metres. Guidance to MHOR 1992 suggests that distances over 10 metres present a growing risk.

Extremes of temperature, high humidity or windy conditions

Injury risk is increased:

- By high temperatures/humidity where fatigue will occur quicker and perspiration may cause hands to become wet and reduce grip.
- Low temperatures can cause fingers to get cold and loose grip, feet to become numb and balance more difficult to achieve. Extra clothing may restrict movement and in extreme conditions floors may become slippery, packages will be harder to grip and may even be stuck together or to the ground by ice.
- Working in high winds will increase the difficulty of control of loads of large area, often compounded by dust entering the worker's eyes reducing visibility.

Poor lighting

Poor lighting prevents a person from seeing the condition of floors and noting any trip or slip potential. Damaged packaging may go unnoticed until the load drops apart during lifting or carrying.

Where handling takes place in an area used by mechanical/electrical transport, the lighting is poor and persons appear from behind racking or stacks, then drivers may not have time to take avoiding action. The sudden appearance of the vehicle may throw the load-carrying person off-balance and result in muscle injury.

Similarly, where loads have to be carried from a bright area into a dark area, for example, from bright sunlight into a warehouse, collision or trip accidents will happen because the eyes take up to 5 seconds to adjust to the lower lighting conditions indoors.

Appropriate lighting should be provided particularly when there is a change of level or direction to prevent slips, falls or dropping of the load, which may result in musculoskeletal injuries.

Good lighting is essential for manual handling operations to enable workers to read packaging to identify any warnings such as weight or unusual distribution of the load within a package. Regulation 4(1)(b) MHOR 1992 requires employers to:

"Take appropriate steps to provide any of those employees who are undertaking any such manual handling operations with general indications and, where it is reasonably practicable to do so, precise information on -

(a) The weight of each load.

(b) The heaviest side of any load whose centre of gravity is not positioned centrally."

Figure B9-16: Regulation 4(1)(b) of MHOR 1992.   *Source: Manual Handling Operations Regulations (MHOR) 1992.*

ASSESSING EQUIPMENT

The equipment available to a worker when conducting manual handling should be considered as part of the assessment. Equipment provided to give mechanical assistance when moving loads will reduce the risk of musculoskeletal injury. The assessment should consider if equipment is in good working order. Equipment that is hard to use, does not work well or requires significant effort to operate will not provide an assured reduction in risk.

ASSESSING INDIVIDUAL CAPABILITY

When assessing the individual it should be done in conjunction with the assessment of the task and the load.

People have varying degrees of capabilities that may not be obvious when they are not working; factors to consider include:

- Levels of medical fitness.
- Level of manual handling training received, and how recently that training was undertaken.
- Recent past history of work which will help indicate a person's abilities.

Care must be taken when assessing persons who are:

- Overweight.
- Frail.
- Young, for example, under 18 years.
- Relatively inactive types.
- Women (who may not be capable of lifting/handling the same loads as men of similar size).
- Pregnant women.

Depending on the task, exclusion or inclusion might depend upon:

- Physical size.
- Height.
- Willingness (often a useful measure).

Assessment of risk - display screen equipment

EQUIPMENT

Display screen

- Characters on the screen should be well-defined and clearly formed, of adequate size and with adequate spacing between the characters and lines.
- Image on the screen should be stable, with no flickering or other forms of instability.
- Brightness and the contrast between the characters and the background shall be easily adjustable by the user, and also be easily adjustable to ambient conditions.
- Screen must swivel and tilt easily and freely to suit the needs of the user.
- A separate base for the screen or an adjustable table should be provided.
- Screen should be free of reflective glare and reflections liable to cause discomfort to the user.

Keyboard

- It should be possible to tilt the keyboard and to use it separately from the screen so as to allow the user to find a comfortable working position, avoiding fatigue in the arms or hands.
- Space in front of the keyboard should be sufficient to provide support for the hands and arms of the user.
- Keyboard should have a matt surface to avoid reflective glare.
- Arrangements of the keyboard and the characteristics of the keys should be such as to facilitate the use of the keyboard.
- The symbols on the keys should be adequately contrasted and legible from the design working position.

Work desk or work surface

- The work desk or work surface should have a sufficiently large, low-reflectance surface and allow a flexible arrangement of the screen, keyboard, documents and related equipment.
- The document holder should be stable and adjustable and shall be positioned so as to minimise the need for uncomfortable head and eye movements.
- There should be adequate space for users to find a more comfortable position.

Work chair

- The work chair should be stable and allow the operator or user easy freedom of movement and a comfortable position.
- The seat should be adjustable in height.
- The seat back should be adjustable in both height and tilt.
- A footrest should be made available to any user who wishes one.

Many of the above factors could be a consideration when assessing the risks of workstations other than those related to display screen equipment.

ENVIRONMENT

- Lighting levels should be appropriate.
- Poor lighting prevents documents from being read easily.
- Excessive lighting results in eye strain.
- Reflective surfaces may cause glare.
- Control daylight with curtains or roller blinds.
- Levels of noise should not be distractive. A maximum limit of 60dBA is recommended for average display screen work, but where the work demands a particularly high degree of concentration 55dBA should not be exceeded.
- A temperature range of 20°C- 23°C for Display Screen Equipment operation is recommended.
- Low humidity, less than 40% Relative Humidity (RH) may cause sore eyes and facial acne.
- Adequate space for working.
- The floor should be level for both the workstation and the user's chair, to ensure the display screen and the chair are in correct alignment and additional postural problems are avoided.

INTERFACE BETWEEN COMPUTER AND USER

- Software should be easy to use, font size and colour definition should be considered.
- The work rate of the user should not be governed by the software, as this can lead to rapid work without rest.
- Monitoring of the user work rate should be open and with the knowledge of the user. The prospect of hidden monitoring methods may cause the operator to take unnecessary risk to deliver the work rate they perceive is necessary.

1) Adequate lighting.

2) Adequate contrast, no glare or distracting reflections.

3) Distracting noise minimised.

4) Leg room and clearances to allow postural changes.

5) Window covering if needed to minimise glare.

6) Software: appropriate to task, adapted to user, providing feedback on system status, no undisclosed monitoring.

7) Screen: stable image, adjustable, readable, glare/reflection-free.

8) Keyboard: usable, adjustable, detachable, and legible.

9) Work surface: with space for flexible arrangement of equipment and documents; glare-free.

10) Chair: stable and adjustable.

11) Footrest if user needs one.

Figure B9-17: Interface between computer and user.

Source: UK HSE, L26 Work with display screen equipment.

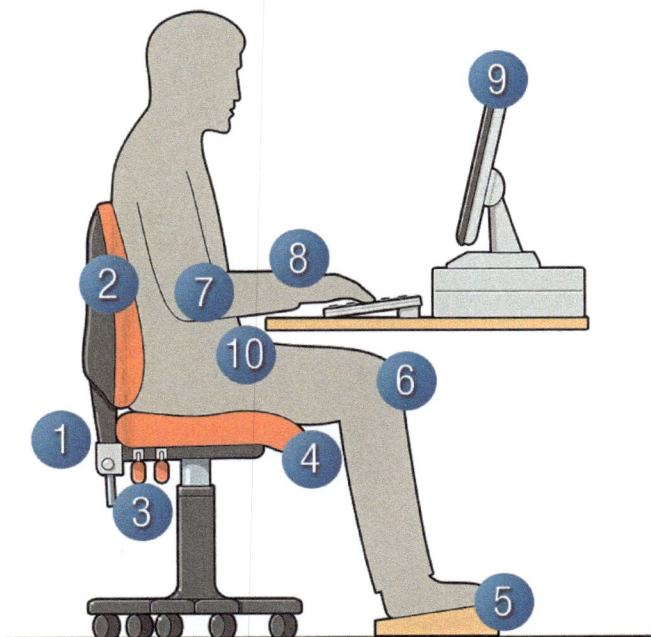

1) Seat back adjustable.

2) Good lumbar support.

3) Seat height adjustable.

4) No excess pressure on underside of thighs and backs of knees.

5) Foot support if needed.

6) Space for postural change, no obstacles under desk.

7) Forearms approximately horizontal.

8) Wrists not excessively bent (up, down or sideways).

9) Screen height and angle to allow comfortable head position.

10) Space in front of keyboard to support hands/wrists during pauses in keying.

Figure B9-18: Seating and posture for typical office tasks.

Source: UK HSE, L26 Work with display screen equipment.

Poor posture when carrying out activities like repetitive physical movement, manual handling and work with DSE equipment can result in back pain. In addition, mobile machine operators and drivers often report back pain. Back pain may be made worse by driving for a long time in a poorly adjusted seat, jolting and jarring from rough roads (whole-body vibration) and by manual handling.

Poor posture should be assessed to determine the level of risk; factors to consider include:

- Layout of the workstation, causing the worker to have to reach to a far point.
- The relationship of the task to the individual, for example, is the person right or left handed, is it possible to alter the layout to best suit their needs?
- Other factors will include individual size, length of legs arms and trunk.
- Constraints of the environment, panels of equipment or parts of the workstation that prevent the worker bending their knees.
- Whether the work is best carried out standing or sitting.
- The duration of postures held and opportunities for rest or adjustment of position.
- Inadequate workstation lighting, design and positioning, that causes the person to lean over to see the work to be done more easily.

Assessment to avoid postural injury		Comment
Vehicle	<ul><li>Can the seat be adjusted to a comfortable position?</li><li>Does the seat provide lower back support?</li><li>Is there good visibility without the need to stretch, twist or lean to one side?</li><li>Are controls within easy reach without the need to stretch, twist or lean excessively to one side?</li><li>Does the seat have a driver's weight adjustment?</li></ul>	
Use of vehicle	<ul><li>Have drivers been trained to:</li><li>Avoid bad postures, for example, slumping in the seat, constantly leaning forward or sideways or driving with the back twisted?</li><li>Drive more slowly to avoid bumping and jolting on poor road surfaces?</li><li>Steer, brake, accelerate, shift gears and operate any attachments, for example, excavator buckets, smoothly?</li><li>Not jump off the vehicle, or make other awkward movements that could jar the back?</li><li>Take regular breaks to avoid sitting in the same position for too long?</li><li>Report any back pain to the employer?</li></ul>	
Journey	<ul><li>Are routes chosen to avoid rough surfaces, rocks or potholes?</li></ul>	

Figure B9-19: Assessment to avoid postural injury.   Source: RMS.

Use of assessment tools

There are a range of different assessment tools:

- HSE manual handling assessment tool (MAC).
- HSE assessment tool for repetitive tasks of the upper limbs (ART).
- HSE variable manual handling assessment chart (V-MAC).
- Appendices 3 and 4 to the HSE's manual handling guidance (L23).
- Appendix 5 (VDU checklist) from HSE guidance (L26).
- NIOSH manual material handling (MMH) checklist.
- Rapid Upper Limb Assessment (RULA).

In addition, as mentioned above - with regards to manual handling of people the Rapid Entire Body Assessment (REBA) tool, details of which can be found within the HOP6 - The Guide to the Handling of People.

HSE MANUAL HANDLING ASSESSMENT TOOL

The Manual Handling Assessment Charts (MAC) were originally designed by the Health and Safety Executive (HSE) to help health and safety inspectors assess the most common risk factors in lifting, lowering, carrying and team-handling operations. They are now made available for open access via their website. It is designed to help the user to understand, interpret and categorise the level of manual handling risk associated with work activities. The tool evaluates the common manual handling risk factors considered when conducting risk assessment and provides guidance on the level of risk related to the factor.

For each type of assessment there is an assessment guide, a flow chart and a score sheet. It is not suitable for some operations such as pushing and pulling and the HSE make it clear that it is not a full risk assessment.

To use the charts the task needs to be observed for some time in detail, speaking to the workers and watching various workers who have to perform it. Following the assessment guide, the flow chart is used to determine the level of risk. The risks are classified as:

- Green - low level of risk, although the vulnerability of certain workers, such as pregnant workers and young people needs to be considered.
- Amber - medium level of risk.
- Red - high level of risk. Prompt action is needed as a significant number of the workforce may be at risk.
- Purple - very high level of risk - serious risk of injury especially if load is supported by one person. Requires close scrutiny.

The HSE MAC tool web site provides a series of videos to enable assessment techniques to be practiced and comparison to be made between the user and the HSE findings for each.

For further details, see 'http://www.hse.gov.uk/msd/mac/'.

HSE ART TOOL

The Assessment of Repetitive Tasks (ART) tool is designed to help the risk assessment of tasks that require repetitive movement of the upper limbs (arms and hands).

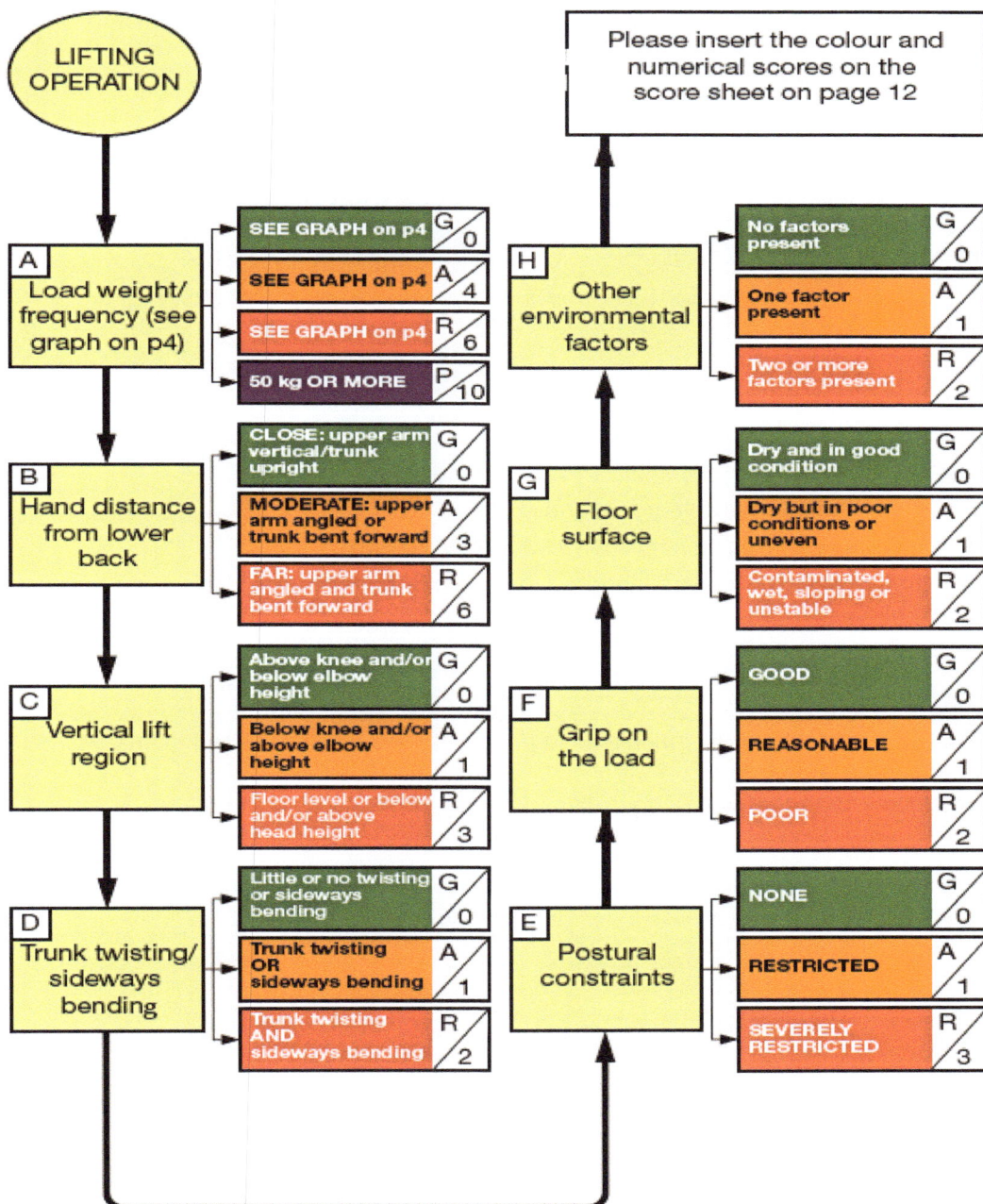

Figure B9-20: MAC tool.

Source: UK, HSE.

It helps in assessing some of the common risk factors in repetitive work that could contribute to the development of Upper Limb Disorders (ULDs).

The ART tool is a method that helps to:

- Identify repetitive tasks that have significant risks and where to focus risk reduction measures.
- Prioritise repetitive tasks for improvement.
- Consider possible risk reduction measures.
- Meet legal requirements to ensure the health and safety of employees who perform repetitive work.

As with the MAC tool, the ART tool uses a numerical score and a traffic light (green, amber, red) approach to indicate the level of risk for twelve factors. These factors are grouped into four stages:

- A: Frequency and repetition of movements.
- B: Force.
- C: Awkward postures of the neck, back, arm, wrist and hand.
- D: Additional factors, including breaks and duration.

The factors are presented on a flow chart, that leads, step-by-step, to evaluate and grade the degree of risk. The tool is supported by an assessment guide, providing instruction to help score the repetitive task being observed. There is also a worksheet to record the assessment.

C1 Head/neck posture

The neck is considered to be bent or twisted if an obvious angle between the neck and back can be observed as a result of performing the task.

The head or neck is:

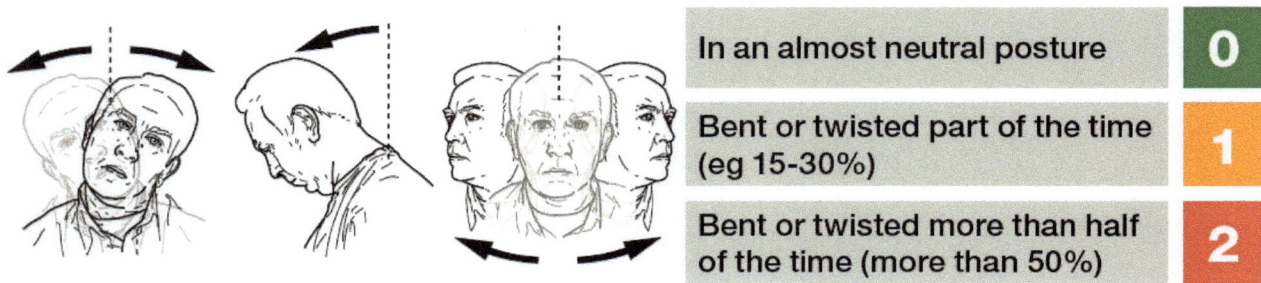

In an almost neutral posture	0
Bent or twisted part of the time (eg 15-30%)	1
Bent or twisted more than half of the time (more than 50%)	2

Figure B9-21: ART tool.   *Source: UK, HSE.*

For further details, see 'http://www.hse.gov.uk/msd/uld/art/'.

HSE VARIABLE MANUAL HANDLING ASSESSMENT (V-MAC)

The V-MAC is a tool for assessing manual handling operations where load weights vary. It should be used in conjunction with the MAC tool.

The MAC tool was designed for assessing handling operations where the same weight is handled over the workday/shift.

However, in practice, load weights are often variable (such as in order picking, parcel sorting, trailer loading/unloading, and parts delivery in manufacturing). The V-MAC was developed to help assess these kinds of jobs.

For further details, see 'http://www.hse.gov.uk/msd/mac/vmac/'.

APPENDICES 3 AND 4 TO THE HSE'S MANUAL HANDLING GUIDANCE (L23)

In the UK, the Manual Handling Operations Regulations 1992, as amended, introduces the need to conduct risk assessments for manual handling operations and to include consideration of the task, individual, load, and environment. The following checklist is adapted from the guidance on the Regulations L23.

No. operator/user		Assessed by	Date
Activities		**Techniques used**	
		Good handling technique per HSE guidance	

Factor to be considered	n/a Y N	Level of risk			Comments What is already being done to control the risk? What further controls are needed?
		H	M	L	
The tasks - do they involve					
• Holding loads away from trunk?					
• Twisting?					
• Stooping?					
• Reaching upwards?					
• Large vertical movement?					
• Long carrying distances?					
• Strenuous pushing or pulling?					
• Unpredictable movement of loads?					
• Repetitive handling?					
• Insufficient rest or recovery?					
• A work rate imposed by a process?					
The loads - are they:					
• Heavy?					
• Bulky/unwieldy?					
• Difficult to gasp?					
• Unstable/unpredictable?					
• Intrinsically harmful (for example, sharp/hot)?					
The working environment - are there:					
• Constraints on posture?					
• Poor floors?					
• Variations in level?					
• Hot/cold/humid conditions?					
• Strong air movements?					
• Poor lighting conditions?					
Individual capability - does the job:					
• Require unusual capability?					
• Hazard those with a health problem?					
• Hazard those who are pregnant?					
• Call for special information/training?					
Other factors:					
• PPE?					
•					
•					
Action					

Figure B9-22 Manual handling checklist.

Source: UK, HSE L23.

APPENDIX 5 (VDU CHECKLIST) FROM HSE GUIDANCE (L26)

In the UK, the Health and Safety (Display Screen Equipment) Regulations 1992, as amended, introduces the concept of ergonomics in the use of display screens. The Regulations identify users who may be at risk, and the regulations require that a risk assessment is conducted. The following assessment form has been adapted from L26 from the guidance to the Regulations

No. operator/user	Assessed by	Date
Activity/description of use		Time per day

Factor to be considered	Tick ans.		Comments
	Yes	No	
DISPLAY SCREENS			
Are the characters clear and readable?			
Is the text size comfortable to read?			
Is the image stable, i.e. free of flicker and jitter?			
Is the screen's specification suitable for its intended use?			
Are the brightness and/or contrast adjustable?			
Does the screen swivel and tilt?			
Is the screen free from glare and reflections?			
Are adjustable window coverings provided and in adequate condition?			
KEYBOARDS			
Is the keyboard separate from the screen?			
Does the keyboard tilt?			
Is it possible to find a comfortable keying position?			
Does the user have good keyboard technique?			
Are the characters on the keys easily readable?			
MOUSE, TRACKBALL, ETC.			
Is the device suitable for the tasks it is used for?			
Is the device positioned close to the user?			
Is there support for the device user's wrist and forearm?			
Does the device work smoothly at a speed that suits the user?			
Can the user easily adjust software settings?			
SOFTWARE			
Is the software suitable for the task?			
FURNITURE			
Is the work surface large enough for all the necessary equipment, papers etc.?			
Can the user comfortably reach all the equipment and papers they need?			
Are surfaces free from glare and reflection?			
Is the small of the back supported by the chair's backrest?			
Are forearms horizontal and eyes at same height as the top of the VDU?			
Are feet flat on the floor, without too much pressure from the seat on the backs of the legs?			
ENVIRONMENT			
Is there enough room to change position and vary movement?			
Is the lighting suitable, for example, not too bright or too dim to work comfortably?			
Does the air feel comfortable?			
Are levels of heat comfortable?			
Are levels of noise comfortable?			
Action			

Figure B9-23: Display screen equipment assessment checklist.   *Source: UK, HSE L26.*

NIOSH MANUAL MATERIAL HANDLING (MMH) CHECKLIST

NIOSH is the National Institute of Occupational Safety and Health, based in the USA. They have developed guidance on manual handling, which is available in a booklet format. There is a substantial amount of guidance on safer handling techniques and plenty of photographs showing how they work. It is not unlike UK guidance documents and the checklist is similar to that in the Manual Handling Operations Regulations 1992 guidance.

The weight stated as being the maximum a healthy person can lift with two hands in ideal conditions, i.e., 'Recommended Weight Limit' (RWL) is 51lbs (23kg). The RWL is defined as the weight of the load that nearly all healthy workers can lift over about 8hrs without developing back pain. Unlike the UK guidance, it does not appear to explicitly differentiate between male and female lifting capabilities, but the UK's rough equivalent of a RWL is 25kg.

For further details, see 'http://www.cdc.gov/niosh/docs/2007-131/pdfs/2007-131.pdf'.

Rapid Upper Limb Assessment (RULA)

Rapid Upper Limb Assessment (RULA) is a survey method developed for use in ergonomic investigations of workplaces where work related upper limb disorders are reported. RULA is a screening tool that assesses biomechanical and postural loading on the whole body, with particular attention to the neck, trunk and upper limbs.

Reliability studies have been conducted using RULA on groups of display screen equipment users and sewing machine operators. A RULA assessment requires little time to complete and the scoring generates an action list that indicated the level of intervention required to reduce the risks of injury due to physical loading on the worker. RULA is intended to be used as part of a broader ergonomic study.

RULA is based on three steps:

- Observing and selecting the postures to assess. There are pictures of different postures available (see opposite) on both the right and left sides to choose from.
- Scoring and recording the posture. There is software available for the scoring and calculation of risk to each upper limb.
- Action level. The score is compared to an action level list. The action level will usually lead to a more detailed investigation.

For further details, see 'http://ergo-plus.com/wp-content/uploads/RULA-A-Step-by-Step-Guide1.pdf'.

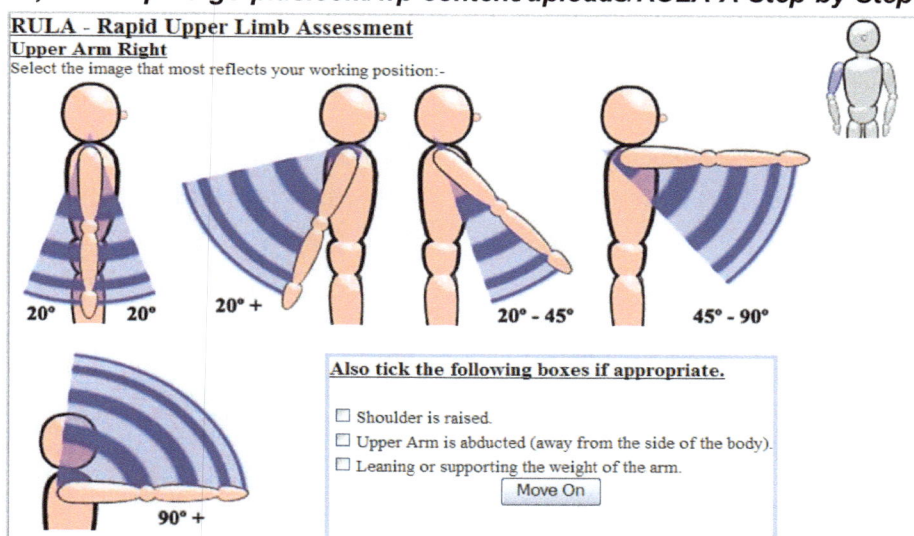

Figure B9-24: RULA. *Source: University of Nottingham.*

Practical control measures to avoid or minimise risks

ELIMINATION

Having completed the assessment of the activity it is necessary to take appropriate steps to reduce the risk of injury to the lowest level reasonably practicable. The MHOR 1992 and accompanying guidance provide guidelines to prevent the risk of injury to employees whilst carrying out manual handling operations. The first step identified is to avoid hazardous manual handling, i.e. elimination or do not carry out the process or activity. If elimination is not possible, it should be considered whether the process could be automated or mechanical assistance be provided.

AUTOMATION/MECHANICAL ASSISTANCE

Could the process be automated or mechanised, for example by providing equipment to feed materials into process machinery instead of handling it? If the whole of a process cannot be automated it may be possible to automate some of the high risk parts.

Mechanical assistance involves the use of handling aids, such as sack trucks, hand pallet trucks, hoists, slings and fork lift trucks. Although this may retain some elements of manual handling, bodily forces are applied more efficiently. Examples are:

Levers	Reduces bodily force to move a load. Can avoid trapping fingers.
Hoists	Can support weights, allowing handler to position load.
Trolley, Sack Truck Roller Conveyer	Reduces effort to move loads horizontally.
Chutes	A way of using gravity to move loads from one place to another.
Handling Devices	Hand-held hooks or suction pads can help when handling a load that is difficult to grasp.

Figure B9-25: Mechanical aids. *Source: UK, HSE Manual Handling (Manual Handling Operations Regulations 1992) Guidance L23.*

ALTERNATIVE WORK METHODS/JOB DESIGN

Consider alternative ways to carry out the task, such as the use of bulk storage and transfer of materials.

Designers should establish health and safety features that reduce musculoskeletal risk at the design stage. Consideration should be given to all aspects of work. In addition, design should minimise risks at all phases of the life of the work including:

- Construction.
- Transport.
- Installation.
- Commissioning.
- Use.
- De-commissioning.
- Dismantling.
- Disposal/recycling.

ERGONOMIC DESIGN OF TOOLS, EQUIPMENT, WORKSTATIONS AND WORKPLACES

Tools and equipment

Tools are designed to extend human physical capabilities of reach, force application and precision of movement and to enhance performance and capability. Harm arising from excessive forces and poor postures are frequently the result of poor tool design or inappropriate use. In some cases if a tool slips, breaks or loses purchase acute injuries can occur.

The forces required to grip tools during use should be minimal and prevent slippage, particularly where gloves are required. Tools should have handles that have the proper shape, thickness and length to prevent pressure on the soft tissues of the hands and to allow a good firm grip.

Where the task requires the large use of force or extended periods of use, the tool should enable the more powerful muscles of the arms and shoulders to be used. Power operated tools should be used instead of muscle power where possible. In addition, the following factors should be considered:

- Tool weight should be minimal. Where tools are heavy, counterbalancing devices can reduce the weight.
- There should be a definite and fixed place for all tools and materials.
- The tool should be easy to set down and pick up.
- Tools, materials and controls should be positioned close in and directly in front of the operator.
- Materials and tools should be located to permit the best sequence of operations.
- Tools and materials should be pre-positioned wherever possible.

- Handles, such as those used on cranks and screwdrivers, should be designed to permit as much of the surface of the hand to come into contact with the handle as possible.
- Levers, crossbars and hand wheels should be located in such positions that the operator can manipulate them with the least change in body position and with the greatest mechanical advantage.

Workstation design

When designing workstations, such as consoles and workbenches, the following factors need to be taken into account in order to accommodate the user:

- Horizontal work area.
- Working position.
- Work height (the height at which the hands are working).
- Viewing distances and angles.
- Reach distances, access and clearance.

Horizontal work area

Horizontal work areas need to include the use of materials and work equipment most frequently used in the foreground, the first and second work areas as illustrated opposite; this will enable ease of access, particularly for repetitive work tasks.

That which is seldom used should be positioned furthest away from the operator, in the third zone.

Working position

A sitting position is generally preferred for fine manipulation, light manual work and where foot controls are regularly used. There should be enough space between the underside of the work surface and the seat, when sitting, for the legs and to allow movement.

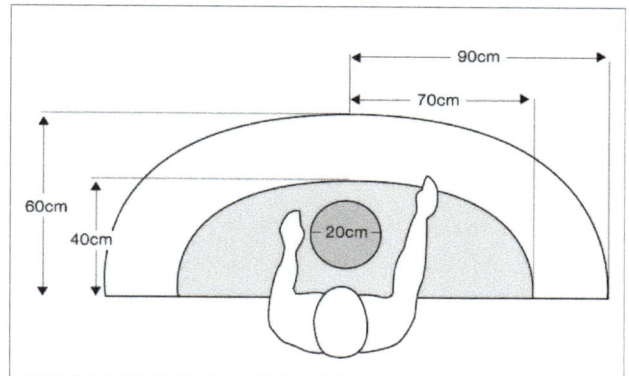

Figure B9-26: Horizontal Work Area showing primary and secondary work spaces. *Source: McPhee, 2005.*

A standing position is preferred where heavier manual handling work is performed, where there is limited or no leg room under equipment or where there are many controls and displays over a wide area that have to be accessed or monitored.

Standing work requires even floor surfaces, preferably with anti-slip coatings to reduce the risks of slipping.

Work height

Work height is concerned with the height at which the hands are working. The nature of the task will determine the preferred work heights. Factors to consider include the need for visual and manual precision, as well as the handling of heavy components. In most manual tasks the work height should be at a level just below the elbow, with the upper arm held in a vertical position close to the body.

For fine work involving close visual distances, the work height should be raised to achieve this with minimal bending of the neck and arm supports should be provided where appropriate.

Viewing distances and angles

Viewing distances will be dependent upon the size of the work object. The smaller the object the shorter the viewing distance and the higher the work surface required.

The most frequently viewed object should be centred in front of the worker. Recommended viewing angles vary depending on the work posture required, for example, when leaning forward such as at a desk or leaning backward such as in a control room.

Distances to the work task should allow the operator to see properly without strain on the eyes or the muscles and joints. Strained neck or shoulder postures should be kept to a minimum and avoided by design whenever possible.

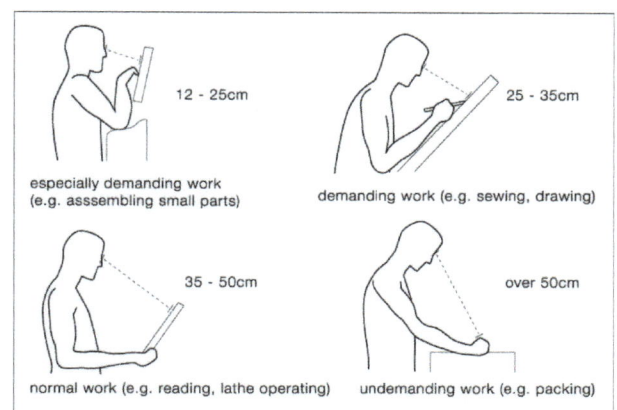

Figure B9-27: Functional viewing distances. *Source: McPhee, 2005.*

Reach distances access and clearance

Arm and leg reach should be adjustable for the needs of all sizes of user, taking into account the work position and postural requirements of the task. Space allowances, including access and clearance to machines and equipment used by operators and for maintenance personnel should be incorporated into the design of the work stations.

Workplaces

The work areas of the workplace should be suitable to accommodate the number of people required to do the tasks employed. Consideration will need to be given to the nature of the work, including any specific hazards from height or space restrictions. All sizes of users should be considered in the design of the work areas, in particular when considering the adequacy of access and visibility for maintenance workers and routine checks. There should be a logical flow of raw materials products or components, with easy access for operators and the ability for adjustments to be made for the needs of the individual as necessary.

Walkways and stairs should be level, even and free from obstructions. Walkways should be wide enough to enable access to the equipment. Where access at different levels is required, slip-resistant surfaces on all steps should be provided. Steps, footholds and stairwells should be fitted with handrails as appropriate to the access requirements of the machinery and equipment.

Care should be taken to eliminate sharp edges and protruding obstructions to prevent injuries while manual handling or conducting repetitive physical activities.

JOB ROTATION/WORK ROUTINE

Job rotation

Job rotation will often be necessary where the work is repetitive or physically demanding, leading to fatigue. Job rotation usually involves two or more individuals changing their patterns of work regularly during the working day to ensure the musculoskeletal risks they are exposed to are minimised and there is time for recovery.

Work routine

Tasks should be designed in accordance with ergonomic principles so that the limitations to human performance can be taken into account. Matching the job to the worker will ensure that they are not overloaded, physically or mentally.

Physical matching includes the whole workplace environment as well as considerations of such things as strength, reach, freedom of movement etc. Mental matching involves the individual's ability to absorb information and make decisions as well as their perception of the task. Any mismatch between the requirements of the job and the worker's abilities provides potential for human error.

The work routine should naturally provide opportunities to control and reduce the musculoskeletal effects of the activities being conducted; for example, a change of task of a display screen equipment user causing them to leave their workstation and providing relaxation of muscles used, including eye muscles.

EYE AND EYESIGHT TESTING

Regulation 5 of the DSE Regulations 1992, relating to eyes and eyesight, states that:

"For DSE users, the employer shall provide, on request, an eyesight test carried out by a competent person".

Figure B9-28: Regulation 5 of DSER 1992.   Source: Display Screen Equipment Regulations (DSER) 1992.

A basic eye examination will include an external and intra-ocular examination; with whatever additional examinations appear clinically necessary.

The tests to be performed are not specified in detail, and the Optometrist is expected to use clinical judgment. Issuing a prescription for corrective lenses, where one is needed, is generally an integral part of the sight test. The sight test could also include the history and any symptoms experienced by the patient.

Examination of the eye and its external area should include eyelids and lacrimal ducts and the front of the eye. Internal examination of the eye function; objective refraction; subjective monocular refraction; subjective binocular refraction; and such other tests as may be necessary.

This is an important control measure that ensures the user can see the screen and associate items clearly and that they are able to adopt the correct posture while working.

TRAINING AND INFORMATION

Employers should ensure that all workers who carry out manual handling operations receive the necessary training to enable them to carry out the task in a safe manner.

A training programme should include:

- How potentially hazardous loads may be recognised.
- How to deal with unfamiliar loads.
- The proper use of handling aids.
- The proper use of PPE.
- Features of the working environments that contribute to safety.
- The importance of good housekeeping.
- Factors affecting individual capability.
- Good handling techniques.

For training to be effective it should be ongoing to reflect improved techniques developed by experienced workers and be supported by periodic refresher training and supervision.

Furthermore, every employer shall ensure that operators and users at work are provided with adequate information about all aspects of health and safety relating to their workstations, to them, and to their work.

Training and information will enable the worker to understand the musculoskeletal risks of work activities they do and apply the control measures to minimise the risk. This is important as some of these measures rely on the worker to take appropriate action, for example, identifying the weight of an item and getting help when needed; lifting in an efficient and effective way and taking breaks from repetitive at the appropriate time. Regulation 6 of the DSE Regulations 1992, relating to provision of training, states:

"The employer shall ensure that the employee is provided with adequate health and safety training in the use of any workstation upon which he may be required to work".

Figure B9-29: Regulation 6 of DSER 1992. *Source: Display Screen Equipment Regulations (DSER) 1992.*

Further, Regulation 7 of the DSE Regulations 1992, relating to provision of information, states:

"Every employer shall ensure that operators and users at work are provided with adequate information about all aspects of health and safety relating to their workstations, to them, and to their work".

Figure B9-30: Regulation 7 of DSER 1992. *Source: Display Screen Equipment Regulations (DSER) 1992.*

See also - Relevant Statutory Provisions - Health and Safety (Display Screen Equipment) Regulations (DSE) 1992 - as amended by the Health and Safety (Miscellaneous Amendments) Regulations 2002.

EFFICIENT MOVEMENT PRINCIPLES

Lifting techniques using kinetic handling principles

Kinetic lifting is a method of lifting that takes into account the body's characteristics and how it works most effectively. It ensures that lifting is carried out without placing unnecessary strain and pressure on vulnerable body parts. The position of the back during normal manual handling activities should ideally include 'moderate flexion (slight bending) of the back, hips and knees', as set out in the Health and Safety Executive's (HSE)

Manual Handling Operations Regulations 1992 Guidance on Regulations L23, and not the out-dated principle of 'knees bent, back straight'.

The principles of good lifting and moving techniques have been established for some considerable time. The basic principles are:

a) ***Think before handling/lifting.*** Plan the lift/handling activity. Where is the load going to be placed? Use appropriate handling aids where possible. Will help be needed with the load? Remove obstructions, such as discarded wrapping materials. For long lifts, such as from floor to shoulder height, consider resting the load mid-way on a table or bench to change grip.

b) ***Keep the load close to the waist.*** Keep the load close to the waist for as long as possible while lifting. The distance of the load from the spine at waist height is an important factor in the overall load on the spine and back muscles. Keep the heaviest side of the load next to the body. If a close approach to the load is not possible, try to slide it towards the body before attempting to lift it.

c) ***Adopt a stable position.*** The feet should be apart with one leg slightly forward to maintain balance (alongside the load if it is on the ground). The worker should be prepared to move their feet during the lift to maintain a stable posture. Wearing over-tight clothing or unsuitable footwear may make this difficult.

d) ***Ensure a good hold on the load.*** Where possible hug the load as close as possible to the body. This may be better than gripping it tightly only with the hands.

e) ***Moderate flexion*** (slight bending) of the back, hips and knees at the start of the lift is preferable to either fully flexing the back (stooping) or fully flexing the hips and knees (full/deep squatting).

f) ***Don't flex the back any further while lifting.*** This can happen if the legs begin to straighten before starting to raise the load.

g) ***Avoid twisting the back*** or leaning sideways especially while the back is bent. Keep shoulders level and facing in the same direction as the hips. Turning by moving the feet is better than twisting and lifting at the same time.

h) ***Keep the head up when handling.*** Look ahead, not down at the load once it has been held securely.

i) ***Move smoothly.*** Do not jerk or snatch the load as this can make it harder to keep control and can increase the risk of injury.

j) ***Don't lift or handle more than can be easily managed.*** There is a difference between what people can lift and what they can safely lift. If in doubt, seek advice or get help.

k) ***Put down, then adjust.*** If precise positioning of the load is necessary, put it down first, then slide it into the desired position.

PERSONAL CONSIDERATIONS

Personal considerations relate to the size of the individual and their capabilities. The size of the individual should be considered when designing and setting up workstations, including display screen equipment. It is a significant aspect if postural and repetitive movement musculoskeletal disorders are to be prevented.

An important aspect of personal considerations is individual capability. The person's capability must be taken into account when considering manual handling operations, if serious musculoskeletal injuries are to be prevented. Pay particular attention to those more vulnerable to injury as a result of manual handling activities, for example, pregnant women, those who have back trouble, or suffer other health problems.

The individual's state of health, fitness and strength can significantly affect the ability to perform a task safely. Some individuals' physical capacity can also be age-related, typically climbing until the early 20's and declining gradually from the mid 40's. It is clear then that an individual's condition and age could significantly affect the ability to perform a task safely.

Studies though have shown no close correlation between individual capacity and capability and injury incidence. There is therefore insufficient evidence for reliable selection of individuals for safe manual handling on the basis of such criteria. It is however recognised that there is often a degree of self-selection for work that is physically demanding.

Work environment risks and controls

On completion of this element, candidates should be able to demonstrate understanding of the content through the application of knowledge to familiar and unfamiliar situations. In particular, they should be able to:

B10.1   Explain the need for, and factors involved in, the provision and maintenance of temperature in both moderate and extreme thermal environments.

B10.2   Explain the need for suitable and sufficient lighting in the workplace, units of measurement of light and the assessment of lighting levels in the workplace.

B10.3   Explain the need for welfare facilities and arrangements in fixed and temporary workplaces.

B10.4   Explain the requirements and provision for first-aid in the workplace.

Content

Relevant statutory provisions

Health and Safety (First-Aid) Regulations (FAR) 1981 (as amended)

Workplace (Health, Safety and Welfare) Regulations (WHSWR) 1992

Sources of reference

Reference information provided, in particular web links, was correct at time of publication, but may have changed.

First aid at work, Guidance on Regulations, L74, HSE Books, ISBN: 978-0-7176-6560-0,
http://www.hse.gov.uk/pUbns/priced/l74.pdf

Lighting at Work, HSG38, second edition 1997, HSE Books, ISBN: 978-0-7176-1232-1,
http://www.hse.gov.uk/pubns/books/hsg38.htm

Workplace health, safety and welfare, Workplace (Health, Safety and Welfare) Regulations 1992, ACOP, L24, HSE Books,
ISBN: 978-0-7176-0413-5, http://www.hse.gov.uk/pubns/books/l24.htm

The above web links along with additional sources of reference, which are additional to the NEBOSH syllabus, are provided on the RMS Publishing website for ease of use - www.rmspublishing.co.uk.

B10.1 - The provision and maintenance of temperature in both moderate and thermal environments

The importance of maintaining heat balance in the body

The human body's optimal core temperature is between 36.1°C (97°F) to 37.2°C (99°F). A significant change in the core temperature will have potentially serious health implications. It is essential the body maintains its optimal core temperature. The 'control centre' in the base of the brain responsible for this regulation is the hypothalamus. Sensors in the skin and organs relay information about body temperature to the hypothalamus.

If there is change in the heat balance; for example, by heat transfer *(see in the following section)* then the individual may suffer with heat exhaustion/stroke or mild/severe hypothermia.

HEAT BALANCE EQUATION = M=K±C±R±E±S

The heat balance equation measures the heat equilibrium in the person. The heat equilibrium is dependent on the heat the body absorbs, the heat given off by evaporation, the body's metabolic rate and the calories (heat energy) used.

Heat equilibrium is where the body can keep the core temperature. If heat generation and inputs are greater than heat outputs, the core body temperature will rise and if the heat outputs are greater, the core body temperature will fall.

The human heat balance equation involves consideration of the following heat processes:

- Heat generation (within the body).
- Heat storage (within the body).
- Heat transfer (to/from the body).

The heat balance equation can be presented in a number of forms one of which is: M=K±C±R±E±S.

Where:

- M = Rate of metabolic heat production.
- S = Rate of heat storage - heat gained or lost by the body.
- K = Heat exchange via conduction - where the body is in contact with (i.e. touching) objects, for example, flooring. The process depends on the temperature difference between the part of the body and the object it is in contact with. In average thermal comfort conditions this is negligible.
- C = Heat exchange via convection - where heat is transferred by a current of moving air. The process of convection depends on the difference between the temperature of the skin and the air temperature, as well as the rate of air movement. In average thermal comfort conditions this is 30 percent.
- R = Heat exchange via radiation - where heat is transferred by radiant energy into the environment surrounding the body through the medium of air. This depends on the surface temperature of the skin combined with the mean radiant temperature of the surroundings. If the surroundings contain a source of high radiant heat such as a furnace, then heat will be transferred to the person's body, which is relatively cooler than the furnace. In average thermal comfort conditions this is 45 percent.
- E = Heat exchanged via evaporation - where heat is transferred to sweat, which takes up heat energy in order to become a vapour. This depends on the amount of sweat on the skin, and its rate of evaporation, which, in turn, depends on air temperature, air velocity and relative humidity. In average thermal comfort conditions this is 25 percent.

M is always a positive value representing a heat creation. For S, C and R, a positive value is heat loss and a negative value is heat gained.

E is the heat exchanged through evaporation of moisture from the skin and respiratory tract so, in reality, this is always negative (heat loss) and not positive (heat gain).

For the body to be in heat balance, at a constant temperature, the rate of heat storage (S) is zero. If there is a net heat gain, storage is positive and the body temperature will rise, but if there is a net heat loss, storage is negative and the body temperature will fall.

Whilst the units of rates of heat production or loss are Js^{-1} or watts (W), it is traditional to standardise over persons of different sizes by using units per square metre of the total body surface area (Wm^2).

The effects of working in high and low temperatures and humidity

UNIT OF HEAT

The basic unit of heat is the joule (J). This is a very small quantity of heat. A typical small electric fire gives out 1,000 J every second. It is more convenient to talk about the *rate* of heat emission i.e. joules per second (J/s) which is the watt (W). (1 kW is 1,000 W).

THE BODY

All warm blooded animals need to maintain the temperature of their body within a narrow range. In human beings this means that we need to maintain a core body temperature of around 37°C (normal limits range from 36.4°C to 37.2°C). The body creates heat during the process of metabolism. Foodstuffs (proteins, carbohydrates and fat) are assimilated by the body and their energy released in the form of heat.

This heat production is influenced by environmental conditions, such as temperature and humidity, and balanced by control mechanisms such as sweating and shivering, which attempt to maintain the core body temperature within the normal range. The 'control centre' in the base of the brain responsible for this regulation is the hypothalamus. When the function of the hypothalamus is affected, for example, by fever, then overheating results, which can lead to heat stress or a fatal heat-stroke if left untreated. Active muscles metabolise food faster than muscles at rest, giving off more heat in the process; thus physical activity increases body temperature. If body temperature is too high for the hypothalamus to balance by sweating then biological function of the body can become impaired, leading to cellular damage. If the body temperature is too low, the rate at which foodstuffs are metabolised decreases; therefore the amount of heat energy produced is less.

HEAT TRANSFER

The heat transfer takes place by means of four methods:

1) **Conduction** - where the body is in contact with (i.e. touching) objects, for example, flooring. The process depends on the temperature difference between the part of the body and the object it is contacting.
2) **Convection** - where heat is transferred by a current of moving air. The process of convection depends on the difference between the temperature of the skin and the air temperature, as well as the rate of air movement.
3) **Radiation** - where heat is transferred by radiant energy into the environment surrounding the body through the medium of air. This depends on the surface temperature of the skin combined with the mean radiant temperature of the surroundings. If the surroundings contain a source of high radiant heat such as a furnace, then heat will be transferred to the person's body, which is relatively cooler than the furnace.
4) **Evaporation** - where heat is transferred to sweat, which takes up heat energy in order to become a vapour. This depends on the amount of sweat on the skin, and its rate of evaporation, which, in turn, depends on air temperature, air velocity and relative humidity.

The contribution of each of the above methods to heat loss, in average thermal comfort conditions, is approximately:

- Radiation   -   45%.
- Convection   -   30%.
- Evaporation   -   25%.
- Conduction   -   Negligible.

EFFECTS OF HIGH AND LOW TEMPERATURE

High temperature

Body temperature is regulated by the rate at which the skin radiates heat and by the evaporation of water vapour (perspiration). The radiant process occurs through vasodilatation. Skin temperature is raised due to an increase in blood flow. If this process is insufficient then the blood vessel size is increased further, in order to induce sweat.

A further increase in skin temperature will occur, which leads to the process of sweating. Sweating (or perspiring) is the evaporation of water, combined with various salts in the body, through special glands and pores in the skin. This is similar to panting which is the evaporation of water vapour through pores and the mouth. Excessive sweating can lead to muscle fatigue, followed by cramps and pains, due to the loss of body salts (the water/salt balance). Overheating can lead to a range of health effects, ranging from heat oedema through to heat stroke, which can lead to coma and death.

These are summarised as:

Condition	Remarks
Heat oedema	Swelling, due to the accumulation of body fluids in an area, for example, the ankles.
Heat syncope	Blood vessel dilation.
Cramps	Due to water/salt balance affected by sweating.
Hidromeiosis	Sweat glands cease to function.
Prickly Heat	Blocked sweat glands, causing sweat to escape into the dermis.
Anhydrosis	Excess radiant heat.
Heat exhaustion	Raised body temperature (39-40°C), where the levels of water and salt in the body begin to drop, causing nausea, faintness and heavy sweating.

Condition	Remarks
Heat stoke	Raised body temperature above 40°C, cells in the body begin to break down and important parts of the body stop working. The person may show signs of mental confusion, hyperventilation and loss of consciousness. There can be signs of neurological failure, leading to coma.

Figure B10-1: Human body regulatory mechanisms.

Source: RMS.

Low temperature

When the body is cold then the blood vessels in the skin contract and the muscles under the skin vibrate (shiver) in an attempt to generate heat and maintain the core temperature. Extreme cold or lowering of body temperature can lead to numbness, frostbite and hypothermia.

Frostbite affects the tissues of the extremities of the body, the toes and fingers. Usually the blood in a person's body carries oxygen to parts of the body so that the body tissue is able to carry out its normal function. When exposed to extreme cold the blood vessels contract so that blood (and oxygen) is diverted from the extremities to vital organs. Early stages may show as signs of blanching, then become mottled and blue. In superficial cases the skin and affected tissues can recover. After some time, the lack of blood supply (and oxygen) to the skin can lead to damage to the cells of the tissue.

In severe cases, as ice crystals form and the blood vessels become irreversibly damaged this can lead to permanent loss of sensation and eventual tissue death (gangrene). Hypothermia is defined as a fall in body temperature to below 35°C. Severe hypothermia is where the body temperature falls below 30°C, which is often fatal. Hypothermia can cause disorientation, apathy and irrational behaviour. Sustained low body temperature can lead to lethargy, weak breathing, weak pulse and unconsciousness. If the person does not receive assistance the heart may stop.

HUMIDITY

The primary effect of high humidity is that it limits the ability of the human body to regulate its heat through sweating; high air humidity limits the amount of sweat taken up by the air from the surface of the skin. This can lead to the body core temperature rising and the effects discussed earlier.

The effects of low humidity are that any available moisture from the body is readily taken up by the low humidity air, which leads to dry and cracked skin, including the lips. In addition, the moisture lubricating the eyes and throat can become reduced, leading to difficulty in blinking and breathing.

Hazard	Typical exposures	Possible effects
High air temperature	Outdoor physical work in hot weather (e.g. road construction). Indoor physical work in a hot working environment (e.g. foundry, bakery).	Discomfort, sweating, flushed skin, fatigue, dizziness, muscle cramps, nausea, vomiting, dehydration, and excessive or erratic pulse. **Severe exposure**: heat stroke, hyperthermia, loss of consciousness, death.
Low air temperature	Prolonged exposure to low air temperatures while wearing clothing inadequate for cold conditions. Outdoor work in cold weather, indoor work in cold environments.	Discomfort, shivering, loss of motor co-ordination, slurred speech. **Severe exposure:** irrational behaviour, frostbite, hypothermia, loss of consciousness, death.
Humidity	Work with plant or processes, which generate humidity (e.g. brick curing, steam presses).	Discomfort, flushed skin, sweating, fatigue, headaches, dizziness, nausea, vomiting, excessive or erratic pulse. **Severe exposure:** collapse, heat stroke, hyperthermia.
Air movement (high)	Prolonged outdoor activity in cold, wet and windy conditions, work in wet clothing in cold wind.	(In cold conditions) discomfort, shivering, cold-related illnesses. **Severe exposure:** hypothermia, loss of consciousness.
Air movement (low)	Work in enclosed area with inadequate ventilation during hot weather.	(In hot conditions) discomfort, flushed skin, sweating, fatigue, headaches, dizziness and excessive or erratic pulse. **Severe exposure:** nausea, vomiting, collapse, heat stroke.
Radiant heat	Exposure to UV radiation from the sun, exposure to radiant or conducted heat from plant (dryer, oven, furnace) or processes such as smelting, molten metals.	Discomfort, sweating, fatigue, dizziness, nausea and vomiting, radiation burns to exposed skin. **Severe exposure:** severe burns, heat stroke, collapse, loss of consciousness.

Figure B10-2: Exposures to heat and cold and possible effects. *Source: WorkCover NSW September 2001.*

Typical work situations likely to lead to thermal discomfort

WORK PRESENTING A HIGH RISK OF THERMAL DISCOMFORT

Work situations that expose employees to extremes of heat or cold that are likely to lead to thermal discomfort include:

- Work in direct sunlight (for example, bitumen laying, construction, horticulture and agriculture workers). The risks increase when combined with high temperatures, high humidity and low air movement.
- Work requiring high physical work rate in humid conditions (for example, laundries, kitchens).
- Work in cold weather (for example, horticulture, power line maintenance, breakdown service engineer). The risks increase when combined with low temperatures, wet and windy conditions.
- Work requiring prolonged physical inactivity in low temperatures or wet conditions (security staff, marshals or police).
- Work with plant that becomes hot (for example, ovens, dryers, boiler room, furnaces, metal forging) or cold (for example, freezers).
- Work in hot or cold conditions where limited space reduces the ability to move about, which could increase the effects of heat or cold (for example, work in confined spaces, plant maintenance workers).
- Work in a workplace with inadequate temperature control or ventilation.
- Work requiring protective clothing that inhibits loss of body heat and causes a build-up of humidity within it (those wearing chemical or asbestos protection suits).

The environmental parameters affecting thermal comfort

ENVIRONMENTAL PARAMETERS

The human body can regulate its temperature to remain fairly constant, despite the surrounding air temperature and other sources of radiated heat and cold, provided that exposure is neither too extreme nor too prolonged.

The efficiency of the body's cooling and heating mechanisms can be affected by a number of environmental factors:

- Air velocity/wind speed.
- The mean radiant temperature (i.e. the temperature of the surroundings).
- The relative humidity of the surrounding air (which affects the rate of sweat evaporation).
- Air temperature.

These factors influence the rate of heat transfer between a person's body and its surrounding environment.

AIR TEMPERATURE

Air temperature is the temperature of the air surrounding the person, given in degrees Celsius °C. The Workplace (Health, Safety and Welfare) Regulations 1992 (L24) Approved Code of Practice states:

- For workplaces where the activity is mainly sedentary, for example offices, the temperature should normally be at least 16°C.
- If work involves physical effort it should be at least 13°C.

The air temperature will give an indication of the thermal environment, but only in the vicinity of where the measurement is taken. It does not show where the temperature is coming from, such as drafts and/or radiant heat; nor will it take the relative humidity into account, all of which are factors necessary to assess the body's ability to remain in equilibrium. However, it is an important first step in most workplaces to measure the air temperature. Air temperature is the temperature of the air.

The ambient air temperature will give an indication of the thermal environment, but only in the vicinity of where the measurement is taken. It does not show where the temperature is coming from, such as drafts and/or radiant heat; nor will it take the relative humidity into account, all of which are factors necessary to assess the body's ability to remain in equilibrium.

However, it is an important first step in most workplaces to measure the air temperature. WHSWR 1992, Regulation 7 requires that a sufficient number of thermometers be provided to enable persons at work to determine the temperature in any workplace inside a building. Measurement of air temperature is usually made with a mercury thermometer. *For further information on measuring equipment, see 'Equipment for measuring environmental parameters' later in this Element.*

RADIANT TEMPERATURE

Thermal radiation is the heat that radiates from a warm object, for example, the sun, ovens and furnaces; hence the term radiant. Radiant heat may therefore be present in the workplace where there is a heat source.

The radiant temperature potentially has a greater effect than air temperature on heat balance. Measurement of radiant temperature is by a globe thermometer. *For further information on measuring equipment, see 'Equipment for measuring environmental parameters' later in this Element.*

RELATIVE HUMIDITY

In hot environments, humidity is important because less sweat evaporates from a person's skin when the air humidity is high, for example, greater than 85%. The evaporation of sweat is a significant method of heat loss and cooling in humans.

Humidity is the mass of water vapour in air per unit volume of air/water vapour mixture and has units of kg/m^2. There is always some moisture present in the air, in the form of water vapour; however the maximum amount that can be held by the air depends upon the air temperature.

Relative humidity (RH) is defined as the ratio of the actual vapour pressure to the saturation vapour pressure, it is commonly expressed as a percentage (unit symbol %rh).

$$\text{Relative Humidity} = \frac{\text{Actual Vapour Density}}{\text{Saturated Vapour Density}} \times 100$$

The amount of vapour that will saturate the air increases with a rise in temperature:

- At 4.4°C, 454kg of moist air contains a maximum of 2kg of water vapour.
- At 37.8°C, 454kg of moist air contains a maximum of 18kg of water vapour.

The level of thermal discomfort is high when the air is saturated with water, because the evaporation of sweat, with its attendant cooling effect, is impossible. The term 'absolute humidity' is used to describe the weight of water vapour contained in a particular volume of air and is expressed in kilograms of water per kilogram of dry air.

One way of measuring the RH is using a ***psychrometer hygrometer***. It consists of a wet and dry bulb thermometer mounted side by side in a frame, which is rotated by hand. When it is rotated, air is forced to flow over the bulbs. A small reservoir supplies the wick surrounding the bulb of the wet bulb thermometer with distilled water. ***For further information on measuring equipment, see 'Equipment for measuring environmental parameters' later in this Element.***

AIR VELOCITY

Air velocity describes the speed of the air moving across the worker and it may help to cool the worker if it is cooler than the environment. Air velocity has been shown to be an important factor in thermal comfort, because people are sensitive to it. Still or stagnant air in indoor environments that are artificially heated may cause people to feel stuffy. It may also lead to a build-up in odour. Moving air in warm or humid conditions can increase a person's heat loss through convection, without any need to change the air temperature. Small air movement in cool or cold environments may be perceived as a draught and if the air temperature is less than skin temperature it will significantly increase convective heat loss. On a larger scale, this is referred to as the wind chill factor.

Physical activity also increases air movement, so air velocity provided in a workplace may be adjusted to account for a person's level of physical activity. Air velocity can be measured by using a vane anemometer; or kata thermometer. ***For further information on measuring equipment, see 'Equipment for measuring environmental parameters' later in this Element.***

The meaning of thermal comfort

Thermal comfort is defined in British Standard BS EN ISO 7730 as: 'that condition of mind which expresses satisfaction with the thermal environment'. The HSE defines thermal comfort as: 'a person's state of mind in terms of whether they feel too hot or too cold'.

Thermal comfort or more typically, thermal discomfort, is that which is typically vocalised as complaints by a group of workers. The HSE accepts that it is unlikely that everyone within a work group will be satisfied with their thermal environment and considers that if 80% of occupants are comfortable, then the employer has done that which is 'reasonable'.

The most commonly used indicator of thermal comfort is the traditional use of the thermometer to measure and display workplace temperature, but air temperature is only one factor and thermal comfort should always be considered in relation to other environmental and personal factors.

ENVIRONMENTAL FACTORS

Environmental factors include:

- Air temperature, measured in degrees Celsius.
- Radiant temperature, heat that radiates from a warm object, for example, sun, fires, furnaces and molten metal.
- Air velocity, because individuals are susceptible to it relatively small movements of air will more often cool the skin.
- Humidity, the higher the humidity (amount of water in the air) the warmer people will feel. At high humidity, above 85%, sweat will be slow to evaporate, which will result in a rise in the body's temperature.

PERSONAL FACTORS

Personal factors include:

- Clothing, thermal comfort is very much dependent on the insulating effect of clothing on the wearer; this may provide insulation from a cold environment or prevent heat being released, causing heat stress.
- Metabolic heat: the heat that is produced within the body, the greater the physical activity the more heat the person produces and the hotter they will feel.

Legal requirements

The Workplace (Health, Safety and Welfare) Regulations (WHSWR) 1992 (as amended), Regulation 7 requires that:

> "(1) During working hours, the temperature in all workplaces inside buildings shall be reasonable.
>
> (2) A method of heating or cooling shall not be used which results in the escape into a workplace of fumes, gas or vapour of such character and to such extent that they are likely to be injurious or offensive to any person.
>
> (3) A sufficient number of thermometers shall be provided to enable persons at work to determine the temperature in any workplace inside a building."

Figure B10-3: Regulation 7 of WHSWR 1992.   Source: The Workplace (Health, Safety and Welfare) Regulations (WHSWR) 1992.

THERMAL COMFORT

The WHSWR 1992 do not refer to managing 'thermal comfort', but thermal comfort is implied in the requirement to ensure a reasonable temperature. The term 'comfort' is introduced by the Approved Code of Practice (ACOP) to the regulations in relation to reasonable temperature. The ACOP states that 'the temperature in workrooms should provide reasonable comfort without the need for special clothing'. Appropriate ordinary personal clothing may be required to establish thermal comfort in some workrooms and in others special clothing may be required. The ACOP emphasises that where a temperature that provides thermal comfort without the use of special clothing is impracticable, because of the hot or cold processes conducted in the workroom, 'all reasonable steps should be taken to achieve a temperature which is as close as possible to comfortable'. This establishes a clear requirement to manage thermal comfort by the means of all reasonable steps.

'REASONABLE' TEMPERATURE INSIDE WORKPLACES

The WHSWR 1992 establishes the general duty to ensure that the temperature is reasonable. The term 'reasonable' in this setting is not further defined by the regulations, but the ACOP to the regulations says the temperature in a workroom should normally be at least 16 degrees Celsius (16°C), unless the work involves severe effort, in which case the temperature should be at least 13 degrees Celsius (13°C). The ACOP accepts that it may be impractical to maintain the required temperature where rooms are open to the outside or where the room is deliberately chilled, for example, to store food. A workroom, in the context of the ACOP, means 'a room where people normally work for more than short periods'. The ACOP also recognises that the required minimum temperatures may not ensure reasonable thermal comfort, depending on factors like air movement and humidity.

Workplaces can vary greatly:

- Very cold and exposed workplaces, such as a frozen food storage warehouse, where temperatures may be minus 25° Celsius (C).
- General warehousing may be at ambient temperature, 5°C to 30°C.
- General offices, where the nature of the work is sedentary, may be at temperatures between16°C to 24°C.
- High temperature processing, such as a laundry or bakery, where temperatures may be 30°C to 38°C.
- High temperature manufacturing, such as with glass, steel or ceramics, where very high temperatures of 30°C to 45°C may arise as a result of the production process.

Figure B10-4: Cool areas.   Source: RMS.

Regulation 7(1A)(a) requires that "a workplace shall be adequately thermally insulated where it is necessary, having regard to the type of work carried out and the physical activity of the persons carrying out the work." This would be particularly important where mobile workplaces are provided as a rest area or for training purposes as they could easily get too cold in the winter and too hot in the summer, for example, in construction activities and temporary school classrooms.

The ACOP does not establish a specific reasonable maximum temperature, but where the temperature might be uncomfortably high, for example, where hot processes take place, then all reasonable steps should be taken to achieve a 'reasonable temperature' through the design of the building.

For example:

- Insulating hot plant or pipes.
- Providing air-cooling plant.
- Shading windows.
- Location of workstations away from places subject to radiant heat.

The WHSWR 1992 specifically requires that the *"excessive effects of sunlight on temperature should be avoided"*.

Where reasonably comfortable temperatures cannot be achieved throughout the workplace, local heating or cooling should be provided. In practice, the level of heating or cooling should be appropriate to provide physical comfort. The nature of the work and the working environment will need to be assessed to achieve the correct level. Several factors should be considered in order to establish a suitable workplace temperature, such as personal capability, mobility of the work, degree of hot or cold, wind speed and humidity. Whenever possible the individual should be able to adjust their workplace temperature to achieve physical comfort.

Where workers remain exposed to temperatures which do not give reasonable comfort, suitable protective clothing and rest facilities should be provided. When appropriate, systems of work should consider alternate work routines, for example, task rotation to reduce exposure to uncomfortable temperatures.

Equipment for measuring environmental parameters

AIR TEMPERATURE

A ***dry bulb thermometer*** is most commonly used to measure air temperature.

In its most basic form this would be the mercury-in-glass type, which consists of a uniform diameter glass capillary that opens into a mercury filled bulb at one end.

The assembly is sealed to preserve a partial vacuum in the capillary. If the temperature increases, the mercury expands and rises inside and the resulting temperature may then be read on an adjacent scale.

Figure B10-5: Mercury thermometer.   *Source: RMS.*

Mercury is most commonly used to measure ordinary air temperature ranges - however for more diverse ranges substances such as alcohol and ether are used. A wide variety of devices are used as thermometers. The primary requirement is that one easily measured property, such as the length of the mercury column, should change markedly and predictably with changes in temperature. The variation of that property should also stay fairly linear with variations in temperature. A unit change in temperature should lead to a unit change in the property to be measured at all points of the scale.

The electrical resistance of conductors and semi-conductors increases with an increase in temperature. This phenomenon is the basis of the resistance thermometer in which a constant voltage, or electrical potential is applied across the thermistor or sensing element. For a thermistor of a given composition the measurement of a specific temperature will induce a specific resistance across the thermistor. This resistance is measured by a galvanometer and becomes the measure of the temperature. Various thermistors made of oxides of nickel, manganese, or cobalt are used to sense temperatures between 46 and 150°C. Similarly, thermistors employing other metals and alloys are designed to be used at higher temperatures. With proper circuitry, the current reading can be converted to a direct digital display of the temperature.

Very accurate temperature measurements can be made using ***thermocouples*** in which a small voltage difference (measured in mV) arises when 2 wires of dissimilar metals are joined to form a loop, and the 2 junctions have different temperatures. To increase the voltage signal, several thermocouples can be connected in series to form a thermopile. Since the voltage depends on the difference of the junction temperatures, one junction must be maintained at a known temperature, otherwise an electronic compensation circuit must be built into the device to measure the actual temperature of the sensor. Thermistors and thermocouples often have sensing units less than a ¼ cm in length, which permits them to respond rapidly to temperature changes and makes them ideal for many environmental, biological and engineering uses.

RADIANT TEMPERATURE

Radiant temperature has a greater influence than air temperature on how we lose or gain heat in the environment. This is because the human body absorbs radiant heat; as radiant heat is absorbed more easily by dark objects more radiant heat will be absorbed if dark clothes are worn. Hence the use of the black globe on the thermometer used to measure radiant temperature.

The most common way of measuring mean radiant temperature is by using a globe thermometer, which has a bulb encased in the centre of a 15 cm diameter copper sphere which has been painted matt black.

This is known as a **vernon globe thermometer** or a **black globe thermometer** and is suspended at the point of measurement for normally at least 20 minutes.

Radiant heat is absorbed into the globe without being influenced by air currents. Electronic, rather than liquid in glass bulb, thermometers fitted with a black globe may also be used to measure radiant temperature.

RELATIVE HUMIDITY

Hygrometers are used for measuring humidity, they take a number of forms and use different mechanisms to provide a measurement; common instruments include pyschrometers and dew point hygrometers.

Figure B10-6: Electronic black globe thermometer.
Source: Extech.com.

Relative humidity (RH) is the ratio between the actual vapour content of the air and the vapour content of air at the same time saturated with water vapour. If the temperature of the atmosphere rises and there is no change in the vapour content of the atmosphere, the absolute humidity remains the same, but the relative humidity is lowered. This is because the amount of water vapour in air at saturation increases as the temperature of the air increases.

The following 2 values are important in order to calculate the RH at any particular air temperature:

1) The measured amount of water vapour in the air.
2) The amount of water vapour in the air at saturation point for that temperature.

As these values are difficult to measure, the technique for measuring RH relies on using a standard chart called a psychrometric chart and also obtaining RH from the chart using 2 temperature measurements relating to the above values.

The 2 temperature measurements made are:

1) The dry bulb measurement of the air.
2) The wet bulb measurement of the air.

Psychrometer hygrometers are used for measuring relative humidity and these comprise a dry bulb thermometer and a wet bulb thermometer, though digital thermometers that use the wet and dry principle may also be used. The **dry bulb temperature** is a measurement taken using a standard liquid in glass bulb thermometer.

The **wet bulb temperature** is measured using a specialised liquid in glass bulb thermometer, in which the bulb is covered with a wick that has been wetted with distilled water. When the water evaporates from the wick, heat is removed from the bulb by the evaporating cooling effect and the thermometer will show a reading that is lower than that of the dry bulb reading.

The rate of water evaporation is directly related to the amount of water in the surrounding air. When the air is saturated with water vapour no evaporation will take place and the two readings will be the same - thus the RH value will be 100%. Both the wet and dry bulb temperature readings can be plotted onto a **psychrometric chart**, which enables the relative humidity reading to be derived.

Whirling (psychrometer) hygrometer - this measuring instrument, used for measuring relative humidity, resembles an old fashioned football rattle and is a simple but effective design.

It consists of a wet and dry bulb thermometer mounted side by side in a frame, which is rotated by hand. When it is rotated, air is forced to flow over the bulbs. A small reservoir supplies the wick surrounding the bulb of the wet bulb thermometer with distilled water.

Although very effective, it is quite limited in its uses as it only gives a reading for a particular spot where the sample was taken and it needs to be checked very frequently to ensure that the distilled water reservoir is not empty.

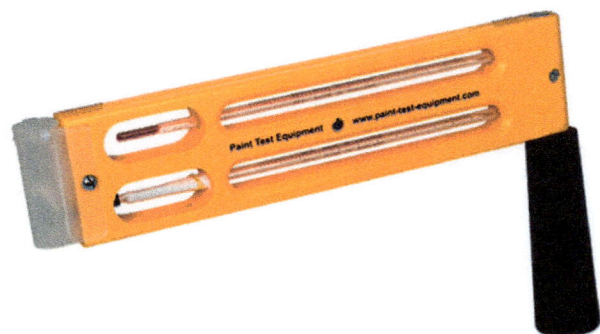

Figure B10-7: Whirling (psychrometer) hygrometer.
Source: Paint test equipment.

Static (psychrometer) hygrometer - this is the static hygrometer or masons hygrometer. This is not very accurate and must be positioned in an area free from draughts as it relies on natural ventilation to induce evaporation of the wet bulb's wick.

Forced draught (psychrometer) hygrometer - this is also known as an Assmann Hygrometer and is extremely accurate. It is of a similar design to the static hygrometer in that it contains **wet** and **dry** bulb thermometers mounted in a frame, but it has a fan that forces a flow of air across the bulbs rather than relying upon natural ventilation. The Assmann hygrometer is insulated, which makes it less susceptible to the effects of nearby radiant heat sources that could influence measurements.

Mirror (dew point) hygrometer - the water vapour held in air will start to settle out on a surface at a specific temperature, called the dew point. The temperature at which this occurs depends on the amount of water vapour held in the air, the humidity. A mirror hygrometer slowly lowers the temperature of a mirror until it reaches the dew point. By measuring the dew point the hygrometer can calculate the humidity.

AIR VELOCITY

Air velocity is measured with a Kata thermometer, the measurements of the cooling effect of air movement are translated into air speed, or an anemometer, which measures the actual speed of the air movement.

The *Kata thermometer* looks like a glass liquid bulb thermometer with a large bulb. The bulb is placed in warm water until the alcohol, the liquid in the bulb, rises to an upper reservoir. The thermometer is wiped dry and suspended in the air. The time taken for the column of alcohol to fall between two marks is measured. This rate of fall is a function of air movement, which can be calculated via nomograms. This instrument is less convenient than the thermo-anemometer and relies on there being a source of hot water. It comes in several sizes, but is accurate at low wind speeds.

There are a number of forms of *anemometer* that fall into two categories, ones that have mechanical movement and those that rely on the cooling effects of the air on a hot wire. One of the most familiar kinds of anemometer is one that consists of three or four cups attached to short rods that are connected at right angles to a vertical shaft. As the air moves, it pushes the cups, which in turn rotates the shaft. The number of turns per minute is translated into wind speed by a system of gears similar to the speedometer on a car. The *rotary vane anemometer* uses a similar mechanical principle, but utilises a multiple vane rotation head to detect the air movement and translate this into mechanical movement of the vane, which can provide a scalar reading or be determined by electrical sensors and translated into a direct digital reading.

Thermo-anemometers rely on the rate of cooling of a heated resistance wire or thermocouple, caused by the air movement. The rate of cooling is dependent on the amount of air movement, which causes circuit balance changes that are translated into a velocity rate. They are portable, battery powered instruments that give instantaneous readings. It is important to appreciate that air movement seldom comes from only one direction. It is therefore important to ensure that air movement from all directions is measured.

ELECTRONIC INSTRUMENTS

Integrated instruments

Electronic instruments are available to provide direct reading of a number of parameters affecting thermal comfort, temperature, humidity and air velocity, integrated into a single instrument. They provide direct reading and are portable. They can also take repeated time based readings and log data for later analysis.

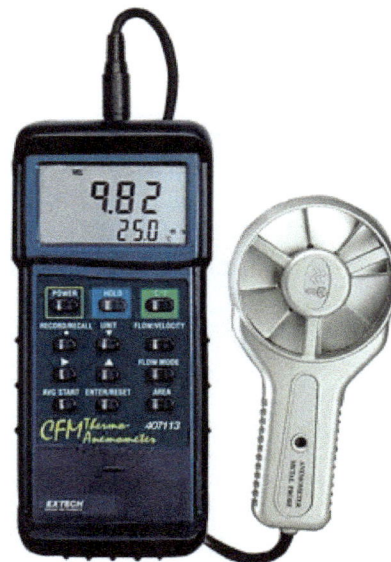

Figure B10-8: Rotary vane anemometer. *Source: Extech Instruments.*

Personal thermal stress monitor

Workers exposed to hot working environments can be susceptible to heat stress, when the internal core temperature of the body rises to dangerous levels. This can result in physiological symptoms like heat cramps, nausea, palpitations, stroke and even, ultimately, death. Typical applications where this can be a hazard include power stations, foundries, steel works, bakeries, glass and many manufacturing processes. Heat stress monitors can be either static instruments sited in areas of risk or data may be collected from personal dose meters that are worn by workers and provided more representative data for the individual. Personal dose meters, fitted with an alarm, should be used when there is a foreseeable risk of heat stress, such as work in confined spaces that may be subject to high ambient temperatures, for example, work on bakery ovens or power generator furnaces.

The personal heat stress monitor is a miniature data-logging instrument for monitoring individual body stress in hot working conditions. The wearer receives information on their physical condition, so that exposure to heat stress is kept within safe and acceptable limits.

The lightweight monitor measures body temperature and heart beat and can be slipped into a shirt pocket or clipped onto a belt worn around the chest.

A two-stage alert with both visual and audio alarms notifies the worker, first, that body temperature, heart rate or both are approaching unsafe levels. The second stage tells the worker to stop work immediately.

Data can be transferred directly to a printer or computer, typically with real-time, minute by minute readings of heart rate, temperature and alert status.

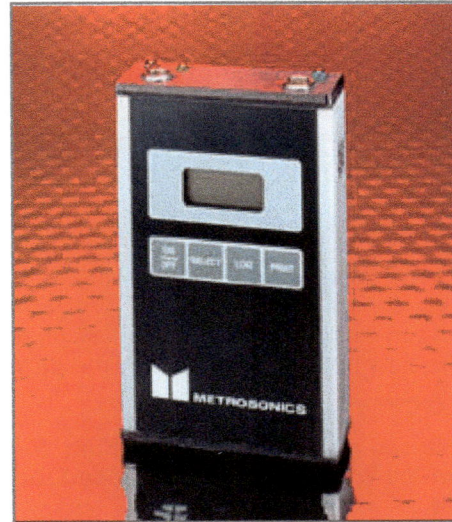

Figure B10-9: Personal heat stress monitor. *Source: Ashtead Technology.*

Workers exposed to cold temperatures that put them at foreseeable risk of cold stress could use personal monitors of a similar type, to monitor how they are affected and their level of cold stress.

Other parameters affecting thermal comfort

METABOLIC RATE

The metabolic rate is the rate at which a person's body burns calories to maintain bodily functions, such as breathing, digesting foods, and beating of the heart. The Basal Metabolic Rate (BMR) is the number of calories used by an inactive person.

For most individuals, about 60% of their calories go towards maintaining the normal body functions, use of muscles and processes for cell reproduction.

Work rates can be calculated by considering the metabolic rate for various activities and are given in W/m^2 or W/person, where the average male adult has a body area of 1.8m^2. As with the clothing index, given later, these are widely available in tables, for example:

Activity	Rate (Watts/m² of body surface)	Watts/Person 1.8m²
Sleeping	43	60
Resting	47	80
Sitting	60	100
Standing	70	120
Strolling (1.5 mph)	107	190
Walking (3 mph)	154	300
Running (10 mph)	600	1080
Sprinting (15 mph)	2370	4270
Shivering	330	600

Figure B10-10: Metabolic rates. *Source: RMS.*

CLOTHING

The tog is a measure of thermal resistance of a unit area, also known as thermal insulance, commonly used in the textile industry, and often seen quoted on, for example, duvets and carpet underlay. The basic unit of insulation coefficient is the R$_{SI}$ (1m^{2}K/W). 1 tog = 0.1 R$_{SI}$.

Therefore, 1 tog = 0.1m^{2}K/W. The thermal resistance in togs is equal to ten times the temperature difference (in °C) between the two surfaces of a material, when the flow of heat is equal to one watt per square metre.

Clothing indices are designed to indicate the insulating properties of clothing. An example of this is the Clo value:

- 1 Clo = 0.155 m^2 °C/Watt - Measurement of resistance.
- 1 tog = 0.1m^2 & °C/Watt.

Naked	0
Shorts	0.1
Light summer clothing	0.5
Indoor clothing	1.0
Heavy suit	1.5
Polar clothing	3.4
Practical maximum	5

Figure B10-11: Typical Clo values *(from Fanger)*.   *Source: RMS.*

SWEAT RATE

Sweat rate is a measure of the amount of sweat produced per unit time. One formula for calculating the sweat rate of an athlete is:

- Sweat rate = (P − A + F − U)/H.

Where:

- P = pre-exercise (naked) body weight.
- A = post-exercise (naked dry) body weight.
- F = fluid intake.
- U = urine volume.
- H = exercise time in hours.

As 1,000 ml of water weigh 1 kg, sweat production can be calculated in units of milliliters of sweat produced per hour.

DURATION OF EXPOSURE

Whilst the temperature and other factors are important, the duration of exposure to heat or cold will influence greatly the amount of heat transferred to, or from, the worker. The longer the exposure, the more heat that can be lost or gained. This can cause the worker to move from a position of comfort to one of heat or cold stress.

Work patterns may vary naturally or by planned rotation, causing the worker to have a varied duration of exposure and this can influence the thermal stress developed by the worker. If the duration is relatively short, it may enable the worker to equalise the thermal effects and reduce the overall thermal stress they experience. However, rapid changes of extreme thermal variance may cause supplementary effects, such as thermal shock.

Workers must be able to function efficiently, both physically and mentally, to sustain work practices that will not place them at risk. If exposure to heat or cold leads to fatigue or discomfort, this could impair decision-making and affect the ability to follow safe working procedures. Prolonged exposure can lead to fatigue, lowered concentration, slowed reflexes and loss of physical co-ordination. Any one of these things increases the possibility of an injury occurring. For example, if an employee should faint as a result of heat stress, there is a possibility of an injury from falling or striking objects. Vibration from tools and equipment also presents increased risk of hand-transmitted or whole body vibration injury to the operator in cold conditions.

The purpose of the heat stress index (WBGT)

The wet bulb globe temperature (WBGT) was originally developed for military use for personnel on active service in the desert. WBGT was brought in during the 1950s as an easier and more practical method to quantify thermal stress. It is now enshrined in BS EN 27243 'Hot Environments - Estimation of the Heat Stress on Working Man'. It takes into account the three main thermal factors affecting heat stress: humidity, the air temperature and radiant heat. The WBGT is calculated using the following equations:

Activity	Equation
Outdoor work with a solar load:	WBGT = 0.7WB + 0.2GT + 0.1DB
Indoor work/outdoor work without a solar load:	WBGT = 0.7WB + 0.3GT

Figure B10-12: WBGT calculations.   *Source: RMS.*

WB	Natural wet bulb temperature.	DB	Dry-bulb temperature (this is the air temperature).	GT	Globe thermometer temperature (this is the radiant temperature).

The *natural wet bulb temperature* means that the wet bulb temperature reading on the whirling hygrometer is taken without rotating the instrument; thus it is naturally, rather than forcibly, ventilated. The readings are then compared against a psychrometric chart to obtain a WBGT reading in °C. As the metabolic rate increases, the value for WBGT °C will increase.

Metabolic rates are classified in accordance with the amount of energy used per second (W). Example guideline WBGT °C limits for work-rest regimes are given in the following table:

Example only, not for field use	Metabolic rate class (work load)		
Work compared to rest regime	Light (Metabolic rate 180 W)	Moderate (Metabolic rate 300 W)	Heavy (Metabolic rate 415 W)
Continuous work	30.0 WBGT °C	26.7 WBGT °C	25 WBGT °C
75% Work : 25% Rest each hour	30.6 WBGT °C	28.0 WBGT °C	25.9 WBGT °C
50% Work : 50% Rest each hour	31.4 WBGT °C	29.4 WBGT °C	27.9 WBGT °C
25% Work : 75% Rest each hour	32.2 WBGT °C	31.1 WBGT °C	30.0 WBGT °C

Figure B10-13: Maximum WBGT readings °C.   Source: RMS.

The WBGT °C value obtained from measurement is compared with reference heat stress index WBGT °C values that consider the metabolic rate class related to the work done and work-rest regimes, for example, those in BS EN 27243.

It is important to note that the reference values generally assume a worker is physically fit, in good health, normally clothed, with adequate salt and water intake and, if conditions stay within limits, are able to work effectively without exceeding a body core temperature of 38°C.

Conclusions are then reached on the heat stress risk related to the thermal conditions and work under consideration. Thus, considering the table in **figure ref B10-13**, a worker in an environment of 30°C WBGT could:

- Perform continuous light work.
- Carry out light to moderate work for 30 minutes and rest for 30 minutes of each hour (i.e. do slightly less than moderate work).
- Perform heavy work for 15 minutes and rest for 45 minutes of each hour.

The capacity for a worker to work, and the effects of heat stress, will differ according to the level of acclimatisation of each worker. Acclimatisation is the 'long-term' adjustment of the individual to the heat stresses placed upon them. It is best achieved by gradually increasing exposure and physical activity in hot conditions.

It enables the body to sweat more and therefore cool down quickly. It improves the cardio-vascular function, which allows more blood to be moved to the skin's surface for cooling. It is important to consider acclimatisation with new workers and for those who have moved to work in a hotter environment.

The BS EN 27243 considers an acclimatised worker as one that has been gradually exposed to heat stress over a seven day period. A limitation of the WBGT index is that the reference values are representative of the mean effect of heat, over a long period of work. It does not provide a reference for those instances where workers are exposed to heat for very short periods of time, for example, a few minutes. These exposures could be as a result of exposure to a very hot environment or a short period of intense physical activity.

When this occurs the reference values may not be exceeded even though the heat stress may exceed the permissible value. In order to take account of this, the highest metabolic rate value is used as the reference value, when there is uncertainty about the metabolic rate that is to be adopted. The most widely used and accepted index for the assessment of heat stress in industry is the Wet Bulb Globe Temperature (WBGT) index.

The measurements that are taken to determine the WBGT index:

- Natural wet-bulb temperature.
- Globe temperature.
- Air temperature.

The principle of operation of the instruments that should be used to make the measurements:

- A thermometer with the bulb wrapped in a wetted cloth or sock is used.
- The water evaporates from the bulb causing it to cool.
- The bulb cools to below the ambient temperature giving a measured wet bulb temperature.
- Air temperature is measured using an alcohol/mercury or digital thermometer.
- The liquid in the tube expands as the temperature rises and a reading obtained from a marked scale or digital display.
- A black globe thermometer is used to measure radiant temperature.
- A mercury filled thermometer is encased in a black painted copper sphere and the radiant heat is absorbed without being influenced by air currents.

| Metabolic Rate class | Metabolic rate, M | | WBGT Reference value | | | |
	Related to a unit skin surface area W/m-2	Total (for a mean skin surface area of 1.8m2) W	Person acclimatised to heat °C		Person not acclimatised to heat °C	
0 (resting)	$M \leq 65$	$M \leq 117$	33		32	
1	$65 < M \leq 130$	$117 < M \leq 234$	30		29	
2	$130 < M \leq 200$	$234 < M \leq 360$	28		26	
3	$200 < M \leq 260$	$360 < M \leq 468$	No sensible air movement 25	Sensible Air movement 26	No sensible air movement 22	Sensible air movement 23
4	$M > 260$	$M > 468$	23	25	18	20

Figure B10-14: Reference values of WBGT heat stress index.   *Source: HSE/ISO 7243.*

Measuring thermal comfort using predicted mean vote (PMV) and percentage people dissatisfied (PPD)

Thermal comfort is 'that condition of mind which expresses satisfaction with the thermal environment'. Due to individual differences, it is extremely hard to establish a thermal environment that will satisfy all workers.

The goal therefore tends to be to establish a thermal environment where there are as few dissatisfied as possible.

Measuring thermal comfort to determine the level of satisfaction can be conducted on a qualitative basis using the Health and Safety Executive (HSE) checklist, which covers the six basic parameters for thermal comfort - air temperature, radiant temperature, humidity, air movement, metabolic rate and the effects of clothing.

This could provide sufficient information for improvements to be made to thermal comfort in the workplace. If there is a need to measure thermal comfort more accurately then methods like predictive mean vote (PMV) and percentage people dissatisfied (PPD) may be used.

BS EN ISO 10551:2001 'Ergonomics of the thermal environment: assessment of the influence of the thermal environment using subjective judgement scales' and BS EN ISO 7730 'Ergonomics of the thermal environment: analytical determination and interpretation of thermal comfort using calculation of the PMV and PPD indices and local thermal comfort criteria' relate to the assessment of the influence of the thermal environment, using subjective judgement scales based on the PMV and PPD of a large group of people.

The predicted mean vote is concerned with the thermal sensation of a large population of people exposed to a certain environment. PMV establishes a thermal strain based on steady-state heat transfer between the body and the environment and assigns a comfort vote to that amount of strain. A seven point scale (hot, warm, slightly warm, neutral, slightly cool, cool, and cold) is used to predict thermal sensation.

ASHRAE psycho-physical scale 'seven point thermal sensation scale'	
Cold	- 3
Cool	- 2
Slightly cool	- 1
Neutral	0
Slightly warm	+ 1
Warm	+ 2
Hot	+ 3

Figure B10-15: Seven point thermal sensation scale.   *Source: Fanger.*

The PMV is established from the **comfort equation**, which is derived from the heat balance equation and takes account of mean skin temperature and sweat loss. This can be calculated using a formula, tables or an instrument that measures thermal comfort and replicates the sensation felt by a human. The comfort equation to calculate PMV and the equation to calculate the PPD are too complicated for this level of study; however it is important to understand the relationship between PMV and PPD.

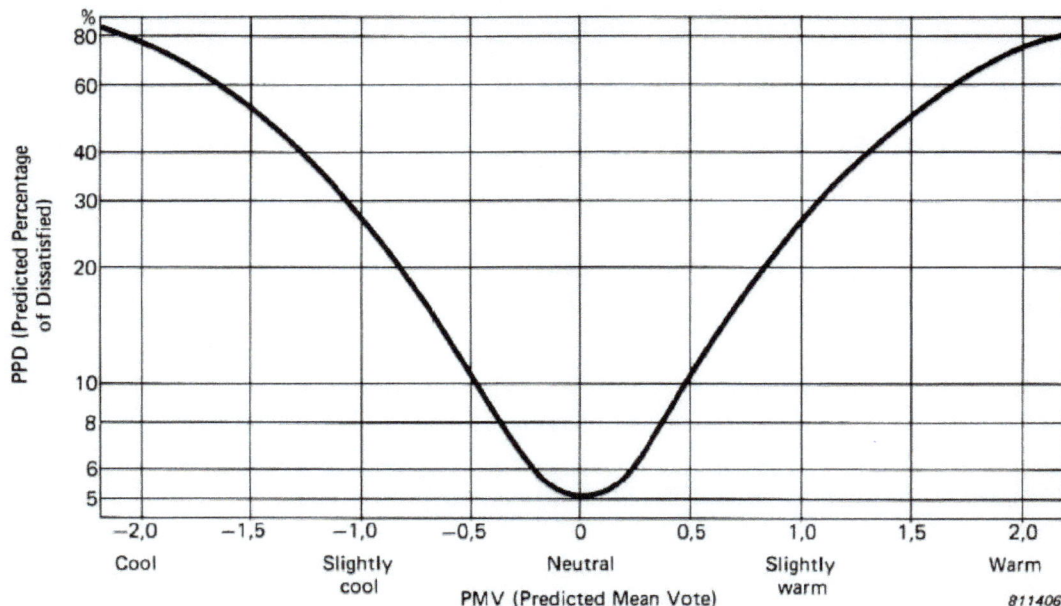

Figure B10-16: Relationship between PMV and PPD.   Source: B W Olesen.

The predicted percentage of people dissatisfied is determined at each stage of the PMV. When PMV changes, away from zero, in the positive or negative direction, PPD increases. Studies have shown that even when the PMV is at zero the minimum percentage of people dissatisfied is 5%, confirming that thermal comfort conditions will not usually be universally acceptable. However, a dissatisfaction rate of 5% is considered reasonable. The BS EN ISO 7730 suggests reasonable limits of thermal comfort based on PMV and PPD; PMV should be limited to plus or minus 0.5 and PPD less than 10%.

The assessment of heat stress

ROLE OF HEAT INDICES

Heat stress indices provide information for assessing hot environments and predicting likely thermal strain on the body.

Heat strain limit values, based upon heat stress indices, indicate when that strain is likely to become unacceptable.

See figure ref B10-17, which shows the variation of heat strain with increasing heat stress. In zone B, the deep body temperature is held constant by the increasing sweat rate. In zone C, the environmentally driven zone, sweat rate can no longer increase and the body temperature rises. The transition between zone B and C is termed the upper limit of the prescriptive zone.

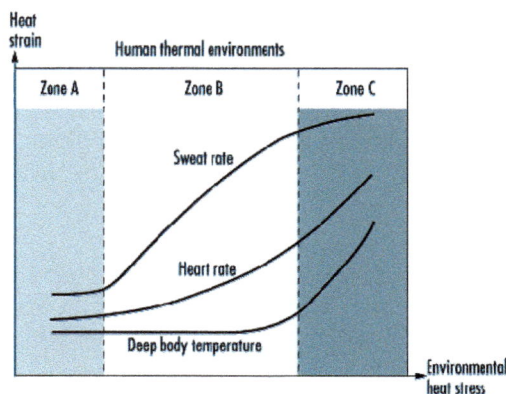

Figure B10-17: Heat strain with increasing heat stress.
Source: World Health Organisation (WHO).

DIFFERENCE BETWEEN EMPIRICAL, DIRECT AND RATIONAL INDICES

There are three types of method used for the assessment of hot environments. These methods are used to estimate or establish the physiological responses of an individual to their environment. The indices provide a numerical comparison between environments, different work situations and different types of clothing to be determined.

Empirical

The empirical indices have been developed from field experimental data of test groups of people to predict the likely effects an environment will have on a human, enabling predictions to be made to estimate work rate, metabolic rate, clothing factors etc.

Direct

The direct indices use standardised instruments to measure environmental parameters. These indices are commonly used to construct safety regulations or an organisation's thermal standards, by providing a simplified approach to obtain an estimate of worker acceptable thermal balance, for example, the use of globe temperature thermometers to measure and establish work place temperature guidelines.

Rational

The rational indices use calculations of the heat exchanges, for example, sweat rate, between the employee and the work environment. It provides a method for predicting employees' responses to different work scenarios.

HEAT STRESS INDICES

The indices most commonly used to assess and manage the risk of heat stress are:

- Effective temperature (ET).
- Corrected effective temperature (CET).
- Heat stress index (HSI).
- Predicted 4-hour sweat rate (P4SR).
- Wet bulb globe temperature (WBGT).

Effective temperature (ET) and corrected effective temperature (CET)

The effective temperature (ET) and corrected effective temperature were early attempts to quantify thermal stress dating back to the 1920s. The effective temperature combines the effects of air temperature, humidity and air movement into one scale. The effective temperature uses both wet and dry bulb temperatures.

The corrected effective temperature (CET) uses the same criteria, but corrected for radiant heat. It is calculated using nomograms that are available for different clothed states, for example, semi-nude and light summer clothed. The following measurements are therefore required in order to calculate the CET.

These are:

- Wet bulb temperature.
- Wind velocity.
- Globe temperature or air temperature if they are the same.

The World Health Organisation recommended that the following limits should apply to corrected effective temperatures:

Work	Temperature
Sedentary work	30°C CET
Light work	28°C CET
Heavy work	26.5°C CET

Heat stress index (HSI)

The heat stress index (HSI) is based on heat exchange and is the comparison of evaporation required to maintain heat balance. It is calculated by using a formula or by charts (nomograms). HSI is obtained by estimating two values:

1) The required evaporative heat loss (by sweating) to achieve heat balance (E_{req}).
2) The maximum evaporative heat loss possible in that environment (E_{max}).

$$HIS = E_{req} / E_{max} \times 100$$

The data used to establish values for and were derived from fit, young, well acclimatised men, which means their applicability to any individual worker being considered may be limited. However, the derived HSI will provide an indicative limit on which an organisation's own standards may be established. The heat strain likely to be experienced due to a given level of heat stress varies; severe heat strain may be experienced in the HIS range 40-60, with increasing severity above this. The upper limit for safety is an HSI that that does not exceed 100. If E_{req} is less than or equal to E_{max} then work can continue with consideration of limiting precautions, such as fluid intake. When E_{req} is greater than E_{max} then there will be heat build-up in the person's body and working time should be limited.

The following measurements are required to establish a heat stress index:

- Metabolic work rate.
- Air velocity.
- Air temperature.
- Wet-bulb temperature.
- Globe temperature.

Predicted 4-hour sweat rate (P4SR)

The predicted 4 hour sweat rate (P4SR) measures sweat rate as a function of heat stress. It is one of the few indices that takes into account all environmental and personal factors affecting thermal comfort.

It therefore takes into account work rates and the clothing worn. The P4SR is obtained from charts (nomograms). The preferred standard is a sweat rate of below 2.7 litres in 4 hours. In any event the maximum sweat rate should not exceed 4.5 litres in 4 hours.

Index	Parameters measured	Other factors	Comment
Effective Temperature (ET)	Dry bulb, wet bulb, air velocity	Two clothing levels, basic and normal	No account of radiant heat or allowance for metabolic rate
Corrected Effective Temperature (CET)	Globe thermometer, wet bulb, air velocity	Two clothing levels, basic and normal	No allowance for metabolic rate
Heat Stress Index	Globe thermometer, dry bulb, wet bulb, air velocity, metabolic rate	Limited assumptions made about clothing	100 max, >70 very severe heat strain, >40 severe heat strain
Predicted 4-hour Sweat Rate (P4SR)	Globe thermometer or dry bulb, wet bulb, air velocity	Corrections applied for metabolic rate and clothing	Absolute max 4.5 litres in 4 hours
Wet Bulb Globe Temperature (WBGT)	Globe thermometer, wet bulb, (dry bulb if outside)	Corrections applied for clothing, different metabolic rate	Two formulae - inside and outside, two limits and action levels

Figure B10-18: Summary of main heat stress indices.   *Source: OH learning.*

Wind chill index (WCI)

The wind chill index (WCI) is relevant to work in low temperatures and takes both air velocity and temperature into account. This can be obtained from charts. The WCI represents heat lost from the body in k cal h^{-1} m^2.

Practical control measures to minimise the risks when working in extreme thermal environment

FACTORS THAT AFFECT THERMAL COMFORT

In order improve the thermal environment it is first necessary to consider what constitutes good thermal comfort. The recommended conditions for thermal comfort are:

1) Temperature:
■ The mean radiant temperature should preferably be equal to or slightly higher than the air temperature.
■ Excessive radiant heat should not fall on the heads of occupants.
■ There should not be large asymmetrics in the radiant fields (for example, cold window, hot surfaces).
■ Temperature gradients should not exceed approximately 3°C over the height of a person.

2) Air movement:
■ Should be approximately 0.1 - 0.2 m/s. Higher values may be desirable in summer. Lower values are likely to cause 'stuffiness'.
■ Air movement should be variable rather than uniform.

3) Humidity:
■ Relative humidity should be in the range of 40-70%.

Alternatively the conditions that cause discomfort are:

■ Thermal differences, for example, under heating in winter and overheating in summer.
■ Localised high radiant conditions, particularly uncomfortable if to the head (an extreme being 'sunstroke').
■ Conditions not matched to activity and clothing.
■ Cold surfaces (for example, windows give localised radiation loss, for example, to the neck).
■ Cold draughts, which affect feet in particular as cold air drops (for example, from windows).
■ Low air movement, causing stuffiness, oppressiveness.
■ Asymmetric conditions, for example, air or radiant temperature gradients.
■ Dry air (<30% RH), causing sore throats, inflamed sinuses etc.
■ Moist air (>70%), causing stuffiness. Particularly uncomfortable when activity rate increased.

CONTROL MEASURES TO IMPROVE UNSATISFACTORY THERMAL ENVIRONMENT PARAMETERS

There are a limited number of ways in which control measures can be applied. The variables which must be taken into consideration are:

- **Control the heat/cold source**, for example, radiant; convection and conduction; screens; ventilation; insulation.
- **Control the environment**, for example, replace hot air with cold, or cold with hot, humidity, air movement, and reduce draught discomfort.
- **Separate heat/cold source from the worker**, for example, the use of screens or barriers, restrict access, use remote working.
- **Workplace design**, for example, for hot work insulating hot plant and pipework, providing air cooling, shading windows, locate workstations away from radiant heat. For cold work, such as food production, storage and distribution; enclosing or insulating the product; pre-chilling the product; keeping chilled areas as small as possible, rest facilities.
- **Job design**, for example, consider physically demanding tasks, work rate (particularly repetitive tasks), use job rotation to reduce exposure, regular rest breaks, make use of mechanical aids where physically demanding work is carried out in a hot environment.
- **Hot/cold drinks**, dehydration effects can be minimised in heat stress situations by encouraging workers to frequently drink cool water (rather than tea, coffee or carbonated drinks) in small volumes to compensate for losses due to sweating. Workers should be made aware that thirst is not a good indicator of dehydration as if they are thirsty they are already starting to suffer from the effects of dehydration. When working hard or at a high rate in heat stress conditions employees should consume around 250 ml (half a pint) every 15 minutes. Where this is not practical an alternative is to drink 500 ml of water per hour before work commences and to encourage the drinking of 500 ml of water during their rest periods.
- **Clothing/PPE**, for example, for hot work light-weight and loose; furnace work use of metallic reflective overalls, make sure that workers are not wearing more (or less) clothing than required. Multiple layers of clothing enable personal adjustments to be made. Dress codes should allow workers to adapt their clothing where possible.
- **Health surveillance**, for example, regular checks for those who work in extreme temperature environments, consider specific surveillance for those that are pregnant, ill, disabled or are on medication. The surveillance should include monitoring of dehydration.
- **Training**, for example, ensure training includes workers being able to recognise the early symptoms of exposure to extreme temperatures, including the symptoms and prevention of dehydration.

Case Study - Too much water can be a bad thing

A health and safety practitioner, working in Nigeria, was not feeling well and was advised to drink plenty of water. He did to the extent that this reduced his sodium concentration in the blood (causing hyponatraemia); this resulted in him being admitted to hospital, followed by a full recovery after being given sodium and other minerals that were low.

B10.2 - Suitable and sufficient lighting in the workplace

Legal requirements

Legal requirements for the provision of lighting in the workplace are set out in the WHSWR 1992 (as amended) and CDM 2015.

WORKPLACE (HEALTH, SAFETY AND WELFARE) REGULATIONS (WHSWR) 1992, REG. 8

"Every workplace shall have suitable and sufficient lighting. The lighting shall, so far as is reasonably practicable, be by natural light. Suitable and sufficient emergency lighting shall be provided in any room in circumstances in which persons at work are specially exposed to danger in the event of failure of artificial lighting."

Figure B10-19: Regulation 8 of WHSWR 1992.   *Source: Workplace (Health, Safety and Welfare) Regulations (WHSWR) 1992.*

CONSTRUCTION (DESIGN AND MANAGEMENT) REGULATIONS (CDM) 2015, REG. 35

"(1)   Each construction site and approach and traffic route to that site must be provided with suitable and sufficient lighting, which must be, so far as is reasonably practicable, by natural light.

(2)   The colour of any artificial lighting provided must not adversely affect or change the perception of any sign or signal provided for the purposes of health or safety.

(3)   Suitable and sufficient secondary lighting must be provided in any place where there would be a risk to the health or safety of a person in the event of the failure of primary artificial lighting."

Figure B10-20: Regulation 35 of CDM 2015.   *Source: Construction (Design and Management) Regulations (CDM) 2015.*

The necessity for lighting in workplaces

The health and safety problems associated with inadequate/unsuitable lighting or lighting arrangements fall into two main categories:

1) Deterioration of visual acuity.
2) Increased likelihood of accidents caused by incorrect perception.

Lighting in workplaces is necessary in order to see things and to be seen. Lighting makes it easier to identify hazards and to respond to them in good time. Workers moving around the workplace need to be seen in order to prevent contact with them, and good lighting helps to ensure this. Poor lighting can also lead to eyestrain, headaches, migraine, tiredness and poor concentration.

Impact of lighting levels on safety issues

Light levels will affect individuals in a number of ways, in particular if they have problems of sight impairment. Older people often develop sight problems of long sight (reduction in the ability to read small print) from the age of 40 to 50 years. Some common issues are discussed in the following section.

INCORRECT PERCEPTION

Deterioration of visual acuity. For example, visual fatigue, glare, falls resulting from level changes which are not apparent.

FAILURE TO PERCEIVE

Increased likelihood of accidents caused by incorrect perception. For example, slips, trips and falls, vehicle collision, etc.

STROBOSCOPIC EFFECTS

Stroboscopic effect is not common with modern lighting systems, but where it does occur, it can be dangerous. Earlier types of fluorescent light gave the impression that machinery was stationary or moving in a different way. Wiring adjacent tubes to different supply phases overcomes this problem.

COLOUR ASSESSMENT

Colour rendition refers to the fact that colours appear different under different light sources. Choice of lamp is important if a 'warm' or 'cool' effect is desired. Colours will appear different when viewed under sodium light compared to natural daylight; this is a consideration when working outside at night using high visibility clothing in areas illuminated by sodium lights, when some colours used for high visibility will not be observed. Good colour discrimination is required for some electrical work and this may be affected by the ambient light available to the worker. At very low levels of illuminance colour vision fails and all colours are seen as shades of grey.

EFFECT ON ATTITUDES

The eye is sensitive to the blue band of natural light and many individuals are depressed when light levels are low or diffused by heavy cloud, reducing that portion of light. Natural daylight is preferred, but this will often need to be supplemented throughout the day for many tasks, for example, when work of a detailed nature is carried out, such as with watch repairs or copy typing small print. Low levels of light can lead to lethargy, poor concentration and irritability. Poor or very bright lighting will cause fatigue and reduce willingness to work; light levels should be assessed and maintained at an appropriate level for the tasks to be done.

'Suitable and sufficient lighting'

The approved code of practice (ACOP) to the WHSWR 1992 emphasises that lighting should be sufficient to enable people to work, use facilities and move from place to place safely and without experiencing eye-strain.

Where risks are higher, then a sufficient amount of lighting should be provided to take account of this; this may include specific lighting of hazards or provision of task lighting to enable hazards to be avoided or reduce eye-strain.

Lighting should also be suitable, such that it provides the right level of luminance, does not create unnecessary additional hazards (such as glare, flicker or explosion hazard) and where reasonably practicable is provided by natural light. To ensure lighting is suitable and sufficient, several aspects of lighting and the workplace need to be considered.

These include:

- Lighting design.
- Type of work.
- The work environment.
- Health aspects.
- Individual requirements.
- Lighting maintenance, replacement and disposal.
- Emergency lighting.

LIGHTING DESIGN

Lighting design should ensure that the right amount of illumination is provided for the work being carried out, the finer the detail the higher the illumination required. It should also achieve a uniform illumination across any task area and take account of the effects of shadow that may be created. Lighting may be split into general, localised or local.

General lighting provides uniform illumination over the whole working area, localised lighting provides different levels of lighting to match the needs of specific tasks and local lighting provides specific lighting to a specific work area or activity, particularly where a high level of detailed work is required. Lighting design should take account of and avoid glare, particularly when providing temporary lightings, for example, in construction and maintenance operations.

TYPE OF WORK

As previously outlined when considering lighting design, the type of work carried out in the workplace greatly influences what may be considered to be suitable and sufficient. Where the task requires fine detailed work, such as in electronic assembly, local lighting may be needed in addition to good general lighting. Tasks that require clear colour rendition will need to be provide with suitable lighting that enables this to take place without eye strain or error. *For recommended illuminance levels for different work activities, see 'Assessment of lighting levels' later in this Element.*

THE WORK ENVIRONMENT

Where possible, natural lighting should be used. This may provide a basis of general lighting but can be of limited use at night and in winter months; it may therefore need to be supplemented with artificial lighting, which can be used when required. Control of natural lighting from windows will often be necessary where it may affect the use of display screen equipment.

The interior design aspect of the work environment should not be forgotten; aspects like colour of walls, floors and ceilings can influence light. The layout of the workplace may lead to shadows or reduced illuminance because of the positioning of luminaries; additional provision may have to be made to compensate for the layout. It is important to remember that the work conditions, which may be dusty, flammable or wet, have to be taken account of when deciding suitable lighting.

HEALTH ASPECTS

Suitability and sufficiency of light is not just about having enough light at point of design. Lighting that flickers, provides reduced illuminance because of its age or creates veiling reflections can create eye strain. These factors must be considered when determining and maintaining the suitability and sufficiency of light. Light flicker may also trigger epileptic seizures in some people, and lighting that flickers should be replaced to maintain standards to a suitable and sufficient level.

INDIVIDUAL REQUIREMENTS

Different individuals have different levels of visual acuity and may need/prefer different levels in order to do work. Where possible, individual requirements should be taken into account, and this may mean providing local lighting for a small number of people.

LIGHTING MAINTENANCE, REPLACEMENT AND DISPOSAL

The requirement for suitable and sufficient lighting is not limited to its initial provision. It is important that lighting is kept clean, and replaced as illumination of the workplace declines with the age of the equipment.

Requirements for this will depend on the type of lighting and the workplace conditions it is in; a maintenance programme to retain a good standard of lighting should be established. Consideration of the need for maintenance should be made when determining the suitability of lighting selected. Some lamps contain mercury, sodium or phosphorous and need to be disposed of in a safe and environmentally responsible way.

EMERGENCY LIGHTING

Suitable and sufficient lighting requirements relate to normal and emergency conditions. It is important that emergency lighting be provided where artificial lighting may fail and light levels might not be sufficient to ensure safe escape or general safety in conditions that loss of light may accentuate hazards. Work situations such as working with machinery, at a height or in confined spaces should be particularly considered.

NATURAL LIGHTING

Most people prefer to work in natural daylight, which also represents an energy efficient/environmentally friendly source, and therefore it is important to make best use of it. Legislation reflects this by requiring, where reasonably practicable, lighting provided from natural sources.

However, natural light will not usually provide sufficient illuminance throughout the working day or in all parts of the workplace. Where natural light is sourced from windows and skylights they should be cleaned regularly and kept free from obstruction.

ARTIFICIAL LIGHTING

Artificial lighting may be used where natural lighting is not reasonably practicable, for example, due to features of the architectural layout or design of a building/structure used as a workplace or because specific tasks require directional light of a given luminance that natural light cannot reliably deliver. A wide range of lamps (or luminaries) are available; different types produce light with different properties.

The choice of lamp depends on factors like the type of workplace, luminous efficiency and service life. Lamps and luminaries should:

- Support and protect the light source.
- Direct the light, where required.
- Avoid glare.
- Provide filtering to reduce electromagnetic radiation.

Types include:

Type	Lamp prefix letter	Construction	Use
High pressure mercury	MBF	An electric discharge in high pressure mercury atmosphere, contained in an arc tube within a glass envelope with fluorescent coating	Exterior large space lighting
Tubular fluorescent	MCF	An electric discharge in a low pressure mercury atmosphere contained in a glass tube internally coated with fluorescent material	General office lighting
Low pressure sodium	SOX/SLI	An electric discharge in a low pressure sodium atmosphere in a glass arc tube contained in a glass envelope	Road and security

Figure B10-21: Types of artificial lighting.   Source: HSG38 Lighting at work.

Type	Typical lumen output
Tubular fluorescent 30 watt	2,400
High pressure sodium 400 watt	47,500

Figure B10-22: Types of artificial lighting.   Source: www.ndlight.com.

LIGHTING APPROACHES

Lighting the task and workstation

In general, for each visual task we require a certain minimum quantity of light arriving on each unit area of the object in view (i.e. a minimum 'planar' illuminance).

The value of this minimum illuminance depends primarily on the size of the detail which must be perceived, but will also depend on the visual contrast that the task makes with the background against which it is seen, the duration of the task, whether or not errors may have serious consequences and the presence or absence of daylight.

Ideally, the workstation should be designed in such a way as to make the task the brightest part of the field of view. Research has shown that favourable conditions exist when the task has a luminance, which is about three times that of its immediate surrounds, and when the immediate surrounds, again, have about three times the luminance of the general surrounds to the workstation.

These conditions can be achieved by a combination of general and local lighting used to illuminate the work surfaces, which have appropriately chosen reflectances. For example, when a desk lamp provides local lighting for white paper seen against a grey-blotting pad placed on a desktop served by a general installation of ceiling-mounted fluorescent lighting, an approximation to these desirable conditions is easily obtained.

Not all problems caused by poor lighting can be overcome simply by the provision of more light. When tasks are visually demanding, the quality of the available light is at least as important as its quantity. Daylight has qualities that are difficult to imitate artificially.

Moreover, the provision of a distant view seen through a window will provide welcome relief to eyes that must focus at their 'near point' while work is in progress.

Aspects of lighting quality, which should receive attention when lighting the workstation, are the control of glare, the provision of adequate modelling, *(see also 'Lighting the interior of the workplace' on the next page)* and, where necessary, good colour rendering.

Figure B10-23: Reflected light from window.   *Source: RMS.*

Figure B10-24: Light from windows controlled by blinds. *Source: RMS.*

Lighting the interior of the workplace

The basic need of lighting the interior of the workplace is to provide sufficient and suitable light in the circulation areas to allow movement of personnel, materials and equipment between workstations to take place conveniently and in safety. The aim should be to provide conditions that remain comfortable to the eye as it passes from one zone of the workplace to another, and to make available all information relevant to the well-being and safety of the workforce that can be received through the sense of sight. All the aspects of quantity and quality considered under task lighting would assume some significance in this larger scale application and so contribute to the achievement of these desirable conditions.

Attention should be drawn to accident 'black-spots', such as changes in floor level or flights of stairs, with increased levels of illuminance served by luminaires that are carefully positioned to provide good three-dimensional modelling, while preventing any direct sight of the unshielded source, so as to avoid disabling glare. Machinery that makes fast cyclic movements, should be carefully illuminated to prevent the occurrence of stroboscopic effects.

Sudden changes in lighting levels should not occur between adjacent workplaces. Levels should be graded to allow time for the eye to adapt. The processes of dark adaptation can take several minutes (more than half an hour in extreme cases), during which time the efficiency of the eye is severely reduced, making accidents more likely.

Differences in the colour characteristics of so-called 'white light' sources are sometimes apparent to the eye and although its ability to colour-adapt is considerable, the occurrence of frequent noticeable changes due to the use of various kinds of sources in the same interior may accelerate the onset of fatigue as well as making fine colour judgements impossible.

Effects of brightness contrast

Shadow will affect the amount of illumination. Its effect will depend on the task being performed. The answer is to use more powerful lights or provide more of them.

DISABLING AND DISCOMFORT GLARE

Glare occurs when one part of the visual field is much brighter than the average to which the eye is adapted. Glare causes discomfort or impairment of vision.

Disabling glare

When there is direct interference with vision the effect is known as disabling glare. Disabling glare prevents the correct vision of things we want to see. This can occur in a direct way, where the glare reduces the contrast between the thing we want to see and the background. Indirect disabling glare affects the eye rather than the visual task.

This effect occurs where bright light causes dazzle, where the light hitting the retina is scattered and there is reduced retinal image contrast. Disability is further prolonged by the time the eye takes to re-adapt to ambient light after exposure to glare. Dazzling lights and glare should be avoided, especially when people move from brightly to dimly lit areas and vice versa. Providing diffusers or screens will reduce the effect of glare from a lamp. Blinds or curtains will reduce glare from sunlight.

Discomfort glare

Discomfort from glare can be thought of as a protective mechanism of the body, the light produces an aversive response in order to encourage withdrawal from the danger presented by the light. At low levels of glare the effect may be one of annoyance, but at higher levels of glare physical discomfort can occur from the action of having to persist with work that involves exposure to glare.

Discomfort is increased as:

- Luminance of the glare source is increased.
- Luminance of the background is reduced.
- Size of the glare is increased.
- Glare source is moved closer to what the person is looking at.

Discomfort glare is related to symptoms of visual fatigue.

TISSUE DAMAGE FROM LIGHT EXPOSURE

Exposure to visible light can lead to thermal, mechanical or photopic effects. Exposure to very bright visible light sources, such as exhibition/stage lighting or image projectors, may cause permanent damage to the light sensing region of the eye, the retinal macula, leading to photoretinitis. Sustained exposure to optical ultraviolet light radiation will cause photokeratitis (inflammation of the cornea), infrared will cause cataracts. Natural light (sunlight) exposure will cause skin changes ranging from tanning or darkening, burns through to formation of melanomas (skin cancers). *See also 'Effects of exposure to non-ionising radiation' in 'Element B7 - Radiation'.*

VISUAL FATIGUE

Visual fatigue occurs when the eye is involuntarily distracted from the task, for example, in data entry the eye may be distracted to the sharp edge of the reflected image. This unconscious eye movement will add to visual fatigue. The eye is particularly sensitive to flicker in a light source, particularly where it occurs at the edges of the person's field of vision. This can be a source of discomfort and fatigue. The use of planned work breaks or changes in work task - before fatigue becomes evident - is an important factor to consider when managing visual fatigue; this is particularly important when work is of a detailed nature.

If the work involves use of visual display screen reflections need to be minimised to reduce visual fatigue. Care should be taken to reduce reflections by elimination of bright light sources or strong reflected images onto the screen. These sources will often include incandescent bulbs, fluorescent lights, vertical or venetian blinds and even reflection of the user themselves, particularly if they are wearing light clothing.

Instrumentation, units (Lux) and measurement of light

The common measuring techniques and terms that are used by the lighting industry are as follows:

LUMEN (LM)

A unit of light flow or luminous flux. Light sources are labelled with an output rating (or luminous flux) in lumens. The lumen rating of a lamp is a measure of the total light output of the lamp, for example, a R30 65-Watt indoor spot lamp may have a rating of 750 lumens. Lamps are usually rated in both initial and mean lumens. Initial lumens indicate how much light is produced once the lamp has stabilised, for fluorescent and high-intensity discharge (HID) lamps this is typically 100 hours.

360⁰ of all light output = one total lumen number measured in a light

Figure B10-25: Lumen.   Source: RMS.

Mean lumens indicate the average light output over the lamp's rated life, which reflects the gradual deterioration of performance due to the rigors of continued operation. For fluorescent lamps, this is usually determined at 40% of the initial rated lumen value.

LUX (LX)

The lux is the metric unit of measurement of illuminance of a surface. One lux is equal to one lumen per square metre. Though the light output of a light unit is important, a ready and useful indicator of the effectiveness of a light unit(s) in the workplace is gained by measuring the lux at a point in the workplace where the light is needed, for example, at a work bench or desk. Digital light meters that give a direct reading of the amount of lux falling on a surface are available and easy to use.

ASSESSING LIGHTING IN THE WORKPLACE

It is important when assessing lighting in the workplace to take account of the following factors:

- Provides sufficient illuminance level for the task.
- Allows people to notice hazards and assess risks.
- Glare.
- Flicker.
- Stroboscopic effects.
- Excessive differences in illuminance level of adjacent areas.
- Positioning for adjustment, maintenance, cleaning or replacement.
- Allows people to see properly and discriminate colours, for example, colour coding of gas cylinders.

- Suitability for the environment and task, for example, flammability, robustness.
- Reflections.
- Hazards arising from the lighting are controlled and risk minimised.
- The needs of each individual.
- Emergency needs.

ASSESSMENT OF LIGHTING LEVELS

When assessing light levels, it is important to take measurements of the general light level, localised light levels and local light levels to ensure that this can be related to the relevant light level standards for the work being done.

It is important to identify where maximum and minimum light levels exist in order to establish the average light levels and ensure levels do not fall below minimum recommended levels. The assessment should seek to measure light levels where there is too much light as well as too little. When lighting levels rely on natural lighting it is important to take measurements when the conditions appear to be average, when bright sunlight is evident and when natural light is limited by weather or the hour of the day.

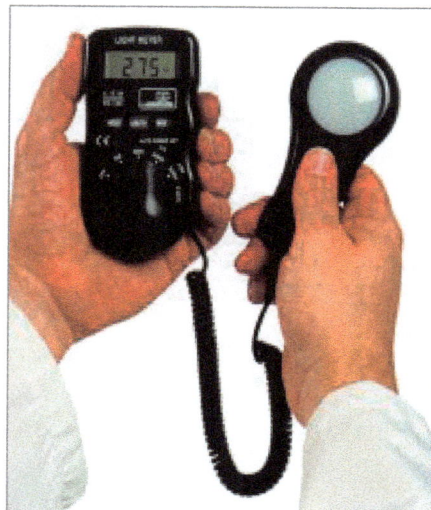

Figure B10-26: Light meter.   *Source: Etl Ltd.*

LIGHTING STANDARDS

Workplace light levels

Workplace light level standards are indicated in the Health and Safety Executive (HSE) guidance on lighting, HSG38. The table in *figure ref B10-27* is based on the HSE recommended levels of illuminance. The HSE advises that reliance on just the average illuminance level might lead to some work areas receiving very low illuminance levels, therefore they advise a minimum illuminance level for the work activities described. For more detailed information on lighting standards the Chartered Institute of Building Services Engineers' (CIBSE) codes for lighting and lighting guides should be consulted.

Activity	Typical locations/types of work	Average illuminance (lux)	Minimum measured illuminance (lux)
Movement of people and vehicles	Depot and marshalling yards, vehicle parks, playgrounds, corridors, aisles and passageways	20	5
Movement of people and vehicles in hazardous areas Rough work not requiring any perception of detail	Construction site clearance, excavation and soil work, loading bays, processing or utilities plant, for example, bottling, food, petro-chemical, power, warehouses involving search and retrieval, stairs Rough fettling, stamping and casting, milking animals	50	20
Work requiring limited perception of detail	Kitchens, factories assembling large components, ceramic work	100	50
Work requiring perception of detail	Offices, sheet metal work, book binding, some food preparation, laboratories	200	100
Work requiring perception of fine detail	Design offices, fine machine work, electronic assembly, some aspects of textile and printing production, some health care activities, jewellery and watch repair	500	200

Figure B10-27: Illuminance levels for different activities.   *Source: HSE, HSG38/RMS.*

Emergency light levels

- *Standby lighting* allows critical activities to continue after a power failure or emergency. It is used to illuminate facilities such as hospitals, fire stations, and other vital services when normal power supplies are unavailable. To ensure these facilities can operate, their standby lighting systems are fuelled by generators or battery packs, which automatically switch on when the main power fails. It can be between 5 and 100% of the illuminance produced by ambient lighting.

- **Safety lighting**, also known as high-risk task lighting, is used to light areas that must be accessed in an emergency. This type of lighting is most commonly found in industrial or manufacturing facilities, and allows workers to shut down equipment or machines that could pose a hazard to evacuees or the public. Safety lighting also allows equipment operators to safely stop work and exit the area in the event of an emergency.
- **Emergency escape lighting** is used to help minimise injury and deaths in the event of an emergency. These fixtures provide light to allow occupants to safely evacuate the building, and help reduce panic and confusion. Powered by either batteries or a generator, the lighting should reach the required luminance within 15 seconds. Battery powered lighting should last for between one and three hours.
- **Anti-panic, or open air, lighting** is designed to allow occupants to quickly flee an area, or to reach a marked escape path. These lights are commonly found in large facilities like sports arenas, stadiums, or auditoriums. They may consist of emergency floodlights or similar fixtures, which are powered by battery packs or generators. Anti-panic lights help keep large crowds calm and safe during an emergency, and reduce lives lost due to panic.

Amenity value and productivity

The lighting levels within a workplace affect the aesthetics of an area. As stated above this can affect attitudes within the workforce. It should be noted that the amenity value of appropriate lighting may enhance productivity.

Further to this appropriate lighting may be essential for certain tasks, for example, job tasks involving close work.

B10.3 - Welfare facilities and arrangements in fixed and temporary workplaces

Legal requirements for welfare facilities and arrangements in workplaces

Most workplaces are regulated by Workplace (Health, Safety and Welfare) Regulations (WHSWR) 1992 (as amended). Regulations 20-25 deal specifically with welfare. They are supported by the Approved Code of Practice and Guidance to the Regulations.

The WHSWR 1992 were amended by the Health and Safety (Miscellaneous Amendments) Regulations 2002 to take account of changes to disability legislation. Regulation 25A sets out a requirement that: *"Where necessary, those parts of the workplace (including in particular doors, passageways, stairs, showers, washbasins, lavatories and workstations) used or occupied directly by disabled persons at work shall be organised to take account of such persons."*

Construction sites are regulated by the regulations for health, safety and welfare in the Construction (Design and Management) Regulations (CDM) 2015, Schedule 2. CDM 2015 has essentially the same requirements as WHSWR 1992, but with the additional option that the facilities do not necessarily have to be provided, they can be 'made available'. This allows for circumstances where the facilities can be made available by other means, such as by accessing those provided by a client on their premises.

When determining welfare provision for works of a temporary nature, the nature of the work to be carried out and the health and safety risks associated with it should be considered. For small scale, low risk work the provision of access to welfare facilities at the workplace may be appropriate. For larger scale work that has specific risks it may be necessary to ensure separate provision of welfare facilities.

For example, showers may be needed if the work involves hazardous substances or very dirty work. This will ensure those carrying out the work are provided with appropriate facilities and ensure there is no contamination of the facilities at the workplace.

This would be particularly important for work that involved hazardous substances like asbestos. Welfare facilities to consider would include those for first-aid, washing, toilets, rest areas, storage of clothing not used at work and drinking water.

The arrangements for welfare provision should include consideration of:
- Duration of the work.
- Number of people who will use facilities.
- Distance workers will have to travel to the welfare facilities.
- Different locations required, including whether the facilities will need to be relocated during phases of the work.
- Cleaning and maintenance of the welfare facilities.

Specific legal requirements for welfare facilities are considered in the following section.

Provision of toilet, washing and changing facilities

SANITARY CONVENIENCES - WHSWR 1992 REGULATION 20

Readily accessible, suitable and sufficient sanitary conveniences must be provided. The conveniences must be adequately lit, kept clean and maintained in an orderly fashion. Separate conveniences for male and female workers must be provided, except where the convenience is in a separate room and the door of the room is capable of being locked from the inside.

The Approved Code of Practice (ACOP) to the WHSWR defines the following as minimum provision:

Number of people at work	Number of water closets	Number of wash stations
1 to 5	1	1
6 to 25	2	2
26 to 50	3	3
51 to 75	4	4
76 to 100	5	5
For every 25 above 100 (or fraction of)	+1	+1

Figure B10-28: Provision for toilet and washing facilities. *Source: ACOP for the Workplace (Health, Safety and Welfare) Regulations (WHSWR) 1992.*

In the case of sanitary conveniences for use only by men, where urinals are provided, the ACOP allows a small adjustment to the number of water closets required.

WASHING FACILITIES - WHSWR 1992 REGULATION 21

Readily accessible, suitable and sufficient washing facilities must be provided. This includes showers if they are required by the nature of the work or for health reasons. Washing facilities must:

■ Be provided in the immediate vicinity of every sanitary convenience, whether or not provided elsewhere as well.

■ Be provided in the vicinity of any changing rooms required by the Regulations, whether or not provided elsewhere as well.

■ Include a supply of clean hot and cold, or warm, water, which must be running water so far as is practicable.

■ Include soap or other suitable means of cleaning.

■ Include towels or other suitable means of drying.

The washing facilities must be adequately lit, kept clean and maintained in an orderly fashion. Separate washing facilities for male and female workers must be provided, except where the washing facilities are in a separate room for one person at a time and the door of the room is capable of being locked from the inside.

The CDM 2015 makes allowance to have mixed washing facilities if they are for the purposes of washing hands, forearms and face only.

FACILITIES FOR CHANGING CLOTHING - WHSWR 1992 REGULATION 24

Where special clothing must be worn at work or for reasons of health or propriety a person cannot change in another room, then suitable and sufficient changing facilities must be provided.

Separate facilities or separate use of facilities for male and female workers must be taken into account. The ACOP recommends that changing facilities should be readily accessible to workrooms (and eating facilities if provided) and should contain adequate seating arrangements.

The facilities provided should be sufficiently large to enable the maximum number of workers to use them comfortably and quickly at any one time. The CDM 2015 expressly requires that changing rooms be provided with seating that has a back.

Storage of clothing - WHSWR Regulation 23

Suitable and sufficient accommodation must be provided for personal clothing not worn at work and special clothing worn at work, but not taken home. Such clothing accommodation must be suitably secure when personal clothing not worn at work is being stored.

Separate storage must be provided for work clothing and other clothing where it is necessary to avoid health risks or damage. It must be sited in a suitable location and, so far as is reasonably practicable, include drying facilities.

The CDM 2015 expressly requires that facilities provide the ability to lock away clothing and that they should provide for personal effects.

Facilities for eating, rest rooms - WHSWR Regulation 25

Readily accessible, suitable and sufficient rest facilities must be provided. The rest facilities must be provided in one or more rest rooms (new and modified, etc. workplaces) or in rest rooms or rest areas (existing workplaces). Where food eaten in the workplace is liable to become contaminated, suitable facilities for eating meals must be included in the rest facilities.

Suitable rest facilities must also be provided for pregnant women and nursing mothers. Where meals are regularly eaten in the workplace suitable and sufficient facilities must be provided for their consumption.

The ACOP to WHSWR 1992 recommends that rest facilities should include suitable and sufficient seats and tables for the number of workers likely to use them at any one time. Work seats in offices or other clean environments may be acceptable as rest facilities provided workers are not subjected to excessive disturbance during rest periods.

Eating facilities should include a facility for preparing or obtaining a hot drink, and where hot food cannot be readily obtained, means should be provided to enable workers to heat their own food. Canteens, etc. may be used as rest facilities providing there is no obligation to buy food. These are mandatory requirements of the CDM 2015.

Facilities for pregnant women and nursing mothers

Suitable rest facilities must also be provided for pregnant women and nursing mothers. This could be a dedicated rest room or when space is short, the first aid room could also be used. In either case, there should be a bed or couch where the pregnant worker can rest. The nursing mother will need a clean area where breast milk can be expressed and a dedicated fridge to keep the containers until she goes home. The shared toilet facilities are not considered appropriate.

Provision of facilities for smokers

SMOKE-FREE PREMISES

The Health Act 2006, Part 1, designated premises used as a workplace or open to the public as smoke-free. This requirement extends to part or whole of the smoke-free premises that are enclosed or substantially enclosed. No-smoking signs must be displayed in smoke-free premises. It is the duty of those concerned with the management of smoke-free premises to stop people smoking in them.

In addition, a person who smokes in a smoke-free premise commits an offence. Provision of facilities for smokers is not a requirement in the workplace, but some organisations provide an external area for the smokers, which can be covered to protect against rain, generally known as 'bus shelters'.

For smoking to be allowed, the shelter must not be partially enclosed at the sides. Some organisations provide containers of sand to ensure the safe extinguishing of smoking materials and to prevent the area being littered.

COMPANY VEHICLES

Company vehicles are also workplaces and therefore smoking in them is banned if the driver carries passengers. A driver of a company vehicle can smoke if they are alone, but they must not smoke when driving as that may be an offence under different legislation. Smoke-free vehicles must display a no smoking sign.

CARE HOMES AND FLATS WHERE RESIDENTS SMOKE

Smoking rooms can be provided in any premises where a person has their home or is living permanently or temporarily, such as care homes. Arrangements should be made with people who live in these premises not to smoke when workers are present.

People can smoke in their own home. In flats, the communal areas, such as lifts and stairwells must be smoke-free if they are open to the public or used by someone at work, such as the postman.

The need to take account of disabled persons

Suitable toilet facilities will need to be provided for those who have physical disability and who rely on equipment to move about (crutches, walking sticks, frames and wheel chairs).

The facility, typically a defined cubicle, will need to have a wider than usual access door and sufficient room to turn around, in relation to the equipment fitted, i.e. water closet, sink, refuse bins etc. located within. Means to summon assistance will also be required.

The WHSWR 1992 were amended to take account of disability and to ensure that the workplace, including doors, passageways, stairs, showers, washbasins, lavatories and workstations used or occupied directly by disabled persons is suitable for them.

B10.4 - Requirements and provision for first-aid in the workplace

Legal requirements for first-aid in the workplace

THE HEALTH AND SAFETY (FIRST-AID) REGULATIONS

The Health and Safety (First-Aid) Regulations (FAR) 1981 set out the following requirements:

Reg 2	First-aid provides treatment for the purpose of preserving life and minimising the consequences of injury or illness until medical (doctor or nurse) help can be obtained. Also, it provides treatment of minor injuries which would otherwise receive no treatment, or which do not need the help of a medical practitioner or nurse.
Reg 3	Requires that every employer must provide equipment and facilities that are adequate and appropriate in the circumstances for administering first-aid to his employees.
Reg 4	An employer must inform employees about the first-aid arrangements, including the location of equipment, facilities and identification of trained personnel.
Reg 5	Self-employed people must ensure that adequate and suitable provision is made for administering first-aid while at work.

PRINCIPLES OF FIRST-AID

The principles of first-aid, for which the FAR 1981 are established to ensure, are to:

- **Preserve** life.
- **Prevent** the condition requiring first-aid getting worse/minimise its consequences until medical help arrives.
- **Promote** recovery of the person requiring first-aid.
- **Provide** treatment where medical attention is not required.

ASSESSMENT OF FIRST-AID NEEDS

To ensure compliance with Regulation 3, an employer must make an assessment to determine the first-aid needs. The assessment will determine the organisation's requirements for first-aiders (quantity and competence) and first-aid equipment/facilities (quantity and type). When making the assessment, consideration should be given to a number of factors, including:

- The size of the organisation - in particular, the number and distribution of workers that need to be covered by the first-aid provision.
- The type of workforce - including the special needs of trainees, younger and older workers, those with known medical conditions and the disabled.
- The need to provide first-aid at all times when workers are on site - considering shift patterns and shift change-over periods, any work done outside normal working hours and cover for first-aiders due to their sickness or other absences.
- Different work activities - first-aid needs will depend on the type of work being done. Some activities, such as office work, have relatively few hazards and low levels of risk and will need less first-aid provisions than other activities that have more hazards or more specific hazards (construction or chemical sites).
- Past experience - previous accidents and ill-health should be considered, including their type, location and the type of harm caused.
- Ease of access to medical treatment - where external emergency medical services are not easily available, more first-aid provision may be required, for example, a first-aid room.
- Workers working away from the employer's premises - first-aid arrangements may have to be made for some work activities, for example, mobile teams of workers, particularly those working in remote locations.
- Workers of more than one employer working together - first-aid facilities may be shared, for example, when a number of contractors work together on a construction site they could share first-aid facilities or when individual workers go to an employer's site to do maintenance work they could use the employer's first-aid facilities.
- Provisions for people who are not the employer's workers - under these regulations employers do not have to make first-aid provision for any person other than their employees. However, civil liability and any interpretation placed on the Health and Safety at Work etc. Act (HASAWA) 1974 may persuade an organisation to provide first-aid to third parties that enter their premises.
- Absences of first-aiders through illness or similar reasons.

The first-aid assessment should be used to guide the employer in providing first-aid arrangements. A sufficient number of competent first-aiders should be readily available. Sometimes first-aiders will need specific competencies related to the risks of the work activities, for example, competencies in the use of a defibrillator or the treatment of chemical burns. There should be a sufficient number and size of first-aid boxes/containers to provide ready access to facilities. Additional equipment and facilities may also be appropriate, for example, a stretcher, defibrillator or first-aid room. When first-aid arrangements have been established, the employer must inform employees where first-aid equipment is and who might be responsible for performing first-aid duties.

The basis of provision of first-aiders

TRAINING AND COMPETENCY OF FIRST-AIDERS

People appointed as first-aiders should be trained and competent in providing first-aid to the level expected of them by their appointment. Some first-aiders may only be trained to provide first-aid for common life-threatening emergencies, others may be able to provide a wider range of first-aid and some may be trained to provide first-aid for specific reasons, for example, the provision of antidotes to chemical poisons.

In the UK, training of first-aiders for the workplace focuses on two main levels of competence:

1) Emergency first-aid at work (EFAW) - EFAW training enables a first-aider to give emergency first-aid to someone who is injured or becomes ill while at work. This involves a minimum of 6 hours training over one day covering competencies in sections (a) and (b) below.

2) First-aid at work (FAW) - FAW training includes EFAW and also equips the first-aider to apply first-aid to a range of specific injuries and illnesses. This involves a minimum of 18 hours training over 3 days covering competencies in sections (a), (b), (c) and (d) below.

On completion of training candidates receiving first-aid training should be able to:

a) Understand the role of the first-aider including reference to:

- The importance of preventing cross-infection.
- The need for recording incidents and actions.
- Use of available equipment.

b) Provide emergency first-aid at work as set out in the following lists.

- Assess the situation and circumstances in order to act safely, promptly and effectively in an emergency.
- Administer first-aid to a casualty who is unconscious (including seizure).
- Administer cardiopulmonary resuscitation (CPR).
- Administer first-aid to a casualty who is choking.
- Administer first-aid to a casualty who is wounded and bleeding.
- Administer first-aid to a casualty who is suffering from shock.
- Provide appropriate first-aid for minor injuries.

The competencies listed in a) and b) are the full extent required for an individual carrying out EFAW. Candidates receiving training to FAW level require these competencies and the additional competencies shown in c) and d).

c) Administer first-aid to a casualty with:

- Injuries to bones, muscles and joints, including suspected spinal injuries.
- Chest injuries.
- Burns and scalds.
- Eye injuries.
- Sudden poisoning.
- Anaphylactic shock.

d) Recognise the presence of major illness and provide appropriate first-aid.

In addition, first-aiders may receive specialist training to enable them to deal with specific emergencies, for example, to enable them to provide medication or chemical antidote by syringe or intravenous drip. Some may be trained in specialist equipment, such as defibrillation equipment. All first-aid training should be provided by a suitable training provider and certificates of qualification for first-aiders are generally valid for three years. A refresher course, followed by examination, is usually required before recertification. To help keep their basic skills up to date, it is recommended that first-aiders undertake annual refresher training. In appropriate, low risk circumstances an employer may provide an 'appointed person' instead of a first-aider. The 'appointed person' is someone appointed by the employer to take charge of the situation (for example, to call an ambulance) if an injury or sudden illness occurs in workplaces where a first-aider has not been appointed. They are also expected to look after the first-aid equipment and facilities provided by the employer. An appointed person may not receive formal training, but may be competent because of their knowledge and other qualities that make them suitable for the position.

NUMBERS OF EMPLOYEES

The number of employees that an employer has, and their distribution, is an important influence on the amount of first-aid provision required. In small organisations the provision may be minimal, particularly if the workplace is located for easy access of emergency services and the provision of someone trained to provide emergency first-aid at work (EFAW) may be all that is needed.

The smallest organisations will only need to establish an 'appointed person' to take charge and ensure that the emergency services are alerted. Typically, this would be where there were low hazard activities with fewer than 25 people or higher hazard activities with fewer than 5 people. They may have basic, emergency first-aid knowledge to help sustain life. They are not classed as a 'first-aider'. Where the number of employees at a given workplace increases then a fully qualified first-aider is required, for example, when an office has 50 employees.

In larger workplaces, with more employees, there would be a need for more than one first-aider, for example, when an office has more than 100 employees the requirement for first-aiders becomes one for every 100 employees.

WORKPLACE RISKS

The workplace risks are a significant factor on the number of first-aiders needed. Any first-aid provision made needs to reflect the workplace risks, both in the level of risk and the type. Different work activities, for example, offices have relatively few hazards and low levels of risk. Others have a higher level of risk or more specific hazards, for example, construction or chemical sites. For low risk workplaces, where hazards are low and reasonably well controlled, the threshold that necessitates a first-aider trained in first-aid at work (FAW) is 50 employees. For higher risk workplaces, where hazards are higher, it is 5 employees.

Category	Number of workers at location	Suggested number and type of suitable person
Low hazard For example, offices, shops, libraries	Less than 25	At least one appointed person
	25-50	At least one first-aider trained in EFAW
	More than 50	At least one first-aider trained in FAW for every 100 employed (or part thereof)
Higher hazard For example, light engineering and assembly work, food processing, warehousing, extensive work with dangerous machinery or sharp instruments, construction, chemical manufacture	Less than 5	At least one appointed person
	5-50	At least one first-aider trained in EFAW or FAW depending on the type of injuries that might occur
	More than 50	At least one first-aider trained in FAW for every 50 employed (or part thereof)

Figure B10-29: Suggested numbers of first-aid personnel to be available at all times people are at work.   *Source: HSE.*

In the case of construction sites they may anticipate a need for rapid local provision of first-aid to deal with a wide range of injuries, particularly impact and fracture injuries relating to falls of people or materials. The chemical site may anticipate exposure to specific chemicals that may need an antidote to be administered promptly or people being overcome by chemicals needing oxygen. In addition, the site may be large and the first-aid provision may have to travel large distances to respond to a request for first-aid.

Requirements for first-aid equipment

The employer should, taking account of the number and location of first-aiders, ensure that an adequate quantity of suitable first-aid equipment is provided. The employer's assessment of first-aid needs, discussed earlier in this section, should indicate the number and type of first-aid boxes/containers and other equipment required. This may be a small personal first-aid pouch, first-aid boxes located at various points and on vehicles as necessary or a first-aid room.

The number of first-aid boxes/containers provided and their contents will usually be influenced by the level of hazard related to the workplace and the number of workers the first-aid box/container is to cover. British Standard BS 8599-1:2011 'Workplace first-aid kits. Specification for the contents of workplace first-aid kits' (BS 8599) specifies three sizes of first-aid box/container for workplaces - small, medium and large. BS 8599 outlines a guide for employers to decide the most suitable size and number for their workplace, presuming there are no special risks in the workplace. BS 8599 also specifies the minimum contents for the three sizes of workplace first-aid boxes/containers.

The contents of a large first-aid box/container would be:

- A leaflet giving general guidance on first-aid.
- 8 medium sized (approximately 12cm x 12cm) individually wrapped sterile un-medicated wound dressings.
- 2 large (approximately 18cm x 18cm) sterile individually wrapped un-medicated wound dressings.
- 4 individually wrapped triangular bandages (preferably sterile).
- 2 other bandages (approximately 7.5cm x 4.5m).
- 4 sterile finger dressings.
- 4 sterile eye pads.
- 2 burn dressings.
- 100 individually wrapped sterile adhesive dressings (assorted sizes), appropriate to the type of work, for example, waterproof.
- 40 sterile saline wipes.
- 1 microporous tape (approximately 2.5cm x 10m).
- 24 safety pins.

- 12 pairs of disposable gloves (Nitrile).
- 2 resuscitation face shields.
- 1 pair of scissors (able to cut through clothing).
- 3 foil blankets.

Additional items for first-aid boxes/containers may be required to deal with specific risks. It is common good first-aid practice not to keep tablets or medicines in the first-aid box/container.

Size of first-aid box	Small	Medium	Large
Low hazard workplaces (offices, shops, etc.)	Less than 25 employees	25 - 100 employees	Over 100 employees, 1 large per 100 employees
Higher hazard workplaces (light engineering and assembly work, food processing, warehousing, extensive work with dangerous machinery or sharp instruments, construction, chemical processing, etc.)	Less than 5 employees	5 - 25 employees	Over 25 employees, 1 large per 25 employees

Figure B10-30: Suggested size of first-aid box/container for different workplaces and numbers of employees.   *Source: UK, BS 8599.*

The minimum level of first-aid equipment should be one suitably stocked and properly identified first-aid box/container for each workplace. In many cases more than one first-aid box/container will have to be provided to ensure they are readily accessible to first-aiders when they are needed.

If the first-aid boxes/containers are placed at fixed locations in the workplace, the employer should position them so that a first-aider only needs to travel a reasonable distance to obtain the first-aid equipment. This might mean providing a first-aid box/container in each work room where there are a number of workers. Alternatively, a transportable first-aid box/container could be provided for each first-aider.

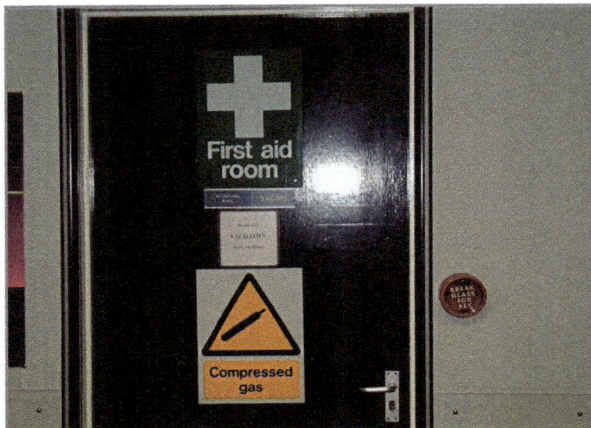

Figure B10-31: First-aid room.   *Source: RMS.*

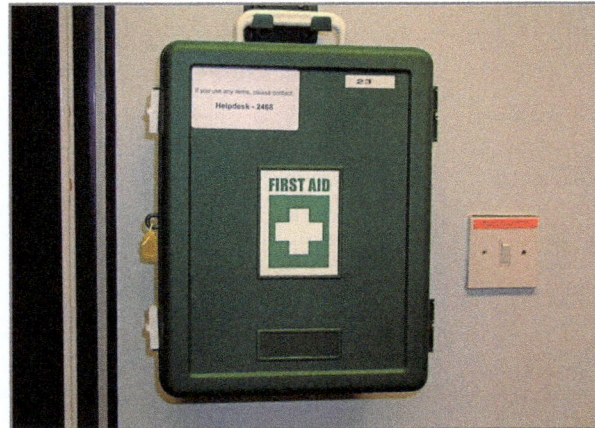

Figure B10-32: First-aid box.   *Source: RMS.*

First-aid boxes/containers should be made of suitable materials and protected from heat, humidity, dust and abuse. The box/container should be clearly identified as containing first-aid material and be marked with a white cross on a green background in accordance with ISO 7010 'Graphical symbols - safety colours and safety signs - registered safety signs'.

PROXIMITY OF EMERGENCY SERVICES

Where access to treatment is difficult first-aiders will usually be required, such as where work activities are a long distance from accident and emergency facilities, for example, when conducting forestry work. Where access to comprehensive treatment is difficult, for example, where a chemical processing site is located away from towns and therefore away from emergency services, an equipped first-aid room may be required to offset possible delay in treatment.

Typical arrangements

LOCAL ARRANGEMENTS

It is necessary for organisations to implement practical arrangements that suit their local conditions and circumstances, for example:

- Sharing first-aiders - arrangements can be made to share the expertise of personnel. Usually, as on a multi-contractor site, one contractor supplies the personnel.
- Workers regularly working away from the employer's premises.
- The numbers of the workers, including fluctuations caused by shift patterns. The more workers there are, the higher the probability of injury.

- Absence of first-aiders through illness or annual leave.
- Shift patterns.

FIRST-AID EQUIPMENT

The employer must ensure that an adequate quantity of suitable first-aid equipment is provided; taking account of how many first-aiders will be using them. In addition, there is the extra consideration that some people may need equipment to administer their own first-aid.

There is no standard list of items to put in a first-aid box. It depends on the risks that workers are exposed to and what the first-aid need is assessed to be. This may be influenced by the proximity of other emergency medical services; where these are not readily available requirements may be higher.

The new British Standard for Workplace First-aid Kits was published on 30th June 2011. These first-aid kits are compliant with the new British Standard for Workplace First-aid Kits BS8599.

- 1 x Guidance leaflet.
- 1 x List of contents.
- 4 x Med HSE dressing.
- 1 x Large HSE dressing.
- 2 x Triangular bandage.
- 6 x Safety pins.
- 2 x No.16 eyepad.
- 40 x Adhesive dressings.
- 20 x Alcohol free wipes.
- 1 x Adhesive tape.
- 6 x Nitrile gloves (pairs).
- 2 x Sterile finger dressing.
- 1 x Resuscitation faceshield.
- 1 x Foil blanket.
- 1 x Burn dressing.
- 1 x Bandage and clothing shears.
- 1 x Conforming bandage.

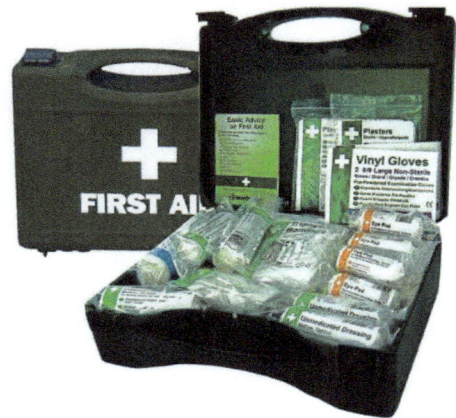

Figure B10-33: First-aid box. *Source: www.aspli.com.*

No drug or remedy used for treating illness (either in liquid or tablet form) should be dispensed by the first-aider or kept in the first-aid box.

Where any employer decides to provide a defibrillator in the workplace, those who may use it should be trained and competent to use it. Any first-aid room provided should be easily accessible to stretchers and to any other equipment needed to convey patients to and from the room and the room should be signposted to enable its location to be identified in an emergency.

Coverage in relation to shift work and geographical location

Additional staff will be necessary to cover for out of hours, shift working or overtime. In particular, where there is a specific legal duty to provide such coverage - for example, first-aid provisions in the UK. The person with overall control of the site must ensure that coverage remains in place throughout the period work is occurring.

Particular care must be paid to high risk work being conducted outside normal working hours. It may be possible to maintain good emergency cover for these times by ensuring, where there is permanent security staff, that these are appropriately trained.

If the area of work is geographically large, for example, gas, electrical or telecommunication field work, all staff may need to be trained and equipped with first-aid equipment in order that they may self-administer first-aid. This may be by means of a full first-aid kit in a vehicle or a personal provision in a pouch. Particular attention should be given to the likelihood and type of injuries when equipping people in this way.

Shared facilities and arrangements

Where a site has multiple occupancy or groups of contracting employers exist within a site, arrangements need to be in place to identify and inform people of where first-aid equipment is and who might be responsible for performing first-aid duties. It is possible for an agreement to be made such that each occupier or employer does not have to make separate arrangements.

This can be particularly useful in providing cover for each other's first-aiders and will avoid the need for small contractors to provide their own first-aid if they can obtain it from a main contractor's facilities. Where small works are going on in a host's premises, the contractor may find an acceptable agreement may be made with the host to provide first-aid and other facilities.

It is important that such agreements be made formally, preferably with a written agreement that will substantiate the existence of such an arrangement if it is challenged or people change their mind as the work unfolds.

TRAINING OF FIRST-AIDERS

People appointed as first-aiders should be trained and competent in providing first-aid to the level required. Training in the UK focuses on two levels of competence:

1) Emergency first-aid at work (EFAW) - EFAW training enables a first-aider to give emergency first-aid to someone who is injured or becomes ill while at work - minimum of six hours training.

2) First-aid at work (FAW) - FAW training includes EFAW and also equips the first-aider to apply first-aid to a range of specific injuries and illness - minimum of 18 hours training.

Additional training may be required; for example:

■ How to use an Automated External Defibrillator (AED).
■ Resuscitation for H_2S (hydrogen sulphide) victims.
■ First-aid treatment for children and babies.
■ First-aid treatment for cyanide poisoning (no longer recommended in UK for first-aid), snake bites etc.

Resuscitation Guidelines 2015 Guidance

Every 5 years the European Resuscitation Council (ERC), the Resuscitation Council (UK) and the International Liaison Committee on Resuscitation (ILCOR) review the latest research and evidence in resuscitation, and then release updated guidelines. In addition to this, for the first time in history, the European Resuscitation Council (ERC) have also produced guidelines for first-aid.

Relevant statutory provisions

RMS Publishing's technical authors regularly review examiners reports for all NEBOSH awards to ensure that the specific publication content is in keeping with the level of study required for the award.

The review considers the core training materials, assessment criteria and relevant legislation. At each stage care is taken to pitch the level of the content to the examination requirements and, in particular, knowledge of the legislation required at this level.

The syllabus does not require knowledge of all legislation to the same depth, this is reflected in the 'relevant statutory provisions' section of the study book. Relevant statutory provisions abstracts are designed to focus on the specific aspects (often popular with the examiners) of the legislation required to meet the syllabus. The study book provides guidance in the form of 'outline of main points' enabling students to focus on the critical learning points and avoid over studying. In addition, legislation is considered in context in the relevant elements of the study book.

Students are advised to obtain or gain access to statutory documents, approved codes of practice and guidance related to the relevant statutory provisions as part of their personal development programme, for the purpose of the examination and future career development. These documents may be purchased on line or downloaded free of charge at www.hse.gov.uk

NEBOSH do not examine on legislation until it has been in force for 6 months. Students may show knowledge of new legislation in their answers until that point, students referring to the former legislation will not lose marks until the 6 month period has passed.

Control of Artificial Optical Radiation at Work Regulations (CAOR) 2010

Considered in context in Element B7.

Arrangement of Regulations

1) Citation, commencement and interpretation.
2) Application of these Regulations.
3) Assessment of the risk of adverse health effects to the eyes or skin created by exposure to artificial optical radiation at the workplace.
4) Obligations to eliminate or reduce risks.
5) Information and training.
6) Health surveillance and medical examinations.
7) Extension outside Great Britain.

Outline of main points

The Regulations came into force on 27 April 2010. The employer has duties to employees and any other person at work who may be affected by the work carried out.

ASSESSMENT OF THE RISK OF ADVERSE HEALTH EFFECTS TO THE EYES OR SKIN

The employer must make a suitable and sufficient assessment of risk for the purpose of identifying the measures it needs to take to meet the requirements of these Regulations where:

(a) The employer carries out work which could expose any of its employees to levels of artificial optical radiation that could create a reasonably foreseeable risk of adverse health effects to the eyes or skin of the employee.
(b) That employer has not implemented any measures to either eliminate or, where this is not reasonably practicable, reduce to as low a level as is reasonably practicable, that risk based on the general principles of prevention set out in Schedule 1 to the Ionising Radiation (IRR) Regulations 1999.

OBLIGATIONS TO ELIMINATE OR REDUCE RISKS

An employer must ensure that any risk of adverse health effects to the eyes or skin of employees as a result of exposure to artificial optical radiation which is identified in the risk assessment is eliminated or, where this is not reasonably practicable, reduced to as low a level as is reasonably practicable.

INFORMATION AND TRAINING

If the risk assessment indicates that employees could be exposed to artificial optical radiation which could cause adverse health effects to the eyes or skin of employees, the employer must provide its employees and representatives with suitable and sufficient information and training relating to the outcome of the risk assessment, and this must include the following:

(a) The technical and organisational measures taken in order to comply with the requirements of regulation 4.
(b) The exposure limit values (see note 1).
(c) The significant findings of the risk assessment, including any measurements taken, with an explanation of those findings.
(d) Why and how to detect and report adverse health effects to the eyes or skin.
(e) The circumstances in which employees are entitled to appropriate health surveillance.
(f) Safe working practices to minimise the risk of adverse health effects to the eyes or skin from exposure to artificial optical radiation.
(g) The proper use of personal protective equipment.

Note 1

Regulation 1(2) defines exposure limit values as being those set out in Annexes I and II to the Directive (2006/25/EC), as amended from time to time, and these Annexes provide for exposure limit values for non-coherent radiation and laser radiation respectively. The employer must ensure that any person, whether or not that person is an employee, who carries out work in connection with the employer's duties under these Regulations has suitable and sufficient information and training.

HEALTH SURVEILLANCE AND MEDICAL EXAMINATIONS

If the risk assessment indicates that there is a risk of adverse health effects to the skin of employees, as a result of exposure to artificial optical radiation, the employer must ensure that such employees are placed under suitable health surveillance.

Classification Labelling and Packaging of Substances (CLP) EC Regulation

Considered in context in Elements B2 and B3.

European Regulation (EC) No 1272/2008 on classification, labelling and packaging of substances and mixtures came into force on 20 January 2009 in all EU Member States, including the UK. It is known by its abbreviated form, 'the CLP Regulation' or 'CLP'.

The CLP Regulation adopts the United Nations' Globally Harmonised System (GHS) on the classification and labelling of chemicals across all European Union countries, including the UK. As the GHS is a voluntary agreement rather than a law, it has to be adopted through a suitable national or regional legal mechanism to ensure it becomes legally binding. That is what the CLP Regulation does.

As GHS was heavily influenced by the old EU system, the CLP Regulation is very similar in many ways. The duties on suppliers are broadly the same and cover classification, labelling and packaging. The process suppliers have to follow when they are classifying substances and mixtures has changed and a new set of hazard pictograms are used.

Explosive	Flammable	Corrosive	Toxic	Human health	Gas bottles	Environmental	Oxidizing

Figure RSP-1: GHS hazard pictograms.

Source: RMS.

Arrangement of Regulation

TITLE I - GENERAL ISSUES

Article 1   Purpose and scope

Article 2   Definitions

Article 3   Hazardous substances and mixtures and specification of hazard classes

Article 4   General obligations to classify, label and package

TITLE II - HAZARD CLASSIFICATION

Chapter 1 - Identification and examination of information

Article 5   Identification and examination of available information on substances

Article 6   Identification and examination of available information on mixtures

Article 7   Animal and human testing

Article 8   Generating new information for substances and mixtures

Chapter 2 - Evaluation of hazard information and decision on classification

Article 9   Evaluation of hazard information for substances and mixtures

Article 10   Concentration limits and M-factors for classification of substances and mixtures

Article 11   Cut-off values

Article 12   Specific cases requiring further evaluation

Article 13   Decision to classify substances and mixtures

Article 14   Specific rules for the classification of mixtures

Article 15   Review of classification for substances and mixtures

Article 16   Classification of substances included in the classification and labelling inventory

TITLE III - HAZARD COMMUNICATION IN THE FORM OF LABELLING

Chapter 1 - Content of the label

Article 17   General rules

Article 18   Product identifiers

Article 19   Hazard pictograms

Article 20   Signal words

Article 21   Hazard statements

Article 22   Precautionary statements

Article 23   Derogations from labelling requirements for special cases

Article 24   Request for use of an alternative chemical name

Article 25   Supplemental information on the label

Article 26   Principles of precedence for hazard pictograms

Article 27   Principles of precedence for hazard statements

Article 28   Principles of precedence for precautionary statements

Article 29   Exemptions from labelling and packaging requirements

Article 30   Updating information on labels

ANNEX I - CLASSIFICATION AND LABELLING REQUIREMENTS FOR HAZARDOUS SUBSTANCES AND MIXTURES

This annex sets out the criteria for classification in hazard classes and in their differentiations and sets out additional provisions on how the criteria may be met.

PART 1 - GENERAL PRINCIPLES FOR CLASSIFICATION AND LABELLING

1.1. Classification of substances and mixtures

1.2. Labelling (General rules for the application of labels required by Article 31)

1.3. Derogations from labelling requirements for special cases

1.4. Request for use of an alternative chemical name

1.5. Exemptions from labelling and packaging requirements

PART 2 - PHYSICAL HAZARDS

Including explosive and flammable hazards

PART 3 - HEALTH HAZARDS

Including acute toxicity and respiratory/skin sensitisation hazards

PART 4 - ENVIRONMENTAL HAZARDS

Including hazards to the aquatic environment

PART 5 - ADDITIONAL HAZARDS

Including hazards to the ozone layer

Annex ii - Special rules for labelling and packaging of certain substances and mixtures

Annex iii - List of hazard statements, supplemental hazard information and supplemental label elements

Annex iv - List of precautionary statements

Annex v - Hazard pictograms

Annex vi - Harmonised classification and labelling for certain hazardous substances

Annex vii - Translation table from classification under directive 67/548/EEC to classification under this regulation

Outline of main points

The CLP Regulation ensures that the hazards presented by chemicals are clearly communicated to workers and consumers in the European Union, including the UK, through classification and labelling of chemicals. Before placing chemicals on the market, the industry must establish the potential risks to human health and the environment of such substances and mixtures, classifying them in line with the identified hazards. The hazardous chemicals also have to be labelled according to a standardised system so that workers and consumers know about their effects before they handle them. Suppliers must label a substance or mixture contained in packaging according to CLP before placing it on the market. CLP defines the content of the label and the organisation of the various labelling elements. This process ensures the hazards of chemicals are communicated through standard statements and pictograms on labels.

For example, when a supplier identifies a substance as 'acute toxicity category 1 (oral)', the labelling will include the hazard statement 'fatal if swallowed', the word 'Danger' and a pictogram with a skull and crossbones.

CLP provides certain exemptions for substances and mixtures contained in packaging that is small (typically less than 125ml) or is otherwise difficult to label.

The exemptions allow the supplier to omit the hazard and/or precautionary statements or the pictograms from the label elements normally required under CLP. The label includes:

- The name, address and telephone number of the supplier.
- The nominal quantity of a substance or mixture in the packages made available to the general public (unless this quantity is specified elsewhere on the package).
- Product identifiers.
- Where applicable, hazard pictograms, signal words, hazard statements, precautionary statements and supplemental information required by other legislation.

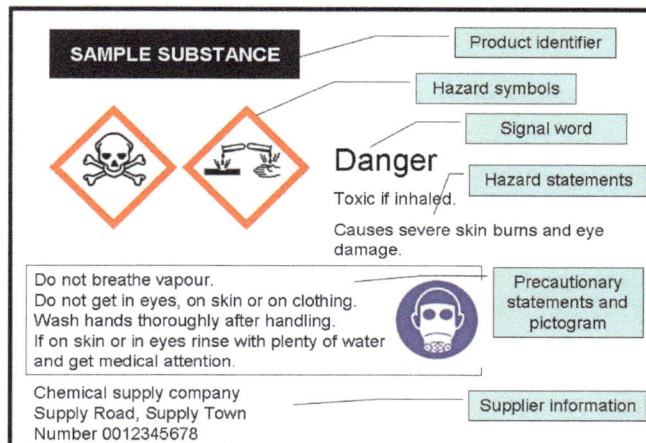

Figure RSP-2: Sample GHS chemical labels. *Source: RMS.*

Control of Asbestos Regulations (CAR) 2012

See also REACH. Considered in context in Elements B3 and B4.

The sixth edition of this Approved Code of Practice and guidance provides practical advice to help duty holders comply with the requirements of the COSHH Regulations. It also takes account of regulatory changes following the introduction of the EU Regulations for REACH (Registration, Evaluation, Authorisation and restriction of Chemicals) and CLP (European Regulation (EC) No 1272/2008 on classification, labelling and packaging of substances and mixtures).

Arrangement of Regulations

PART 1 - PRELIMINARY

1) Citation and commencement.
2) Interpretation.
3) Application of these Regulations.

PART 2 - GENERAL REQUIREMENTS

4) Duty to manage asbestos in non-domestic premises.
5) Identification of the presence of asbestos.
6) Assessment of work which exposes employees to asbestos.
7) Plans of work.
8) Licensing of work with asbestos.
9) Notification of work with asbestos.

10) Information, instruction and training.
11) Prevention or reduction of exposure to asbestos.
12) Use of control measures etc.
13) Maintenance of control measures etc.
14) Provision and cleaning of protective clothing.
15) Arrangements to deal with accidents, incidents and emergencies.
16) Duty to prevent or reduce the spread of asbestos.
17) Cleanliness of premises and plant.
18) Designated areas.
19) Air monitoring.
20) Standards for air testing.
21) Standards for analysis.
22) Health records and medical surveillance.
23) Washing and changing facilities.
24) Storage, distribution and labelling of raw asbestos and asbestos waste.

PART 3 - PROHIBITIONS AND RELATED PROVISIONS

25) Interpretation of prohibitions.
26) Prohibitions of exposure to asbestos.
27) Labelling of products containing asbestos.
28) Additional provisions in the case of exceptions and exemptions.

PART 4 MISCELLANEOUS

29) Exemption certificates.
30) Exemptions relating to the Ministry of Defence.
31) Extension outside Great Britain.
32) Existing licences and exemption certificates.
33) Revocations and savings.
34) Defence.
35) Review.

SCHEDULE 1
Particulars to be included in a notification.

SCHEDULE 2
Appendix 7 to Annex XVII of the REACH Regulation – special provisions on the labelling of articles containing asbestos.

SCHEDULE 3
Amendments.

Outline of main points

SUMMARY

The **Control of Asbestos Regulations (CAR) 2012** place emphasis on assessment of exposure; exposure prevention, reduction and control; adequate information, instruction and training for employees; monitoring and health surveillance. The regulations also apply to incidental exposure. The section on prohibitions is now covered by REACH. The amendments in these regulations have introduced an additional category of work with asbestos. The three categories are: Licensed, Non-Licensed and a new category of Notifiable Non-Licensed (NNLW).

A summary of the requirements of each category is detailed below:

Non-licenced work	Notifiable non-licenced work	Licenced work
■ Carry out and comply with a risk assessment ■ Control exposure ■ Provide training and information	■ Notify before work starts ■ Provide medical examinations every three years ■ Keep health records of employees ■ Carry out and comply with a risk assessment. ■ Control exposure ■ Provide training and information	■ Licencing ■ Notify fourteen days in advance ■ Develop emergency arrangements ■ Designate of asbestos areas ■ Provide medical examination every two years ■ Keep health records of all employees ■ Carry out and comply with a risk assessment. ■ Control exposure ■ Provide training and information

In order to achieve the required changes the regulations provide a separate definition of licensable work and set out the scope of the work which is exempt from the various requirements as now. Several other amendments have also been necessary and as a result there are changes to the notification requirements and those relating to health records and medical surveillance to distinguish between licensed and non-licensed work and amendments to permit a wider range of medical professionals to carry out the required medical examinations. The work for which a licence is required is defined as 'Licensable work with asbestos' and is work:

(a) Where the exposure to asbestos of employees is not sporadic and of low intensity; or

(b) In relation to which the risk assessment cannot clearly demonstrate that the control limit will not be exceeded; or

(c) On asbestos coating; or

(d) On asbestos insulating board or asbestos insulation for which the risk assessment:

  (i) Demonstrates that the work is not sporadic and of low intensity, or

  (ii) Cannot clearly demonstrate that the control limit will not be exceeded, or

  (iii) Demonstrates that the work is not short duration work.

Regulation 3(2) sets out the exemptions for non-licensable work as follows:

Regulations 9 (notification of work with asbestos), 18(1)(a) (designated areas) and 22 (health records and medical surveillance) do not apply where:

(a) The exposure to asbestos of employees is sporadic and of low intensity; and

(b) It is clear from the risk assessment that the exposure to asbestos of any employee will not exceed the control limit; and

(c) The work involves:

  (i) Short, non-continuous maintenance activities in which only non-friable materials are handled, or

  (ii) Removal without deterioration of non-degraded materials in which the asbestos fibres are firmly linked in a matrix, or

  (iii) Encapsulation or sealing of asbestos-containing materials which are in good condition, or

  (iv) Air monitoring and control, and the collection and analysis of samples to ascertain whether a specific material contains asbestos.

Whether a type of asbestos work is either licensable, NNLW or non-licensed work has to be determined in each case and will depend on the type of work being done, the type of material being worked on and its condition. The identification of the type of asbestos-containing material (ACM) to be worked on and an assessment of its condition are important parts of your risk assessment, which needs to be completed before work starts. It is the responsibility of the person in charge of the job to assess the ACM to be worked on and decide if the work is NNLW or non-licensed work. This will be a matter of judgement in each case, dependent on consideration of the above factors. A decision flow chart is available from the HSE at www.hse.gov.uk/asbestos/essentials/index.htm

Duty to manage asbestos in non-domestic premises (regulation 4)

The Regulations include the 'duty to manage asbestos' in non-domestic premises. Guidance on the duty to manage asbestos can be found in the Approved Code of Practice, Work with Materials Containing Asbestos, L143, ISBN 9780717662067. At the time of going to print this ACOP is being reviewed as part of a wider review of HSE's guidance and ACOPs and therefore has not been updated to reflect the changes made in the revised regulations.

Information, instruction and training (regulation 10)

The Regulations require mandatory training for anyone liable to be exposed to asbestos fibres at work. This includes maintenance workers and others who may come into contact with or who may disturb asbestos (e.g. cable installers) as well as those involved in asbestos removal work.

Prevention or reduction of exposure to asbestos (regulation 11)

When work with asbestos or which may disturb asbestos is being carried out, the Asbestos Regulations require employers and the self-employed to prevent exposure to asbestos fibres. Where this is not reasonably practicable, they must make sure that exposure is kept as low as reasonably practicable by measures other than the use of respiratory protective equipment. The spread of asbestos must be prevented. The Regulations specify the work methods and controls that should be used to prevent exposure and spread.

Control limits

Worker exposure must be below the airborne exposure limit (Control Limit). The Asbestos Regulations have a single Control Limit for all types of asbestos of 0.1 fibres per cm^3. A Control Limit is a maximum concentration of asbestos fibres in the air (averaged over any continuous 4 hour period) that must not be exceeded. In addition, short term exposures must be strictly controlled and worker exposure should not exceed 0.6 fibres per cm^3 of air averaged over any continuous 10 minute period using respiratory protective equipment if exposure cannot be reduced sufficiently using other means.

Respiratory protective equipment

Respiratory protective equipment is an important part of the control regime but it must not be the sole measure used to reduce exposure and should only be used to supplement other measures. Work methods that control the release of fibres such as those detailed in the **Asbestos Essentials task sheets** (available on the HSE website) for non-licensed work should be used. Respiratory protective equipment must be suitable, must fit properly and must ensure that worker exposure is reduced as low as is reasonably practicable.

Asbestos removal work undertaken by a licensed contractor

Most asbestos removal work must be undertaken by a licensed contractor but any decision on whether particular work is licensable is based on the risk. Work is only exempt from licensing if:

■ The exposure of employees to asbestos fibres is sporadic and of low intensity (but exposure cannot be considered to be sporadic and of low intensity if the concentration of asbestos in the air is liable to exceed 0.6 fibres per cm3 measured over 10 minutes).

■ It is clear from the risk assessment that the exposure of any employee to asbestos will not exceed the control limit.

■ The work involves:

- Short, non-continuous maintenance activities. Work can only be considered as short, non-continuous maintenance activities if any one person carries out work with these materials for less than one hour in a seven-day period. The total time spent by all workers on the work should not exceed a total of two hours*.
- Removal of materials in which the asbestos fibres are firmly linked in a matrix. Such materials include: asbestos cement; textured decorative coatings and paints which contain asbestos; articles of bitumen, plastic, resin or rubber which contain asbestos where their thermal or acoustic properties are incidental to their main purpose (e.g. vinyl floor tiles, electric cables, roofing felt) and other insulation products which may be used at high temperatures but have no insulation purposes, for example gaskets, washers, ropes and seals.
- Encapsulation or sealing of asbestos-containing materials which are in good condition.
- Air monitoring and control, and the collection and analysis of samples to find out if a specific material contains asbestos.

It is important that the amount of time employees spend working with asbestos insulation, asbestos coatings or asbestos insulating board (AIB) is managed to make sure that these time limits are not exceeded. This includes the time for activities such as building enclosures and cleaning.

Under the Asbestos Regulations, anyone carrying out work on asbestos insulation, asbestos coating or AIB needs a licence issued by HSE unless they meet one of the exemptions above.

Although you may not need a licence to carry out a particular job, you still need to comply with the rest of the requirements of the Asbestos Regulations.

Licensable work - additional duties

If the work is licensable there are a number of additional duties. The need to:

- Notify the enforcing authority responsible for the site where you are working (for example HSE or the local authority).
- Designate the work area (see regulation 18 for details).
- Prepare specific asbestos emergency procedures.
- Pay for your employees to undergo medical surveillance.

Non-notifiable licensable work - additional duties

If work is determined to be NNLW, the duties are:

- To notify the enforcing authority responsible for the site where the work is before work starts. (There is no minimum period.)
- By 2015 all employees will have to undergo medical examinations which are repeated every three years.
- To have prepared procedures which can be put into effect should an accident, incident or emergency occur.
- To keep a register of all NNLW work for all employees.
- To record the significant findings of and comply with a risk assessment.
- To prevent or reduce exposure so far as is reasonably practicable and to take reasonable steps that all control measures are used.
- To ensure that adequate information, instruction and training is given to employees.

Air monitoring (regulation 19)

The Asbestos Regulations require any analysis of the concentration of asbestos in the air to be measured in accordance with the 1997 WHO recommended method.

Standards for air testing and site clearance certification (regulation 20)

From 06 April 2007, a clearance certificate for re-occupation may only be issued by a body accredited to do so. At the moment, such accreditation can only be provided by the United Kingdom Accreditation Service (UKAS). You can find more details of how to undertake work with asbestos containing materials, the type of controls necessary, what training is required and analytical methods in the following HSE publications:

- Approved Code of Practice Work with Materials containing Asbestos, L143, ISBN 978 0 7176 6206 7.
- Asbestos: the Licensed Contractors Guide, HSG 247, ISBN 978 0 7176 2874 2.
- Asbestos: The analysts' guide for sampling, analysis and clearance procedures, HSG 248, ISBN 978 0 7176 2875 9.
- Asbestos Essentials, HSG 210, ISBN 978 0 7176 6263 0. (See also the 'Asbestos Essentials task sheets' available on the HSE website).

Other health and safety legislation must be complied with.

Source: HSE Website: www.hse.gov.uk.

Control of Lead at Work Regulations (CLAW) 2002

Considered in context in Element B4.

Arrangement of Regulations

1) Citation and commencement.
2) Interpretation.
3) Duties under these Regulations.
4) Prohibitions.
5) Assessment of the risk to health created by work involving lead.
6) Prevention or control of exposure to lead.
7) Eating, drinking and smoking.
8) Maintenance, examination and testing of control measures.
9) Air monitoring.

10) Medical surveillance.
11) Information, instruction and training.
12) Arrangements to deal with accidents, incidents and emergencies.
13) Exemption certificates.
14) Extension outside Great Britain.
15) Revocation and savings.

Schedule 1 - Activities in which the employment of young persons and women of reproductive capacity is prohibited.

Schedule 2 - Legislation concerned with the labelling of containers and pipes.

Outline of main points

The **Control of Lead at Work Regulations (CLAW) 2002** aims to protect people at work exposed to lead by controlling that exposure. The Regulations, which are summarised below, apply to any work which exposes people to lead.

Exposure to lead must be assessed by employers so that they may take adequate measures to protect both employees and anyone else who may be exposed to lead at work. Once the level of exposure has been assessed, then adequate measures can be taken ranging from simple maintenance of good washing facilities through to the provision of control measures such as respiratory equipment and constant medical surveillance. The Regulations prohibit the employment of young persons and women of reproductive capacity from some manufacturing, smelting and refining processes (specified in Schedule 1).

WORK WITH LEAD

The Regulations apply to any work which exposes employees or others to lead. In practical terms, this means any work from which lead arises:

a) In the form of lead dust, fume or vapour in such a way as it could be inhaled.
b) In any form which is liable to be ingested such as powder, dust, paint or paste.
c) In the form of lead compounds such as lead alkyls and compounds of lead, this could be absorbed through the skin.

Employers' duties under the 2002 Regulations extend to any other people at work on the premises where work with lead is being carried on.

LEAD ASSESSMENT

Before employers (or a self-employed person) can take adequate measures to protect people from lead at work, they need to know exactly what the degree of risk of lead exposure is. The level of risk dictates the measures to be taken. The employer's first duty, therefore, is to assess whether the exposure of any employee is liable to be significant. The next step is to determine the nature and degree of exposure. The assessment must be made before the work is commenced and revised where there is a reason to suspect that it is incorrect.

The purpose of the assessment is to determine whether or not exposure to lead is significant. Where exposure is significant then the employer must, so far as is reasonably practicable, ensure the prevention or adequate control of exposure by means other than the provision of personal protective equipment (PPE). Where control measures are not sufficient on their own and PPE is issued then it must comply with the PPE Regulations or be of a type approved by the Health and Safety Executive (HSE). When deciding controls the employer must take reasonable steps to ensure that they are being used and employees are under a duty to make full and proper use of control measures, PPE or any other measures dictated by the Regulations.

CONTROL MEASURES

Employers must, so far as is reasonably practicable, provide such control measures for materials, plant and processes as will adequately control the exposure of their employees to lead otherwise than by the use of respiratory protective equipment or protective clothing by those employees. Again, personal protective equipment and clothing should be used as a last resort. Employers are under a duty to restrict access to areas to ensure that only people undertaking necessary work are exposed. If other control measures are inadequate, respiratory protective equipment must be provided for employees exposed to airborne lead. Employees must also be provided with protective clothing where they are significantly exposed to lead. Respiratory protective equipment (RPE) or protective clothing should comply with any UK legislation which implements relevant EU 'design and manufacture' Directives.

Employers must also carry out an assessment before selecting RPE or protective clothing to ensure it will satisfy the necessary requirements and provide adequate protection. The assessment should define the characteristics required by the RPE or protective clothing in order to be suitable, and compare these characteristics against those of the protective equipment actually available. RPE must be examined and tested at appropriate intervals and records kept for a minimum period of 5 years.

Control measures, respiratory equipment and protective clothing must be maintained in an efficient state, in efficient working order and good repair. Employers should ensure that employees use the measures provided properly and employees must make full and proper use of all respiratory protective equipment or protective clothing provided, report defects immediately to the employer and take all reasonable steps to ensure RPE or protective clothing is returned to its storage accommodation after use. Eating, drinking and smoking are prohibited in any place that is, or is liable to be, contaminated with lead.

OCCUPATIONAL EXPOSURE LIMITS

Control limits for exposure to lead in atmosphere are:

- For lead other than lead alkyls, a concentration of lead in air which any employee is exposed of 0.15 mg per m^3 (8 hour TWA).
- For lead alkyls a concentration of lead of 0.10 mg per m^3 (8 hour TWA).

Air monitoring in relevant areas must be carried out at least every 3 months. This interval can be increased to 12 months providing that there are no material changes to the workplace and lead in air concentrations have not exceeded 0.10 mg per m^3 on two previous consecutive occasions.

MEDICAL SURVEILLANCE

Employees subject to, or liable to be, significantly exposed (or for whom a relevant doctor has certified that they should be) must be placed under medical surveillance by an employment medical adviser or appointed doctor. The Regulations set down action levels for blood-lead concentrations, which are:

- 20 µg/dl for women of reproductive capacity.
- 35 µg/dl for any other employee.

Levels for urinary lead concentration are also specified. The adviser or doctor can certify that employees should not be employed on work which exposes them to lead or can only be employed under certain conditions. An investigation must be made when blood-lead action levels are exceeded.

Employees exposed to lead at work are under a duty to present themselves, in normal working hours, for medical examination or such biological tests as may be required. Employers and employees have a right to appeal against decisions made by relevant doctors.

INFORMATION, INSTRUCTION AND TRAINING

Every employer must ensure that adequate information, instruction and training is given to employees who are liable to be exposed to lead so that they are aware of the risks from lead and the precautions which should be observed. Information must also be given about the results relating to air monitoring and health surveillance and their significance. Adequate information, instruction and training must also be given to anyone who is employed by the employer to carry out lead assessments, air monitoring, etc.

RECORDS

Adequate records must be kept of assessments, examination and testing of controls, air monitoring, medical surveillance and biological tests. Those records should be made available for inspection by employees (although not health records of identifiable individuals). Specific recording requirements are made in respect of female employees who are, or who are likely to be, exposed to significant levels of lead. Air monitoring records must be kept for at least 5 years and individual medical records for 40 years.

Control of Noise at Work Regulations (CNWR) 2005

Considered in context in Element B6.

The implementation of the European Physical Agents (Noise) Directive as the Control of Noise at Work Regulations 2005 came into force on 06 April 2006. These regulations replaced the Noise at Work Regulations 1989. The main changes are the reduction by 5dB of the exposure levels at which action has to be taken, and the introduction of a new exposure limit value and a specific requirement on health surveillance. In practice this means that there is now a requirement to ensure that noise exposure levels are reduced to meet the 2005 regulations.

Arrangement of Regulations

1) Citation and commencement.
2) Interpretation.
3) Application.
4) Exposure limit values and action values.
5) Assessment of the risk to health and safety created by noise at the workplace.
6) Elimination or control of exposure to noise at the workplace.
7) Hearing protection.
8) Maintenance and use of equipment.
9) Health surveillance.
10) Information, instruction and training.
11) Exemption certificates from hearing protection.
12) Exemption certificates for emergency services.
13) Exemptions relating to the Ministry of Defence.
14) Extension outside Great Britain.
15) Revocations and amendments.

Outline of main points

CHANGES TO THE ACTION LEVELS

The values of the actions levels associated with noise at work have been lowered and their names have been changed. The first action level is reduced from *85 dB(A) down to 80 dB(A)* and will be known as the *lower exposure action value.* Meanwhile, the section level is reduced from *90 dB(A) down to 85 dB(A)* and will be known as the *upper exposure action value*. The Regulations also allow the employer to average out the exposure to noise over a one week period instead of the previous normal eight hour period, in situations where the noise exposure varies on a day-to-day basis. When determining noise levels for the purposes of determining exposure action levels, the noise exposure reducing effects of hearing protection may not be taken in to account.

Where exposure is at, or above, the *lower exposure action value* (80 dB(A)) the employer has a duty to provide hearing protection to those employees that request it. The employer also has a duty to information, instruction and training on the risks posed by exposure to noise and the control measures to be used.

Where the exposure it at, or above, the *upper exposure action value* (85 dB(A)) the employer is also required to introduce a formal programme of control measures. The measures to be taken as part of this programme of control measures will depend on the findings of the noise risk assessment (see below).

The Control of Noise at Work Regulations 2005 also introduces a new value known as the **exposure limit value**. When evaluating the risks to employees from noise, the employer needs to take account of the exposure limit values. These are limits set both in terms of daily (or weekly) personal noise exposure (*LEP,d* of 87 dB) and in terms of peak noise (*LCpeak* of 140 dB). The exposure action values, take account of the protection provided by personal hearing protection. **If an employee is exposed to noise at or above the exposure limit value, then the employer must take immediate action to bring the exposure down below this level.**

Summary of exposure limit values and action values

The lower exposure action values are:	A daily or weekly personal noise exposure of 80 dB (A-weighted)
	A peak sound pressure of 135 dB (C-weighted)
The upper exposure action values are:	A daily or weekly personal noise exposure of 85 dB (A-weighted)
	A peak sound pressure of 137 dB (C-weighted)
The exposure limit values are:	A daily or weekly personal exposure of 87 dB (A-weighted)
	A peak sound pressure of 140 db (C-weighted)

NOISE RISK ASSESSMENT AND CONTROL MEASURES

The requirement for a noise risk assessment carries through from the Noise at Work Regulations 1989 into the Control of Noise at Work Regulations 2005. Employers are required (in accordance with the general risk assessment and control measure hierarchy contained in Schedule 1 to the Management of Health and Safety at Work Regulation 1999) to ensure that the risks associated with employees' exposure to noise are eliminated where this is reasonably practicable. Where elimination is not reasonably practicable, then the employer must reduce the risks down to as low a level as is reasonably practicable.

Regulation 6(2) of the proposed Control of Noise at Work Regulations 2005 introduces the requirement for a formal programme of control measures and states: If any employee is likely to be exposed to noise at or above an upper exposure action value, the employer shall reduce exposure to a minimum by establishing and implementing a programme of organisational and technical measures, excluding the provision of personal hearing protectors, which is appropriate to the activity and consistent with the risk assessment, and shall include consideration of:

(a) Other working methods which eliminate or reduce exposure to noise.

(b) Choice of appropriate work equipment emitting the least possible noise, taking account of the work to be done.

(c) The design and layout of workplaces, work stations and rest facilities.

(d) Suitable and sufficient information and training for employees, such that work equipment may be used correctly, in order to minimise their exposure to noise.

(e) Reduction of noise by technical means including:

    (i) In the case of airborne noise the use of shields, enclosures, and sound-absorbent coverings.

    (ii) In the case of structure-borne noise by damping and isolation.

(f) Appropriate maintenance programmes for work equipment, the workplace and workplace systems.

(g) Limitation of the duration and intensity of exposure to noise.

(h) Appropriate work schedules with adequate rest periods.

If the risk assessment indicates an employee is likely to be exposed to noise at or above an upper exposure action value, the employer shall ensure that:

- The area is designated a Hearing Protection Zone;
- The area is demarcated and identified by means of the sign specified for the purpose of indicating "ear protection must be worn" (to be consistent with the Health and Safety (Safety Signs and Signals) Regulations 1996)
- The sign shall be accompanied by text that indicates that the area is a Hearing Protection Zone and that employees must wear personal hearing protectors while in that area;
- Access to the area is restricted where this is technically feasible and the risk of exposure justifies it and shall make every effort to ensure that no employee enters that area unless they are wearing personal hearing protectors.

MAINTENANCE

There is a duty on the employer to maintain the control introduced to protect employees. This will include maintenance of acoustic enclosures, etc as well as the maintenance of machinery (as required under the Provision and Use of Work Equipment Regulations 1998) to control noise at source.

HEALTH SURVEILLANCE

Under the Control of Noise at Work Regulations 2005, employees who are regularly exposed to noise levels of 85 dB(A) or higher must be subject to health surveillance, including audiometric testing. This constitutes a big change from the previous Regulations that only required an employer to carry out health surveillance where the employee was subject to noise levels of 95 dB(A) or higher. Where exposure is between 80 dB and 85 dB, or where employees are only occasionally exposed above the upper exposure action values, health surveillance will only be required if information comes to light that an individual may be particularly sensitive to noise induced hearing loss.

SUMMARY

The Control of Noise at Work Regulations 2005 became part of UK health and safety law in April 2006. They introduce levels for employees to control exposure down to, including a new exposure limit value, above which employers are obliged to take immediate action to reduce exposure. These new lower limits mean that about a further million workers will be afforded protection by these new Regulations. The requirements for risk assessments, control measures and health surveillance have been updated, but are broadly similar to previous requirements.

(Source: www.lrbconsulting com and www.hse.gov.uk).

Control of Substances Hazardous to Health Regulations (COSHH) 2002 (as amended)

Considered in context in Elements B2, B3, B4 and B5.

Amendments to these Regulations were made by the Control of Substances Hazardous to Health (Amendment) Regulations 2004. The main change being that MELs and OESs were replaced by workplace exposure limits (WELs).

Arrangement of Regulations

1) Citation and commencement.
2) Interpretation.
3) Duties under these Regulations.
4) Prohibitions on substances.
5) Application of regulations 6 to 13.
6) Assessment of health risks created by work involving substances hazardous to health.
7) Control of exposure.
8) Use of control measures etc.
9) Maintenance of control measures.
10) Monitoring exposure.
11) Health surveillance.
12) Information etc.
13) Arrangements to deal with accidents, incidents and emergencies.
14) Exemption certificates.
15) Extension outside Great Britain.
16) Defence in proceedings for contravention of these Regulations.
17) Exemptions relating to the Ministry of Defence etc.
18) Revocations, amendments and savings.
19) Extension of meaning of "work".
20) Modification of section 3(2) of the Health and Safety at Work etc Act (HASAWA) 1974.

SCHEDULES

Schedule 1 - Other substances and processes to which the definition of "carcinogen" relates.

Schedule 2 - Prohibition of certain substances hazardous to health for certain purposes.

Schedule 3 - Special provisions relating to biological agents.

Schedule 4 - Frequency of thorough examination and test of local exhaust ventilation plant used in certain processes.

Schedule 5 - Specific substances and processes for which monitoring is required.

Schedule 6 - Medical surveillance.

Schedule 7 - Legislation concerned with the labelling of containers and pipes.

Schedule 8 - Fumigations excepted from regulation 14.

Schedule 9 - Notification of certain fumigations.

Appendix 1 - Control of carcinogenic substances.

Annex 1 - Background note on occupational cancer.

Annex 2 - Special considerations that apply to the control of exposure to vinyl chloride.

Appendix 2 - Additional provisions relating to work with biological agents.

Appendix 3 - Control of substances that cause occupational asthma.

NOTE the main impact to the latest version of the COSHH Regulations concern the control of substances that cause occupational asthma.

Outline of main points

REGULATIONS

Reg.2 **Interpretation**

"Substance hazardous to health" includes:

1) Substances which under The Chemicals Labelling and Packaging Regulations (CLP) 2009 are in categories of very toxic, toxic, harmful, corrosive or irritant.
2) A substance listed in Schedule 1 to the Regulations or for which the HSE *(formerly HSC)* have approved a maximum exposure limit or an occupational exposure standard.
3) A biological agent.
4) Dust in a concentration in air equal to or greater than:
- 10 mg/m3 inhalable dust as an 8hr TWA.

- 4 mg/m3 respirable dust as an 8hr TWA.

5) Any other substance which creates a health hazard comparable with the hazards of the substances in the other categories above.

Reg. 3

Duties

Are on employer to protect:

- Employees.

Any other person who may be affected, except:

- Duties for health surveillance do not extend to non-employees.
- Duties to give information may extend to non-employees if they work on the premises.

Reg. 4

Prohibitions on substances

Certain substances are prohibited from being used in some applications. These are detailed in Schedule 2.

Reg. 5

Application of regulations 6-13

Regulations 6-13 are made to protect a person's health from risks arising from exposure. They do not apply if:

The following Regulations already apply:

- The Control of Lead at Work Regulations (CLAW) 2002.
- The Control of Asbestos at Work Regulations (CAWR) 2002.

The hazard arises from one of the following properties of the substance:

- Radioactivity, explosive, flammable, high or low temperature, high pressure.
- Exposure is for medical treatment.
- Exposure is in a mine.

Reg. 6

Assessment

Employers must not carry out work that will expose employees to substances hazardous to health unless they have assessed the risks to health and the steps that need to be taken to meet the requirements of the Regulations. The assessment must be reviewed if there are changes in the work and at least once every 5 years.

A suitable and sufficient assessment should include:

- An assessment of the risks to health.
- The practicability of preventing exposure.
- Steps needed to achieve adequate control.

An assessment of the risks should involve:

- Types of substance including biological agents.
- Where the substances are present and in what form.
- Effects on the body.
- Who might be affected?
- Existing control measures.

Reg. 7

Control of exposure

1) Employer shall ensure that the exposure of employees to substances hazardous to health is either prevented or, where this is not reasonably practicable, adequately controlled.

2) So far as is reasonably practicable (1) above except to a carcinogen or biological agent shall be by measures other than personal protective equipment (PPE).

3) Where not reasonably practicable to prevent exposure to a carcinogen by using an alternative substance or process, the following measure shall apply:

- Total enclosure of process.
- Use of plant, process and systems which minimise generation of, or suppress and contain, spills, leaks, dust, fumes and vapours of carcinogens.
- Limitation of quantities of a carcinogen at work.
- Keeping of numbers exposed to a minimum.
- Prohibition of eating, drinking and smoking in areas liable to contamination.
- Provision of hygiene measures including adequate washing facilities and regular cleaning of walls and surfaces.
- Designation of areas/installations liable to contamination and use of suitable and sufficient warning signs.
- Safe storage, handling and disposal of carcinogens and use of closed and clearly-labelled containers.

4) If adequate control is not achieved, then employer shall provide suitable PPE to employees in addition to taking control measures.

5) PPE provided shall comply with The Personal Protective Equipment at Work Regulations (PPER), 2002 (dealing with the supply of PPE).

6&7) For substances which have a maximum exposure limit (MEL), control of that substance shall, so far as inhalation is concerned, only be treated if the level of exposure is reduced as far as is reasonably practicable and in any case below the MEL.

Where a substance has an occupational exposure standard (OES), control of that substance shall, so far as inhalation is concerned, only be treated as adequate if the OES is not exceeded or if it is, steps are taken to remedy the situation as soon as reasonably practicable.

8) Respiratory protection must be suitable and of a type or conforming to a standard approved by the HSE.

9) In the event of failure of a control measure which may result in the escape of carcinogens, the employer

shall ensure:

Only those who are responsible for repair and maintenance work are permitted in the affected area and are provided with PPE.

Employees and other persons who may be affected are informed of the failure forthwith.

Reg. 8   Employer shall take all reasonable steps to ensure control measures; PPE, etc. are properly used/applied.

Employee shall make full and proper use of control measures, PPE etc. and shall report defects to employer.

Reg. 9   Maintenance of control measures

Employer providing control measures to comply with Reg.7 shall ensure that it is maintained in an efficient state, in efficient working order and in good repair and in the case of PPE in a clean condition, properly stored in a well-defined place checked at suitable intervals and when discovered to be defective repaired or replaced before further use.

- Contaminated PPE should be kept apart and cleaned, decontaminated or, if necessary destroyed.
- Engineering controls - employer shall ensure thorough examination and tests.
- Local exhaust ventilation (LEV) - Once every 14 months unless process specified in Schedule 4.
- Others - At suitable intervals.
- Respiratory protective equipment - employer shall ensure thorough examination and tests at suitable intervals.
- Records of all examinations, tests and repairs kept for 5 years.

Reg. 10   Monitoring exposure

Employer shall ensure exposure is monitored if:

- Needed to ensure maintenance of adequate control.
- Otherwise needed to protect health of employees.
- Substance/process specified in Schedule 5.

Records kept if:

- There is an identified exposure of identifiable employee - 40 years.
- Otherwise - 5 years.

Reg. 11   Health surveillance

1) Where appropriate for protection of health of employees exposed or liable to be exposed, employer shall ensure suitable health surveillance.
2) Health surveillance is appropriate if:
- Employee exposed to substance/process specified in Schedule 6.
- Exposure to substance is such that an identifiable disease or adverse health effect can result, there is a reasonable likelihood of it occurring and a valid technique exists for detecting the indications of the disease or effect.
3) Health records kept for at least 40 years.
4) If employer ceases business, HSE notified and health records offered to HSE.
5) If employee exposed to substance specified in Schedule 6, then health surveillance shall include medical surveillance, under Employment Medical Adviser (EMA) at 12 monthly intervals - or more frequently if specified by EMA.
6) EMA can forbid employee to work in process, or specify certain conditions for him to be employed in a process.
7) EMA can specify that health surveillance is to continue after exposure has ceased. Employer must ensure.
8) Employees to have access to their own health record.
9) Employee must attend for health/medical surveillance and give information to EMA.
10) EMA entitled to inspect workplace.
11) Where EMA suspends employee from work exposing him to substances hazardous to health, employer of employee can apply to HSE in writing within 28 days for that decision to be reviewed.

Reg. 12   Information etc

Employer shall provide suitable and sufficient information, instruction and training for him to know:

- Risks to health.
- Precautions to be taken.

This should include information on:

- Results of monitoring of exposure at workplace.
- Results of collective health surveillance.

If the substances have been assigned a maximum exposure limit, then the employee/Safety Representative must be notified forthwith if the MEL has been exceeded.

Reg. 13   Arrangements to deal with accidents, incidents and emergencies

To protect the health of employees from accidents, incidents and emergencies, the employer shall ensure that:

- Procedures are in place for first aid and safety drills (tested regularly).
- Information on emergency arrangements is available.
- Warning, communication systems, remedial action and rescue actions are available.
- Information made available to emergency services: external and internal.
- Steps taken to mitigate effects, restore situation to normal and inform employees.
- Only essential persons allowed in area.

These duties do not apply where the risks to health are slight or measures in place Reg 7(1) are sufficient to control the risk. The employee must report any accident or incident which has or may have resulted in the release of a biological agent which could cause severe human disease.

Note: The main impact to the latest version of the COSHH Regs concern the control of substances that cause occupational asthma.

APPENDIX 3 CONTROL OF SUBSTANCES THAT CAUSE OCCUPATIONAL ASTHMA

This relates certain regulations specifically to substances with the potential to cause asthma.

- Regulation 6 - assessment of risk to health created by work involving substances hazardous to health, (i.e. substances that may cause asthma).
- Regulation 7 - prevention or control of exposure to substances hazardous to health, (i.e. substances that may cause occupational asthma).
- Regulation 11 - health surveillance, (for employees who are or may be exposed to substances that may cause occupational asthma).
- Regulation 12 - information, instruction and training for persons who may be exposed to substances hazardous to health, to include: typical symptoms of asthma, substances that may cause it, the permanency of asthma and what happens with subsequent exposures, the need to report symptoms immediately and the reporting procedures.

Training should be given, including induction training before they start the job.

SCHEDULE 3 ADDITIONAL PROVISIONS RELATING TO WORK WITH BIOLOGICAL AGENTS

Regulation 7(10)

Part I Provision of general application to biological agents.

1 **Interpretation**

2 **Classification of biological agents**

The HSC shall approve and publish a "Categorisation of Biological Agents according to hazard and categories of containment" which may be revised or re-issued. Where no approved classification exists, the employer shall assign the agent to one of four groups according to the level of risk of infection.

Group 1 - unlikely to cause human disease.
Group 2 - can cause human disease.
Group 3 - can cause severe disease and spread to community.
Group 4 - can cause severe disease, spread to community and there is no effective treatment.

3 **Special control measures for laboratories, animal rooms and industrial processes**

Every employer engaged in research, development, teaching or diagnostic work involving Group 2, 3 or 4 biological agents; keeping or handling laboratory animals deliberately or naturally infected with those agents, or industrial processes involving those agents, shall control them with the most suitable containment.

4 **List of employees exposed to certain biological agents**

The employer shall keep a list of employees exposed to Group 3 or 4 biological agents for at least 10 years. If there is a long latency period then the list should be kept for 40 years.

5 **Notification of the use of biological agents**

Employers shall inform the HSE at least 20 days in advance of first time use or storage of Group 2, 3 or 4 biological hazards. Consequent substantial changes in procedure or process shall also be reported.

6 **Notification of the consignment of biological agents**

The HSE must be informed 30 days before certain biological agents are consigned.

Part II Containment measures for health and veterinary care facilities, laboratories and animal rooms.

Part III Containment measures for industrial processes.

Part IV Biohazard sign.

The biohazard sign required by regulation 7(6) (a) shall be in the form shown.

Part V Biological agents whose use is to be notified in accordance with paragraph 5(2) of Part I of this Schedule.

- Any Group 3 or 4 agent.
- Certain named Group 2 agents.

Figure RSP-3: Biohazard sign. *Source: COSHH 2002.*

Control of Substances Hazardous to Health (Amendment) Regulations 2004

See also - COSHH 2002, CLP 2009, PPER 1992 and CLAW Regulations 2002, REACH 2006.

These Regulations make minor amendments to The Control of Substances Hazardous to Health Regulations (COSHH) 2002 and came into force on 17th January 2005 and 6th April 2005.

Arrangement of Regulations

1) Citation and commencement.

2) Amendments of the Control of Substances Hazardous to Health (Amendment) Regulations 2004.

3) Amendment of the Chemicals (Hazard Information and Packaging for Supply) Regulations 2002.

4) Amendments of the Control of Lead at Work Regulations 2002.

Outline of main points

The main changes are that maximum exposure limits (MEL) and occupational exposure standards (OES) have been replaced by the new workplace exposure limits (WELs) and schedule 2a has been introduced.

Regulation 2(a)(v) workplace exposure limit for a substance hazardous to health means the exposure limit approved by the Health and Safety Commission for that substance in relation to the specified reference period when calculated by a method approved by the Health and Safety Commission, as contained in HSE publication 'EH/40 Workplace Exposure Limits 2005' as updated from time to time.

Regulation 7(7) Schedule 2a - Principles of good practice for the control of exposure to substances hazardous to health.

(a) Design and operate processes and activities to minimise emission, release and spread of substances hazardous to health.

(b) Take into account all relevant routes of exposure - inhalation, skin absorption and ingestion - when developing control measures.

(c) Control exposure by measures that are proportionate to the health risk.

(d) Choose the most effective and reliable control options which minimise the escape and spread of substances hazardous to health.

(e) Where adequate control of exposure cannot be achieved by other means, provide, in combination with other control measures, suitable personal protective equipment.

(f) Check and review regularly all elements of control measures for their continuing effectiveness.

(g) Inform and train all employees on the hazards and risks from the substances with which they work and the use of control measures developed to minimise the risks.

(h) Ensure that the introduction of control measures does not increase the overall risk to health and safety.

Control of Vibration at Work Regulations (CVWR) 2005

Considered in context in Element B6.

Hand-arm vibration (HAV) and whole body vibration (WBV) are caused by the use of work equipment and work processes that transmit vibration into the hands, arms and bodies of employees in many industries and occupations. Long-term, regular exposure to vibration is known to lead to permanent and debilitating health effects such as vibration white finger, loss of sensation, pain, and numbness in the hands, arms, spine and joints. These effects are collectively known as hand-arm or whole body vibration syndrome. These Regulations introduce controls, which aim substantially to reduce ill-health caused by exposure to vibration. These Regulations came into force on 06 July 2005.

Arrangement of Regulations

1) Citation and commencement.
2) Interpretation.
3) Application.
4) Exposure limit values and action values.
5) Assessment of the risk to health created by vibration at the workplace.
6) Elimination or control of exposure to vibration at the workplace.
7) Health surveillance.
8) Information, instruction and training for persons who may be exposed to risk from vibration.
9) Exemption certificates for emergency services.
10) Exemption certificates for air transport.
11) Exemption relating to the Ministry of Defence etc.
12) Extension outside Great Britain.
13) Amendment of the Offshore Installations and Wells (Design and Construction, etc.) Regulations 1996.

Outline of main points

Regulation 4 states the personal daily exposure limits and daily exposure action values normalised over an 8-hour reference period.

	Daily exposure limits	Daily exposure action values
Hand arm vibration	5 m/s^2	2.5 m/s^2
Whole body vibration	1.15 m/s^2	0.5 m/s^2

Regulation 5 requires the employer to make a suitable and sufficient assessment of the risk created by work that is liable to expose employees to risk from vibration. The assessment must observe work practices, make reference to information regarding the magnitude of vibration from equipment and if necessary measurement of the magnitude of the vibration.

Consideration must also be given to the type, duration, effects of exposure, exposures limit / action values, effects on employees at particular risk, the effects of vibration on equipment and the ability to use it, manufacturers' information, availability of replacement equipment, and extension of exposure at the workplace (e.g. rest facilities), temperature and information on health surveillance. The risk assessment should be recorded as soon as is practicable after the risk assessment is made and reviewed regularly.

Regulation 6 states that the employer must seek to eliminate the risk of vibration at source or, if not reasonably practicable, reduce it to a minimum. Where the personal daily exposure limit is exceeded the employer must reduce exposure by implementing a programme of organisational and technical measures. Measures include the use of other methods of work, ergonomics, maintenance of equipment, design and layout, information, instruction and training, limitation by schedules and breaks and the provision of personal protective equipment.

Regulation 7 states that health surveillance must be carried out if there is a risk to the health of employees liable to be exposed to vibration. This is in order to diagnose any health effect linked with exposure to vibration. A record of health shall be kept of any employee who undergoes health surveillance. If health surveillance identifies a disease or adverse health effect, considered by a doctor to be a result of exposure to vibration, the employer shall ensure that a qualified person informs the employee and provides information and advice. In addition the employer must also review risk assessments and the health of any other employee who has been similarly exposed and consider alternative work.

Regulation 8 states that employers must provide information, instruction and training to all employees who are exposed to risk from vibration. This includes any organisational and technical measures taken, exposure limits and values, risk assessment findings, why and how to detect injury, health surveillance entitlement and safe working practices. The requirement for information, instruction and training extends to persons whether or not an employee, but who carries out work in connection with the employers duties.

Criminal Law Act (CLA) 1967, Section 3, (reasonable force)

Considered in context in Element B8.

The Criminal Law Act 1967 is an Act to amend the law of England and Wales by abolishing the division of crimes into felonies and misdemeanours and to amend and simplify the law in respect of matters arising from or related to that division or the abolition of it; to do away (within or without England and Wales) with certain obsolete crimes together with the torts of maintenance and champerty; and for purposes connected therewith.

Arrangement of Act

The Act has three parts.

Part I - abolished the distinction between felony and misdemeanour and makes consequential provisions.

Part II - abolished a number of obsolete crimes.

Part III - contains supplementary provisions.

Outline of main points

Part I Section 3 replaces the common law rules on self-defence, such as the duty to retreat. It simply requires that any force used must be "reasonable in the circumstances". It is still in force today and states:

"3. (1) A person may use such force as is reasonable in the circumstances in the prevention of crime, or in effecting or assisting in the lawful arrest of offenders or suspected offenders or of persons unlawfully at large.

(2) Subsection (1) above shall replace the rules of the common law on the question when force used for a purpose mentioned in the subsection is justified by that purpose".

(Further provision about when force is "reasonable" was made by section 76 of the Criminal Justice and Immigration Act 2008).

Employment Rights Act (ERA) 1996

Considered in context in Element B8.

The Employment Rights Act 1996 deals with rights that most employees can get when they work, including unfair dismissal, reasonable notice before dismissal, time off rights for parenting, redundancy and more. It was amended substantially by the Labour government since 1997, to include the right to request flexible working time.

Arrangement of Act

1) Employment particulars.
2) Protection of wages.
3) Guarantee payments.
4) Sunday working for shop and betting workers.
4A) Protected disclosures.
5) Protection from suffering detriment in employment.
6) Time off work.
7) Suspension from work.
8) Maternity leave, adoption leave, parental leave, paternity leave and flexible working.
9) Termination of employment.
10) Unfair dismissal.
11) Redundancy payments etc.
12) Insolvency of employers.
13) Miscellaneous.
14) Interpretation.
15) General and supplementary.

Schedule 1 - Consequential amendments.

Schedule 2 - Transitional provisions, savings and transitory provisions.

Schedule 3 - Repeals and revocations.

Outline of main points

EMPLOYMENT PROTECTION

All employees are protected by the Employment Rights Act 1996, as amended, against suffering any harm because of any reasonable actions they take on health and safety grounds. This applies regardless of their length of service.

NEW AND EXPECTANT MOTHERS

The Employment Rights Act requires employers, when offering suitable alternative, to ensure that the work is:

- Suitable and appropriate for her to do in the circumstances.
- On terms and conditions no less favourable than her normal terms and conditions.

TIME OFF WORK

The Employment Rights Act 1996 also covers the time off from work to which an employee is entitled. One such entitlement is maternity leave. An employee cannot be dismissed for taking maternity leave. If an employer were to do so, they would be guilty of pregnancy discrimination. Other provisions related to time off include those concerning training, public duties and medical problems.

DISMISSAL AND REDUNDANCY

Two major topics covered by the Employment Rights Act 1996 are dismissal and redundancy. The Act makes unfair dismissal illegal. Unfair dismissal occurs when an employer terminates an employee without a legitimate and lawful reason for doing so. It is also the Employment Rights Act 1996 that provides for a redundancy payment.

EMPLOYMENT TRIBUNAL

The process by which disputes related to a violation of your employment rights are resolved is outlined within the Employment Rights Act 1996. It provides aggrieved employees with the right to take their case before an Employment Tribunal. The cases which entitle you to make an application to an Employment Tribunal are outlined on what is known as a "jurisdiction list." A jurisdiction list can be obtained from the local tribunal office.

Equality Act (EA) 2010

Considered in context in Element B1.

Arrangement of relevant Parts, Chapters and Sections of Act

Parts of the Act

1) Socio-economic inequalities.
2) Equality: key concepts.
3) Services and public functions.
4) Premises.
5) Work.
6) Education.
7) Association.
8) Prohibited conduct: ancillary.
9) Enforcement.
10) Contracts.
11) Advancement of equality.
12) Disabled person: transport.
13) Disability: miscellaneous.
14) General exemptions.
15) Family property.
16) General and miscellaneous.
17) Schedules 1-28.

Part 2, Chapters

1) Protected characteristics.
2) Prohibited conduct.

Discrimination.

Adjustments for disabled.

Discrimination: supplementary.

Outline of main points

PART 2 EQUALITY: KEY CONCEPTS

Chapter 1 Protected Characteristics

Section 4 the following characteristics are protected characteristics:

- Age.
- Disability.
- Gender reassignment.
- Marriage and civil partnership.
- Pregnancy and maternity.
- Race.
- Religion or belief.
- Sex.
- Sexual orientation.

Section 5 Age

(1) In relation to the protected characteristic of age:

(a) A reference to a person who has a particular protected characteristic is a reference to a person of a particular age group.

(b) A reference to persons who share a protected characteristic is a reference to persons of the same age group.

(2) A reference to an age group is a reference to a group of persons defined by reference to age, whether by reference to a particular age or to a range of ages.

Section 6 Disability

(1) A person (P) has a disability if:

(a) P has a physical or mental impairment.

(b) The impairment has a substantial and long-term adverse effect on P's ability to carry out normal day-to-day activities.

(2) A reference to a disabled person is a reference to a person who has a disability.

(3) In relation to the protected characteristic of disability:

(a) A reference to a person who has a particular protected characteristic is a reference to a person who has a particular disability.

(b) A reference to persons who share a protected characteristic is a reference to persons who have the same disability.

(4) This Act (except Part 12 and section 190) applies in relation to a person who has had a disability as it applies in relation to a person who has the disability; accordingly (except in that Part and that section).

(a) A reference (however expressed) to a person who has a disability includes a reference to a person who has had the disability.

(b) A reference (however expressed) to a person who does not have a disability includes a reference to a person who has not had the disability.

(5) A Minister of the Crown may issue guidance about matters to be taken into account in deciding any question for the purposes of subsection (1).

(6) Schedule 1 (disability: supplementary provision) has effect.

Section 9 Race

(1) Race includes:

(a) Colour.

(b) Nationality.

(c) Ethnic or national origins.

(2) In relation to the protected characteristic of race:

(a) A reference to a person who has a particular protected characteristic is a reference to a person of a particular racial group.

(b) A reference to persons who share a protected characteristic is a reference to persons of the same racial group.

(3) A racial group is a group of persons defined by reference to race; and a reference to a person's racial group is a reference to a racial group into which the person falls.

(4) The fact that a racial group comprises two or more distinct racial groups does not prevent it from constituting a particular racial group.

(5) A Minister of the Crown may by order:

(a) Amend this section so as to provide for caste to be an aspect of race.

(b) Amend this Act so as to provide for an exception to a provision of this Act to apply, or not to apply, to caste or to apply, or not to apply, to caste in specified circumstances.

(6) The power under section 207(4)(b), in its application to subsection (5), includes power to amend this Act.

Section 11 Sex

In relation to the protected characteristic of sex:

(a) A reference to a person who has a particular protected characteristic is a reference to a man or to a woman.

(b) A reference to persons who share a protected characteristic is a reference to persons of the same sex.

Part 2 Prohibited Conduct

Section 13 Direct discrimination

(1) A person (A) discriminates against another (B) if, because of a protected characteristic, A treats B less favourably than A treats or would treat others.

(2) If the protected characteristic is age, A does not discriminate against B if A can show A's treatment of B to be a proportionate means of achieving a legitimate aim.

(3) If the protected characteristic is disability, and B is not a disabled person, A does not discriminate against B only because A treats or would treat disabled persons more favourably than A treats B.

(4) If the protected characteristic is marriage and civil partnership, this section applies to a contravention of Part 5 (work) only if the treatment is because it is B who is married or a civil partner.

(5) If the protected characteristic is race, less favourable treatment includes segregating B from others.

(6) If the protected characteristic is sex:

(a) Less favourable treatment of a woman includes less favourable treatment of her because she is breast-feeding.

(b) In a case where B is a man, no account is to be taken of special treatment afforded to a woman in connection with pregnancy or childbirth.

(7) Subsection (6)(a) does not apply for the purposes of Part 5 (work).

(8) This section is subject to sections 17(6) and 18(7).

Section 15 Discrimination arising from disability

(1) A person (A) discriminates against a disabled person (B) if:

(a) A treats B unfavourably because of something arising in consequence of B's disability.

(b) A cannot show that the treatment is a proportionate means of achieving a legitimate aim.

(2) Subsection (1) does not apply if A shows that A did not know, and could not reasonably have been expected to know, that B had the disability.

Section 19 Indirect discrimination

(1) A person (A) discriminates against another (B) if A applies to B a provision, criterion or practice which is discriminatory in relation to a relevant protected characteristic of B's.

(2) For the purposes of subsection (1), a provision, criterion or practice is discriminatory in relation to a relevant protected characteristic of B's if:

(a) A applies, or would apply, it to persons with whom B does not share the characteristic.

(b) It puts, or would put, persons with whom B shares the characteristic at a particular disadvantage when compared with persons with whom B does not share it.

(c) It puts, or would put, B at that disadvantage.

(d) A cannot show it to be a proportionate means of achieving a legitimate aim.

Section 20 Duty to make adjustments

(1) Where this Act imposes a duty to make reasonable adjustments on a person, this section, sections 21 and 22 and the applicable Schedule apply; and for those purposes, a person on whom the duty is imposed is referred to as A.

(2) The duty comprises the following three requirements.

(3) The first requirement is a requirement, where a provision, criterion or practice of A's puts a disabled person at a substantial disadvantage in relation to a relevant matter in comparison with persons who are not disabled, to take such steps as it is reasonable to have to take to avoid the disadvantage.

(4) The second requirement is a requirement, where a physical feature puts a disabled person at a substantial disadvantage in relation to a relevant matter in comparison with persons who are not disabled, to take such steps as it is reasonable to have to take to avoid the disadvantage.

(5) The third requirement is a requirement, where a disabled person would, but for the provision of an auxiliary aid, be put at a substantial disadvantage in relation to a relevant matter in comparison with persons who are not disabled, to take such steps as it is reasonable to have to take to provide the auxiliary aid.

(6) Where the first or third requirement relates to the provision of information, the steps which it is reasonable for A to have to take include steps for ensuring that in the circumstances concerned the information is provided in an accessible format.

(7) A person (A) who is subject to a duty to make reasonable adjustments is not (subject to express provision to the contrary) entitled to require a disabled person, in relation to whom A is required to comply with the duty, to pay to any extent A's costs of complying with the duty.

(8) A reference in section 21 or 22 or an applicable Schedule to the first, second or third requirement is to be construed in accordance with this section.

(9) In relation to the second requirement, a reference in this section or an applicable Schedule to avoiding a substantial disadvantage includes a reference to:

(a) Removing the physical feature in question.

(b) Altering it.

(c) Providing a reasonable means of avoiding it.

(10) A reference in this section, section 21 or 22 or an applicable Schedule (apart from paragraphs 2 to 4 of Schedule 4) to a physical feature is a reference to:

(a) A feature arising from the design or construction of a building.

(b) A feature of an approach to, exit from or access to a building.

(c) A fixture or fitting, or furniture, furnishings, materials, equipment or other chattels, in or on premises.

(d) Any other physical element or quality.

(11) A reference in this section, section 21 or 22 or an applicable Schedule to an auxiliary aid includes a reference to an auxiliary service.

(12) A reference in this section or an applicable Schedule to chattels is to be read, in relation to Scotland, as a reference to moveable property.

(13) The applicable Schedule is, in relation to the Part of this Act specified in the first column of the Table, the Schedule specified in the second column.

Section 26 Harassment

(1) A person (A) harasses another (B) if:

(a) A engages in unwanted conduct related to a relevant protected characteristic.

(b) The conduct has the purpose or effect of:

(i) Violating B's dignity.

(ii) Creating an intimidating, hostile, degrading, humiliating or offensive environment for B.

(2) A also harasses B if:

(a) A engages in unwanted conduct of a sexual nature.

(b) The conduct has the purpose or effect referred to in subsection (1)(b).

(3) A also harasses B if:

(a) A or another person engages in unwanted conduct of a sexual nature or that is related to gender reassignment or sex.

(b) The conduct has the purpose or effect referred to in subsection (1)(b).

(c) Because of B's rejection of or submission to the conduct, A treats B less favourably than A would treat B if B had not rejected or submitted to the conduct.

(4) In deciding whether conduct has the effect referred to in subsection (1)(b), each of the following must be taken into account:

(a) The perception of B.

(b) The deriving from these hazards and the health and safety policy of their employer and the organisation and arrangements for fulfilling that policy. other circumstances of the case.

(c) Whether it is reasonable for the conduct to have that effect.

Section 27 Victimisation

(1) A person (A) victimises another person (B) if A subjects B to a detriment because:

(a) B does a protected act.

(b) A believes that B has done, or may do, a protected act.

(2) Each of the following is a protected act:

(a) Bringing proceedings under this Act.

(b) Giving evidence or information in connection with proceedings under this Act.

(c) Doing any other thing for the purposes of or in connection with this Act.

(d) Making an allegation (whether or not express) that A or another person has contravened this Act.

(3) Giving false evidence or information, or making a false allegation, is not a protected act if the evidence or information is given, or the allegation is made, in bad faith.

(4) This section applies only where the person subjected to a detriment is an individual.

(5) The reference to contravening this Act includes a reference to committing a breach of an equality clause or rule.

Health and Safety (Display Screen Equipment) Regulations (DSE) 1992 (as amended)

See also - Health and Safety (Miscellaneous Amendments) Regulations (MAR) 2002.

Considered in context/more depth in Element B9.

Arrangement of Regulations

2) Every employer shall carry out suitable and sufficient analysis of workstations.

3) Employers shall ensure that equipment provided meets the requirements of the schedule laid down in these Regulations.

4) Employers shall plan activities and provide such breaks or changes in work activity to reduce employees' workload on that equipment.

5) For display screen equipment (DSE) users, the employer shall provide, on request, an eyesight test carried out by a competent person.

6 & 7) Provision of information and training.

Outline of main points

WORKSTATION ASSESSMENTS (REGULATION 2)

Should take account of:

■ Screen	-	positioning, character definition, character stability etc.
■ Keyboard	-	tilt able, character legibility etc.
■ Desk	-	size, matt surface etc.
■ Chair	-	adjustable back and height, footrest available etc.
■ Environment	-	noise, lighting, space etc.
■ Software	-	easy to use, work rate not governed by software.

INFORMATION AND TRAINING (REGULATIONS 6 & 7)

Should include:

■ Risks to health.
■ Precautions in place (e.g. the need for regular breaks).
■ How to recognise problems.
■ How to report problems.

Health and Safety (First-Aid) Regulations (FAR) 1981 (as amended)

See also Health and Safety (Miscellaneous Amendments) Regulations 2002

Law considered in context/more depth in Element B10.

Arrangement of Regulations

1) Citation and commencement.
2) Interpretation.
3) Duty of employer to make provision for first-aid.
4) Duty of employer to inform his employees of the arrangements.
5) Duty of self-employed person to provide first-aid equipment.
6) Power to grant exemptions.
7) Cases where these Regulations do not apply.
8) Application to mines.
9) Application offshore.
10) Repeals, revocations and modification.
Schedule 1 - Repeals.
Schedule 1 - Revocations.

Outline of main points

2) Regulation 2 defines first aid as: '…treatment for the purpose of preserving life and minimising the consequences of injury or illness until medical (doctor or nurse) help can be obtained. Also, it provides treatment of minor injuries which would otherwise receive no treatment, or which do not need the help of a medical practitioner or nurse'.
3) Requires that every employer must provide equipment and facilities that are adequate and appropriate in the circumstances for administering first-aid to his employees.
4) The employer must inform their employees about the first-aid arrangements, including the location of equipment, facilities and identification of trained personnel.
5) Self-employed people must ensure that adequate and suitable provision is made for administering first-aid while at work.

Health and Safety (Miscellaneous Amendments) Regulations (MAR) 2002

See also - DSE 1992, MHOR 1992, PPER 1992 and WHSWR 1992.

These Regulations made minor amendments to UK law to come into line with the requirements of the original Directives and came into force on 17th September 2002. The Regulations that are affected by the amendments are detailed below.

Arrangement of Regulations

1) Citation and commencement.
2) Amendment of the Health and Safety (First-Aid) Regulations 1981.
3) Amendment of the Health and Safety (Display Screen Equipment) Regulations 1992.
4) Amendment of the Manual Handling Operations Regulations 1992.
5) Amendment of the Personal Protective Equipment at Work Regulations 1992.
6) Amendment of the Workplace (Health, Safety and Welfare) Regulations 1992.
7) Amendment of the Provision and Use of Work Equipment Regulations 1998.
8) Amendment of the Lifting Operations and Lifting Equipment Regulations 1998.
9) Amendment of the Quarries Regulations 1999.

Outline of main points

REGULATION 3 - AMENDMENT OF THE HEALTH AND SAFETY (FIRST AID) REGULATIONS 1981

The Health and Safety (First Aid) Regulations 1981 are amended by adding the additional requirements that any first-aid room provided under requirements of these regulations must be easily accessible to stretchers and to any other equipment needed to convey patients to and from the room and that the room be sign-posted by use of a sign complying with the Health and Safety (Safety Signs and Signals) Regulations 1996.

REGULATION 3 - AMENDMENT OF THE HEALTH AND SAFETY (DISPLAY SCREEN EQUIPMENT) REGULATIONS 1992

The Health and Safety (Display Screen Equipment) Regulations 1992, Regulation 3 – Requirements of workstations, was amended to remove transitional arrangements relating to the introduction of the regulations and provide a simpler requirement for workstations, removing reference to who may use the workstation:

"Regulation 3. Every employer shall ensure that any workstation which may be used for the purposes of his undertaking meets the requirements laid down in the Schedule to these Regulations, to the extent specified in paragraph 1 thereof."

The requirement "used for the purposes of his undertaking" includes workstations provided by the employer and others and to workstations provided by the employer to others (including members of the public) as part of the undertaking.

The Health and Safety (Display Screen Equipment) Regulations 1992, Regulation 5 – Eyes and eyesight tests, was amended to clarify the requirement for eye and eyesight tests to ensure that those who may become employed as users are provide for and receive eye and eyesight tests before they become users.

The Health and Safety (Display Screen Equipment) Regulations 1992 Regulation 6 – Provision of training, was similarly amended with regard to health and safety training for those who may become employed as users, such that they receive training before they become a user.

REGULATION 4 - AMENDMENT OF THE MANUAL HANDLING OPERATIONS REGULATIONS 1992

Regulation 4 of the Manual Handling Operations Regulations 1992 were amended by adding the requirement to, when determining whether manual handling operations at work involve a risk of injury and the appropriate steps to reduce that risk, have regard to:

- Physical suitability of the employee to carry out the operations.
- Clothing, footwear or other personal effects they are wearing.
- Knowledge and training.
- Results of any relevant risk assessment conducted for the Management of Health and Safety at Work Regulations.
- Whether the employee is within a group of employees identified by that assessment as being especially at risk.
- Results of any health surveillance provided under the Management of Health and Safety Regulations.

REGULATION 5 - AMENDMENT OF THE PERSONAL PROTECTIVE EQUIPMENT AT WORK REGULATIONS 1992

The Personal Protective Equipment at Work Regulations 1992 were amended so that personal protective equipment (PPE) must also be suitable for the period for which it is worn and account is taken of the characteristics of the workstation of each person.

Provision of personal issue of PPE needs to take place in situations where it is necessary to ensure it is hygienic and free of risk to health.

Where an assessment of PPE is made this must consider whether it is compatible with other personal protective equipment that is in use and which an employee would be required to wear simultaneously.

The amendments require that information provided to satisfy regulation 9 for the provision of information, instruction and training must be kept available to employees. A new, additional duty is created requiring the employer, where appropriate, and at suitable intervals, to organise demonstrations in the wearing of PPE.

REGULATION 6 - AMENDMENT OF THE WORKPLACE (HEALTH, SAFETY AND WELFARE) REGULATIONS 1992

The Workplace (Health, Safety and Welfare) Regulations 1992 have been amended to improve clarity, include additional regulations and make provision for the disabled.

An additional regulation (4A) sets out a requirement where a workplace is in a building, the building shall have stability and solidity appropriate to the nature of the use of the workplace. The range of things requiring maintenance under these regulations is extended to equipment and devices intended to prevent or reduce hazards. A new duty requires workplaces to be adequately thermally insulated where it is necessary, having regard to the type of work carried out and the physical activity of the persons carrying out the work. In addition, excessive effects of sunlight on temperature must be avoided.

The regulations were amended with regard to facilities for changing clothing in that the facilities need to be easily accessible, of sufficient capacity and provided with seating. Requirements were amended such that rest rooms and rest areas must include suitable arrangements to protect non-smokers from discomfort caused by tobacco smoke. They also must be equipped with an adequate number of tables and adequate seating with backs for the number of persons at work likely to use them at any one time and seating which is adequate for the number of disabled persons at work and suitable for them.

A new regulation (25A) was added requiring, where necessary, those parts of the workplace (including in particular doors, passageways, stairs, showers, washbasins, lavatories and workstations) used or occupied directly by disabled persons at work to be organised to take account of such persons.

AMENDMENT OF THE PROVISION AND USE OF WORK EQUIPMENT REGULATIONS 1998

The regulations have a small number of amendments affecting 3 main regulations. Regulation 10, which deals with equipment's conformity with community requirements, is amended such that the requirement to conform to 'essential requirements' is no longer limited to the point at which the equipment was designed and constructed - equipment must now conform at 'all times'. The 'essential requirements' are those that were applicable at the time it was put into first service. Regulation 11 was amended to remove the opportunity of reliance on information, instruction, training and supervision as a separate option in the hierarchy of control of dangerous parts of machinery. The requirement to provide information, instruction, training and supervision is now amended to apply to each stage of the dangerous parts of machinery control hierarchy. Regulation 18, which deals with control systems, carries a small but important amendment which means the requirement that all control systems of work equipment are "chosen making due allowance for the failures, faults and constraints to be expected in the planned circumstances of use" is modified from an absolute duty to one of so far as is reasonably practicable.

AMENDMENT OF THE LIFTING OPERATIONS AND LIFTING EQUIPMENT REGULATIONS 1998

Minor changes to the definitions in the Lifting Operations and Lifting Equipment Regulations 1998 were made by these regulations.

(a) In the definition of "accessory for lifting" in regulation 2(1), by substituting for the word "work" the word "lifting".
(b) In regulation 3(4), by substituting for the words "(5)(b)" the words "(3)(b)".

Health and Safety (Miscellaneous Repeals, Revocations and Amendments) Regulations (MRRA) 2013

See also - PPER 1992 and WHSWR 1992.

These Regulations repeal one Act and revoke twelve instruments (plus a related provision in the Factories Act 1961) and came into force on 6th April 2013. The Regulations, applicable to this unit, that are affected by the amendments are detailed below.

Arrangement of Regulations

1) Citation and commencement.
2) Repeals and revocations.
3) Consequential amendments to the Dangerous Substances (Notification and Marking of Sites) Regulations 1990.
4) Consequential amendments to the Workplace (Health, Safety and Welfare) Regulations 1992.

Outline of main points

CONSEQUENTIAL AMENDMENTS TO THE PERSONAL PROTECTIVE EQUIPMENT AT WORK REGULATIONS 1992

The Personal Protective Equipment at Work Regulations 1992 have been amended so that they cover the provision and use of head protection on construction sites thus maintaining the level of legal protection when the Construction (Head Protection) Regulations were revoked as part of The Health and Safety (Miscellaneous Repeals, Revocations and Amendments) Regulations 2013. These measures are being removed because they have either been overtaken by more up to date Regulations, are redundant or do not deliver the intended benefits.

AMENDMENT OF THE WORKPLACE (HEALTH, SAFETY AND WELFARE) REGULATIONS 1992

The Workplace (Health, Safety and Welfare) Regulations 1992 have been amended to include the requirement for adequate lighting and safe access for workers on ships in a shipyard or harbour undergoing construction, repair or maintenance.

The Health and Safety at Work Act (HASAWA) 1974

Considered in context/more depth in Element B7.

HASAWA is amended by Article 2 of the Legislative Reform (Health and Safety Executive) Order. LRHSEO 2008 abolishes the Health and Safety Commission and Executive; Article 4 establishes the new Health and Safety Executive and amends reference to the Commission in the HASWA to "the Executive".

Arrangement of Act

PRELIMINARY

1) Preliminary.

GENERAL DUTIES

2) General duties of employers to the employees.

3) General duties of employers and self-employed to persons other than their employees.

4) General duties of persons concerned with premises to persons other than their employees.

5) [repealed].

6) General duties of manufacturers etc. as regards articles and substances for use at work.

7) General duties of employees at work.

8) Duty not to interfere with or misuse things provided pursuant to certain provisions.

9) Duty not to charge employees for things done or provided pursuant to certain specific requirements.

THE HEALTH AND SAFETY EXECUTIVE

10) Establishment of the Executive.

11) Functions of the Executive.

12) Control of the Executive by the Secretary of State.

13) Other powers of the Executive.
14) Power of the Executive to direct investigations and Inquiries.

HEALTH AND SAFETY REGULATIONS AND APPROVED CODES OF PRACTICE

15) Health and safety regulations.
16) Approval of codes of practice by the Executive.
17) Use of approved codes of practice in criminal proceedings.

ENFORCEMENT

18) Authorities responsible for enforcement of the relevant statutory provisions.
19) Appointment of inspectors.
20) Powers of inspectors.
21) Improvement notices.
22) Prohibition notices.
23) Provisions supplementary toss. 21 and 22.
24) Appeal against improvement or prohibition notice.
25) Power to deal with cause of imminent danger.
26) Power of enforcing authorities to indemnify their inspectors.

OBTAINING AND DISCLOSURE OF INFORMATION

27) Obtaining of information by the Executive, enforcing authorities etc.
28) Restrictions on disclosure of information.

SPECIAL PROVISIONS RELATING TO AGRICULTURE

29-32) [repealed].

PROVISIONS AS TO OFFENCES

33) Offences.
34) Extension of time for bringing summary proceedings.
35) Venue.
36) Offences due to fault of other person.
37) Offences by bodies corporate.
38) Restriction on institution of proceedings in England and Wales.
39) Prosecutions by inspectors.
40) Onus of proving limits of what is practicable etc.
41) Evidence.
42) Power of court to order cause of offence to be remedied or, in certain cases, forfeiture.

FINANCIAL PROVISION

43) Financial provisions.

MISCELLANEOUS AND SUPPLEMENTARY

44) Appeals in connection with licensing provisions in the relevant statutory provisions.
45) Default powers.
46) Service of notices.
47) Civil liability.

Section 69 of the Enterprise and Regulatory Reform Act 2013 (ERRA), which came into force on 1 October 2013, amends the Health and Safety at Work etc Act 1974 (HSWA) Section 47 to the effect that claims for compensation for workplace incidents can only be made when negligence can be proved.

48) Application to Crown.
49) Adaptation of enactments to metric units or appropriate metric units.
50) Regulations under the relevant statutory provisions.
51) Exclusion of application to domestic employment.
52) Meaning of work and at work.
53) General interpretation of Part I.
54) Application of Part I to Isles of Scilly.

THE EMPLOYMENT MEDICAL ADVISORY SERVICE

55) Functions of, and responsibility for maintaining, employment medical advisory service.
56) Functions of authority responsible for maintaining the service.
57) Fees.
58) Other financial provisions.

59) Duty of responsible authority to keep accounts and to report.

60) Supplementary.

Outline of main points

OVERALL AIMS OF THE ACT
1) To protect people.
2) To protect the public from risks which may arise from work activities.

THE MAIN PROVISIONS - SECTION 1
a) Securing the health, safety and welfare of people at work.
b) Protecting others against risks arising from workplace activities.
c) Controlling the obtaining, keeping, and use of explosive and highly flammable substances.
d) Controlling emissions into the atmosphere of noxious or offensive substances (Repealed).

Duties are imposed on:

a) The employer.
b) The self-employed.
c) Employees.
d) Contractors and subcontractors.
e) Designers, manufacturers, suppliers, importers and installers.
f) Specialists - architects, surveyors, engineers, personnel managers, health and safety practitioners, and many more.

EMPLOYER'S DUTIES - [TO EMPLOYEES]

Section 2(1)

To ensure, so far as *reasonably practicable*, the health, safety and welfare at work of employees.

Section 2(2)

Ensuring health, safety and welfare at work through:

- Safe plant and systems of work e.g. provision of guards on machines.
- Safe use, handling, storage and transport of goods and materials e.g. good manual handling of boxes.
- Provision of information, instruction, training and supervision e.g. provision of induction training.
- Safe place of work including means of access and egress e.g. aisles kept clear.
- Safe and healthy working environment e.g. good lighting.

Further duties are placed on the employer by:

Section 2(3)

Prepare and keep up to date a written safety policy supported by information on the organisation and arrangements for carrying out the policy. The safety policy has to be brought to the notice of employees. If there are fewer than five employees, this section does not apply.

Section 2(4)

Recognised Trade Unions have the right to appoint safety representatives to represent the employees in consultations with the employer about health and safety matters.

Section 2(6)

Employers must consult with any safety representatives appointed by recognised Trade Unions.

Section 2(7)

To establish a safety committee if requested by two or more safety representatives.

EMPLOYER'S DUTIES - [TO PERSONS NOT HIS EMPLOYEES]

Section 3
a) Not to expose them to risk to their health and safety e.g. contractor work barriered off.
b) To give information about risks which may affect them e.g. location induction for contractors.

SELF EMPLOYED DUTIES

Section 3
a) Not to expose themselves to risks to their health and safety e.g. wear personal protection.
b) Not to expose other persons to risks to their health and safety e.g. keep shared work area tidy.

Some of the practical steps that an organisation might take in order to ensure the safety of visitors to its premises are:

- Identify visitors by signing in, badges etc.
- Provide information regarding the risks present and the site rules and procedures to be followed, particularly in emergencies.
- Provide escorts to supervise visitors throughout the site.
- Restrict access to certain areas.

CONTROL OF PREMISES

Section 4

This section places duties on anyone who has control to any extent of non-domestic premises used by people who are not their employees. The duty extends to the provision of safe premises, plant and substances, e.g. maintenance of a boiler in rented out property.

Figure RSP-4: Risks from roadside work.   *Source: RMS.*

Figure RSP-5: Risks from street light repairing or tree felling.
Source: RMS.

MANUFACTURERS, DESIGNERS, SUPPLIERS, IMPORTERS, INSTALLERS

Section 6

This section places specific duties on those who can ensure that articles and substances are as safe and without risks as is reasonably practicable. The section covers:

- Safe design, installation and testing of equipment (including fairground equipment).
- Safe substances tested for risks.
- Provision of information on safe use and conditions essential to health and safety.
- Research to minimise risks.

EMPLOYEES' DUTIES

Section 7

a) To take reasonable care for themselves and others that may be affected by their acts/omissions e.g. wear eye protection, not obstructs a fire exit.
b) To co-operate with the employer or other to enable them to carry out their duty and/or statutory requirements e.g. report hazards or defects in controls, attend training, provide medical samples.

Additional duties created by the Management of Health and Safety at Work Regulations 1999 employees' duties:

- Every employee shall use any equipment, material or substance provided to them in accordance with any training and instruction.
- Every employee shall inform (via supervisory staff) their employer of any (a) risk situation or (b) shortcoming in the employer's protection arrangements.

OTHER DUTIES

Section 8

No person to interfere with or misuse anything provided to secure health and safety - e.g. wedge fire door open, remove first aid equipment without authority, breach lock off systems.

Section 9

Employees cannot be charged for anything done or provided to comply with a specific legal obligation e.g. personal protective equipment, health surveillance or welfare facilities.

OFFENCES COMMITTED BY OTHER PERSONS

Section 36

- Where the commission by any person of the breach of legislation is due to the act or default of some other person, that other person shall be guilty of the offence and may be charged with and convicted of the offence whether or not proceedings are taken against the first mentioned person.
- Case law indicates that 'other person' refers to persons lower down the corporate tree than mentioned in section 37, e.g. middle managers, safety advisors, training officers; and may extend to people working on contract e.g. architects, consultants or a planning supervisor.

OFFENCES COMMITTED BY THE BODY CORPORATE

Section 37

- Where there has been a breach of legislation on the part of a body corporate (limited company or local authority) and the offence can be proved to have been committed with the consent or connivance of or to be attributable to any neglect on the part of any director, manager, secretary or similar officer of the body corporate, he, as well as the body corporate, can be found guilty and punished accordingly.

ONUS OF PROOF

Section 40

In any proceedings for an offence under any of the relevant statutory involving a failure to comply with a duty or requirement:

- To do something so far as is practicable.
- To do something so far as is reasonably practicable.
- It shall be for the accused to prove that the requirements were met rather than for the prosecution to prove that the requirements were not met.

THE EMPLOYMENT MEDICAL ADVISORY SERVICE

Section 50

The HSE operates the Employment Medical Advisory Service (EMAS), which assists in the investigation of work-related ill-health.

Through the HSE Info Line, EMAS also offers advice on general enquiries regarding fitness for work, medical aspects of employment, health in the workplace, and general and specific complaints of ill-health attributable to work.

EMAS' main function is to assist in the investigation of ill-health attributable to work, and help to bring enforcement, action or prosecution of employers, if appropriate.

Ionising Radiations Regulations (IRR) 1999

Considered in context in Elements B4 and B7.

Arrangement of Regulations

Part I	*Interpretation and General*	
1)	Citation and commencement.	
2)	Interpretation.	
3)	Application.	
4)	Duties under the Regulations.	
Part II	*General Principles and Procedures*	
5)	Authorisation of specified practices.	
6)	Notification of specified work.	
7)	Prior risk assessment etc.	
8)	Restriction of exposure.	
9)	Personal protective equipment.	
10)	Maintenance and examination of engineering controls etc. and personal protective equipment.	
11)	Dose limitation.	
12)	Contingency plans.	
Part III	*Arrangements for the Management of Radiation Protection*	
13)	Radiation protection adviser.	
14)	Information, instruction and training.	
15)	Co-operation between employers.	
Part IV	*Designated Areas*	
16)	Designation of controlled or supervised areas.	
17)	Local rules and radiation protection supervisors.	
18)	Additional requirements for designated areas.	
19)	Monitoring of designated areas.	
Part V	*Classification and monitoring of persons*	
20)	Designation of classified persons.	
21)	Dose assessment and recording.	
22)	Estimated doses and special entries.	
23)	Dosimetry for accidents etc.	
24)	Medical surveillance.	
25)	Investigation and notification of overexposure.	
26)	Dose limitation for overexposed employees.	

Part VI	*Arrangements for the Control of Radioactive Substances, Articles and Equipment*
27)	Sealed sources and articles containing or embodying radioactive substances.
28)	Accounting for radioactive substances.
29)	Keeping and moving of radioactive substances.
30)	Notification of certain occurrences.
31)	Duties of manufacturers etc. of articles for use in work with ionising radiation.
32)	Equipment used for medical exposure.
33)	Misuse of or interference with sources of ionising radiation.
Part VII	***Duties of Employees and Miscellaneous***
34)	Duties of employees.
35)	Approval of dosimetry services.
36)	Defence on contravention.
37)	Exemption certificates.
38)	Extension outside Great Britain.
39)	Transitional provisions.
40)	Modifications relating to the Ministry of Defence.
41)	Modification, revocation and saving.
Schedule 1.	Work not required to be notified under regulation 6.
Schedule 2.	Particulars to be provided in a notification under regulation 6(2).
Schedule 3.	Additional particulars that the Executive may require.
Schedule 4.	Dose limits.
Schedule 5.	Matters in respect of which radiation protection adviser must be consulted by a radiation employer.
Schedule 6.	Particulars to be entered in the radiation passbook.
Schedule 7.	Particulars to be contained in a health record.
Schedule 8.	Quantities and concentrations of radionuclides.
Schedule 9.	Modifications.

Outline of main points

These Regulations supersede and consolidate the Ionising Radiations Regulations 1985 and the Ionising Radiation (Outside Workers) Regulations 1993.

IRR99 apply to a large range of workplaces where radioactive substances and electrical equipment emitting ionising radiation are used. They also apply to work with natural radiation, including work in which people are exposed to naturally occurring radon gas and its decay products. Any employer who undertakes work with ionising radiation must comply with IRR99. Under IRR99 employers who work with ionising radiation are called radiation employers.

IRR99 requires employers to keep exposure to ionising radiations as low as reasonably practicable. Exposures must not exceed specified dose limits. Restriction of exposure should be achieved first by means of engineering control and design features. Where this is not reasonably practicable employers should introduce safe systems of work and only rely on the provision of personal protective equipment as a last resort.

Physical Agents (Electromagnetic Fields) Directive

Considered in context in Elements B7.

A directive, entitled "Physical Agents (Electromagnetic Fields) Directive 20th June 2013, is the 20th individual Directive within the meaning of Article 16(1) of Directive 89/391/EEC (repeals Directive 2004/40/EC), and lays down minimum requirements for the protection of workers from risks to their health and safety arising, or likely to arise, from exposure to electromagnetic fields during their work.

The Directive deals only with health and safety at work, and applies to work activities where workers are exposed to risks from electromagnetic fields.

For the purposes of this Directive, the following definitions shall apply:

(a) 'electromagnetic fields' (EMF) means static electric, static magnetic and time-varying electric, magnetic and electromagnetic fields with frequencies up to 300 GHz;

(b) 'direct biophysical effects' means effects in the human body directly caused by its presence in an electromagnetic field, including:

    (i) thermal effects, such as tissue heating through energy absorption from electromagnetic fields in the tissue;

    (ii) non-thermal effects, such as the stimulation of muscles, nerves or sensory organs. These effects might have a detrimental effect on the mental and physical health of exposed workers. Moreover, the stimulation of sensory organs may lead to transient symptoms, such as vertigo or phosphenes (a phenomenon characterised by the experience of seeing light without light actually entering the eye). These effects might create temporary annoyance or affect cognition or other brain or muscle functions, and may thereby affect the ability of a worker to work safely (i.e. safety risks); and

    (iii) limb currents;

(c) 'indirect effects' means effects, caused by the presence of an object in an electromagnetic field, which may become the cause of a safety or health hazard, such as:

(i) interference with medical electronic equipment and devices, including cardiac pacemakers and other implants or medical devices worn on the body;

(ii) the projectile risk from ferromagnetic objects in static magnetic fields;

(iii) the initiation of electro-explosive devices (detonators);

(iv) fires and explosions resulting from the ignition of flammable materials by sparks caused by induced fields, contact currents or spark discharges; and

(v) contact currents;

(d) 'exposure limit values (ELVs)' means values established on the basis of biophysical and biological considerations, in particular on the basis of scientifically well-established short-term and acute direct effects, i.e. thermal effects and electrical stimulation of tissues;

(e) 'health effects ELVs' means those ELVs above which workers might be subject to adverse health effects, such as thermal heating or stimulation of nerve and muscle tissue;

(f) 'sensory effects ELVs' means those ELVs above which workers might be subject to transient disturbed sensory perceptions and minor changes in brain functions;

(g) 'action levels (ALs)' means operational levels established for the purpose of simplifying the process of demonstrating the compliance with relevant ELVs or, where appropriate, to take relevant protection or prevention measures specified in the Directive.

The AL terminology used in Annex II is as follows:

(i) for electric fields, 'low ALs' and 'high ALs' means levels which relate to the specific protection or prevention measures specified in the Directive; and

(ii) for magnetic fields, 'low ALs' means levels which relate to the sensory effects ELVs and 'high ALs' to the health effects ELVs.

IN SUMMARY

The Directive places a number of duties on employers, main ones being that it:

- Places a duty on the employer to conduct a risk assessment and calculate EMF strengths.
- Places a duty on the employer to eliminate or reduce as low as possible the risk of exposure; and where risk can't be eliminated that measures are devised by the employer to reduce the risk of exposure below ELV.
- Requires the employer to provide: the risk assessment to the nominated person responsible for health surveillance.
- Requires an investigation and medical examination where an employee is 'detected' as having been exposed.
- Records of Health surveillance activities are kept.

http://eur-lex.europa.eu/LexUriServ/LexUriServ.do?uri=OJ:L:2013:179:0001:0021:EN:PDF

Management of Health and Safety at Work Regulations (MHSWR) 1999 (as amended)

Considered in context in Elements B7 and B9.

Arrangement of Regulations

1) Citation, commencement and interpretation.
2) Disapplication of these Regulations.
3) Risk assessment.
4) Principles of prevention to be applied.
5) Health and safety arrangements.
6) Health surveillance.
7) Health and safety assistance.
8) Procedures for serious and imminent danger and for danger areas.
9) Contacts with external services.
10) Information for employees.
11) Co-operation and co-ordination.
12) Persons working in host employers' or self-employed persons' undertakings.
13) Capabilities and training.
14) Employees' duties.
15) Temporary workers.
16) Risk assessment in respect of new or expectant mothers.
17) Certificate from a registered medical practitioner in respect of new or expectant mothers.
18) Notification by new or expectant mothers.
19) Protection of young persons.
20) Exemption certificates.
21) Provisions as to liability.
22) Exclusion of civil liability.

23) Extension outside Great Britain.

24) Amendment of the Health and Safety (First-Aid) Regulations 1981.

25) Amendment of the Offshore Installations and Pipeline Works (First-Aid) Regulations 1989.

26) Amendment of the Mines Miscellaneous Health and Safety Provisions Regulations 1995.

27) Amendment of the Construction (Health, Safety and Welfare) Regulations 1996.

28) Regulations to have effect as health and safety regulations.

29) Revocations and consequential amendments.

30) Transitional provision.

Schedule 1 - General principles of prevention.

Schedule 2 - Consequential amendments.

Outline of main points

Management of Health and Safety at Work Regulations (MHSWR) 1999 set out some broad general duties that apply to almost all kinds of work. They are aimed mainly at improving health and safety management. The 1999 regulation is amended by the Management of Health and Safety at Work Regulations (MHSWR) 2006 the principal Regulations are discussed below.

RISK ASSESSMENT (REGULATION 3)

The regulations require employers (and the self-employed) to assess the risk to the health and safety of their employees and to anyone else who may be affected by their work activity. This is necessary to ensure that the preventive and protective steps can be identified to control hazards in the workplace.

Where an employer is employing or about to employ young persons (under 18 years of age) he must carry out a risk assessment which takes particular account of:

- The inexperience, lack of awareness of risks and immaturity of young persons.
- The layout of the workplace and workstations.
- Exposure to physical, biological and chemical agents.
- Work equipment and the way in which it is handled.
- The extent of health and safety training to be provided.
- Risks from agents, processes and work listed in the Annex to Council Directive 94/33/EC on the protection of young people at work.

Where 5 or more employees are employed, the significant findings of risk assessments must be recorded in writing (the same threshold that is used in respect of having a written safety policy). This record must include details of any employees being identified as being especially at risk.

PRINCIPLES OF PREVENTION TO BE APPLIED (REGULATION 4)

Regulation 4 requires an employer to implement preventive and protective measures on the basis of general principles of prevention specified in Schedule 1 to the Regulations. These are:

1) Avoiding risks.
2) Evaluating the risks which cannot be avoided.
3) Combating the risks at source.
4) Adapting the work to the individual, especially as regards the design of workplaces, the choice of work equipment and the choice of working and production methods, with a view, in particular, to alleviating monotonous work and work at a predetermined work-rate and to reducing their effect on health.
5) Adapting to technical progress.
6) Replacing the dangerous by the non-dangerous or the less dangerous.
7) Developing a coherent overall prevention policy which covers technology, organisation of work, working conditions, social relationships and the influence of factors relating to the working environment.
8) Giving collective protective measures priority over individual protective measures.
9) Giving appropriate instructions to employees.

HEALTH AND SAFETY ARRANGEMENTS (REGULATION 5)

Appropriate arrangements must be made for the effective planning, organisation, control, monitoring and review of preventative and protective measures (in other words, for the management of health and safety).

Again, employers with five or more employees must have their arrangements in writing.

HEALTH SURVEILLANCE (REGULATION 6)

In addition to the requirements of other specific regulations, consideration must be given to carrying out health surveillance of employees, where there is a disease or adverse health condition identified in risk assessments.

HEALTH AND SAFETY ASSISTANCE (REGULATION 7)

The employer must appoint one or more competent persons to assist him in complying with the legal obligations imposed on the undertaking. The number of persons appointed should reflect the number of employees and the type of hazards in the workplace.

If more than one competent person is appointed, then arrangements must be made for ensuring adequate co-operation between them. The competent person(s) must be given the necessary time and resources to fulfil their functions. This will depend on the size the undertaking, the risks to which employees are exposed and the distribution of those risks throughout the undertaking.

The employer must ensure that competent person(s) who are not employees are informed of the factors known (or suspected) to affect the health and safety of anyone affected by business activities.

Competent people are defined as those who have sufficient training and experience or knowledge and other qualities to enable them to perform their functions.

Persons may be selected from among existing employees or from outside. Where there is a suitable person in the employer's employment, that person shall be appointed as the 'competent person' in preference to a non-employee.

PROCEDURES FOR SERIOUS AND IMMINENT DANGER AND FOR DANGER AREAS (REGULATION 8)

Employers are required to set up emergency procedures and appoint **competent persons** to ensure compliance with identified arrangements, to devise control strategies as appropriate and to limit access to areas of risk to ensure that only those persons with adequate health and safety knowledge and instruction are admitted.

The factors to be considered when preparing a procedure to deal with workplace emergencies such as fire, explosion, bomb scare, chemical leakage or other dangerous occurrence should include:

- The identification and training requirements of persons with specific responsibilities.
- The layout of the premises in relation to escape routes etc.
- The number of persons affected.
- Assessment of special needs (disabled persons, children etc.).
- Warning systems.
- Emergency lighting.
- Location of shut-off valves, isolation switches, hydrants etc.
- Equipment required to deal with the emergency.
- Location of assembly points.
- Communication with emergency services.
- Training and/or information to be given to employees, visitors, local residents and anyone else who might be affected.

CONTACTS WITH EXTERNAL SERVICES (REGULATION 9)

Employers must ensure that, where necessary, contacts are made with external services. This particularly applies with regard to first-aid, emergency medical care and rescue work.

INFORMATION FOR EMPLOYEES (REGULATION 10)

Employees must be provided with relevant information about hazards to their health and safety arising from risks identified by the assessments. Clear instruction must be provided concerning any preventative or protective control measures including those relating to serious and imminent danger and fire assessments. Details of any competent persons nominated to discharge specific duties in accordance with the regulations must also be communicated as should risks arising from contact with other employer's activities (see Regulation 11).

Before employing a child (a person who is not over compulsory school age) the employer must provide those with parental responsibility for the child with information on the risks that have been identified and preventative and protective measures to be taken.

CO-OPERATION AND CO-ORDINATION (REGULATION 11)

Employers who work together in a common workplace have a duty to co-operate to discharge their duties under relevant statutory provisions. They must also take all reasonable steps to inform their respective employees of risks to their health or safety which may arise out of their work. Specific arrangements must be made to ensure compliance with fire legislation.

PERSONS WORKING IN HOST EMPLOYERS' OR SELF EMPLOYED PERSONS' UNDERTAKINGS (REGULATION 12)

This regulation extends the requirements of regulation 11 to include employees working as sole occupiers of a workplace under the control of another employer. Such employees would include those working under a service of contract and employees in temporary employment businesses under the control of the first employer.

CAPABILITIES AND TRAINING (REGULATION 13)

Employers need to take into account the capabilities of their employees before entrusting tasks. This is necessary to ensure that they have adequate health and safety training and are capable enough at their jobs to avoid risk. To this end, consideration must be given to recruitment including job orientation when transferring between jobs and work departments. Training must also be provided when other factors such as the introduction of new technology and new systems of work or work equipment arise.

Training must:

- Be repeated periodically where appropriate.
- Be adapted to take account of any new or changed risks to the health and safety of the employees concerned.
- Take place during working hours.

EMPLOYEES' DUTIES (REGULATION 14)

Employees are required to follow health and safety instructions by using machinery, substances, transport etc. in accordance with the instructions and training that they have received.

They must also inform their employer (and other employers) of any dangers or shortcoming in the health and safety arrangements, even if there is no risk of imminent danger.

TEMPORARY WORKERS (REGULATION 15)

Consideration is given to the special needs of temporary workers. In particular to the provision of particular health and safety information such as qualifications required to perform the task safely or any special arrangements such as the need to provide health screening.

RISKS ASSESSMENT IN RESPECT OF NEW OR EXPECTANT MOTHERS (REGULATION 16)

Where the work is of a kind which would involve risk to a new or expectant mother or her baby, then the assessment required by regulation 3 should take this into account.

If the risk cannot be avoided, then the employer should take reasonable steps to:

- Adjust the hours worked.
- Offer alternative work.
- Give paid leave for as long as is necessary.

CERTIFICATE FROM A REGISTERED MEDICAL PRACTITIONER IN RESPECT OF NEW OR EXPECTANT MOTHERS (REGULATION 17)

Where the woman is a night shift worker and has a medical certificate identifying night shift work as a risk then the employer must put her on day shift or give paid leave for as long as is necessary.

NOTIFICATION BY NEW OR EXPECTANT MOTHERS (REGULATION 18)

The employer need take no action until he is notified in writing by the woman that she is pregnant, has given birth in the last six months, or is breastfeeding.

PROTECTION OF YOUNG PERSONS (REGULATION 19)

Employers of young persons shall ensure that they are not exposed to risk as a consequence of their lack of experience, lack of awareness or lack of maturity.

No employer shall employ young people for work which:

- Is beyond his physical or psychological capacity.
- Involves exposure to agents which chronically affect human health.
- Involves harmful exposure to radiation.
- Involves a risk to health from extremes of temperature, noise or vibration.
- Involves risks which could not be reasonably foreseen by young persons.

This regulation does not prevent the employment of a young person who is no longer a child for work:

- Where it is necessary for his training.
- Where the young person will be supervised by a competent person.
- Where any risk will be reduced to the lowest level that is reasonably practicable.

Note: Two HSE publications give guidance on these topics. HSG122 - New and expectant mothers at work: a guide for employers and HSG165 - Young people at work: a guide for employers.

EXEMPTION CERTIFICATES (REGULATION 20)

The Secretary of State for Defence may, in the interests of national security, by a certificate in writing exempt the armed forces, any visiting force or any headquarters from certain obligations imposed by the Regulations.

PROVISIONS AS TO LIABILITY (REGULATION 21)

Employers cannot submit a defence in criminal proceedings that contravention was caused by the act or default either of an employee or the competent person appointed under Regulation 7.

EXCLUSION OF CIVIL LIABILITY (REGULATION 22)

As amended by Health and Safety at Work etc. Act 1974 (Civil Liability) (Exceptions) Regulations 2013):

Regulation 22 specifies:

"(1) Breach of a duty imposed by regulation 16, 16A, 17 or 17A shall, so far as it causes damage, be actionable by the new or expectant mother.

(2) Any term of an agreement which purports to exclude or restrict any liability for such a breach is void."

Manual Handling Operations Regulations (MHOR) 1992 (as amended)

See also - Health and Safety (Miscellaneous Amendments) Regulations (MAR) 2002.

Considered in context in Element B9.

Arrangement of Regulations

1) Citation and commencement.
2) Interpretation.
3) Disapplication of Regulations.
4) Duties of employers.
5) Duty of employees.
6) Exemption certificates.
7) Extension outside Great Britain.
8) Repeals and revocations.

Outline of main points

1)	**Citation and commencement**
2)	**Interpretation**

"Injury" does not include injury caused by toxic or corrosive substances which:

- Have leaked/spilled from load.
- Are present on the surface but not leaked/spilled from it.
- Are a constituent part of the load.

"Load" includes any person or animal.

"Manual Handling Operations" means transporting or supporting a load including:

- Lifting and putting down.
- Pushing, pulling or moving by hand or bodily force.
- Shall as far as is reasonably practicable.

3)	**Disapplication of regulations**
4)	**Duties of employers**
4) (1)(a)	**Avoidance of manual handling**

The employer's duty is to avoid the need for manual handling operations which involve a risk of their employees being injured - as far as is reasonably practicable.

4) (1)(b)(i)   Assessment of risk

Where not reasonably practicable make a suitable and sufficient assessment of all such manual handling operations.

4) (1)(b)(ii)   Reducing the risk of injury

Take appropriate steps to reduce the risk of injury to the lowest level reasonably practicable.

4) (1)(b)(iii)   The load - additional information

Employers shall provide information on general indications or where reasonably practicable precise information on:

- The weight of each load.
- The heaviest side of any load whose centre of gravity is not central.

4) (2)   Reviewing the assessment

Assessment review:

- Where there is reason to believe the assessment is no longer valid.
- There is sufficient change in manual handling operations.

5)   Duty of employees

Employees shall make full and proper use of any system of work provided for his use by his employer.

6)	**Exemption certificates**
7)	**Extension outside Great Britain**
8)	**Repeals and revocations**

Schedule 1	Factors to which the employer must have regard and questions he must consider when making an assessment of manual handling operations.
Schedule 2	Repeals and revocations.
Appendix 1	Numerical guidelines for assessment.
Appendix 2	Example of an assessment checklist.

Thus the Regulations establish a clear hierarchy of measures:

1) Avoid hazardous manual handling operations so far as is reasonably practicable.
2) Make a suitable and sufficient assessment of any hazardous manual handling operations that cannot be avoided.
3) Reduce the risk of injury so far as is reasonably practicable.

Personal Protective Equipment at Work Regulations (PPER) 1992

See also - Health and Safety (Miscellaneous Amendments) Regulations (MAR) 2002 and Health and Safety (Miscellaneous Repeals, Revocations and Amendments) Regulations (MRRA) 2002.

Considered in context in Elements B3 and B6.

Arrangement of Regulations

1) Citation and commencement.
2) Interpretation.
3) Disapplication of these Regulations.
4) Provision of personal protective equipment.

5) Compatibility of personal protective equipment.

6) Assessment of personal protective equipment.

7) Maintenance and replacement of personal protective equipment.

8) Accommodation for personal protective equipment.

9) Information, instruction and training.

10) Use of personal protective equipment.

11) Reporting loss or defect.

12) Exemption certificates.

13) Extension outside Great Britain.

14) Modifications, repeal and revocations directive.

Schedule 1   Relevant Community.

Schedule 2   Modifications.

Part I   Factories Act 1961.

Part II   The Coal and Other Mines (Fire and Rescue) Order 1956.

Part III   The Shipbuilding and Ship-Repairing Regulations 1960.

Part IV   The Coal Mines (Respirable Dust) Regulations 1975.

Part V   The Control of Lead at Work Regulations 1980.

Part VI   The Ionising Radiations Regulations 1985.

Part VII   The Control of Asbestos at Work Regulations 1987.

Part VIII   The Control of Substances Hazardous to Health Regulations 1988.

Part IX   The Noise at Work Regulations 1989.

Part X   The Construction (Head Protection) Regulations 1989.

Schedule 3   Revocations.

Outline of main points

2) Personal protective equipment (PPE) means all equipment (including clothing provided for protection against adverse weather) which is intended to be worn or held by a person at work and which protects him against risks to his health or safety.

3) These Regulations do not apply to:
- Ordinary working clothes/uniforms.
- Offensive weapons.
- Portable detectors which signal risk.
- Equipment used whilst playing competitive sports.
- Equipment provided for travelling on a road.

The Regulations do not apply to situations already controlled by other Regulations i.e.
- Control of Lead at Work Regulations 1998.
- Ionising Radiation Regulations 1999.
- Control of Asbestos at Work Regulations 1987 (and as amended 1999).
- CoSHH Regulations 1999.
- Noise at Work Regulations 1989.
- Construction (Head Protection) Regulations 1989.

4) Suitable PPE must be provided when risks cannot be adequately controlled by other means. Regulation 4 shall ensure suitable PPE:
- Appropriate for the risk and conditions.
- Ergonomic requirements.
- State of health of users.
- Correctly fitting and adjustable.
- Complies with EEC directives.

5) Equipment must be compatible with any other PPE which has to be worn.

6) Before issuing PPE, the employer must carry out a risk assessment to ensure that the equipment is suitable.
- Assess risks not avoided by other means.
- Define characteristics of PPE and of the risk of the equipment itself.
- Compare characteristics of PPE to defined requirement.
- Repeat assessment when no longer valid or significant change has taken place.

7) PPE must be maintained.
- In an efficient state and working order.
- In good repair.

8) Accommodation must be provided for equipment when it is not being used.

9) Information, instruction and training must be given on:
- The risks PPE will eliminate or limit.
- Why the PPE is to be used.

- How the PPE is to be used.
- How to maintain the PPE.

Information and instruction must be comprehensible to the wearer/user.

10) Employers shall take reasonable steps to ensure PPE is worn.
- Every employee shall use PPE that has been provided.
- Every employee shall take reasonable steps to return PPE to storage.

11) Employees must report any loss or defect.

The Guidance on the Regulations points out:

"Whatever PPE is chosen, it should be remembered that, although some types of equipment do provide very high levels of protection, none provides 100%".

PPE includes the following when worn for health and safety reasons at work:

- Aprons.
- Adverse weather gear.
- High visibility clothing.
- Gloves.
- Safety footwear.
- Safety helmets.
- Eye protection.
- Life-jackets.
- Respirators.
- Safety harness.
- Underwater breathing gear.

PPE comes almost last of the measures to be taken to control hazards:

Eliminate

Reduce by substitution

Isolate

Control

Personal Protection

Discipline

There are some good reasons why PPE is used as a last resort:

- PPE only protects the wearer and not others who may be in the area and also at risk.
- The introduction of PPE may bring another hazard such as impaired vision, impaired movement or fatigue.

Protection will depend upon fit in many cases, i.e. respirators, breathing apparatus, noise protection devices. Not only does this presuppose that adequate training and instruction have been given, it also presupposes that other individual conditions have been assessed, such as:

- Long hair.
- Wearing of spectacles.
- Stubble growing on men's faces which may have an adverse effect on protection.

PPE will not be suitable unless:

- It is appropriate to the risk involved and the prevailing conditions of the workplace.
- It is ergonomic (user-friendly) in design and takes account of the state of health of the user.
- It fits the wearer correctly, perhaps after adjustment.
- It is effective in preventing or controlling the risk(s) [without increasing the overall risk], so far as is practicable.
- It must comply with EEC directives or other specific standards.

It must be compatible with other equipment where there is more than one risk to guard against.

Training and instruction will include both:

- *Theoretical training -* where the reasons for wearing the PPE, factors which may affect the performance, the cleaning, maintenance and identification of defects are discussed.
- *Practical training -* wearing, adjusting, removing, cleaning, maintenance and testing are practised.

Public Order Act (POA) 1986

Considered in context in Element B8.

Arrangement of Regulations

1) New offences.
2) Processions and assemblies.
3) Racial hatred.
4) Exclusion orders.
5) Miscellaneous and General.

Schedule 1 - Sporting events.

Schedule 2 - Other amendments.

Schedule 3 - Repeals.

Outline of main points

The 1986 Act extends police controls over public processions and marches and creates for the first time controls over public assemblies.

It abolishes a number of common law offences, including riot, violent disorder and affray and replaces them with a wide range of statutory public order offences. Included as "sections" of the Act are:

Section 1 - Riot.

Section 2 - Violent Disorder.

Section 3 - Affray.

Section 4 - Threatening Behaviour.

Section 5 - Causing Harassment, Alarm and Distress (sometimes referred to as "disorderly conduct").

This act makes it an offence for a person to use threatening, abusive or insulting words or behaviour, or to display any written material which is threatening, abusive or insulting, if:

(a)   He intends thereby to stir up racial hatred.

(b)   Having regard to all the circumstances racial hatred is likely to be stirred up thereby.

The Act covers the crime of Incitement to Racial Hatred.

Registration, Evaluation, Authorisation and Restriction of Chemicals (REACH) Regulation, EC 1907/2006)

Considered in context in Element B2.

REACH came into force on 1st June 2007 and replaces a number of European Directives and Regulations (see table below for details) with a single system to gather hazard information, assess risks, classify, label, and restrict the marketing and use of individual chemicals and mixtures. This is known as the REACH system:

R egistration of basic information of substances to be submitted by companies, to a central database.

E valuation of the registered information to determine hazards and risks.

A uthorisation requirements imposed on the use of high-concern substances.

CH emicals.

REACH covers both "new" and "existing" substances and puts the onus on Industry to prove that chemicals it uses are safe.

REACH only applies to chemicals manufactured in or imported into the EU. It does not apply to the use of chemicals in finished products. So a product like a television, or computer or shampoo made outside the EU could contain chemicals that are not registered under REACH - providing they are not banned under specific safety regulations (such as lead).

REACH entered into force on 1st June 2007. The pre-registration period for eligible chemicals was 1st June - 1st December 2008, and chemicals that are pre-registered benefit from phased registration deadlines up to June 2018.

The new regime creates the European Chemicals Agency (ECHA) with a central coordination and implementation role in the overall process. (See http://reach.jrc.it/guidance for more information). Industry will be able to use "substances of very high concern" only if they have authorisation from ECHA. Authorisation will be granted under specific conditions, and will have to be regularly renewed, encouraging companies to seek safer alternatives.

"Substances of very high concern" are defined by REACH, these are chemicals that:

- Cause cancer or mutation or interfere with the body's reproductive function (CMRs).
- Take a long time to break down, accumulate in the body and are toxic (PBTs).
- Take a very long time to break down and accumulate in the body (vPvBs).
- Have serious and irreversible effects on humans and the environment, for example substances that disturb the body's hormone system.

However, polymers, a group of chemicals that includes plastics, will be exempted. But monomers - the basic building block of an individual polymer do have to be registered and evaluated.

All manufacturers and importers of chemicals must identify and manage risks linked to the substances they manufacture and market. For substances produced or imported in quantities of 1 tonne or more per year per company, manufacturers and importers need to demonstrate that they have appropriately done so by means of a registration dossier, which shall be submitted to the Agency.

Once the registration dossier has been received, the Agency may check that it is compliant with the Regulation and shall evaluate testing proposals to ensure that the assessment of the chemical substances will not result in unnecessary testing, especially on animals.

Where appropriate, authorities may also select substances for a broader substance evaluation to further investigate substances of concern.

Manufacturers and importers must provide their downstream users with the risk information they need to use the substance safely. This will be done via the classification and labelling system and Safety Data Sheets (SDS), where needed.

ENFORCEMENT

REACH requires each Member State to appoint a Competent Authority (CA) and maintain an appropriate control system with respect to enforcement.

The authorities given enforcement responsibility by the REACH Enforcement Regulations 2008 are those with existing remits to protect human health, consumer safety, and the environment:

- The Health and Safety Executive (HSE).
- The Health and Safety Executive for Northern Ireland (HSENI).
- The Environment Agency (EA).
- The Scottish Environment Protection Agency (SEPA).
- The Northern Ireland Environment Agency (NIEA).
- The Department of Energy and Climate Change (DECC).

- Local Authorities (LAs), as regards health and safety and consumer protection (trading standards).

Regulation 3 and Schedule 1 of the REACH Enforcement Regulations 2008 sets out which enforcing authority is responsible for enforcing the listed REACH provisions, though broadly speaking:

- HSE, in its capacity as UK REACH CA, will enforce those duties in REACH concerning registration.
- HSE in Great Britain and HSENI in Northern Ireland will enforce supply chain related duties up to the point of retail sale, and for retail sale local authority trading standards departments are responsible.
- A wide range of enforcing authorities will enforce use related duties, as per existing arrangements for enforcing health, safety and environmental legislation. HSE will enforce use-related duties relating to occupational safety and health in Great Britain.

TABLE OF AMENDMENTS

Citation	Extent of revocation
The Notification of New Substances (Amendment) Regulations 2001.	The whole instrument.
The Chemicals (Hazard Information and Packaging for Supply) Regulations 2002.	In regulation 8, paragraphs (4), (5) and (6). Regulation 12(1)(d). Schedule 4.
The Control of Substances Hazardous to Health Regulations 2002.	Regulation 4(4). In Schedule 2, entries numbered 11, 12 and 13. Regulation 2(g).
The Chemicals (Hazard Information and Packaging for Supply) (Amendment) Regulations 2005.	Regulation 2(3).
The Control of Asbestos Regulations 2006.	Regulations 25(3), 27, 28, 29 and 32(2).

Main sources: www.hse.gov.uk/reach

At the time of writing further amendments to REACH have been made. The amendments made until July 2013 are listed below. The latest consolidated version of REACH together with related Legislation and Case Law is available on the legislation page on website of the European Chemicals Agency link to external website.

Please note that the legislation has been amended on a number of occasions since its original publication. The amendments made until the end of February 2012 are listed below. The latest consolidated version of REACH (December 2011) is available on the legislation page on website of the European Chemicals Agency link to external website.

Amendments and Corrigenda (corrections) to REACH

Below is a list of REACH amendments and corrigenda followed by a short explanation of the changes they introduced.

- Amendments
- Regulation (EC) No 1354/2007: Adaptation to account for accession of Bulgaria and Romania
- Regulation (EC) No 987/2008: Amendments to Annexes IV & V
- Regulation (EC) No 1272/2008: Adaptations to REACH to account for the CLP 2009 regulation.
- Regulation (EC) No 134/2009: Amendment to Annex XI
- Regulation (EC) No 552/2009: Amendments to Annex XVII to account for repeal of Marketing & Use Directive (76/769/EEC)
- Regulation (EU) No 276/2010: Amendments Annex XVII (dichloromethane, lamp oils and grill lighter fluids and organostannic compounds)
- Regulation (EU) No 453/2010: Phased Amendments to Annex II (Safety Data Sheet requirements) to account for CLP 2009
- Regulation (EU) No 143/2011: 1st list of substances added to Annex XIV
- Regulation (EU) No 207/2011: Annex XVII (Diphenylether, pentabromo derivative and PFOS)
- Regulation (EU) No 252/2011: Amendments to Annex I
- Regulation (EU) No 253/2011: Amendments to Annex XIII
- Regulation (EU) No 366/2011: Amendment to Annex XVII (Acrylamide)
- Regulation (EU) No 494/2011: Amendment to Annex XVII (Cadmium)
- Regulation (EU) No 109/2012: Annex XVII (CMR substances)
- Regulation (EU) No 125/2012: 2nd list of substances added to Annex XIV
- Regulation (EU) No 412/2012: Amendment to Annex XVII (Dimethylfumarate)
- Regulation (EU) No 835/2012: Amendment to Annex XVII (Cadmium)
- Regulation (EU) No 836/2012: Amendment to Annex XVII (Lead)
- Regulation (EU) No 126/2013: Amendment to Annex XVII (editorial corrections & clarifications)
- Regulation (EU) No 348/2013: 3rd list of substances added to Annex XIV
- Regulation (EU) No 517/2013: Amendment to account for Croatia joining EU
- Corrigenda
- Corrigendum, OJ L 141, 31.5.2008, p. 22 & Corrigendum, OJ L 36, 5.2.2009, p. 84: Corrections of definition of phase-in substances for no-longer polymer substances
- Corrigendum, OJ L 49, 24.2.2011, p. 52: Sunset Date corrections for 1st Annex XIV substances
- Corrigendum, OJ L 136, 24.5.2011, p. 105: Date correction for Cadmium restriction.

Outline of main points

Scope and exemptions

REACH applies to substances manufactured or imported into the EU in quantities of 1 tonne or more per year. Generally, it applies to all individual chemical substances on their own, in preparations or in articles (if the substance is intended to be released during normal and reasonably foreseeable conditions of use from an article).

Some substances are specifically excluded:

- Radioactive substances.
- Substances under customs supervision.
- The transport of substances.
- Non-isolated intermediates.
- Waste.
- Some naturally occurring low-hazard substances.

Some substances, covered by more specific legislation, have tailored provisions, including:

- Human and veterinary medicines.
- Food and foodstuff additives.
- Plant protection products and biocides.

Other substances have tailored provisions within the REACH legislation, as long they are used in specified conditions:

- Isolated intermediates.
- Substances used for research and development.

REACH places the burden of proof of compliance on companies. To comply with the Regulation, companies must identify and manage the risks linked to the substances they manufacture and supply in the EU. They have to demonstrate to ECHA how the substance can be safely and healthily used, and they must communicate the risk management measures to the users.

Registration

Companies have the responsibility of collecting information on the properties and the uses of substances that they manufacture or import at or above one tonne per year. They also have to make an assessment of the hazards and potential risks presented by the substance. They have to register this by submitting a dossier to ECHA, based in Helsinki. The dossier contains details of the substance's properties, other relevant information about risks and how these risks can be managed. They will not be able to manufacture or import a substance within the EU, or import an article that intentionally releases a substance, unless the substance has been registered.

Evaluation

ECHA and the Member States evaluate the information submitted by companies to examine the quality of the registration dossiers and the testing proposals and to clarify if a given substance constitutes a risk to human health or the environment. REACH promotes the use of alternative methods for the assessment of the hazardous properties of substances in order to reduce the number of tests on animals, for example, quantitative structure-activity relationships (QSAR) and read across.

Authorisation

The authorisation procedure aims to assure that the risks from Substances of Very High Concern (SVHC) are properly controlled and that these substances are progressively replaced by suitable alternatives, while ensuring the good functioning of the EU internal market. SVHCs will need to be authorised for specific uses if they appear in Annex XIV of REACH. There will not be a 'blanket' authorisation for a substance to be used generally. Instead, applications for authorisation for a specific use may be made by companies that register the substances, or by those that use them. When a substance is placed on Annex XIV of REACH, a 'sunset date' will be set after which its use will be prohibited, unless an authorisation has been granted.

Restriction

Restrictions are set to protect human health and the environment from unacceptable risks posed by chemicals. Restrictions may limit or ban the manufacture, placing on the market or use of a substance. This is a direct means of controlling the risks associated with any given hazardous substance. Restriction will be used when it is felt that action at the European level is needed. Restriction decisions will also be made by the European Commission based on advice from the ECHA and consultation with EU Member States and others. Restrictions in place already from previous legislation are carried over into REACH in Annex XVII of REACH; further restrictions will be added to this Annex as necessary.

Information

The passage of information up and down the supply chain is one of the important features of REACH - users should be able to understand what manufacturers and importers know about the dangers involved in using chemicals, and they can also pass information back up the supply chain. REACH adopts and builds on the existing system for passing information in a structured way down to chemicals users - the Safety Data Sheet (SDS). This should accompany materials down through the supply chain, providing the information that users need to ensure chemicals are safely managed. REACH will also allow for information on uses of chemicals to be passed back up the supply chain, so that these can be reflected in the SDS.

Users of chemicals - Article 37 - 39 particularly

REACH aims to reduce the use of (and risks from) hazardous chemicals. Certain substances will be identified as substances of very high concern (SVHC), and may become subject to 'authorisation'. Substances that are subject to authorisation cannot be used in the EU unless a company (and their identified users) have been authorised to do so. This will mean that these substances are eventually phased out of all non-essential uses. Other substances are subject to a 'restriction'. Restrictions limit or ban the manufacture, placing on the market or use of substances that pose an unacceptable risk to human health and the environment. REACH may make things better for users of chemicals as it is designed to provide more information on chemicals and increase confidence in their safe and healthy use. In particular, better information on the hazards of chemicals and how to use them safely and healthily will be passed down the supply chain by chemical manufacturers and importers through improved Safety Data Sheets. Users have a duty to apply risk reduction measures identified in Safety Data Sheets provided to them form chemicals (Article 37(5)).

If users of chemicals use a chemical in a novel way that is perhaps not expected, then they will need to consider letting their supplier know. This use will need to be considered for registration by the supplier. If users do not want to let your supplier know about this use (for example, because of commercial concerns) then they do not have to, but it will mean that they will have to let ECHA know about this use and possibly have to submit their own risk assessment. Once the chemical has been registered and they have been provided with a safety data sheet listing the registration number, they have a maximum of 6 months to provide information about their use to ECHA if it is not included in the supplier's registration. The user's supplier should be able to tell them which uses are covered by the registration. Users of hazardous substances should check whether any of the substances they use are subject to a 'restriction'. This information should be communicated to users in the safety data sheet or other communication given to them by their supplier. If users use restricted substances, they will need to ensure their use of the substance meets the conditions of the restriction. Users are able to contribute to the public consultations whenever a new restriction is proposed, or an existing restriction is amended. Users of hazardous substances should check whether any of the substances they use are subject to authorisation, or are likely to become subject to authorisation in the future. If a substance is subject to authorisation, users will only be able to use the substance if they, or someone else further up the supply chain, have been granted an authorisation for that use. Information about whether a substance is subject to authorisation, and whether an authorisation has been granted or refused, should be communicated to users in the safety data sheet or other communication given by the supplier.

Safety data sheets under REACH - Article 31 particularly

Once chemicals are registered, safety data sheets will list registration numbers. Safety data sheets may also include exposure scenarios. An exposure scenario describes the operating conditions and risk management measures that have been identified by the supplier as necessary to use the chemical safely and healthily in the users processes. Under Article 31 of REACH the supplier of a substance or mixture must provide 'the recipient', including end users, with a safety data sheet compiled in accordance with Annex II of REACH. Safety Data Sheets need not be supplied where substances are provided to the general public and they are provided with sufficient information to enable them to take the necessary measures, unless they are requested by a distributor or user. REACH requires users to follow the advice on risk management measures given in the exposure scenario attached to the safety data sheet. If users use different risk management measures to those described in the exposure scenario, then they should be able to justify why their measures offer an equivalent (or better) level of protection for human health and the environment to those described in the exposure scenario. If users need to change the risk management measures used at their workplace in order to comply with the exposure scenario then they have a maximum of 12 months (after they have received a safety data sheet listing a registration number for a chemical) to do this.

REACH, Article 31 and Annex II, specifies the scope of the content of safety data sheets for chemicals:

1) Identification of the substance/mixture and of the company/undertaking.
2) Hazards identification.
3) Composition/information on ingredients.
4) First-aid measures.
5) Fire-fighting measures.
6) Accidental release measures.
7) Handling and storage.
8) Exposure controls/personal protection.
9) Physical and chemical properties.
10) Stability and reactivity.
11) Toxicological information.
12) Ecological information.
13) Disposal considerations.
14) Transport information.
15) Regulatory information.
16) Other information.

Reporting of Injuries, Diseases and Dangerous Occurrences (Amendment) Regulations (RIDDOR) 2013

Considered in context in Element B8.

Arrangement of Regulations

1) Citation and commencement.
2) Interpretation.
3) Responsible person.
4) Non-fatal injuries to workers.
5) Non-fatal injuries to non-workers.
6) Work-related fatalities.
7) Dangerous occurrences.
8) Occupational diseases.
9) Exposure to carcinogens, mutagens and biological agents.

10) Diseases offshore.
11) Gas-related injuries and hazards.
12) Recording and record-keeping.
13) Mines, quarries and offshore site disturbance.
14) Restrictions on the application of regulations 4 to 10.
15) Restriction on parallel requirements.
16) Defence.
17) Certificates of exemption.
18) Revocations, amendments and savings.
19) Extension outside Great Britain.
20) Review.

SCHEDULES

Schedule 1 - Reporting and Recording Procedures.
Schedule 2 - Dangerous Occurrences.
Schedule 3 - Diseases Reportable Offshore.
Schedule 4 - Revocations and Amendments.

Outline of main points

The Reporting of Injuries, Diseases and Dangerous Occurrences Regulations (RIDDOR) 2013 covers the requirement to report certain categories of injury and disease sustained at work, along with specified dangerous occurrences and gas incidents, to the relevant enforcing authority. These reports are used to compile statistics to show trends and to highlight problem areas, in particular industries or companies.

THE MAIN POINTS OF RIDDOR

Reporting

(1) When a person **dies or suffers any specified injury** listed in regulation 4 **(non-fatal injuries to workers)** as a result of a work-related accident or dies as a result of occupational exposure to a biological agent or an incident occurs of the type listed as a dangerous occurrences in Schedule 2 the responsible person must notify the relevant enforcing authority by the quickest practicable means (usually by telephone) without delay and must send them a report in an approved manner (on-line) within 10 days. This therefore includes accidents connected with work where:

- An employee or a self-employed person at work is killed or suffers a specified injury (including as a result of physical violence).
- A member of the public is killed or taken to hospital.

A work-related **'accident'** in the context of RIDDOR 2013 includes an act of non-consensual physical violence done to a person at work.

(2) In cases of work-related diseases that are listed in regulation 8 and 10 the responsible person must send a report of the diagnosis in an approved manner (on-line) to the relevant enforcing authorities without delay. In cases of diseases related to carcinogens, mutagens and biological agents that are listed in regulation 9 the responsible person must notify the relevant enforcing authority in an approved manner.

(3) If personal injury results in **more than 7 days (excluding the day of the accident) incapacity** for routine work, but is not one of the specified non-fatal injuries, the responsible person must send a report to the relevant enforcing authority in an approved manner (on-line) as soon as is practicable and in any event within **15 days** of the accident. The day of the accident is not counted, but any days which would not have been working days are included.

(4) If there is an accident connected with work (including an act of physical violence) and an employee, or a self-employed person at work, suffers an over-three-day injury it must be recorded by the employer.

(5) The enforcing authority for most workplaces is either the Health and Safety Executive or the Local Authority, for railway operations it is the Office of Rail Regulation (ORR).

Responsible person

Reportable event	To	Responsible person
Death, specified non-fatal injury, over 7 day injury, disease.	Employee.	Employer.
	Self-employed person working in someone else's premises.	Person in control of the premises: ■ At the time of the event. ■ In connection with trade, business or undertaking.
Specified injury, over 7 day injury, disease.	Self-employed in own premises.	Self-employed person or someone acting for them.
Death, being taken to hospital or a specified non-fatal injury on hospital premises.	A person not at work.	Person in control of the premises: ■ At the time of the event. ■ In connection with trade, business or undertaking.
Dangerous occurrences - general.		Person in control of the premises where, or in connection with the work going on at which, the dangerous occurrence happened: ■ At the time of the event. ■ In connection with trade, business or undertaking.

Road traffic accidents

Road traffic accidents only have to be reported if:

- Death or injury results from an accident involving a train.
- Death or injury results from exposure to a substance being conveyed by a vehicle.
- Death or injury results from the person being engaged in work connected with loading or unloading of any article/substance or results from another person engaged in these activities.
- Death or injury results from the person being engaged in work on or alongside a road or results from another person engaged in these activities.

Work on or alongside a road means work concerned with the construction, demolition, alteration, repair or maintenance of:

- The road or the markings or equipment on the road.
- The verges, fences, hedges or other boundaries of the road.
- Pipes or cables on, under, over or adjacent to the road.
- Buildings or structures adjacent to or over the road.

Non-employee

The responsible person must not only report non-employee deaths, but also cases that involve major injury or being taken to hospital if caused by an accident out of or in connection with their work.

Employee death

Where an employee dies within one year of the date of an accident, as a result of a reportable injury, as soon as the employer knows they must inform the enforcing authority in writing of the death.

Gas incidents

Specified gas incidents are notified without delay and reported within 14 days to the Health and Safety Executive.

Injury under medical supervision

Reporting and recording requirements do not apply in situations where the injury or death of a person arises out of the conduct of an operation, examination or other medical treatment of that person whilst under the supervision of a registered medical practitioner or dentist.

Self-employed people

If a self-employed person suffers a specified non-fatal injury while working at premises that are owned or occupied by themselves they do not need to notify the enforcing authority immediately. However, they or someone acting for them must report the injury within 10 days.

Where an injury is not a specified non-fatal injury, but causes a self-employed person to be incapacitated from routine work for more than seven consecutive days, the self-employed person or someone acting for them must report it within 15 days of the accident. There is no reporting requirement for situations where a self-employed person suffers a fatal accident or fatal exposure on premises controlled by that self-employed person.

Recording

In the case of an accident at work, the following details must be recorded:

- Date and time.
- Name.
- Occupation.
- Nature of injury.
- Place of accident.
- Brief description of the circumstances in which the accident happened.
- In the case of a person not a work, instead of occupation a record of their status should be made (for example, passenger, customer, visitor or bystander).
- The date on which the accident was first notified or reported to the relevant enforcing authority and the method used.

For non-reportable injuries that incapacitate for more than 3 days a record of notification or reporting is not relevant, but a record of the accident must be maintained.

Similar information should be recorded for dangerous occurrences, except that details of injured persons will not be relevant.

In the case of a diagnosis of a reportable disease, the following details must be recorded:

- The date of diagnosis of the disease.
- The name of the person affected.
- The occupation of the person affected.
- The name or nature of the disease.
- The date on which the disease was first reported to the relevant enforcing authority.
- The method by which the disease was reported.

Records must be kept for at least 3 years and kept at the place where the work it relates to is carried out or at the usual place of business of the responsible person.

Defences

A person must prove that they were not aware of the event requiring reporting and that they had taken all reasonable steps to be made aware, in sufficient time.

SPECIFIED INJURIES (RIDDOR 2013 - REGULATION 4)

The list of specified non-fatal injuries is:

- Any bone fracture diagnosed by a registered medical practitioner, other than to a finger, thumb or toe.

- Amputation of an arm, hand, finger, thumb, leg, foot or toe.
- Any injury diagnosed by a registered medical practitioner as being likely to cause permanent blinding or reduction in sight in one or both eyes.
- Any crush injury to the head or torso causing damage to the brain or internal organs in the chest or abdomen.
- Any burn injury (including scalding) which:
  - Covers more than 10% of the whole body's total surface area.
  - Causes significant damage to the eyes, respiratory system or other vital organs.
- Any degree of scalping requiring hospital treatment.
- Loss of consciousness caused by head injury or asphyxia.
- Any other injury arising from working in an enclosed space which:
  - Leads to hypothermia or heat-induced illness.
  - Requires resuscitation or admittance to hospital for more than 24 hours.

DISEASES (RIDDOR 2013 - REGULATIONS 8, 9 AND 10)

Regulation 8 - Occupational diseases

- Carpal Tunnel Syndrome (CTS), where the person's work involves regular use of percussive or vibrating tools.
- Cramp in the hand or forearm, where the person's work involves prolonged periods of repetitive movement of the fingers, hand or arm.
- Occupational dermatitis, where the person's work involves significant or regular exposure to a known skin sensitiser or irritant.
- Hand Arm Vibration Syndrome (HAVS), where the person's work involves regular use of percussive or vibrating tools, or the holding of materials which are subject to percussive processes, or processes causing vibration.
- Occupational asthma, where the person's work involves significant or regular exposure to a known respiratory sensitizer.
- Tendonitis or tenosynovitis in the hand or forearm, where the person's work is physically demanding and involves frequent, repetitive movements.

Regulation 9 - Exposure to carcinogens, mutagens and biological agents

- Any cancer attributed to an occupational exposure to a known human carcinogen or mutagen (including ionising radiation).
- Any disease attributed to an occupational exposure to a biological agent.

Regulation 10 and Schedule 3 - Diseases offshore

Examples of diseases listed in this Schedule are:

- Chickenpox.
- Cholera.
- Diphtheria.
- Dysentery (amoebic or bacillary).
- Mumps.
- Food poisoning.
- Legionellosis.
- Malaria.
- Measles.
- Meningitis.

DANGEROUS OCCURRENCES (RIDDOR 2013 - SCHEDULE 2)

Dangerous occurrences are events that have the potential to cause death or serious injury and so must be reported whether anyone is injured or not. Examples of dangerous occurrences that must be reported are:

- The failure of any load bearing part of any lifting equipment, other than an accessory for lifting.
- The failure of any pressurised closed vessel or any associated pipework.
- Any unintentional incident in which plant or equipment either:
  - Comes into contact with an uninsulated overhead electric line in which the voltage exceeds 200 volts.
  - Causes an electrical discharge from such an electric line by coming into close proximity to it.
- Electrical short-circuit or overload attended by fire or explosion which results in the stoppage of the plant involved for more than 24 hours.

The schedule also identifies Dangerous Occurrences that are specific to mines, quarries, transport systems, and offshore workplaces.

Working Time Regulations (WTR) 1998

Considered in context in Element B8.

The Working Time Regulations (1998) came into effect on 1st October 1998 to implement the European Working Time Directive into GB law. Since their introduction, the Regulations have been updated and amended through additional legislation to cover an even wider range of workers, and granted additional rights to young workers. While special rules apply in respect of young workers and junior doctors, the core rights at the heart of the Regulations remain the same.

Arrangement of Regulations

Part I - General

Part II - Rights and obligations concerning working time

Part III - Exceptions

Part IV - Miscellaneous

Part V - Special classes of person

Schedules

1) Workforce agreements.
2) Workers employed in agriculture.

Outline of main points

These Regulations protect workers from being forced to work excessive hours. They also make the provision of paid annual leave mandatory, and include rights to rest breaks and uninterrupted periods of rest.

The Regulations apply to "adult workers" (over 18) and to "young workers" (over compulsory school age), and some of the detailed provisions are slightly different for each of these two groups. There are exceptions to some of the Regulations, primarily for employees working in transport (air, road, rail, sea etc.), sea-fishing, work at sea, doctors in training, certain activities of the armed forces, police and civil protection workers.

The Regulations have 4 principal effects which are, in summary:

- To place a 48 hour limit on the average working week.
- To place an 8 hour limit on average night work for each 24 hours.
- To guarantee rest breaks during the day and daily and weekly rest periods.
- To guarantee the right to four weeks paid holiday per year.

Workplace (Health, Safety and Welfare) Regulations (WHSWR) 1992

See also - Health and Safety (Miscellaneous Amendments) Regulations (MAR) 2002 and Health and Safety (Miscellaneous Repeals, Revocations and Amendments) Regulations (MRRA) 2002.

Considered in context in Elements B7, B9 and B10.

Arrangement of Regulations

1) Citation and commencement.
2) Interpretation.
3) Application of these Regulations.
4) Requirements under these Regulations.
5) Maintenance of workplace, and of equipment, devices and systems.
6) Ventilation.
7) Temperature in indoor workplaces.
8) Lighting.
9) Cleanliness and waste materials.
10) Room dimensions and space.
11) Workstations and seating.
12) Condition of floors and traffic routes.
13) Falls or falling objects.
14) Windows, and transparent or translucent doors, gates and walls.
15) Windows, skylights and ventilators.
16) Ability to clean windows etc. safely.
17) Organisation etc. of traffic routes.
18) Doors and gates.
19) Escalators and moving walkways.
20) Sanitary conveniences.
21) Washing facilities.
22) Drinking water.
23) Accommodation for clothing.
24) Facilities for changing clothing.
25) Facilities for rest and to eat meals.
26) Exemption certificates.
27) Repeals, saving and revocations.
Schedule 1   Provisions applicable to factories which are not new workplaces, extensions or conversions.
Schedule 2   Repeals and revocations.

Outline of main points

SUMMARY

The main requirements of the Workplace (Health, Safety and Welfare) Regs 1992 are:

1) **Maintenance** of the workplace and equipment.
2) **Safety** of those carrying out maintenance work and others who might be at risk (e.g. segregation of pedestrians and vehicles, prevention of falls and falling objects etc.).
3) Provision of **welfare** facilities (e.g. rest rooms, changing rooms etc.).
4) Provision of a safe **environment** (e.g. lighting, ventilation etc.).

ENVIRONMENT

Regulation 1   New workplaces, extensions and modifications must comply.

Regulation 4   Requires employers, persons in control of premises and occupiers of factories to comply with the regulations.

Regulation 6   Ventilation - enclosed workplaces should be ventilated with a sufficient quantity of fresh or purified air (5 to 8 litres per second per occupant).

Regulation 7   Temperature indoors - This needs to be reasonable and the heating device must not cause injurious fumes. Thermometers must be provided. Temperature should be a minimum of 16°C or 13°C if there is physical effort.

Regulation 8   Lighting - must be suitable and sufficient. Natural light if possible. Emergency lighting should be provided if danger exists.

Regulation 10   Room dimensions and space - every room where persons work shall have sufficient floor area, height and unoccupied space (min 11 cu.m per person).

Regulation 11   Workstations and seating have to be suitable for the person and the work being done.

SAFETY

Regulation 12   Floors and traffic routes must be of suitable construction. This includes absence of holes, slope, uneven or slippery surface. Drainage where necessary. Handrails and guards to be provided on slopes and staircases.

Regulation 13   Tanks and pits must be covered or fenced.

Regulation 14   Windows and transparent doors, where necessary for health and safety, must be of safety material and be marked to make it apparent.

Regulation 15   Windows, skylights and ventilators must be capable of opening without putting anyone at risk.

Regulation 17   Traffic routes for pedestrians and vehicles must be organised in such a way that they can move safely.

Regulation 18   Doors and gates must be suitably constructed and fitted with any necessary safety devices.

Regulation 19   Escalators and moving walkways shall function safely, be equipped with any necessary safety devices and be fitted with emergency stop.

HOUSEKEEPING

Regulation 5   Workplace and equipment, devices and systems must be maintained in efficient working order and good repair.

Regulation 9   Cleanliness and waste materials - workplaces must be kept sufficiently clean. Floors, walls and ceilings must be capable of being kept sufficiently clean. Waste materials shall not be allowed to accumulate, except in suitable receptacles.

Regulation 16   Windows etc. must be designed so that they can be cleaned safety.

FACILITIES

Regulation 20   Sanitary conveniences must be suitable and sufficient and in readily accessible places. They must be adequately ventilated, kept clean and there must be separate provision for men and women.

Regulation 21   Washing facilities must be suitable and sufficient. Showers if required (a table gives minimum numbers of toilets and washing facilities).

Regulation 22   Drinking water - an adequate supply of wholesome drinking water must be provided.

Regulation 23   Accommodation for clothing must be suitable and sufficient.

Regulation 24   Facilities for changing clothes must be suitable and sufficient, where a person has to use special clothing for work.

Regulation 25   Facilities for rest and eating meals must be suitable and sufficient.

The WHSWR 1992 were amended by the Health and Safety (Miscellaneous Amendments) Regulations (MAR) 2002 to establish specific requirements that rest rooms be equipped with:

- An adequate number of tables and adequate seating with backs for the number of persons at work likely to use them at any one time.
- Seating which is adequate for the number of disabled persons at work and suitable for them.

In addition, WHSWR 1992 were amended to take account of disability arrangements. Where necessary, those parts of the workplace used or occupied directly by disabled persons at work, including in particular doors, passageways, stairs, showers, washbasins, lavatories and workstations, must be organised to take account of such persons.

The Health and Safety (Miscellaneous Repeals, Revocations and Amendments) Regulations (MRRA) 2013 came into force on 6th April 2013 and introduced minor amendments to Regulation 3 of the WHSWR 1992 Regulations.

This page is intentionally blank

This page is intentionally blank

Index

Practical control measures, 184, 205, 283, 306
Practical measures
    minimise exposure, 215
    prevent exposure, 215
Precautionary statements, 30
Predicted 4-hour sweat rate (P4SR), 306
Predicted mean vote (PMV), 303
Pregnant women, 316
Pre-placement assessment, 8
Pre-weighted filters, 102
Primary care, 13
Principles
    bio-psychosocial model, 7
    first-aid, 317
    fitness to work, 8
    kinetic handling, 287
    measurement, 103
    movement, 287
    return to work management, 7
    vocational rehabilitation, 7
Product labels, 39
Prophylaxis, 132
Prospective cohort studies, 50
Protected head samplers, 98
Protons, 190
Protozoa, 121
Provision
    changing facilities, 315
    first-aid, 317
    first-aiders, 318
    temperature, 291
    toilet facilities, 315
    washing facilities, 315
Psittacosis, 123
Psychological effects, 146
    aggression, 242
    stress, 229
    violence, 242
Psycho-social, 4
Public Health England (PHE), 195
Public Order Act (POA) 1986, 358
Pumps, 106
    activated, 105
    compressed chamber, 100
    diaphragm, 100
    piston, 100
    rotary vane, 100
    sampling, 100
Pushing and pulling, 274

Q
Qualitative data, 234
Qualitative dust monitoring, 109
Qualitative/Quantitative structure activity relationship (QSAR), 51
Quantitative data, 234

R
Radiant exposure, 201
Radiant temperature, 294, 297
Radiation, 292
    cosmic, 208
    ionising, 189
    ionising, 207
    methods of measurement, 201
    non-ionising, 189
    radio frequency, 200
    risk assessment, 204
    units of measurement, 201
Radiation dose, 209, 210
Radio frequency radiation, 197, 200
Radio wave, 194
Radioactivity measuring, 210
Radiological protection organisations, 195
Radiometers, 203
Radon, 208
Rapid multiplication, 122
Rapid mutation, 121
Rapid Upper Limb Assessment (RULA), 283
Rational indices, 305
Reach distances, 285
Read-across, 51
Ready-reckoner, 154, 182
Reappraisal, 93
Reduced virulence, 133
Reflection, 166
Registration, Evaluation, Authorisation and Restriction of Chemicals (REACH)
Regulation, EC 1907/2006), 35, 359
Rehabilitation overcoming barriers, 11
Relative humidity, 295, 298
Repeat victimisation, 243
Repetitive physical activities, 267

Reportable diseases, 126
Reporting incidents, 246
Reporting of Injuries, Diseases and Dangerous Occurrences (Amendment)
Regulations (RIDDOR) 2013, 362
Reproductive toxicity, 34
Respirable dust, 30
Respiratory
    ciliary escalator, 28
    nose, 27
    respiratory tract, 27
    sensitisation, 32
Respiratory system, 21
    defensive responses, 27
Respiratory, 27
Retinal photo-chemical injury, 199
Retinal thermal injuries, 199
Retrospective cohort studies, 50
Return to work management – benefits, 7
Return to work management– principles, 7
Return to work policy, 235
Return to work, 7
Risk assessments, 11
    factors, 128
    noise, 149
    prior to return to work, 11
    process, 47
    radiation, 204
    review, 11
    review, 47
    vibration, 179
Risk identification, 246
    incident reporting, 246
    staff surveys, 246
Role
    biological monitoring guidance values, 115
    control measures, 236
    fit note, 9
    hazard and precautionary statements, 30
    health surveillance, 183
    heat indices, 304
    job, 231
    Laser Protection Advisor, 223
    occupational health services, 14
    occupational hygienist, 90
    pre-placement assessment, 8
    support agencies, 13
    toxicological testing, 50
Rotary vane pumps, 100
Routes of entry
    ears, 25
    eyes, 25
    hazardous substances, 21, 25
    physical form, 29
    properties, 29
    skin, 25
Routine monitoring, 93

S
Safety assessment/reports, 43
Safety data sheets (SDS), 41
Samplers
    active personal, 104
    airborne contaminants, 89
    cowl head, 100
    cyclone head, 99
    dust particles, 97
    equipment, 89
    long-term, 94
    methods, 89, 96
    passive personal, 104
    preliminary considerations, 94
    protected head, 98
    short-term, 94
    techniques, 94
Sampling heads, 98
Sampling pumps, 100
Sampling techniques, 94
Sanitary conveniences, 315
Schizophrenia, 228
Scintillation detectors, 214
Scottish Environment Protection Agency (SEPA), 195
Segregation, 133
Sensory organs, 21
Serious eye damage/eye irritation, 32
Sharps control, 136
Shielding, 216
Short-term sampling, 94
Short-term absence/incapacity, 9
Short-term exposure limits, 89
Short-term sickness absence, 9
Silica, 37
Simple exposure, 182

This page is intentionally blank

This page is intentionally blank